Advances in Intelligent Systems and Computing

Volume 1035

The series "Advances in Intelligent Systems and Computing" contains publications on theory, applications, and design methods of Intelligent Systems and Intelligent Computing. Virtually all disciplines such as engineering, natural sciences, computer and information science, ICT, economics, business, e-commerce, environment, healthcare, life science are covered. The list of topics spans all the areas of modern intelligent systems and computing such as: computational intelligence, soft computing including neural networks, fuzzy systems, evolutionary computing and the fusion of these paradigms, social intelligence, ambient intelligence, computational neuroscience, artificial life, virtual worlds and society, cognitive science and systems, Perception and Vision, DNA and immune based systems, self-organizing and adaptive systems, e-Learning and teaching, human-centered and human-centric computing, recommender systems, intelligent control, robotics and mechatronics including human-machine teaming, knowledge-based paradigms, learning paradigms, machine ethics, intelligent data analysis, knowledge management, intelligent agents, intelligent decision making and support, intelligent network security, trust management, interactive entertainment, Web intelligence and multimedia.

The publications within "Advances in Intelligent Systems and Computing" are primarily proceedings of important conferences, symposia and congresses. They cover significant recent developments in the field, both of a foundational and applicable character. An important characteristic feature of the series is the short publication time and world-wide distribution. This permits a rapid and broad dissemination of research results.

** Indexing: The books of this series are submitted to ISI Proceedings, EI-Compendex, DBLP, SCOPUS, Google Scholar and Springerlink **

More information about this series at http://www.springer.com/series/11156

Leonard Barolli · Hiroaki Nishino ·
Hiroyoshi Miwa
Editors

Advances in Intelligent Networking and Collaborative Systems

The 11th International Conference
on Intelligent Networking and
Collaborative Systems (INCoS-2019)

 Springer

Editors
Leonard Barolli
Department of Information
and Communication Engineering
Faculty of Information Engineering
Fukuoka Institute of Technology
Fukuoka, Japan

Hiroaki Nishino
Division of Computer Science
and Intelligent Systems
Faculty of Science and Technology
Oita University
Oita, Japan

Hiroyoshi Miwa
School of Science and Technology
Kwansei Gakuin University
Sanda, Japan

ISSN 2194-5357   ISSN 2194-5365   (electronic)
Advances in Intelligent Systems and Computing
ISBN 978-3-030-29034-4   ISBN 978-3-030-29035-1   (eBook)
https://doi.org/10.1007/978-3-030-29035-1

This Springer imprint is published by the registered company Springer Nature Switzerland AG
The registered company address is: Gewerbestrasse 11, 6330 Cham, Switzerland

Welcome Message from the INCoS-2019 Organizing Committee

Welcome to the 11th International Conference on Intelligent Networking and Collaborative Systems (INCoS-2019), which is held from 5–7 September 2019, at Oita University, Japan.

INCoS is a multidisciplinary conference that covers latest advances in intelligent social networks and collaborative systems, intelligent networking systems, mobile collaborative systems, secure intelligent cloud systems, etc. Additionally, the conference addresses security, authentication, privacy, data trust and user trustworthiness behaviour, which have become crosscutting features of intelligent collaborative systems. With the fast development of the Internet, we are experiencing a shift from the traditional sharing of information and applications as the main purpose of the networking systems to an emergent paradigm, which locates people at the very centre of networks and exploits the value of people's connections, relations and collaboration. Social networks are playing a major role as one of the drivers in the dynamics and structure of intelligent networking and collaborative systems.

Virtual campuses, virtual communities and organizations strongly leverage intelligent networking and collaborative systems by a great variety of formal and informal electronic relations, such as business-to-business, peer-to-peer and many types of online collaborative learning interactions, including the virtual campuses and e-learning and MOOCs systems. Altogether, this has resulted in entangled systems that need to be managed efficiently and in an autonomous way. In addition, the conjunction of the latest and powerful technologies based on cloud, mobile and wireless infrastructures is currently bringing new dimensions of collaborative and networking applications a great deal by facing new issues and challenges. INCoS-2019 conference paid a special attention to cloud computing services, storage, security and privacy, data mining, machine learning and collective intelligence, cooperative communication and cognitive systems, management of virtual organization and enterprises, big data analytics, e-learning, virtual campuses and MOOCs, among others.

The aim of this conference is to stimulate research that will lead to the creation of responsive environments for networking and, at longer-term, the development of adaptive, secure, mobile and intuitive intelligent systems for collaborative work and learning.

This edition the conference received 107 submissions, and based on the review results, we accepted 30 regular papers (28% acceptance ratio). Additionally, five workshops were organized in conjunction with the conference.

The successful organization of the conference is achieved thanks to the great collaboration and hard work of many people and conference supporters. First and foremost, we would like to thank all the authors for their continued support to the conference by submitting their research work to the conference, for their presentations and discussions during the conference days. We would like to thank TPC members and external reviewers for their work by carefully evaluating the submissions and providing constructive feedback to authors. We would like to thank the track chairs for their work on setting up the tracks and the respective TPCs and also for actively promoting the conference and their tracks. We would like to appreciate the work of PC Co-chairs and Workshops' Co-chairs for the successful organization of workshops in conjunction with main conference.

We would like to acknowledge the excellent work and support by the International Advisory Committee. Also, we would like to thank the conference keynote speakers for their interesting and inspiring keynote speeches.

We greatly appreciate the support by Web Administrator Co-chairs and the local organizing committee from Oita University, Japan, for excellent arrangements for the conference. We are very grateful to Springer as well as several academic institutions for their endorsement and assistance.

Finally, we hope that you will find these proceedings to be a valuable resource in your professional, research and educational activities!

Hiroaki Nishino
Leonard Barolli
General Co-chairs

Hiroyoshi Miwa
Flora Amato
Programme Co-chairs

Message from the INCoS-2019 Workshops Chairs

Welcome to the Workshops of the 11th International Conference on Intelligent Networking and Collaborative Systems (INCoS-2019), which is held from 5–7 September 2019, at Oita University, Japan.

In this edition of the conference, there are held five workshops, which complemented the INCoS main themes with specific themes and research issues and challenges, as follows:

1. The 11th International Workshop on Information Network Design (WIND-2019)
2. The 7th International Workshop on Frontiers in Intelligent Networking and Collaborative Systems (FINCoS-2019)
3. The 5th International Workshop on Theory, Algorithms and Applications of Big Data Science (BDS-2019)
4. The 5th International Workshop on Collaborative e-business Systems (e-Business-2019)
5. The 2nd International Workshop on Machine Learning in Intelligent and Collaborative Systems (MaLICS-2019)

We would like to thank the workshop organizers for their great efforts and hard work in proposing the workshop, selecting the papers, the interesting programmes and for the arrangements of the workshop during the conference days. We are grateful to the INCoS-2019 Conference Chairs for inviting us to be the Workshops Co-chairs of the conference.

We hope you will enjoy the workshop programmes.

Masato Tsuru
Jakub Nalepa
INCoS-2019 Workshops Co-chairs

INCoS-2019 Organizing Committee

Honorary Co-chairs

Seigo Kitano	Oita University, Japan
Makoto Takizawa	Hosei University, Japan

General Co-chairs

Hiroaki Nishino	Oita University, Japan
Leonard Barolli	Fukuoka Institute of Technology, Japan

Programme Co-chairs

Hiroyoshi Miwa	Kwansei Gakuin University, Japan
Flora Amato	University of Naples, Italy

Workshops' Co-chairs

Masato Tsuru	Kyushu Institute of Technology, Japan
Jakub Nalepa	Silesian University of Technology, Poland

International Advisory Committee

Vincenzo Loia	University of Salerno, Italy
Amélia Ferreira da Silva	CEOS. PP, Politécnico do Porto, Portugal
Christine Strauss	University of Vienna, Austria
Fang-Yie Leu	Tunghai University, Taiwan
Albert Zomaya	University of Sydney, Australia

International Liaison Co-chairs

Pavel Kromer   Technical Univ. of Ostrava, Czech Republic
Kin Fun Li   University of Victoria, Canada
Ana Azevedo   CEOS.PP, Politécnico do Porto, Portugal
Joseph Tan   McMaster University, Canada

Award Co-chairs

Marek Ogiela   AGH Univ. of Science and Technology, Poland
Vaclav Snasel   Technical Univ. of Ostrava, Czech Republic

Web Administrator Co-chairs

Miralda Cuka   Fukuoka Institute of Technology, Japan
Kevin Bylykbashi   Fukuoka Institute of Technology, Japan
Donald Elmazi   Fukuoka Institute of Technology, Japan

Local Arrangement Co-chairs

Ken'ichi Furuya   Oita University, Japan
Makoto Nakashima   Oita University, Japan

Finance Chair

Makoto Ikeda   Fukuoka Institute of Technology, Japan

Steering Committee Chair

Leonard Barolli   Fukuoka Institute of Technology, Japan

Track Areas and PC Members

Track 1: Data Mining, Machine Learning and Collective Intelligence
Track Co-chairs

Carson K. Leung   University of Manitoba, Canada
Thomas Lenhard   Comenius University in Bratislava, Slovakia

PC Members

Alfredo Cuzzocrea	University of Trieste, Italy
Fan Jiang	University of Northern British Columbia, Canada
Wookey Lee	Inha University, Korea
Oluwafemi A. Sarumi	Federal University of Technology—Akure, Nigeria
Syed K. Tanbeer	University of Manitoba, Canada
Tomas Vinar	Comenius University in Bratislava, Slovakia
Kin Fun Li	University of Victoria, Canada

Track 2: Fuzzy Systems and Knowledge Management
Track Co-chairs

Marek Ogiela	AGH University of Science and Technology, Poland
Morteza Saberi	UNSW Canberra, Australia
Chang Choi	Chosun University, Korea

PC Members

Hsing-Chung (Jack) Chen	Asia University, Taiwan
Been-Chian Chien	National University, Taiwan
Junho Choi	Chosun University, Korea
Farookh Khadeer Hussain	University of Technology Sydney, Australia
Hae-Duck Joshua Jeong	Korean Bible University, Korea
Hoon Ko	Sungkyunkwan University, Korea
Natalia Krzyworzeka	AGH University of Science and Technology, Poland
Libor Mesicek	J. E. Purkinje University, Czech Republic
Lidia Ogiela	Pedagogical University of Cracow, Poland
Su Xi	Hohai University, China
Ali Azadeh	Tehran University, Iran
Jin Hee Yoon	Sejong University, South Korea
Hamed Shakouri	Tehran University, Iran
Jee-Hyong Lee	Sungkyunkwan University, South Korea
Jung Sik Jeon	Mokpo National Maritime University, South Korea

Track 3: Grid and P2P Distributed Infrastructure for Intelligent Networking and Collaborative Systems

Track Co-chairs

Aneta Poniszewska-Maranda	Lodz University of Technology, Poland
Michal Gregus	Comenius University in Bratislava, Slovakia
Takuya Asaka	Tokyo Metropolitan University, Japan

PC Members

Jordi Mongay Batalla	National Institute of Telecommunications, Poland
Nik Bessis	Edge Hill University, UK
Aniello Castiglione	University of Naples Parthenope, Italy
Naveen Chilamkurti	La Trobe University, Australia
Radu-Ioan Ciobanu	University Politehnica of Bucharest, Romania
Alexandru Costan	IRISA/INSA Rennes, France
Vladimir-Ioan Cretu	University Politehnica of Timisoara, Romania
Marc Frincu	West University of Timisoara, Romania
Rossitza Ivanova	Goleva Technical University of Sofia, Bulgaria
Dorian Gorgan	Technical University of Cluj-Napoca, Romania
Mauro Iacono	Seconda Universita' degli Studi di Napoli, Italy
George Mastorakis	Technological Educational Institute of Crete, Greece
Constandinos X. Mavromoustakis	University of Nicosia, Cyprus
Gabriel Neagu	National Institute for Research and Development in Informatics, Romania
Rodica Potolea	Technical University of Cluj-Napoca, Romania
Radu Prodan	University of Innsbruck, Austria
Ioan Salomie	Technical University of Cluj-Napoca, Romania
George Suciu	BEIA International, Romania
Sergio L. Toral Marín	University of Seville, Spain
Radu Tudoran	European Research Center, Germany
Lucian Vintan	Lucian Blaga University, Romania
Mohammad Younas	Oxford Brookes University, UK

Track 4: Nature's Inspired Parallel Collaborative Systems
Track Co-chairs

Mohammad Shojafar	University of Rome, Italy
Zahra Pooranian	University of Padua, Italy
Daichi Kominami	Osaka University, Japan

PC Members

Francisco Luna	University of Málaga, Spain
Sergio Nesmachnow	University La Republica, Uruguay
Nouredine Melab	University of Lille 1, France
Julio Ortega	University of Granada, Spain
Domingo Giménez	University of Murcia, Spain
Gregoire Danoy	University of Luxembourg, Luxembourg
Carolina Salto	University of La Pampa, Argentina
Stefka Fidanova	IICT-BAS, Bulgaria
Michael Affenzeller	Upper Austria University, Austria
Hernan Aguirre	Shinshu University, Japan
Francisco Chicano	University of Malaga, Spain
Javid Tahery	Karlstad University, Sweden
Enrique Domínguez	University of Málaga, Spain
Guillermo Leguizamón	Universidad Nacional de San Luis, Argentina
Konstantinos Parsopoulos	University of Ioannina, Greece
Carlos Segura	CIMAT, Mexico
Eduardo Segredo	Edinburgh Napier University, UK
Javier Arellano	University of Málaga, Spain

Track 5: Security, Organization, Management and Autonomic Computing for Intelligent Networking and Collaborative Systems

Track Co-chairs

Jungwoo Ryoo	Pennsylvania State University, USA
Simon Tjoa	St. Pölten University of Applied Sciences, Austria

PC Members

Nikolaj Goranin	Vilnius Gediminas Technical University, Lithuania
Kenneth Karlsson	Lapland University of Applied Sciences, Finland
Peter Kieseberg	SBA Research, Austria
Hyoungshick Kim	Sungkyunkwan University, Korea
Hae Young Lee	DuDu IT, Korea
Moussa Ouedraogo	Wavestone, Luxembourg
Sebastian Schrittwieser	St. Pölten University of Applied Sciences, Austria
Syed Rizvi	Pennsylvania State University, USA

Track 6: Software Engineering, Semantics and Ontologies for Intelligent Networking and Collaborative Systems

Track Co-chairs

Kai Jander	University of Hamburg, Germany
Giovanni Cozzolino	University of Naples "Federico II", Italy

PC Members

Tsutomu Kinoshita	Fukui University of Technology, Japan
Kouji Kozaki	Osaka University, Japan
Hiroyoshi Miwa	Kwansei Gakuin University, Japan
Burin Rujjanapan	Nation University, Thailand
Hiroshi Kanasugi	Tokyo University, Japan
Takayuki Shimotomai	Advanced Simulation Technology of Mechanics R&D, Japan
Jinattaporn Khumsri	Fukui University of Technology, Japan
Rene Witte	Concordia University, Canada
Amal Zouaq	University of Ottawa, Canada
Jelena Jovanovic	University of Belgrade, Serbia
Zeinab Noorian	Ryerson University, Canada
Faezeh Ensan	Ferdowsi University of Mashhad, Ireland
Alireza Vazifedoost	Sun Life Financial, Canada
Morteza Mashayekhi	Royal Bank of Canada, Canada

Track 7: Wireless and Sensor Systems for Intelligent Networking and Collaborative Systems

Track Co-chairs

Do van Thanh	Telenor & Oslo Metropolitan University, Norway
Salem Lepaja	University of Pristina @ AAB College in Pristina, Kosovo
Shigeru Kashihara	Nara Institute of Science and Technology, Japan

PC Members

Dhananjay Singh	HUFS, Korea
Shirshu Varma	IIIT-Allahabad, India
B. Balaji Naik	NIT-Sikkim, India
Sayed Chhattan Shah	HUFS, Korea, USA
Madhusudan Singh	Yonsei University, Korea

Irish Singh	Ajou University, Korea
Gaurav Tripathi	Bharat Electronics Limited, India
Jun Kawahara	Kyoto University, Japan
Muhammad Niswar	Hasanuddin University, Indonesia
Vasaka Visoottiviseth	Mahidol University, Thailand
Jane Louie F. Zamora	Weathernews Inc., Japan

Track 8: Service-Based Systems for Enterprise Activities Planning and Management

Track Co-chairs

Corinna Engelhardt-Nowitzki	University of Applied Sciences, Austria
Natalia Kryvinska	Comenius University in Bratislava, Slovakia

PC Members

Maria Bohdalova	Comenius University in Bratislava, Slovakia
Ivan Demydov	Lviv Polytechnic National University, Ukraine
Jozef Juhar	Technical University of Košice, Slovakia
Nor Shahniza Kamal Bashah	Universiti Teknologi MARA, Malaysia
Eric Pardede	La Trobe University, Australia
Francesco Moscato	University of Campania, Italy
Tomoya Enokido	Rissho University, Japan
Olha Fedevych	Lviv Polytechnic National University, Ukraine

Track 9: Next-Generation Secure Network Protocols and Components
Track Co-chairs

Xu An Wang	Engineering University of CAPF, China
Mingwu Zhang	Hubei University of Technology, China

PC Members

Fushan Wei	The PLA Information Engineering University, China
He Xu	Nangjing University of Posts and Telecommunications, China
Yining Liu	Guilin University of Electronic Technology, China
Yuechuan Wei	Engineering University of CAPF, China
Weiwei Kong	Xi'an University of Posts & Telecommunications, China

| Dianhua Tang | CETC 30, China |
| Hui Tian | Huaqiao University, China |

Track 10: Big Data Analytics for Learning, Networking and Collaborative Systems

Track Co-chairs

Santi Caballe	Open University of Catalonia, Spain
Francesco Orciuoli	University of Salerno, Italy
Shigeo Matsubara	Kyoto University, Japan

PC Members

Jordi Conesa	Open University of Catalonia, Spain
Soumya Barnejee	National Institute of Applied Sciences, France
David Bañeres	Open University of Catalonia, Spain
Nicola Capuano	University of Salerno, Italy
Nestor Mora	Open University of Catalonia, Spain
Jorge Moneo	University of San Jorge, Spain
David Gañán	Open University of Catalonia, Spain
Isabel Guitart	Open University of Catalonia, Spain
Elis Kulla	Okayama University of Science, Japan
Evjola Spaho	Polytechnic University of Tirana, Albania
Florin Pop	University Politehnica of Bucharest, Romania
Kin Fun Li	University of Victoria, Canada
Miguel Bote	University of Valladolid, Spain
Pedro Muñoz	University of Carlos III, Spain

Track 11: Cloud Computing: Services, Storage, Security and Privacy
Track Co-chairs

| Javid Taheri | Karlstad University, Sweden |
| Shuiguang Deng | Zhejiang University, China |

PC Members

Ejaz Ahmed	National Institute of Standards and Technology, USA
Asad Malik	National University of Science and Technology, Pakistan
Usman Shahid	COMSATS University Islamabad, Pakistan
Assad Abbas	North Dakota State University, USA
Nikolaos Tziritas	Chinese Academy of Sciences, China
Osman Khalid	COMSATS University Islamabad, Pakistan

Kashif Bilal	Qatar University, Qatar
Javid Taheri	Karlstad University, Sweden
Saif Rehman	COMSATS University Islamabad, Pakistan
Inayat Babar	University of Engineering and Technology, Pakistan
Thanasis Loukopoulos	Technological Educational Institute of Athens, Greece
Mazhar Ali	COMSATS University Islamabad, Pakistan
Tariq Umer	COMSATS University Islamabad, Pakistan

Track 12: Intelligent Collaborative Systems for Work and Learning, Virtual Organization and Campuses

Track Co-chairs

| Nikolay Kazantsev | National Research University, Russia |
| Monika Davidekova | Comenius University in Bratislava, Slovakia |

PC Members

Luis Alberto Casillas	University of Guadalajara, Mexico
Nestor Mora	University of Cadiz, Spain
Michalis Feidakis	University of Aegean, Greece
Sandra Isabel Enciso	Fundación Universitaria Juan N. Corpas, Colombia
Nicola Capuano	University of Salerno, Italy
Rafael Del Hoyo	Technological Center of Aragon, Spain
George Caridakis	University of Aegean, Greece
Kazunori Mizuno	Takushoku University, Japan
Satoshi Ono	Kagoshima University, Japan
Yoshiro Imai	Kagawa University, Japan
Takashi Mitsuishi	Tohoku University, Japan
Hiroyuki Mitsuhara	Tokushima University, Japan

Track 13: Social Networking and Collaborative Systems
Track Co-chairs

Nicola Capuano	University of Salerno, Italy
Dusan Soltes	Comenius University in Bratislava, Slovakia
Yusuke Sakumoto	Kwansei Gakuin University, Japan

PC Members

Santi Caballé	Open University of Catalonia, Spain
Thanasis Daradoumis	University of the Aegean, Greece
Angelo Gaeta	University of Salerno, Italy
Christian Guetl	Graz University of Technology, Austria
Miltiadis Lytras	American College of Greece, Greece
Agathe Merceron	Beuth University of Applied Sciences Berlin, Germany
Francis Palma	Screaming Power, Canada
Krassen Stefanov	Sofia University "St. Kliment Ohridski", Bulgaria
Daniele Toti	Roma Tre University, Italy
Jian Wang	Wuhan University, China
Jing Xiao	South China Normal University, China
Jian Yu	Auckland University of Technology, Australia
Aida Masaki	Tokyo Metropolitan University, Japan
Takano Chisa	Hiroshima City University, Japan
Sho Tsugawa	Tsukuba University, Japan

Track 14: Intelligent and Collaborative Systems for E-Health
Track Co-chairs

Massimo Esposito	Institute for High Performance Computing and Networking—National Research Council of Italy, Italy
Mario Ciampi	Institute for High Performance Computing and Networking—National Research Council of Italy, Italy
Giovanni Luca Masala	University of Plymouth, UK

PC Members

Tim Brown	Australian National University, Australia
Mario Marcos do Espirito Santo	Universidad Stadual de Montes Claros, Brazil
Jana Heckenbergerova	University Pardubice, Czech Republic
Zdenek Matej	Masaryk University, Czech Republic
Michal Musilek	University Hradec Kralove, Czech Republic
Michal Prauzek	VSB-TU Ostrava, Czech Republic
Vaclav Prenosil	Masaryk University, Czech Republic
Alvin C. Valera	Singapore Management University, Singapore
Nasem Badr El Din	University of Manitoba, Canada
Emil Pelikan	Academy of Sciences, Czech Republic
Joanne Nightingale	National Physical Laboratory, UK
Tomas Barton	University of Alberta, Canada

Track 15: Mobile Networking and Applications

Track Co-chairs

Miroslav Voznak	VSB-Technical University of Ostrava, Czech Republic
Akihiro Fujihara	Chiba Institute of Technology, Japan
Lukas Vojtech	Czech Technical University in Prague, Czech Republic

PC Members

Nobuyuki Tsuchimura	Kwansei Gakuin University, Japan
Masanori Nakamichi	Fukui University of Technology, Japan
Masahiro Shibata	Kyushu Institute of Technology, Japan
Yusuke Ide	Kanazawa Institute of Technology, Japan
Takayuki Shimotomai	Advanced Simulation Technology of Mechanics R&D, Japan
Dinh-Thuan Do	Ton Duc Thang University, Vietnam
Floriano De Rango	University of Calabria, Italy
Homero Toral-Cruz	University of Quintana Roo, Mexico
Remigiusz Baran	Kielce University of Technology, Poland
Mindaugas Kurmis	Klaipeda State University of Applied Sciences, Lithuania
Radek Martinek	VSB-Technical University of Ostrava, Czech Republic
Mauro Tropea	University of Calabria, Italy
Gokhan Ilk	Ankara University, Turkey
Shino Iwami	Microsoft, Japan

INCoS-2019 Reviewers

Amato Flora
Barolli Admir
Barolli Leonard
Caballé Santi
Capuano Nicola
Chen Xiaofeng
Cui Baojiang
Daradoumis Thanasis
Elmazi Donald
Enokido Tomoya
Esposito Christian
Fenza Giuseppe

Ficco Massimo
Fiore Ugo
Fujihara Akihiro
Fun Li Kin
Funabiki Nobuo
Gañán David
Hori Yoshiaki
Hsing-Chung Chen
Hussain Farookh
Hussain Omar
Ikeda Makoto
Joshua Hae-Duck

Juggapong Natwichai
Kohana Masaki
Kolici Vladi
Köppen Mario
Koyama Akio
Kromer Pavel
Kryvinska Natalia
Kulla Elis
Leu Fang-Yie
Li Yiu
Loia Vincenzo
Ma Kun
Maeda Hiroshi
Mangione Giuseppina Rita
Matsuo Keita
Messina Fabrizio
Miguel Jorge
Miwa Hiroyoshi
Nadeem Javaid
Nalepa Jakub
Nishino Hiroaki
Nowakowa Jana
Ogiela Lidia
Ogiela Marek
Palmieri Francesco

Pardede Eric
Poniszewska-Maranda Aneta
Rahayu Wenny
Rawat Danda
Sakaji Hiroki
Shibata Masahiro
Shibata Yoshitaka
Snasel Vaclav
Spaho Evjola
Suganuma Takuo
Sugita Kaoru
Sukumoto Yusuke
Takizawa Makoto
Terzo Olivier
Tsukamoto Kazuya
Tsuru Masato
Uchida Masato
Uchida Noriki
Uehara Minoru
Wang Xu An
Woungang Isaac
Younas Mohammad
Zhang Mingwu
Zhou Neng-Fa
Zomaya Albert

Welcome Message from WIND-2019 Workshop Organizers

Welcome to the 11th International Workshop on Information Network Design (WIND-2019), which is held in conjunction with the 11th International Conference on Intelligent Networking and Collaborative Systems (INCoS-2019), which is held from 5–7 September 2019, at Oita University, Japan.

Nowadays, the Internet is playing a role of social and economical infrastructure and is expected to support not only comfortable communication and information dissemination but also any kind of intelligent and collaborative activities in a dependable manner. However, the explosive growth of its usage with diversifying the communication technologies and the service applications makes it difficult to manage efficient sharing of the Internet. In addition, an inconsistency between Internet technologies and the human society forces a complex and unpredictable tension among end-users, applications and Internet service providers (ISPs).

It is thought, therefore, that the Internet is approaching a turning point and there might be the need for rethinking and redesigning the entire system composed of the human society, nature and the Internet. To solve the problems across multiple layers on a large-scale and complex system and to design the entire system of systems towards future information networks for human/social orchestration, a new tide of multiperspective and multidisciplinary research is essential. It will involve not only the network engineering (network routing, mobile and wireless networks, network measurement and management, high-speed networks, etc.) and the networked applications (robotics, distributed computing, human–computer interactions, Kansei information processing, etc.), but the network science (providing new tools to understand and control the huge-scale complex systems based on theories, e.g. graph theory, game theory, information theory, learning theory and statistical physics) and the social science (enabling safe, secure and human-centric application principles and business models).

The Information Network Design Workshop aims at exploring ongoing efforts in the theory and application on a wide variety of research fields related to the design of information networks and resource sharing in the networks. The workshop provides an opportunity for academic/industry researchers and professionals to share, exchange and review recent advances on information network design research. Original contribution describing recent modelling, analysis and experiment on network design research with particular, but not exclusive, regard to:

- Large-scale and/or complex networks;
- Cross-layered networks;
- Overlay and/or P2P networks;
- Sensor and/or mobile ad hoc networks;
- Delay-/disruption-tolerant networks;
- Social networks;
- Applications on networks;
- Fundamental theories for network design.

We would like to thank the organizing committee of INCoS-2019 International Conference for giving us the opportunity to organize the workshop. We also like to thank our Programme Committee members and referees and of course, all authors of the workshop for submitting their research works and for their participation.

We wish all participants and contributors to spend an event with high research impact, interesting discussions, exchange of research ideas, to pave future research cooperations.

Masaki Aida
Mario Koeppen
Hiroyoshi Miwa
Masato Tsuru
Masato Uchida
Neng-Fa Zhou
WIND-2019 Workshop Co-chairs

WIND-2019 Organizing Committee

WIND-2019 Workshop Co-chairs

Masaki Aida	Tokyo Metropolitan University, Japan
Mario Koeppen	Kyushu Institute of Technology, Japan
Hiroyoshi Miwa	Kwansei Gakuin University, Japan
Masato Tsuru	Kyushu Institute of Technology, Japan
Masato Uchida	Waseda University, Japan
Neng-Fa Zhou	The City University of New York, USA

Programme Committee

Yoshiaki Hori	Saga University, Japan
Hideaki Iiduka	Meiji University, Japan
Kenichi Kourai	Kyushu Institute of Technology, Japan
Kei Ohnishi	Kyushu Institute of Technology, Japan
Masahiro Sasabe	Nara Institute of Science and Technology, Japan
Kazuya Tsukamoto	Kyushu Institute of Technology, Japan

Welcome Message from FINCoS-2019 Workshop Organizer

Welcome to the 7th International Workshop on Frontiers in Intelligent Networking and Collaborative Systems (FINCoS-2019), which is held in conjunction with the 11th International Conference on Intelligent Networking and Collaborative Systems (INCoS-2019), from 5–7 September 2019, at Oita University, Japan.

The FINCoS-2019 covers the latest advances in the interdisciplinary fields of intelligent networking, social networking, collaborative systems, cloud-based systems and business intelligence, which lead to gain competitive advantages in business and academia scenarios. The ultimate aim is to stimulate research that will lead to the creation of responsive environments for networking and, at longer-term, the development of adaptive, secure, mobile and intuitive intelligent systems for collaborative work and learning.

Industry and academic researchers, professionals and practitioners are invited to exchange their experiences and present their ideas in this field. Specifically, the scope of FINCoS-2019 comprises research work and findings on intelligent networking, cloud and fog distributed infrastructures, security and privacy and data analysis. We would like to thank all authors of the workshop for submitting their research works and their participation. We would like to express our appreciation to the reviewers for their timely review and constructive feedback to authors.

We are looking forward to meeting you again in the forthcoming editions of the workshop.

Leonard Barolli
FINCoS-2019 Workshop Organizer

FINCoS-2019 Organizing Committee

Workshop Organizer

Leonard Barolli   Fukuoka Institute of Technology, Japan

Programme Committee

Santi Caballé	Open University of Catalonia, Spain
Makoto Ikeda	Fukuoka Institute of Technology, Japan
Kin Fun Li	University of Victoria, Canada
Shengli Liu	Shanghai Jiaotong University, China
Janusz Kacpryzk	Polish Academy of Science, Poland
Hiroaki Nishino	University of Oita, Japan
Makoto Takizawa	Hosei University, Japan
David Taniar	Monash University, Australia
Xu An Wang	CAPF Engineering University, P.R. China

Welcome Message from BDS-2019 Workshop Organizers

Welcome to the 5th International Workshop on Theory, Algorithms and Applications of Big Data Science (BDS-2019), which is held in conjunction with the 11th International Conference on Intelligent Networking and Collaborative Systems (INCoS-2019), from 5–7 September 2019, at Oita University, Japan.

Diverse multidisciplinary approaches are being continuously developed and advanced to address the challenges that big data research raises. In particular, the current academic and professional environments are working to produce algorithms, theoretical advance in big data science, to enable the full utilization of its potential, and better applications.

The proposed workshop focuses on the dissemination of original contributions to discuss and explore theoretical concepts, principles, tools, techniques and deployment models in the context of big data. Via the contribution of both academics and industry practitioners, the current approaches for the acquisition, interpretation and assessment of relevant information will be addressed to advance the state-of-the-art big data technology.

The workshop covers the following topics:

- Contributions should focus on (but not limited to) the following topics:
- Statistical and dynamical properties of big data;
- Applications of machine learning for information extraction;
- Hadoop and big data;
- Data and text mining techniques for big data;
- Novel algorithms in classification, regression, clustering and analysis;
- Distributed systems and cloud computing for big data;
- Big data applications;
- Theory, applications and mining of networks associated with big data;
- Large-scale network data analysis;
- Data reduction, feature selection and transformation algorithms;
- Data visualization;
- Distributed data analysis platforms;
- Scalable solutions for pattern recognition;

- Stream and real-time processing of big data;
- Information quality within big data;
- Threat detection in big data.

We would like to thank the organizing committee of INCoS-2019 International Conference for giving us the opportunity to organize the workshop and the Local Arrangement Chairs for facilitating the workshop organization.

We are looking forward to meeting you again in the forthcoming editions of the workshop.

Marcello Trovati
Mark Liptrott
Jeffrey Ray
Workshop Organizers

BDS-2019 Organizing Committee

Workshop Organizers

Marcello Trovati	Edge Hill University, UK
Mark Liptrott	Edge Hill University, UK
Jeffrey Ray	Edge Hill University, UK

Programme Committee

Georgios Kontonatsios	Edge Hill University, UK
Richard Conniss	University of Derby, UK
Ovidiu Bagdasar	University of Derby, UK
Peter Larcombe	University of Derby, UK
Stelios Sotiriadis	University of Toronto, Canada
Jer Hayes	IBM Research, Dublin Lab, Ireland
Xiaolong Xu	Nanjing University of Post and Telecommunications, China
Nan Hu	Nanjing University of Post and Telecommunications, China
Tao Lin	Nanjing University of Post and Telecommunications, China

Welcome Message from e-Business-2019 International Workshop Organizers

Welcome to the 5th International Workshop on Collaborative e-business Systems (e-Business-2019), which is held in conjunction with the 11th International Conference on Intelligent Networking and Collaborative Systems (INCoS-2019), at Oita University, Japan, during 5–7 September 2019.

The rapid expansion of business relationships and processes involved led to the emerging standards and infrastructure for business collaborations. Business large or small can no longer survive alone. The efficient and effective links with the business partners and consumers become critical. Overall, the collaborations occur between the communities of buyers, i.e. service consumers and sellers, i.e. service providers.

As much of the competition occurs between service providers and service consumers along the e-business value chains, the main theme of IWCBS is on collaborative e-business systems through aspects of business-IT alignment, business process integration, mobility, technology and tools, platforms and architectures, and applications.

The workshop aims to address the resource planning, modelling, coordination and integration in order to develop long-term sustainable and beneficial business relationships among all the partners and consumers along the value chains. Development of well-cooperated and coordinated e-business environment is crucial. Information technology has significant roles in supporting more competitive collaborative and integrated e-business systems. For business stakeholders, the long-term sustainability and efficiency are to be increasingly important. Indeed, to appropriately address the balance between the community of buyers and sellers through collaboration and support is becoming urgent. The technology trend of supply chain management and logistics is heading towards all aspects of the integration, coordination and intelligent use of the network-based resources. In practical deployment of the solutions, mobility and handheld devices are to be involved.

We are looking forward to meeting you again in the forthcoming editions of the workshop.

Leonard Barolli
Natalia Kryvinska
e-Business-2019 International Workshop Organizers

e-Business-2019 Organizing Committee

Workshop Organizers

Leonard Barolli	Fukuoka Institute of Technology, Japan
Natalia Kryvinska	Comenius University in Bratislava, Slovakia

Programme Committee

Maria Bohdalova	Comenius University in Bratislava, Slovakia
Ivanna Dronyuk	Lviv Polytechnic National University, Ukraine
Corinna Engelhardt-Nowitzki	University of Applied Sciences, Austria
Olha Fedevych	Lviv Polytechnic National University, Ukraine
Michal Gregus	Comenius University in Bratislava, Slovakia
Ivan Izonin	Lviv Polytechnic National University, Ukraine
Nikolay Kazantsev	University of Manchester, UK
Kamal Bashah Nor Shahniza	Universiti Teknologi MARA, Malaysia
Aneta Poniszewska-Maranda	Lodz University of Technology, Poland
Christine Strauss	University of Vienna, Austria
Do van Thanh	Telenor & Norwegian University of Science & Technology, Norway
Volodymyr Zhezhukha	Lviv Polytechnic National University, Ukraine
Corinna Engelhardt-Nowitzki	University of Applied Sciences, Austria
Max Lackner	University of Applied Sciences, Austria
Olha Fedevych	Lviv Polytechnic National University, Ukraine

Welcome Message from MaLICS-2019 International Workshop Organizer

Welcome to the 2nd International Workshop on Machine Learning in Intelligent and Collaborative Systems (MaLICS-2019), which is held in conjunction with the 11th International Conference on Intelligent Networking and Collaborative Systems (INCoS-2019), from 5–7 September 2019, at Oita University, Japan.

The era of big data is here and now. The amount of data produced every day grows tremendously in most real-life domains, including medical imaging, genomics, text categorization and computational biology. Hence, data-driven machine learning-powered approaches are consistently gaining research attention, and they are applied in multiple fields, with intelligent and collaborative systems not being an exception. In this workshop, we strive to present current advances on novel ideas and practical aspects concerning intelligent and collaborative systems, which benefit from machine learning, and deep learning in particular. Also, we hope to identify and highlight challenges, which are being faced by research and industrial communities in the field.

This workshop covers the latest advances in machine- and deep learning-powered intelligent systems that lead to gain competitive advantages in business and academia scenarios. The ultimate aim is to stimulate research that will lead to the creation of robust intelligent and collaborative systems applicable in a variety of fields (ranging from medical image analysis to smart delivery systems).

The workshop covers the topics of:

- Deep learning and neural Networks: applications, techniques and tools;
- Soft computing techniques for design of intelligent and collaborative systems;
- Computer vision and image processing in intelligent and collaborative systems;
- Bio-inspired algorithms for intelligent and collaborative systems;
- Hybrid algorithms for intelligent and collaborative systems;
- Heuristic and meta-heuristic algorithms in intelligent and collaborative systems;
- Machine learning in intelligent and collaborative systems;
- Advanced data analysis in intelligent and collaborative systems;
- Approaches, techniques and challenges in parallelizing machine learning-powered intelligent systems;

- Practical applications of intelligent and collaborative systems;
- Automated design and auto-tuning of deep learning and machine learning systems;
- Smart delivery systems (incl. autonomous vehicles);
- Medical image analysis in intelligent decision-support systems;
- Learning systems: approaches, techniques and tools;
- Multi- and hyperspectral imaging, analysis and processing.

We are looking forward to meeting you again in the forthcoming editions of the workshop.

Jakub Nalepa
MaLICS-2019 Workshop Organizer

MaLICS-2019 Organizing Committee

Workshop Organizer

Jakub Nalepa — Silesian University of Technology & Future Processing, Poland

Programme Committee

Aneta Poniszewska-Maranda	Lodz University of Technology, Poland
Leonard Barolli	Fukuoka Institute of Technology, Japan
Fatos Xhafa	Technical University of Catalonia, Spain
Shinji Sakamoto	Seikei University, Japan
Makoto Ikeda	Fukuoka Institute of Technology, Japan
Nadeem Javaid	COMSATS University Islamabad, Pakistan
Elis Kulla	Okayama University of Science, Japan
Admir Barolli	Aleksander Moisiu University of Durres, Albania
Donald Elmazi	Fukuoka Institute of Technology, Japan
Evjola Spaho	Polytechnic University of Tirana, Albania
Xu An Wang	Engineering University of CAPF, China

INCoS-2019 Keynote Talks

3D Graphics Applications for Education and Visualization

Yoshihiro Okada

Kyushu University, Kyushu, Japan

Abstract. In this talk, I will introduce the research activities about 3D graphics applications. We have developed environments for 3D graphics applications for many years. We have proposed a new system for 3D graphics applications called IntelligentBox. I will present the IntelligentBox and several applications especially for education and visualization. There are many education applications such as collaborative dental training system, Tai Chi-based physical therapy game and so on. While as visualization applications, we can mention room layout system, Time-tunnel (a visual analytics tool for multi-dimensional data), Treecube (a visualization tool for browsing 3D multimedia data), and so on. I will introduce the development activities of our laboratory for e-learning materials using 3D graphics and VR/AR, such as web-based interactive educational materials for Japanese history and IoT security, and games for medical education.

Secure Resilient Edge Cloud Designed Network

Tarek Saadawi

The City University of New York, City College, New York, USA

Abstract. IoT systems have put forth new requirements in all aspects of their existence: a diverse QoS requirements, resiliency of computing and connectivity, and the scalability to support massive number of end devices in a plethora of envisioned applications. The trustworthy IoT/cyber physical system (CPS) networking for smart and connected communities will be realized by distributed secure resilient Edge Cloud (EC). This distributed EC system will be a network of geographically distributed EC nodes, brokering between end-devices and Backend Cloud (BC) servers. In this talk, I will present three main topics in secure resilient cloud designed network: (1) resource management in mobile cloud computing; (2) information management in dynamic distributed databases; and (3) biological-inspired intrusion detection system (IDS). A focus in the presentation will be on the biological-inspired IDS.

Container-Leveraged Service Realization Challenges for Cloud-Native Computing

JongWon Kim

Gwangju Institute of Science & Technology (GIST), Gwangju, Korea

Abstract. Cloud-native computing, employing container-based microservices architecture, is accelerating its adoption for agile and scalable service deployment over worldwide multi-cloud infrastructure. In order to transparently enable diversified interconnections for container-based cloud-native computing, by leveraging SDN/NFV technology, we need to tie distributed IoT things through multi-site edge clouds to hyper-scale core clouds. Thus, in this talk, I first attempt to relate the open-source-driven development for CNI (Container Networking Interface) and CSI (Container Storage Interface) to the required container-enabled cloud-native computing/storage with end-to-end (i.e., IoT–SDN/NFV–Cloud) interconnections. Then, selected container-leveraged service realization challenges such as multi-tenant/multi-cluster Kubernetes orchestration, pvc (physical+virtual+containerized) harmonization, kernel-friendly accelerated and secured networking, and network-aware service meshes will be briefly discussed.

Contents

The 11th International Conference on Intelligent Networking and Collaborative Systems (INCoS-2019)

A Hierarchical Group of Peers in Publish/Subscribe Systems

Takumi Saito[1](✉), Shigenari Nakamura[1], Tomoya Enokido[2], and Makoto Takizawa[1]

[1] Hosei University, Tokyo, Japan
takumi.saito.3j@stu.hosei.ac.jp, nakamura.shigenari@gmail.com,
makoto.takizawa@computer.org
[2] Rissho University, Tokyo, Japan
eno@ris.ac.jp

Abstract. In the P2P (peer-to-peer) type of a topic-based PS (publish/subscribe) (P2PPS) model, each peer can be a subscriber and publisher and directly communicates with other peers in a distributed manner. Messages which have a common publication topic are considered to be related. In our previous studies, the topic vector is proposed to causally deliver related messages published by peers. In addition, messages are assumed to be broadcast to every peer in a system. In scalable systems, it is not easy, maybe impossible to broadcast messages to every peer. In this paper, we consider a two-layered group of multiple peers, which is composed of subgroups, in order to efficiently publish and receive messages. Here, messages are broadcast to every member peer in each subgroup but messages are unicast to subgroups. We propose a GBLC (Group-Based Linear time Causally ordering) protocol for a hierarchical group of peers in a P2P type of topic-based PS model. In the evaluation, we show the number of pairs of unnecessarily ordered messages can be reduced in the GBLC protocol than the linear time (LT) protocol.

Keywords: Topic-based publish/subscribe system · P2P model ·
Hierarchical P2PPS model · Gateway communication ·
Unnecessarily ordered messages ·
GBLC (Group-Based Linear time Causally ordering) protocol ·
Linear Time (LT) vector

1 Introduction

The publish/subscribe (PS) model is an event-driven model of a distributed system. In topic-based PS models [1,4,13,14], each subscriber process specifies interesting topics named subscription topics and a publisher process publishes a message with publication topics. Each message is delivered to only a subscriber process whose subscription topics include some publication topic of the message. In the P2PPS (P2P (peer-to-peer) type of topic-based PS) system [1,7,8],

L. Barolli et al. (Eds.): INCoS 2019, AISC 1035, pp. 3–13, 2020.
https://doi.org/10.1007/978-3-030-29035-1_1

each process is a peer which can both publish and subscribe messages in a distributed manner [2,5]. Each peer publishes a message with topics which denote the contents of the message. The topics of a message are publication topics. A peer receives only a message, some of whose publication topics of the message are subscription topics of the peer. Here, the peer is a target peer of the message. A pair of messages which have common topics are considered to be *related* with each other in the topic-based PS systems. A message m_1 causally precedes another message m_2 ($m_1 \rightarrow m_2$) if and only if (iff) the publication event of the message m_1 happens before the message m_2 according to the causality theory [6]. Each peer is required to causally deliver related messages in the P2PPS system. In papers [10–12], the TBC (topic-based causally) precedent relation among messages is defined. The OBC (object-based causally) precedent relation among messages is also defined where each message carries same objects which are characterized in terms of topics. Here, a pair of messages are considered to be related if the messages carry at least one common topic.

In order to causally deliver messages related with respect to topics, the topic vector is proposed [8,11]. The size of the topic vector is $O(l)$ for total number l of topics in a system. If topics are scalable, it is not easy for each message to carry the topic vector due to the message overhead. In addition, each peer is assumed to broadcast messages to all the peers. However, it is difficult, maybe impossible for each peer to broadcast messages to all the peers in scalable systems. In this paper, we propose a hierarchical group of peers, i.e. two-layered group of a P2PPS system. Here, a group of peers is composed of exclusive subgroups of peers. Each subgroup is composed of member peers. In each subgroup, messages are broadcast to every member peer. The size of each subgroup is decided to be so small that every message published by a peer can be efficiently broadcast to every member peer. Each subgroup includes a gateway peer. A group of gateway peers is a gateway subgroup. Messages published in a subgroup are forwarded to other subgroups by a gateway peer in the gateway subgroup with unicast communication. In order to causally deliver messages, a linear time (LT) vector is proposed in this paper. The tth element of an LT vector shows linear time [6] used in a subgroup G_t. Messages received by a peer are ordered in the LT vectors of the messages. The message length is $O(m)$ for number m of subgroups. In this paper, we propose a GBLC (Group-Based Linear time Causally ordering) protocol for a two-layered group of peers where related messages are causally delivered to target peers by using the LT vector. We show the number of unnecessarily ordered messages can be reduced in the LT vector compared with the linear time (LT) protocol in the evaluation.

In Sect. 2, we present a system model. In Sect. 3, we propose the GBLC protocol for a two-layered group of peers. In Sect. 4, we evaluate thee GBLC protocol.

2   System Model

A group G of peers is composed of subgroups $G_0, G_1, ..., G_g$ ($g \geq 1$) as shown in Fig. 1. Here, G_0 is a *gateway* subgroup and $G_1, ..., G_g$ are *member* subgroups.

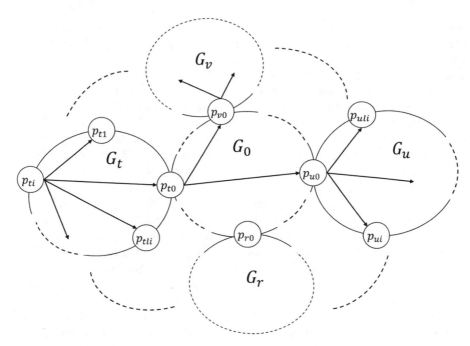

Fig. 1. Two-layered group.

Each member subgroup G_t is composed of member peers $p_{t1}, ..., p_{tl_t}$ $(l_t > 1)$ and a gateway peer p_{t0} $(t = 1, ..., g)$. Each member peer belongs to only one subgroup. The gateway subgroup G_0 is composed of gateway peers $p_{10}, ..., p_{g0}$ of subgroups $G_1, ..., G_g$, respectively. Each gateway peer p_{t0} belongs to both a member subgroup G_t and a gateway subgroup G_0.

In each subgroup G_t, every message is broadcast to all the member peers $p_{t1}, ..., p_{tl_t}$ and the gateway peer p_{t0} by taking advantage of the underlying network service. For example, all the peers of a member subgroup are interconnected in a local area network (LAN) with broadcast communication. In the gateway subgroup G_0, a gateway peer p_{t0} receives messages sent by a member peer p_{ti} in a subgroup G_t and forwards the messages to gateway peers of other subgroups by using the unicast communication. For example, gateway peers of a gateway subgroups communicate with one another by using TCP [3] in a wide area network (WAN). On receipt of a message m from another gateway peer p_{u0}, a gateway peer p_{t0} broadcasts the message m to every member peer in the member subgroup G_t. Here, every pair of messages have to be causally delivered to member peers in member subgroups.

We assume the underlying network of a subgroup is reliable. That is, a pair of messages m_1 and m_2 published by member peers are delivered to every common target member peer in the publication order without any message loss. Let T be a set of topics $t_1, ..., t_l$ $(l \geq 1)$ in a system. Each member peer p_{ti} of a subgroup G_t specifies a set $p_{ti}.S$ $(\subseteq T)$ of interesting topics. Here, the topic set $p_{ti}.S$ is referred to as *subscription* of the member peer p_{ti}. A member peer p_{ti} publishes a message m with publication topics in a subgroup G_t. The publication topics denote the contents of the message. Here, let $m.P$ ($\subseteq T$) be a *publication* of the message m which is a subset of the topic set T. A member peer p_{ti} receives a message m only if the publication $m.P$ of the message m and the subscription $p_{ti}.S$ of the peer p_{ti} include at least one common topic ($p_{ti}.S \cap m.P \neq \phi$). Here, the member peer p_{ti} is a target peer of the message m. If a peer p_{ti} publishes a message m, every target peer p_{tj} receives the message m in the member subgroup G_t. The gateway peer p_{t0} also receives the message m in the member subgroup G_t. Then, the gateway peer p_{t0} forwards the message m to another gateway peer p_{u0} of another member subgroup G_u. The gateway peer p_{u0} publishes the message m in the member subgroup G_u. A pair of messages m_1 and m_2 are *related* with each other iff (if and only if) $m_1.P \cap m_2.P \neq \phi$, i.e. the messages m_1 and m_2 carry at least one common publication topic.

In a member subgroup G_t, the subscription $p_{t0}.S$ of a gateway peer p_{t0} is a set of subscription topics of all the member peers $p_{t1}, ..., p_{tl_t}$, i.e. $p_{t0}.S = \cup_{p_{ti} \in G_t} p_{ti}.S$. We assume each gateway peer p_{t0} knows the subscription topics $p_{u0}.S$ of every other gateway peer p_{u0}. If a gateway peer p_{u0} sends a message m to another gateway peer p_{t0}, the gateway peer p_{t0} receives the message m only if $m.P \cap p_{t0}.S \neq \phi$. On receipt of a message m from a member peer p_{ti}, the gateway peer p_{t0} forwards the message m to every gateway peer p_{u0} where $p_{u0}.S \cap m.P \neq \phi$.

A message m_1 causally precedes a message m_2 ($m_1 \rightarrow m_2$) iff the publication event of the message m_1 happens before the publication event of the message m_2 according to the traditional causality theory [6]. In the P2PPS system, even if $m_1 \rightarrow m_2$, a target member peer of the messages m_1 and m_2 is not required to deliver a messages m_1 before another message m_2 if the messages m_1 and m_2 are not related with each other. A message m_1 has to be delivered before another message m_2 in a common target peer if $m_1 \rightarrow m_2$ and the messages m_1 and m_2 are related. It is decided whether or not a pair of messages are related by checking the publication topics of the message. We discuss how to causally deliver messages to member peers in a scalable system in this paper.

3   A GBLC (Group-Based Linear Time Causally Ordering) Protocol

A group G is a collection of peers which publish and receive messages. We propose a GBLC (Group-Based Linear time Causally ordering) protocol to causally deliver related messages to target peers in a group G of peers.

In this paper, we consider a group G is composed of a gateway subgroup G_0 and member subgroups $G_1, ..., G_g$ ($g > 1$) to efficiently realize a scalable P2PPS system. Each member subgroup G_t supports a linear clock [6] which is manipulated by member peers. Each member peer p_{ti} in a member subgroup G_t manipulates a linear time (LT) variable $p_{ti}.L_t$. Initially, $p_{ti}.L_t = 0$. Each time a peer p_{ti} publishes a message m, the LT variable $p_{ti}.L_t$ is incremented by one in the member peer p_{ti}. The message m carries the linear time $m.L_t$ in a field $m.L$. On receipt of a message m, a member peer p_{ti} changes the variable $p_{ti}.L_t$ with a larger one of $p_{ti}.L_t$ and linear time $m.L_t$ carried by the message m as discussed [6], i.e. $L_t = max(p_{ti}.L_t, m.L_t)$.

In a GBLC group G, each message m carries a linear time (LT) vector $L = \langle L_1, ..., L_g \rangle$. The tth element L_t shows the LT of a member subgroup G_t $(t = 1, ..., g)$. Here, $p_{ti}.L$ stands for a linear time vector $L = \langle L_1, ..., L_g \rangle$ of a member peer p_{ti} in a member subgroup G_t. Initially, $L = \langle 0, ..., 0 \rangle$. $m.L$ shows an LT vector carried by a message m. The length of each message m is $O(g)$ where g is the number of subgroups, since the message m carries the field $m.L$.

Each member peer p_{ti} of a member subgroup G_t publishes and receives a message m by manipulating the LT vectors $p_{ti}.L$ and $m.L$.

First, a member peer p_{ti} publishes a message m in a subgroup G_t as follows:

[Publication]

1. The tth element L_t in the LT vector $p_{ti}.L = \langle L_1, ..., L_g \rangle$ is incremented by one in a peer p_{ti}, i.e. $p_{ti}.L_t = p_{ti}.L_t + 1$;
2. $m.L = p_{ti}.L \ (= \langle L_1, ..., L_g \rangle)$;
3. $m.P =$ publication topics $(\subseteq T)$;
4. The peer p_{ti} **publishes** the message m;

Each member peer p_{ti} first increments the LT variable $p_{ti}.L_t$ by one. Then, the element $m.L_t$ in the LT vector field $m.L$ of a message m is changed with the variable $p_{ti}.L_t$. Then, the member peer p_{ti} publishes the message m.

Messages published by each member peer p_{ti} are delivered to the gateway peer p_{t0} in a subgroup G_t. On receipt of a message m from a member peer p_{ti}, a gateway peer p_{t0} forwards the message m to other gateway peers in the gateway subgroup G_0 by using the unicast communication like TCP.

[Gateway peer receives a message in a member subgroup]

1. A gateway peer p_{t0} receives a message m published by a member peer p_{ti} in a member subgroup G_t;
2. The gateway peer p_{t0} **forwards** the message m to every gateway peer p_{u0} of a member subgroup G_u such that $p_{u0}.S \cap m.P \neq \phi$ in the gateway subgroup G_0;

Messages sent by another gateway peer p_{u0} are delivered to a gateway peer p_{t0} in a gateway subgroup G_0. On receipt of a message m from another gateway peer p_{u0}, a gateway peer p_{t0} of a member subgroup G_t behaves as follows:

[Gateway peer receives a message in a gateway subgroup]

1. A gateway peer p_{t0} receives a message m from another gateway peer p_{u0} in a gateway subgroup G_0;
2. If $p_{t0}.S \cap m.P = \phi$, the gateway peer p_{t0} **neglects** the message m;
3. $p_{t0}.L_u = max(p_{t0}.L_u, m.L_u)$ (for $u = 1, ..., g$);
4. $p_{t0}.L_t = max(p_{t0}.L_1, ..., p_{t0}.L_g)$;

5. The gateway peer p_{t0} **publishes** the message m in the member subgroup G_t;

On receipt of a message m from the gateway subgroup G_0, a gateway peer p_{t0} first changes each LT vector time $p_{t0}.L_u$ with a larger one of $p_{t0}.L_u$ and $m.L_u$ for $u = 1, ..., g$. Then, the largest one of the linear time $p_{t0}.L_1, ..., p_{t0}.L_g$ is taken to be the linear time (LT) variable $p_{t0}.L_t$, i.e. $p_{t0}.L_t = max(p_{t0}.L_1, ..., p_{t0}.L_g)$.

Messages published by member peers in another member subgroup G_u arrives at a member peer p_{ti} in a member subgroup G_t. A member peer p_{ti} receives a message m in a member subgroup G_t as follows:

[Receipt of a message]

1. A message m arrives at a member peer p_{ti} in a member subgroup G_t;
2. $p_{ti}.L_u = max(p_{ti}.L_u, m.L_u)$ (for $u = 1, ..., g$);
3. $p_{ti}.L_t = max(p_{ti}.L_1, ..., p_{ti}.L_g)$;
4. **If** $m.P \cap p_{ti}.S = \phi$, the member peer p_{ti} **neglects** the message m;
5. The member peer p_{ti} **receives** the message m;

On receipt at a message m, a member peer p_{ti} takes a larger one of $p_{ti}.L_u$ and $m.L_u$ as a new value of the variable $p_{ti}.L_u$ for $u = 1, ..., g$. In addition, the variable $p_{ti}.L_t$ is changed with the largest one of $p_{ti}.L_1, ..., p_{ti}.L_g$ in the LT vector $p_{ti}.L$.

The size of a message m is $O(g)$ for number g of member subgroups. In the topic vector [9–11], the size of a message m is $O(l)$ for number l of topics. The number g of member subgroups is smaller than the number l of topics. Furthermore, the topic set T might be changed by adding new topics and deleting absolute topics. It is not easy to change the topic vector as the membership of the topic set T is changed.

In the GBLC protocol, the publication topics $m.P$ of each message m and subscription topics $p_{ti}.S$ of each member peer p_{ti} are implemented in a bitmap to reduce the size of the message m. The bitmap is l bits long for number l of topics.

The following property holds for LT vectors.

[Property] If a message m_1 causally precedes a message m_2 $(m_1 \rightarrow m_2)$, $m_1.L < m_2.L$.

On receipt of messages, a member peer p_{ti} stores the messages in the receipt buffer RB_{ti}. Then, messages in the buffer RB_{ti} are ordered in the LT vector. Messages in the receipt buffer RB_{ti} are delivered in the order of LT vectors, i.e. the message m_1 is delivered before the message m_2 $(m_1 \prec m_2)$ if $m_1.L < m_2.L$. However, even if $m_1.L < m_2.L$, m_1 may not causally precedes m_2.

Suppose every member peer manipulates a linear time variable LT [6]. Each message m carries the linear time $m.LT$ which is linear time of a source member peer. Here, each member peer p_{ti} increments $p_{ti}.LT$ by one and sends a message m with $m.LT$ $(= p_{ti}.LT)$. On receipt of a message m, a member peer p_{ti} changes the variable LT with $max(p_{ti}.LT, m.LT)$. Here, if $m_1 \rightarrow m_2$, $m_1.LT < m_2.LT$.

[Example1] In Fig. 2, there are a pair of member subgroups G_t and G_u which are interconnected in a gateway subgroup G_0. The gateway subgroup G_0 is

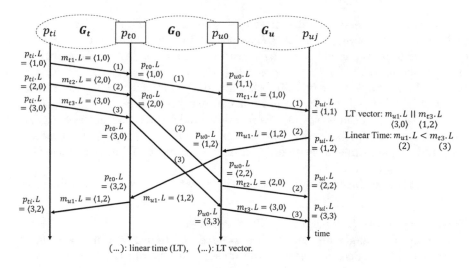

Fig. 2. Linear time (LT) vector.

composed of a pair of gateway peers p_{t0} and p_{u0}. The LT vector L is composed of two elements $\langle L_1, L_2 \rangle$ since there are two member subgroups G_t and G_u. Initially, $L = \langle 0, 0 \rangle$ in every member peer. A member peer p_{ti} in the member subgroup G_t first publishes a message m_{t1} where $m_{t1}.L = \langle 1, 0 \rangle$ by incrementing the first element L_1 by one.

On receipt of the message m_{t1}, a gateway peer p_{t0} forwards the message m_{t1} to the gateway peer p_{u0}. The gateway peer p_{u0} receives the message m in the gateway subgroup G_u. Here, since $m.L_t$ $(= 1) > p_{u0}.L_t$ $(= 0)$, L_t in the LT vector $p_{u0}.L$ is changed with 1, i.e. $p_{u0}.L = \langle 1, 0 \rangle$. Then, the element L_u in the LT vector $p_{u0}.L$ is changed with the maximum value of $p_{u0}.L_t$ $(= 1)$ and $p_{u0}.L_u$ $(= 0)$, i.e. 1. The LT vector $p_{u0}.L = \langle 1, 1 \rangle$. The message m_{t1} carries the LT vector $m_{t1}.L = \langle 1, 0 \rangle$ to target member peers in the member subgroup G_u.

On receipt of the message m_{ti}, a member peer p_{uj} changes the LT vector $p_{uj}.L$ with $\langle 1, 1 \rangle$ in a way similar to the gateway peer p_{u0}. Then, the member peer p_{uj} publishes a message m_{u1} with the LT vectors $m_{u1}.L = \langle 1, 2 \rangle$. The member peer p_{ti} publishes a pair of messages m_{t2} and m_{t3} with $m_{t2}.L = \langle 2, 0 \rangle$ and $m_{t3}.L = \langle 3, 0 \rangle$, respectively.

The gateway peer p_{t0} receives the message m_{u1} with the LT vector $m_{u1}.L = \langle 1, 2 \rangle$ in the gateway subgroup G_0. The LT vector $p_{t0}.L$ is changed with $\langle 3, 2 \rangle$ since $p_{t0}.L_u$ is 0 but $m_{ui}.L_u$ is 2. The gateway peer p_{t0} receives the message m_{u1} in the gateway subgroup G_0 and publishes the message m_{u1} in the member subgroup G_t.

The member peer p_{ti} receives the message m_{u1} with $m_{u1}.L = \langle 1, 2 \rangle$, the LT vector $p_{ti}.L$ of the peer p_{ti} is changed with $\langle 3, 2 \rangle$. The message m_{t1} causally precedes the message m_{u1} $(m_{t1} \rightarrow m_{u1})$. Here, $m_{t1}.L$ $(= \langle 1, 0 \rangle) < m_{u1}.L$ $(= \langle 1, 2 \rangle)$. The member peer p_{ti} delivers the message m_{t1} before the message m_{u1}.

Next, suppose each member peer and gateway peer manipulates linear time (LT) and each message carries LT as shown in Fig. 2. The linear time LT is composed of only one element (LT). Initially, $LT = (0)$ in every member peer. Here, since $m_{u1}.LT \ (= (2)) < m_{t3}.LT \ (= (3))$, the message m_{u1} is delivered before the message m_{t3}. On the other hand, a pair of the LT vectors $m_{u1}.L$ $(= \langle 1, 2 \rangle)$ and $m_{t3}.L \ (= \langle 3, 0 \rangle)$ are not comparable in the LT vector $(m_{u1}.L \parallel m_{t3}.L)$. Hence, the message m_{u1} is not delivered before the message m_{t3} since the message m_{t3} is received before the message m_{u1}. Thus, a pair of messages m_{u1} and m_{t3} are unnecessarily ordered in the LT protocol but not in the GBLC protocol.

The following property [4] holds for the linear time vector L and linear time LT.

[Property] If $m_1.L < m_2, L$, $m_1.LT < m_2.LT$.

A pair of messages m_1 and m_2 are *unnecessarily ordered* in a peer p_{ti} with a protocol iff the peer p_{ti} receives the message m_2 before the message m_1 and the message m_1 is delivered before the message m_2 but $m_1 \rightarrow m_2$ does not hold.

4   Evaluation

We evaluate the GBLC protocol in terms of number of messages unnecessarily ordered compared with the LT protocol. A system includes member peers $p_1, ..., p_n \ (n \geq 1)$. A group G is composed of a gateway subgroup G_0 and member subgroups $G_1, ..., G_g \ (g > 1)$. Let T be a set of topics $t_1, ... \ t_l \ (l \geq 1)$ in a system.

Each member peer p_{ti} randomly publishes messages. That is, the publication time $m.PBT$ of a message m published by a member peer p_{ti} is randomly decided from time 0 to time $maxT$ [time unit (tu)]. In the evaluation, $maxT$ is 500 [tu]. At each time unit, a member peer p_{ti} is randomly taken in all peers. Then, the member peer p_{ti} creates one message m which carries contents and p_{ti} publishes the message m to all member peers and the gateway peer p_{t0} in the member subgroup G_t. The receiving time $m.RVT_j$ of a message m of each member peer p_{tj} is $m.PBT + \delta_{t_{ij}}$. Here, the delay time $\delta_{t_{ij}}$ between a pair of member peers p_{ti} and p_{tj} in the member subgroup G_t is randomly taken from 1 to 10 [tu]. The delay time δ_{tu_0} between a pair of gateway peers p_{t0} and p_{u0} in the gateway subgroup G_0 is also randomly taken from 1 to 10 [tu]. If the delay time of each message is 0, each message arrives at a target peer and gateway peer. If the gateway peer p_{t0} receives a message m from a member peer p_{ti} in a member subgroup G_t, the gateway peer p_{t0} forwards the message m to other gateway peers. If the gateway peer p_{u0} receives the message m from another gateway peer p_{t0}, the gateway peer p_{u0} forwards the message m to all the member peers in the member subgroup G_u.

In the evaluation, publication and subscription topics of each member peer p_{ti} are randomly taken from the topic set T. Each member peer p_{ti} manipulates the linear time LT vector variable $p_{ti}.L$. Each message m also carries the linear time vector $m.L$. Each member peer p_{ti} also manipulates the linear time LT variable $p_{ti}.LT$ and each message m carries the linear time $m.LT$. We show the number of pairs of messages m_1 and m_2 such as $m_1.LT < m_2.LT$ but $m_1.L$ and $m_2.L$ are not comparable. Here, a pair of messages m_1 and m_2 are unnecessarily ordered in the LT protocol but not in the GBLC protocol.

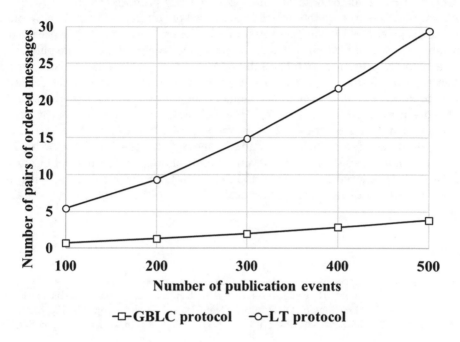

Fig. 3. Evaluation LT and GBLC protocol.

Figure 3 shows the number of pairs of messages ordered in the LT protocol and the GBLC protocol. Here, there are 30 member peers ($n = 30$), 5 member subgroups ($g = 5$) and 30 topics ($l = 30$). Each member subgroup includes six member peers and one gateway peer. Subscription topics of each member peer are randomly selected from all topics. The number of pairs of ordered messages in the GBLC protocol are about 85% smaller than the LT protocol. The number of subscription topics of each member peer is randomly selected out of numbers 1, ..., 10. The difference between the LT and GBLC protocols shows how much the number of pairs of unnecessarily ordered messages in the LT protocol can be reduced in the GBLC protocol. Here, about 86% of pairs of ordered messages in the LT protocol are reduced in the GBLC protocol. For example, about 20 pairs of unnecessarily ordered messages are reduced in the GBLC protocol compared with the LT protocol for 400 publication events.

5   Concluding Remarks

In the P2PPS system, a pair of messages which have a common publication topic are defined to be related. Related messages have to be causally delivered in every peer. Current information systems are scalable. Related messages are required to be efficiently causally delivered in a scalable system. In this paper, we newly proposed the GBLC protocol where a group is composed of subgroups of peers. Here, a vector of linear time (LT) is newly proposed to causally deliver related messages in a system which is composed of subgroups. In each subgroup, messages published by a member peer are broadcast to every member peer. Subgroups are interconnected with gateway peers. On receipt of a message published by a member peer in a member subgroup, a gateway peers forwards the message to other gateway peers by a unicast communication. The target gateway peer broadcasts the message in the member subgroup. If a message m_1 causally precedes a message m_2 ($m_1 \rightarrow m_2$), the LT vector $m_1.L$ of the message m_1 is smaller than the LT vector $m_2.L$ of the message m_2. In the linear time (LT) protocol [6], $m.LT$ shows the linear time of a message m. Here, we showed $m_1.LT < m_2.LT$ if $m_1.L < m_2.L$. In the evaluation, we showed the number of messages unnecessarily ordered can be reduced in the GBLC protocol compared with the linear time LT protocol. We are now evaluating the GBLC protocol in a scalable system.

References

1. Google alert. http://www.google.com/alerts
2. Blanco, R., Alencar, P.: Event models in distributed event based systems. In: Principles and Applications of Distributed Event-Based Systems, pp. 19–42 (2010)
3. Comer, D.E.: Internetworking with TCP/IP, vol. 1. Prentice Hall, Upper Saddle River (1991)
4. Eugster, P., Felber, P., Guerraoui, R., Kermarrec, A.: The many faces of publish/subscribe. ACM Comput. Surv. **35**(2), 114–131 (2003)
5. Hinze, A., Buchmann, A.: Principles and Applications of Distributed Event-Based Systems. IGI Grobal, Hershey (2010)
6. Lamport, L.: Time, clocks, and the ordering of event in a distributed systems. Commun. ACM **21**(7), 558–565 (1978)
7. Nakayama, H., Duolikun, D., Enokido, T., Takizawa, M.: Selective delivery of event messages in peer-to-peer topic-based publish/subscribe systems. In: Proceedings of the 18th International Conference on Network-Based Information Systems (NBiS 2015), pp. 379–386 (2015)
8. Nakayama, H., Duolikun, D., Enokido, T., Takizawa, M.: Reduction of unnecessarily ordered event messages in peer-to-peer model of topic-based publish/subscribe systems. In: Proceedings of IEEE the 30th International Conference on Advanced Information Networking and Applications (AINA 2016), pp. 1160–1167 (2016)
9. Nakayama, H., Ogawa, E., Nakamura, S., Enokido, T., Takizawa, M.: Topic-based selective delivery of event messages in peer-to-peer model of publish/subscribe systems in heterogeneous networks. In: Proceedings of the 18th International Conference on Network-Based Information Systems (WAINA 2017), pp. 1162–1168 (2017)
10. Saito, T., Nakamura, S., Duolikun, D., Enokido, T., Takizawa, M.: Object-based selective delivery of event messages in topic-based publish/subscribe systems. In: Proceedings of the 13th International Conference on Broadband and Wireless Computing, Communication and Applications (BWCCA 2018), pp. 444–455 (2018)

11. Saito, T., Nakamura, S., Duolikun, D., Enokido, T., Takizawa, M.: Evaluation of TBC and OBC precedent relations among messages in P2P type of topic-based publish/subscribe system. In: Proceedings of the Workshops of the 33rd International Conference on Advanced Information Networking and Applications (WAINA 2019), pp. 570–581 (2019)
12. Saito, T., Nakamura, S., Enokido, T., Takizawa, M.: A causally precedent relation among messages in topic-based publish/subscribe systems. In: Proceedings of the 21st International Conference on Network-Based Information Systems (NBiS 2018), pp. 543–553 (2018)
13. Tarkoma, S.: Publish/Subscribe System: Design and Principles, 1st edn. Wiley, Hoboken (2012)
14. Tarkoma, S., Rin, M., Visala, K.: The publish/subscribe internet routing paradigm (PSIRP): designing the future internet architecture. In: Future Internet Assembly, pp. 102–111 (2009)

Performance Analysis of WMNs by WMN-PSOHC-DGA Simulation System Considering Linearly Decreasing Inertia Weight and Linearly Decreasing Vmax Replacement Methods

Admir Barolli[1], Shinji Sakamoto[2(✉)], Seiji Ohara[3], Leonard Barolli[4], and Makoto Takizawa[5]

[1] Department of Information Technology, Aleksander Moisiu University of Durres, L.1, Rruga e Currilave, Durres, Albania
admir.barolli@gmail.com
[2] Department of Computer and Information Science, Seikei University, 3-3-1 Kichijoji-Kitamachi, Musashino-shi, Tokyo 180-8633, Japan
shinji.sakamoto@ieee.org
[3] Graduate School of Engineering, Fukuoka Institute of Technology, 3-30-1 Wajiro-Higashi, Higashi-Ku, Fukuoka 811-0295, Japan
seiji.ohara.19@gmail.com
[4] Department of Information and Communication Engineering, Fukuoka Institute of Technology, 3-30-1 Wajiro-Higashi, Higashi-Ku, Fukuoka 811-0295, Japan
barolli@fit.ac.jp
[5] Department of Advanced Sciences, Faculty of Science and Engineering, Hosei University, Kajino-Machi, Koganei-Shi, Tokyo 184-8584, Japan
makoto.takizawa@computer.org

Abstract. The Wireless Mesh Networks (WMNs) are becoming an important networking infrastructure because they have many advantages such as low cost and increased high speed wireless Internet connectivity. In our previous work, we implemented a Particle Swarm Optimization (PSO) and Hill Climbing (HC) based hybrid simulation system, called WMN-PSOHC, and a simulation system based on Genetic Algorithm (GA), called WMN-GA, for solving node placement problem in WMNs. Then, we implemented a hybrid simulation system based on PSOHC and distributed GA (DGA), called WMN-PSOHC-DGA. In this paper, we analyze the performance of WMNs using WMN-PSOHC-DGA simulation system considering Linearly Decreasing Inertia Weight Method (LDIWM) and Linearly Decreasing Vmax Method (LDVM). Simulation results show that a good performance is achieved for LDIWM compared with the case of LDVM.

© Springer Nature Switzerland AG 2020
L. Barolli et al. (Eds.): INCoS 2019, AISC 1035, pp. 14–23, 2020.
https://doi.org/10.1007/978-3-030-29035-1_2

1   Introduction

The wireless networks and devices are becoming increasingly popular and they provide users access to information and communication anytime and anywhere [2,6–8,10,14,20,25–27]. Wireless Mesh Networks (WMNs) are gaining a lot of attention because of their low cost nature that makes them attractive for providing wireless Internet connectivity. A WMN is dynamically self-organized and self-configured, with the nodes in the network automatically establishing and maintaining mesh connectivity among them-selves (creating, in effect, an ad hoc network). This feature brings many advantages to WMNs such as low up-front cost, easy network maintenance, robustness and reliable service coverage [1]. Moreover, such infrastructure can be used to deploy community networks, metropolitan area networks, municipal and corporative networks, and to support applications for urban areas, medical, transport and surveillance systems.

Mesh node placement in WMN can be seen as a family of problems, which are shown (through graph theoretic approaches or placement problems, e.g. [4,11]) to be computationally hard to solve for most of the formulations [31]. We consider the version of the mesh router nodes placement problem in which we are given a grid area where to deploy a number of mesh router nodes and a number of mesh client nodes of fixed positions (of an arbitrary distribution) in the grid area. The objective is to find a location assignment for the mesh routers to the cells of the grid area that maximizes the network connectivity and client coverage. Node placement problems are known to be computationally hard to solve [9,32]. In some previous works, intelligent algorithms have been recently investigated [3,5,12,15–18,22,23].

In [26], we implemented a Particle Swarm Optimization (PSO) and Hill Climbing (HC) based simulation system, called WMN-PSOHC. Also, we implemented another simulation system based on Genetic Algorithm (GA), called WMN-GA [3,13], for solving node placement problem in WMNs. Then, we designed a Hybrid Intelligent System Based on PSO, HC and DGA, called WMN-PSOHC-DGA [24].

In this paper, we evaluate the performance of WMNs using WMN-PSOHC-DGA simulation system considering Linearly Decreasing Inertia Weight Method (LDIWM) and Linearly Decreasing Vmax Method (LDVM).

The rest of the paper is organized as follows. The mesh router nodes placement problem is defined in Sect. 2. We present our designed and implemented hybrid simulation system in Sect. 3. The simulation results are given in Sect. 4. Finally, we give conclusions and future work in Sect. 5.

2   Node Placement Problem in WMNs

For this problem, we have a grid area arranged in cells we want to find where to distribute a number of mesh router nodes and a number of mesh client nodes of fixed positions (of an arbitrary distribution) in the considered area. The objective is to find a location assignment for the mesh routers to the area that maximizes the network connectivity and client coverage. Network connectivity is measured by Size of Giant Component (SGC) of the resulting WMN graph, while the user coverage is simply the number of mesh client nodes that fall within the radio coverage of at least one mesh router node and is measured by Number of Covered Mesh Clients (NCMC).

An instance of the problem consists as follows.

- N mesh router nodes, each having its own radio coverage, defining thus a vector of routers.
- An area $W \times H$ where to distribute N mesh routers. Positions of mesh routers are not pre-determined and are to be computed.
- M client mesh nodes located in arbitrary points of the considered area, defining a matrix of clients.

It should be noted that network connectivity and user coverage are among most important metrics in WMNs and directly affect the network performance.

In this work, we have considered a bi-objective optimization in which we first maximize the network connectivity of the WMN (through the maximization of the SGC) and then, the maximization of the NCMC.

In fact, we can formalize an instance of the problem by constructing an adjacency matrix of the WMN graph, whose nodes are router nodes and client nodes and whose edges are links between nodes in the mesh network. Each mesh node in the graph is a triple $v = <x, y, r>$ representing the 2D location point and r is the radius of the transmission range. There is an arc between two nodes u and v, if v is within the transmission circular area of u.

3 Proposed and Implemented Simulation System

3.1 WMN-PSOHC-DGA Hybrid Simulation System

Distributed Genetic Algorithm (DGA) has been focused from various fields of science. DGA has shown their usefulness for the resolution of many computationally hard combinatorial optimization problems. Also, Particle Swarm Optimization (PSO) has been investigated for solving NP-hard problem.

PSOHC Part

WMN-PSOHC-DGA decide the velocity of particles by a random process considering the area size. For instance, when the area size is $W \times H$, the velocity is decided randomly from $-\sqrt{W^2 + H^2}$ to $\sqrt{W^2 + H^2}$. Each particle's velocities are updated by simple rule [19].

For HC mechanism, next positions of each particle are used for neighbor solution s'. The fitness function f gives points to the current solution s. If $f(s')$ is better than $f(s)$, the s is updated to s'. However, if $f(s')$ is not better than $f(s)$, the s is not updated. It should be noted that the positions are not updated but the velocities are updated even if the $f(s)$ is better than $f(s')$.

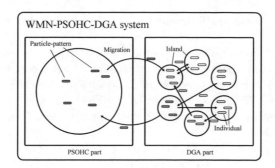

Fig. 1. Model of WMN-PSOHC-DGA migration.

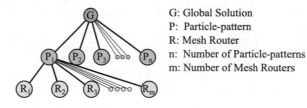

G: Global Solution
P: Particle-pattern
R: Mesh Router
n: Number of Particle-patterns
m: Number of Mesh Routers

Fig. 2. Relationship among global solution, particle-patterns and mesh routers in PSOHC part.

Routers Replacement Method for PSO Part

A mesh router has x, y positions and velocity. Mesh routers are moved based on velocities. There are many router replacement methods. In this paper, we use LDIWM and LDVM.

Linearly Decreasing Inertia Weight Method (LDIWM)
    In LDIWM, C_1 and C_2 are set to 2.0, constantly. On the other hand, the ω parameter is changed linearly from unstable region ($\omega = 0.9$) to stable region ($\omega = 0.4$) with increasing of iterations of computations [29,30].
Linearly Decreasing Vmax Method (LDVM)
    In LDVM, PSO parameters are set to unstable region ($\omega = 0.9$, $C_1 = C_2 = 2.0$). A value of V_{max} which is maximum velocity of particles is considered. With increasing of iteration of computations, the V_{max} is kept decreasing linearly [21,28].

DGA Part

Population of individuals: Unlike local search techniques that construct a path in the solution space jumping from one solution to another one through local perturbations, DGA use a population of individuals giving thus the search a larger scope and chances to find better solutions. This feature is also known as "exploration" process in difference to "exploitation" process of local search methods.

Selection: The selection of individuals to be crossed is another important aspect in DGA as it impacts on the convergence of the algorithm. Several selection schemes have been proposed in the literature for selection operators trying to cope with premature convergence of DGA. There are many selection methods in GA. In our system, we implement 2 selection methods: Random method and Roulette wheel method.

Crossover operators: Use of crossover operators is one of the most important characteristics. Crossover operator is the means of DGA to transmit best genetic features of parents to offsprings during generations of the evolution process. Many methods for crossover operators have been proposed such as Blend Crossover (BLX-α), Unimodal Normal Distribution Crossover (UNDX), Simplex Crossover (SPX).

Mutation operators: These operators intend to improve the individuals of a population by small local perturbations. They aim to provide a component of randomness in the neighborhood of the individuals of the population. In our system, we implemented two mutation methods: uniformly random mutation and boundary mutation.

Escaping from local optima: GA itself has the ability to avoid falling prematurely into local optima and can eventually escape from them during the search process. DGA has one more mechanism to escape from local optima by considering some islands. Each island computes GA for optimizing and they migrate its gene to provide the ability to avoid from local optima.

Convergence: The convergence of the algorithm is the mechanism of DGA to reach to good solutions. A premature convergence of the algorithm would cause that all individuals of the population be similar in their genetic features and thus the search would result ineffective and the algorithm getting stuck into local optima. Maintaining the diversity of the population is therefore very important to this family of evolutionary algorithms.

In following, we present our proposed and implemented simulation sistem called WMN-PSOHC-DGA. We show the fitness function, migration function, particle-pattern, gene coding and client distributions.

Fitness Function

The determination of an appropriate fitness function, together with the chromosome encoding are crucial to the performance. Therefore, one of most important thing is to decide the determination of an appropriate objective function and its encoding. In our case, each particle-pattern and gene has an own fitness value which is comparable and compares it with other fitness value in order to share information of global solution. The fitness function follows a hierarchical approach in which the main objective is to maximize the SGC in WMN. Thus, the fitness function of this scenario is defined as

$$\text{Fitness} = 0.7 \times \text{SGC}(\boldsymbol{x}_{ij}, \boldsymbol{y}_{ij}) + 0.3 \times \text{NCMC}(\boldsymbol{x}_{ij}, \boldsymbol{y}_{ij}).$$

Table 1. WMN-PSOHC-DGA parameters.

Parameters	Values
Clients distribution	Normal distribution
Area size	32.0×32.0
Number of mesh routers	16
Number of mesh clients	48
Number of migrations	200
Evolution steps	9
Number of GA islands	16
Radius of a mesh router	2.0
Selection method	Roulette wheel method
Crossover method	SPX
Mutation method	Boundary mutation
Crossover rate	0.8
Mutation rate	0.2
Replacement method	LDIWM, LDVM

Migration Function

Our implemented simulation system uses Migration function as shown in Fig. 1. The Migration function swaps solutions between PSOHC part and DGA part.

Particle-Pattern and Gene Coding

In ordert to swap solutions, we design particle-patterns and gene coding carefully. A particle is a mesh router. Each particle has position in the considered area and velocities. A fitness value of a particle-pattern is computed by combination of mesh routers and mesh clients positions. In other words, each particle-pattern is a solution as shown is Fig. 2.

A gene describes a WMN. Each individual has its own combination of mesh nodes. In other words, each individual has a fitness value. Therefore, the combination of mesh nodes is a solution.

4   Simulation Results

In this section, we show simulation results using WMN-PSOHC-DGA system. In this work, we analyse the performance of WMNs considering LDIWM and LDVM router replacement methods. The number of mesh routers is considered 16 and the number of mesh clients 48. We conducted simulations 100 times, in order to avoid the effect of randomness and create a general view of results. We show the parameter setting for WMN-PSOHC-DGA in Table 1.

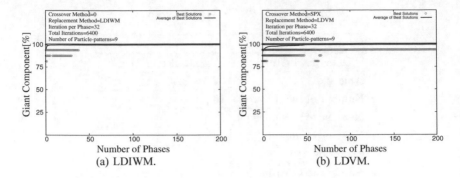

Fig. 3. Simulation results of WMN-PSOHC-DGA for SGC.

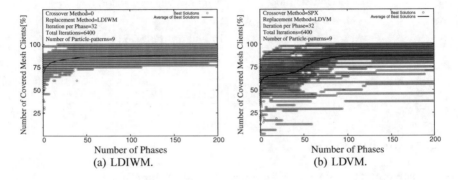

Fig. 4. Simulation results of WMN-PSOHC-DGA for NCMC.

We show simulation results in Figs. 3 and 4. We see that for both SGC and NCMC, the performance of LDIWM is better than LDVM.

5   Conclusions

In this work, we evaluated the performance of WMNs using a hybrid simulation system based on PSOHC and DGA (called WMN-PSOHC-DGA) considering LDIWM and LDVM router replacement methods. Simulation results show that the performance is better for LDIWM compared with the case of LDVM.

In our future work, we would like to evaluate the performance of the proposed system for different parameters and patterns.

References

1. Akyildiz, I.F., Wang, X., Wang, W.: Wireless mesh networks: a survey. Comput. Netw. **47**(4), 445–487 (2005)

2. Barolli, A., Sakamoto, S., Barolli, L., Takizawa, M.: Performance analysis of simulation system based on particle swarm optimization and distributed genetic algorithm for WMNs considering different distributions of mesh clients. In: International Conference on Innovative Mobile and Internet Services in Ubiquitous Computing, pp. 32–45. Springer, Heidelberg (2018)
3. Barolli, A., Sakamoto, S., Ozera, K., Barolli, L., Kulla, E., Takizawa, M.: Design and implementation of a hybrid intelligent system based on particle swarm optimization and distributed genetic algorithm. In: International Conference on Emerging Internetworking, Data & Web Technologies, pp. 79–93. Springer, Heidelberg (2018)
4. Franklin, A.A., Murthy, C.S.R.: Node placement algorithm for deployment of two-tier wireless mesh networks. In: Proceedings of Global Telecommunications Conference, pp. 4823–4827 (2007)
5. Girgis, M.R., Mahmoud, T.M., Abdullatif, B.A., Rabie, A.M.: Solving the wireless mesh network design problem using genetic algorithm and simulated annealing optimization methods. Int. J. Comput. Appl. 96(11), 1–10 (2014)
6. Inaba, T., Elmazi, D., Sakamoto, S., Oda, T., Ikeda, M., Barolli, L.: A secure-aware call admission control scheme for wireless cellular networks using fuzzy logic and its performance evaluation. J. Mob. Multimedia 11(3&4), 213–222 (2015)
7. Inaba, T., Obukata, R., Sakamoto, S., Oda, T., Ikeda, M., Barolli, L.: Performance evaluation of a QoS-aware fuzzy-based CAC for LAN access. Int. J. Space-Based Situated Comput. 6(4), 228–238 (2016)
8. Inaba, T., Sakamoto, S., Oda, T., Ikeda, M., Barolli, L.: A testbed for admission control in WLAN: a fuzzy approach and its performance evaluation. In: International Conference on Broadband and Wireless Computing, Communication and Applications, pp. 559–571. Springer, Heidelberg (2016)
9. Maolin, T., et al.: Gateways placement in backbone wireless mesh networks. Int. J. Commun. Netw. Syst. Sci. 2(1), 44 (2009)
10. Matsuo, K., Sakamoto, S., Oda, T., Barolli, A., Ikeda, M., Barolli, L.: Performance analysis of WMNs by WMN-GA simulation system for two WMN architectures and different TCP congestion-avoidance algorithms and client distributions. Int. J. Commun. Netw. Distrib. Syst. 20(3), 335–351 (2018)
11. Muthaiah, S.N., Rosenberg, C.P.: Single gateway placement in wireless mesh networks. In: Proceedings of 8th International IEEE Symposium on Computer Networks, pp. 4754–4759 (2008)
12. Naka, S., Genji, T., Yura, T., Fukuyama, Y.: A hybrid particle swarm optimization for distribution state estimation. IEEE Trans. Power Syst. 18(1), 60–68 (2003)
13. Sakamoto, S., Kulla, E., Oda, T., Ikeda, M., Barolli, L., Xhafa, F.: A comparison study of hill climbing, simulated annealing and genetic algorithm for node placement problem in WMNs. J. High Speed Netw. 20(1), 55–66 (2014)
14. Sakamoto, S., Kulla, E., Oda, T., Ikeda, M., Barolli, L., Xhafa, F.: A simulation system for WMN based on SA: performance evaluation for different instances and starting temperature values. Int. J. Space-Based Situated Comput. 4(3–4), 209–216 (2014)
15. Sakamoto, S., Kulla, E., Oda, T., Ikeda, M., Barolli, L., Xhafa, F.: Performance evaluation considering iterations per phase and SA temperature in WMN-SA system. Mob. Inf. Syst. 10(3), 321–330 (2014)
16. Sakamoto, S., Lala, A., Oda, T., Kolici, V., Barolli, L., Xhafa, F.: Application of WMN-SA simulation system for node placement in wireless mesh networks: a case study for a realistic scenario. Int. J. Mob. Comput. Multimedia Commun. (IJMCMC) 6(2), 13–21 (2014)

17. Sakamoto, S., Oda, T., Ikeda, M., Barolli, L., Xhafa, F.: An integrated simulation system considering WMN-PSO simulation system and network simulator 3. In: International Conference on Broadband and Wireless Computing, Communication and Applications, pp. 187–198. Springer, Heidelberg (2016)
18. Sakamoto, S., Oda, T., Ikeda, M., Barolli, L., Xhafa, F.: Implementation and evaluation of a simulation system based on particle swarm optimisation for node placement problem in wireless mesh networks. Int. J. Commun. Netw. Distrib. Syst. **17**(1), 1–13 (2016)
19. Sakamoto, S., Oda, T., Ikeda, M., Barolli, L., Xhafa, F.: Implementation of a new replacement method in WMN-PSO simulation system and its performance evaluation. In: The 30th IEEE International Conference on Advanced Information Networking and Applications (AINA-2016), pp. 206–211 (2016). https://doi.org/10.1109/AINA.2016.42
20. Sakamoto, S., Obukata, R., Oda, T., Barolli, L., Ikeda, M., Barolli, A.: Performance analysis of two wireless mesh network architectures by WMN-SA and WMN-TS simulation systems. J. High Speed Netw. **23**(4), 311–322 (2017)
21. Sakamoto, S., Ozera, K., Ikeda, M., Barolli, L.: Performance evaluation of WMNs by WMN-PSOSA simulation system considering constriction and linearly decreasing inertia weight methods. In: International Conference on Network-Based Information Systems, pp. 3–13. Springer, Heidelberg (2017)
22. Sakamoto, S., Ozera, K., Oda, T., Ikeda, M., Barolli, L.: Performance evaluation of intelligent hybrid systems for node placement in wireless mesh networks: a comparison study of WMN-PSOHC and WMN-PSOSA. In: International Conference on Innovative Mobile and Internet Services in Ubiquitous Computing, pp. 16–26. Springer, Heidelberg (2017)
23. Sakamoto, S., Ozera, K., Oda, T., Ikeda, M., Barolli, L.: Performance evaluation of WMN-PSOHC and WMN-PSO simulation systems for node placement in wireless mesh networks: a comparison study. In: International Conference on Emerging Internetworking, Data & Web Technologies, pp. 64–74. Springer, Heidelberg (2017)
24. Sakamoto, S., Barolli, A., Barolli, L., Takizawa, M.: Design and implementation of a hybrid intelligent system based on particle swarm optimization, hill climbing and distributed genetic algorithm for node placement problem in WMNs: a comparison study. In: The 32nd IEEE International Conference on Advanced Information Networking and Applications (AINA-2018), pp. 678–685. IEEE (2018)
25. Sakamoto, S., Ozera, K., Barolli, A., Barolli, L., Kolici, V., Takizawa, M.: Performance evaluation of WMN-PSOSA considering four different replacement methods. In: International Conference on Emerging Internetworking, Data & Web Technologies, pp. 51–64. Springer, Heidelberg (2018)
26. Sakamoto, S., Ozera, K., Ikeda, M., Barolli, L.: Implementation of intelligent hybrid systems for node placement problem in WMNs considering particle swarm optimization, hill climbing and simulated annealing. Mob. Netw. Appl. **23**(1), 27–33 (2018)
27. Sakamoto, S., Ozera, K., Barolli, A., Ikeda, M., Barolli, L., Takizawa, M.: Implementation of an intelligent hybrid simulation systems for WMNs based on particle swarm optimization and simulated annealing: performance evaluation for different replacement methods. Soft. Comput. **23**(9), 3029–3035 (2019)
28. Schutte, J.F., Groenwold, A.A.: A study of global optimization using particle swarms. J. Glob. Optim. **31**(1), 93–108 (2005)
29. Shi, Y.: Particle swarm optimization. IEEE Connect. **2**(1), 8–13 (2004)
30. Shi, Y., Eberhart, R.C.: Parameter selection in particle swarm optimization. In: Evolutionary Programming VII, pp. 591–600 (1998)

31. Vanhatupa, T., Hannikainen, M., Hamalainen, T.: Genetic algorithm to optimize node placement and configuration for WLAN planning. In: Proceedings of the 4th IEEE International Symposium on Wireless Communication Systems, pp. 612–616 (2007)
32. Wang, J., Xie, B., Cai, K., Agrawal, D.P.: Efficient mesh router placement in wireless mesh networks. In: Proceedings of IEEE Internatonal Conference on Mobile Adhoc and Sensor Systems (MASS-2007), pp. 1–9 (2007)

Development and Evaluation of IoT/M2M Application Using Real Object Oriented Model

Hiroyuki Suzuki$^{(\boxtimes)}$, Liyang Zhang, and Akio Koyama

Graduate School of Science and Engineering, Yamagata University,
4-3-16 Jonan, Yonezawa, Yamagata 992-8510, Japan
{shiroyuki,ttf04338,akoyama}@yz.yamagata-u.ac.jp

Abstract. Currently, although IoT/M2M applications are developed and used, the development model is mainly a vertical type model that is used for specific services. This model is difficult to incorporate already developed functions into new services and requires redevelopment. Meanwhile, a horizontal type model can be used even except the specific services because once developed functions can be reused. Consequently, development efficiency becomes high. We have previously proposed a development method for application development that uses a Real Object-Oriented Model of a horizontal type model without being aware of common functions such as communication. In this paper, we compared and verified the number of source code and development time for a crime prevention application and a physical condition management application that developed by using the vertical type model, the horizontal type model and the proposed method. As a result, we confirmed that the second application development was more effective than the first application development.

1 Introduction

Recently, various things are connected to the Internet, and IoT (Internet of Things) which is a technology for collecting, managing, and controlling information, has become widespread. It is said that the number of devices used in IoT (IoT devices) worldwide reached about 27.5 billion in 2017 and is expected to increase to 40.3 billion by 2020 [1].

Applications that use data generated from huge numbers of IoT devices has been studied in various fields such as manufacturing [2], crime prevention [3], medical care [4], agriculture [5, 6], healthcare [7] and energy management [8]. Since these applications use various data, the centralized management with a cloud server is more efficient than the distributed management. Therefore, research on load balancing to efficiently operate cloud servers is also being conducted [9–11].

Such application development methods using IoT include a vertical type model and a horizontal type model. Currently, the vertical type model is mainstream [9, 10]. In the vertical type model, devices such as sensors are independent of each application. The data communication function is also the same. Therefore, it is not possible to use the function that developed by other applications such as data communication function. The function is required redevelopment. In contrast, the horizontal type model provides multiple functions on a common infrastructure such as middleware. By using this

© Springer Nature Switzerland AG 2020
L. Barolli et al. (Eds.): INCoS 2019, AISC 1035, pp. 24–35, 2020.
https://doi.org/10.1007/978-3-030-29035-1_3

common infrastructure, we can use the functions developed once for other applications. Therefore, development efficiency of the horizontal type model is higher than the vertical type model. This is because development using those functions is possible, the horizontal type model can be expected in a shorter time than the vertical model.

However, the horizontal type model has limitations on improving development efficiency because application developers need to grasp the huge number of functions (functions, methods, etc.) included in the common infrastructure and create source code. If the amount of code to be created by the developer can be reduced by automatically generating the source code, the development efficiency can be improved [15, 16]. Therefore, the development period can be reduced. Based on this idea, we previously proposed an application development method using a Real Object-Oriented Model which is one of horizontal models [12, 13].

In this paper, in order to verify the usefulness of the proposed method, based on a real object-oriented model, we developed a crime prevention application and a physical condition management which are often used as IoT applications. Compared with other methods, how much development period and code number can be reduced is evaluated and reported.

The structure of this paper is as follows. Section 2 introduces related works, and Sect. 3 introduces development methods using real object-oriented models. Section 4 describes the development procedure of an application using the real object-oriented model. Section 5 describes the performance evaluation, and Sect. 6 describes the conclusions.

2 Related Works

This chapter introduces related works on IoT application development method and source code automatic generation using a horizontal type model.

2.1 IoT Application Development Using a Horizontal Type Model

Datta et al. has developed an M3 framework that extends oneM2M (horizontal type model framework), which is a framework for IoT, and proposed an application development method for mobile devices using it [14]. The developer stores the application on a website in advance, and the user can use the application by downloading it by the smartphone or tablet from the website. Devices such as sensors used in the application need to be prepared by the developer at the stage of developing the application. In the development of the application, although the developer develops it using the M3 framework, it takes time to grasp the framework, so there is a limit to shortening development time.

2.2 Automatic Generation of Source Code

Ueda et al. has proposed a method to facilitate the development of Web applications using RESTful API and blanco framework [15]. RESTful API can manipulate devices on the network using HTTP Method (GET, POST, PUT, DELETE). After receiving the sent HTTP Method, it is possible to predefine in the device what processing is to be performed, and to perform specific operations. In addition, blanco framework, which is an automatic code generation tool, defines data structures and objects in an Excel sheet, and automatically generates code based on that. In this research, they are able to utilize RESTful API in blanco framework and has realized shortening of application development period by automatically generating communication function. However, communication between clients is excluded because it is premised on communication between client and server. Since application developers need to develop applications using the automatically generated communication function, they need time to grasp the communication function. Therefore, there is a limit to shortening the development period.

3 Development Method Using a Real Object-Oriented Model

We have previously proposed an application development method using a real object-oriented model that is assumed to use in IoT/M2M [12, 13]. This chapter explains the method.

3.1 Real Object-Oriented Model

The Real Object-Oriented Model uses an object-oriented concept. This model manages a device as an object, and objects cooperate and construct an application. As a premise, it is assumed that the device is equipped with a microcomputer and a communication function. It places the object on the microcomputer and operates the device through the object.

Objects have methods (device operations) and fields (device information), and devices and objects have a one-to-one correspondence. There is a common platform to develop this model as an application. Common platform includes communication function and object management function. Communication functions eliminate differences in data types and communication methods handled for each object. Therefore, various devices can be linked. The object management function determines and executes the specified execution method from the other object through communication.

3.2 Automatic Generation and Deployment of Source Code

Automatic code generation has been implemented to reduce the amount of source code generated by application developers. In addition, in the case of IoT applications, the method of manually deploying and executing each device is not realistic because it needs embed into many devices. Therefore, we realized the method of automatically deploying the application from the server to each device and the device automatically executing the deployed application.

An outline of the automatic generation of code is shown in Fig. 1. First, the developer creates an application combining objects and a container embedding a common platform. Once created containers can be used with other applications as well as common platform. An application combining objects is a general program created in an object-oriented language, which is a combination of objects. A container is a program that describes common platform such as communication and GUI, etc., and does not describe only the operation of an application. The program of the object corresponding to each device is automatically generated by the automatic generation tool of the proposed method. Automatic code generation adds application operations to this container to complete the application for each device. Then, the compiled binary data can be automatically distributed and executed on the embedded device.

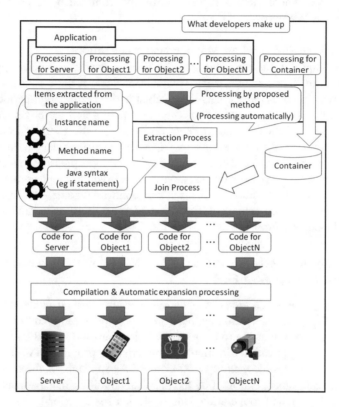

Fig. 1. Outline of automatic code generation process.

4 Development Procedure Using Real Object Orientation

This chapter explains the actual creation and development process of the crime prevention application and the physical condition management application. The application is created by Java and experimented with emulation that makes the laptop PC look like an object. Since the steps 2 to 4 are automatically executed by the tool of the proposed method, the application developer needs to make only the step 1.

4.1 Prototype Application

4.1.1 The Crime Prevention Application

The crime prevention application is an application aimed at detecting the intrusion of a suspicious person from windows or doors. This application is used for nighttime security, and assumes cooperation between a human sensor, GPS, and a smartphone. When there is a reaction in the human sensor due to a suspicious person intrusion, etc., GPS location information is sent to the user to notify of an abnormality. Moreover, the map of the place where there was an abnormality from the position information is displayed on the Web browser.

4.1.2 The Physical Condition Management Application

The physical condition management application is an application that supports user's physical condition management using wearable devices, weight scales, and servers. The weight scale records daily changes in weight and sets the target period and target weight. Also, the weight scale transmits the target period and the target weight to the wearable device 1 (device 1) and presents the user with the result of calculating the daily exercise amount (the number of steps). Device 1 measures sleep time and exercise amount, and wearable device 2 (device 2) measures blood pressure and pulse. If the user's blood pressure or pulse worsens, a warning is given to the user to notify him or her of poor physical condition. The user can confirm whether the target exercise amount per day has been achieved. The server stores daily user information (such as weight and blood pressure).

4.2 Development Procedure 1 (Create Application and Container, IP Address Management File)

Application developers can create applications and use them if containers exist. The programs for the crime prevention application and the physical condition management application (only for the activation part of the object) are shown in Figs. 2 and 3. The source code in this application is a simple object-oriented program in which each object is a device and executes the methods of the object. As an example, the execute method of the weight scale object executed by the physical condition management application is as shown in Fig. 4, and cooperation with the device 1 is described. Similarly, specific processing is described for the user, the device 1 and the server, but this will be omitted due to space concerns. The device 2 is not described in Fig. 4 because it cooperates only with the user.

Next is the container. The container mainly describes the GUI and common platform, and defines the operation start of the application. This time, the application operates when the start button is pressed (Fig. 5). Communication and object management functions are functions of common platform. The communication function defines TCP and UDP communication functions. The object management function is a function to determine which method should be executed when a method operation instruction is received through communication. The application developer embeds it as a container together with the display function of GUI so that these functions can be used.

```
 8 | public static void main (String[] args){
 9 |   /** process start */
10 |   Sensor sensor1 = new Sensor();
11 |   GPS gps1 = new GPS();
12 |   createMap map = new createMap();
13 |   String data = sensor1.check();
14 |   if(data.equals("There is an intruder.")){
15 |     String gps = gps1.Start("3");
16 |     if(gps.contains(",")){
17 |       //Create google map html file
18 |       String gps_list[] = gps.split(",");
19 |       String html = map.createHTML(gps_list[1], gps_list[0]);
20 |
21 |       if(html.equals("done")){
22 |         String url = "C:\\ROP\\test_modify.html";
23 |         try{
24 |         //open googlemap using web
25 |         ProcessBuilder pb = new ProcessBuilder("cmd.exe", "/c", "start", url);
26 |         pb.redirectErrorStream(true);
27 |         Process process = pb.start();
28 |         }
29 |         catch (Exception ec) {
30 |           ec.printStackTrace();
31 |         }
32 |       }
33 |     }
34 |   }
35 | }
36 | /**process end */
37 |}
```

Fig. 2. Program of the crime prevention application

```
 1 |/** process start */
 2 |dia = table.getIP("1", "User" );
 3 |tcpsender.Operate(dia, Operate,"execute" );
 4 |
 5 |dia = table.getIP("1", "WeightDevice" );
 6 |tcpsender.Operate(dia, Operate,"execute" );
 7 |
 8 |dia = table.getIP("1", "Wearable" );
 9 |tcpsender.Operate(dia, Operate,"execute" );
10 |
11 |dia = table.getIP("1", "Server" );
12 |tcpsender.Operate(dia, Operate,"execute" );
13 |/**process end */
```

Fig. 3. Program of the physical condition management application (object starting part)

```
 4 |public void execute() {
 5 | /** process start */
 6 |
 7 | WeightDevice weight1 = new WeightDevice();
 8 | Wearable wearable1 = new Wearable();
 9 | String height = String.valueOf(getHeight());
10 | String myWeight = String.valueOf(getMyWeight());
11 | String targetWeight= String.valueOf(getTargetWeight());
12 | String targetPeriod= String.valueOf(getTargetPeriod());
13 |
14 | wearable1.setHeight(height);
15 | wearable1.setWeight(myWeight);
16 | wearable1.setTargetWeight(targetWeight);
17 | wearable1.setTargetPeriod(targetPeriod);
18 | wearable1.getAllData();
19 |
20 | /** process end */
21 |}
```

Fig. 4. Weight scale execute method (cooperation with device 1)

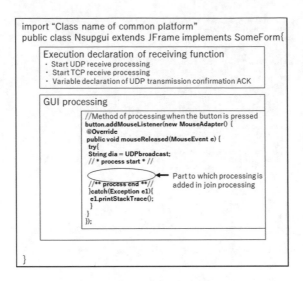

Fig. 5. Container outline

The IP address management file defines what object should be sent to which device when transferring the finally compiled binary file (Fig. 6). If the own object sends a method execution instruction to another object, it is described in the method part of Fig. 6. The communication method is selected from TCP, UDP 0 (without retransmission) and UDP 1 (with retransmission). The format of the method is "Communication method: Instance name: Method name".

The crime prevention application

IP Address	Object Name	Instance Name	Method 1	Method 2
192.168.11.2	Main	Main	TCP:sensor1:check	TCP:gps1:Start1
192.168.11.3	Sensor	sensor1		
192.168.11.4	GPS	gps1		

The physical condition management application

IP Address	Object Name	Instance Name	Method 1	Method 2	...
192.168.11.2	Main	Main	TCP:user1:execute	TCP:weight1:execute	...
192.168.11.3	WeightDevice	weight1	TCP:wearable1:setHeight	TCP:wearable1:setWeight	...
192.168.11.4	Wearable	wearable1	TCP:user1:setSteps	TCP:user1:setTargetSteps	...
192.168.11.5	User	user1	TCP:weight1:setName	TCP:weight1:setAddress	
192.168.11.6	Wearable2	wearable2	TCP:user1:setHeartbeat	TCP:user1:setGPS	...
192.168.11.7	Server	server1			

Fig. 6. IP address management file

4.3   Development Procedure 2 (Extraction Process)

The application developer who has finished creating the application executes the tool of the proposal method and extracts the part to be embedded in the container from the application for each object. The target of extraction process is a range between "process start" and "process end" of extraction range characters (for example, from the 9th to 36th lines in Fig. 2, from the 1st to 13th lines in Fig. 3, and from the 5th to 20th lines in Fig. 4). The items extracted by extraction process are as follows.

- Instance name (Device name on application)
- Method name (Name of method to execute on application)
- Java syntax (Source code surrounded by extracted characters)

4.4   Development Procedure 3 (Combine Process)

In the combine process, the items extracted in the previous section are combined with the container and generates code that operates as an application. The flow of combine process is as follows.

I. Source code is converted to code with communication function of common platform (Fig. 7).

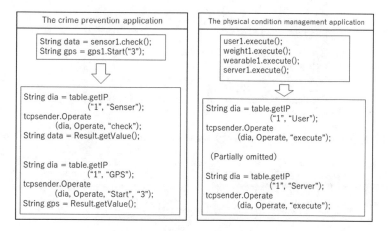

Fig. 7. Example of source code replacement.

II. Create as many folders as the number of objects and copy the container to that folder.

III. Output source code converted by process I at the part between container extraction range characters.

The converted code of each object is stored to between the extraction range characters. This code is output to the part between container and object extraction range characters (Figs. 8, 9 and 10).

```
 1 /** process start */
 2 createMap map = new createMap();
 3
 4 dia = table.getIP("1", "Sensor" );
 5 tcpsender.Operate(dia, Operate,"check" );
 6 String data = Result.getValue();
 7 if(data.equals("There is an intruder.")){
 8   dia = table.getIP("1", "GPS" );
 9   tcpsender.Operate(dia, Operate, "Start", "3");
10   String gps = Result.getValue();
11   if(gps.contains(",")){
12     String gps_list[] = gps.split(",");
13     String html = map.createHTML(gps_list[1], gps_list[0]);
14
15     if(html.equals("done")){
16       String url = "C:¥¥ROP¥¥test_modify.html";
17       try{
18         ProcessBuilder pb = new ProcessBuilder("cmd.exe", "/c", "start", url);
19         pb.redirectErrorStream(true);
20         Process process = pb.start();
21       }
22       catch (Exception ec) {
23         ec.printStackTrace();
24       }
25     }
26   }
27 }
28 /** process end */
29
```

Fig. 8. Completed code of container (crime prevention application)

```
 1 /** process start */
 2 dia = table.getIP("1", "User" );
 3 tcpsender.Operate(dia, Operate,"execute" );
 4
 5 dia = table.getIP("1", "WeightDevice" );
 6 tcpsender.Operate(dia, Operate,"execute" );
 7
 8 dia = table.getIP("1", "Wearable" );
 9 tcpsender.Operate(dia, Operate,"execute" );
10
11 dia = table.getIP("1", "Server" );
12 tcpsender.Operate(dia, Operate,"execute" );
13 /**process end */
```

Fig. 9. Completed code of container (physical condition management application)

```
 1 public void execute() {
 2   /** process start */
 3   String height = String.valueOf(getHeight());
 4   String myWeight = String.valueOf(getMyWeight());
 5   String targetWeight= String.valueOf(getTargetWeight());
 6   String targetPeriod= String.valueOf(getTargetPeriod());
 7
 8   dia = table.getIP("1", "Wearable" );
 9   tcpsender.Operate(dia, Operate,"setHeight",height);
10   dia = table.getIP("1", "Wearable" );
11   tcpsender.Operate(dia, Operate,"setWeight",myWeight);
12   dia = table.getIP("1", "Wearable" );
13   tcpsender.Operate(dia, Operate,"setTargetWeight",targetWeight);
14   dia = table.getIP("1", "Wearable" );
15   tcpsender.Operate(dia, Operate,"setTargetPeriod",targetPeriod);
16   dia = table.getIP("1", "Wearable" );
17   tcpsender.Operate(dia, Operate,"getAllData" );
18 /** process end */
19 }
```

Fig. 10. Execute method of weight scale (after replacement)

4.5   Development Procedure 4 (Compilation and Automatic Deployment)

This tool generates a bat file for compilation. The bat file is described in the file so that it can compile from the file path of each object by the javac command. After application developer creates all the compilation bat files, the developer creates execution bat file and execute the bat file. The execution bat file contains the execution command of the compilation bat file for all objects. When the execution bat file is executed, all compilation bat files are executed. In automatic deployment, based on the IP address management file created in development procedure 1, the instance name and IP address are compared, and executable binary data is sent to any device by FTP. The device needs to execute the monitoring program before sending executable binary data. The monitoring program executes the binary data of the application received by FTP and starts the application.

5   Performance Evaluation

We compare the development time and the number of codes developed with the proposed method, horizontal type model and vertical type model. The development order will be developed in the order of the crime prevention application and the physical condition management application. The crime prevention application develops all source codes in all develop methods. In the physical condition management application, horizontal type model is developed by reusing common platform. In the proposed method, it is developed by reusing container and common platform.

   We compare the amounts of actual development source codes for application development. The amounts mean the number of lines for source codes. The results of the crime prevention application are shown in Fig. 11. In the vertical type model, device processing, communication functions and object management functions are created as processing for each device. In the horizontal type model, communication function and object management function are created as common platform with device processing. In the proposed method, common platform, containers and processing of devices is created. Comparing the amounts of codes of the proposed method, the improvement rate is 52.3% for the vertical type model and 14.3% for the horizontal type model. The development time decreased with the reduction of the amount of codes. Therefore, the improvement rate is 60.7% for the vertical type model and 25% for the horizontal type model.

   Figure 12 shows the results of the physical condition management application. The horizontal type model uses the common platform, so even if the number of devices increases from three for crime prevention application to five for physical condition management application, the amounts of codes do not increase significantly, and the increase rate is about 6%. In contrast, in the proposed method, although the number of devices increased, the amounts of codes for the physical condition management application is smaller than that for the crime prevention application. That's because we reused common platform and containers. The amounts of codes show an improvement of 80.9% for vertical type model and 36.8% for horizontal type model. The development time is improved by 76.5% for the vertical type model and 53.1% for the

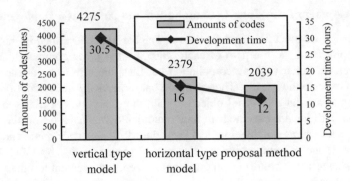

Fig. 11. The crime prevention application development time and amounts of codes.

horizontal type model. From the above, the proposed method can further shorten application development period by reusing common platform and containers.

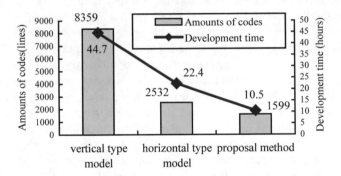

Fig. 12. The physical condition management application development time and amounts of codes.

6   Conclusion

Based on the application development method using the Real Object-Oriented Model that we proposed previously, application development was carried out and verification was performed. As a result, in the crime prevention application, the source code volume difference with the horizontal type model was small. However, in the second physical condition management application development, the proposed method could reduce the amounts of codes to be developed rather than the horizontal type model. In addition, the development time could also be shortened. The reason is that the proposed method can reuse not only common platform but also containers.

In the future, we would like to have another application developer use a Real Object-Oriented Model, and to test if there is a difference in technology.

References

1. Outline of the 2018 White Paper on Information and Communications in Japan: Chapter 1 ICT in Japan and the World, Trend of ICT Marckets in Japan and the World. http://www. soumu.go.jp/johotsusintokei/whitepaper/eng/WP2018_outline.pdf
2. Ikezawa, I., Tsuruhata, K., Hayashi, S.: IoT and system architecture for industries. J. Soc. Instrum. Control Eng. **55**(4), 295–299 (2016)
3. Yamato, Y., Fukumoto, Y., Kumazaki, H.: Proposal of shoplifting prevention service using image analysis and ERP check. IEEJ Trans. Electr. Electron. Eng. **12**(S1), S142–S145 (2017)
4. Yamato, Y.: Experiments of posture estimation on vehicles using wearable acceleration sensors. In: The 3rd IEEE International Conference on Big Data Security on Cloud, pp. 14–17 (2017)
5. Deepak, V., Zerina, K., Jong-ho, W., Xinxin, J., Ranveer, C., Ashish, K., Sudipta, N.S., Madhusudhan, S., Sean, S.: FarmBeats: an IoT platform for data-driven agriculture. In: Proceedings of the 14th USENIX Conference on Networked Systems Design and Implementation, pp. 515–528 (2017)
6. Christopher, B., Ioanna, R., Nikos, K., Kevin, D., Keith, E.: IoT in agriculture designing a Europe-wide large-scale pilot. IEEE Commun. Mag. **55**, 26–33 (2017)
7. Sapna, T., Amit, A., Piyush, M.: A conceptual framework for IoT-based healthcare system using cloud computing. In. Proceeding of the International Conference on Cloud System and Big Data Engineering, pp. 503–507 (2016)
8. Francesco, G.B., Edoardo, P., Anna, O., Matteo, D.G., Niccolò, R., Alexandr, K., Marco, J., Vittorio, V., Elisa, G., Laura, R., Andrea, A.: IoT software infrastructure for energy management and simulation in smart cities. IEEE Trans. Industr. Inf. **13**(2), 832–840 (2016)
9. Yamamoto, Y., Hoshikawa, N., Noguchi, H., Demizu T., Kataoka, M.: A study of optimizing heterogeneous resources for open IoT. Technical report of IEICE, vol. 117, no. 271, pp. 1–6 (2017)
10. Kurosaki, Y., Takefusa, A., Nakada, H., Oguchi, M.: Performance evaluation of load balancing between sensors and a cloudor a real time video streaming analysis application framework. Technical report of IEICE, vol. 115, no. 230, pp. 23–28 (2015)
11. Iwasaki, Y., Matsumono, J., Okamoto, S., Yamanaka, N.: Processing load balancing method for home gateway internet of things platform. Technical report of IEICE, vol. 117, no. 131, pp. 205–210 (2017)
12. Suzuki, H., Koyama, A.: An implementation and evaluation of IoT application development method based on real object-oriented model. Int. J. Space-Based Situated Comput. **8**(3), 151–159 (2018)
13. Suzuki, H., Koyama, A.: IoT/M2M application development method using real object-oriented model. In: Complex, Intelligent, and Software Intensive Systems, vol. 772, pp. 561–572 (2018)
14. Datta, S.K., Gyrard, A., Bonnet, C., Boudaoud, K.: oneM2M architecture based user centric IoT application development. In: International Conference on Future Internet of Things and Cloud, pp. 100–107 (2015)
15. Ueda, T., Matsumono, S., Haino, Y.: A design of a source code generator for RESTful API system. IPSJ SIG Technical report, vol. 2016-SPT-20, no. 2, pp. 1–6 (2016)
16. Kyoya, K., Ito, K., Okuno, T.: An experiment of code generation for CakePHP framework using a prototype development tool. In: Proceedings of the 33rd JSSST Annual Conference, vol. FOSE2, no. 2, pp. 1–10 (2016)

A Fuzzy-Based Decision System for Sightseeing Spots Considering Noise Level as a New Parameter

Yi Liu$^{(\boxtimes)}$ and Leonard Barolli

Department of Languages and Informatics Systems,
Fukuoka Institute of Technology (FIT), 3-30-1 Wajiro-Higashi, Higashi-Ku,
Fukuoka 811-0295, Japan
ryuui1010@gmail.com, barolli@fit.ac.jp

Abstract. Discovering and recommending points of interest are drawing more attention to meet the increasing demand from personalized tours. In this paper, we propose and evaluate a new fuzzy-based system for decision of sightseeing spots considering different conditions. In our system, we considered four input parameters: Ambient Temperature (AT), Air Quality (AQ), Noise Levle (NL) and Current Number of People (CNP) to decide the sightseeing spots Visit or Not Visit (VNV). We evaluate the proposed system by computer simulations. From the simulations results, we conclude that when CNP is increased, the VNV is increased and when the AT is around $18\,°C-26\,°C$, the VNN is the best. But when AQ and NL are increased, the VNV is decreased. The simulation results have shown that the proposed system has a good performance and can choose good sightseeing spots.

1 Introduction

Social image hosting websites have recently become very popular. On these sites, users can upload and tag images for sharing their travelling experiences. The geotagged images are widely used in landmark recognitions and trip recommendations. Large amount of information generated from these location-based social services covers not only popular locations but also obscure ones. Since personalized tours are becoming popular, more attention is focusing on obscure sightseeing locations that are less well-known while still worth visiting. In Fig. 1 are show two dimensions of diverse sightseeing resources. The evaluation can be done using the sightseeing quality and popularity [1–5].

In this work, we use Fuzzy Logic (FL) for decision of sightseeing spots. The FL is the logic underlying modes of reasoning which are approximate rather then exact. The importance of FL derives from the fact that most modes of human reasoning and especially common sense reasoning are approximate in nature [6]. FL uses linguistic variables to describe the control parameters. By using relatively simple linguistic expressions it is possible to describe and grasp very complex problems. A very important property of the linguistic variables is the capability of describing imprecise parameters.

© Springer Nature Switzerland AG 2020
L. Barolli et al. (Eds.): INCoS 2019, AISC 1035, pp. 36–45, 2020.
https://doi.org/10.1007/978-3-030-29035-1_4

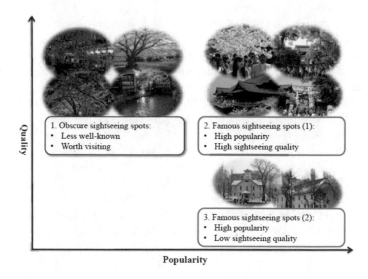

Fig. 1. Two dimensions of diverse sightseeing resources.

The concept of a fuzzy set deals with the representation of classes whose boundaries are not determined. It uses a characteristic function, taking values usually in the interval [0, 1]. The fuzzy sets are used for representing linguistic labels. This can be viewed as expressing an uncertainty about the clear-cut meaning of the label. But important point is that the valuation set is supposed to be common to the various linguistic labels that are involved in the given problem.

The fuzzy set theory uses the membership function to encode a preference among the possible interpretations of the corresponding label. A fuzzy set can be defined by examplification, ranking elements according to their typicality with respect to the concept underlying the fuzzy set [7].

In this paper, we propose and evaluate a fuzzy-based system for decision of sightseeing spots considering noise level as a new parameter. In our system, we considered four input parameters: Ambient Temperature (AT), Air Quality (AQ), Noise Levle (NL) and Current Number of People (CNP) to decide the output parameter Visit or Not Visit (VNV).

The structure of this paper is as follows. In Sect. 2, we introduce FL used for control. In Sect. 3, we present the proposed fuzzy-based system. In Sect. 4, we discuss the simulation results. Finally, conclusions and future work are given in Sect. 5.

2 Application of Fuzzy Logic for Control

The ability of fuzzy sets and possibility theory to model gradual properties or soft constraints whose satisfaction is matter of degree, as well as information pervaded with imprecision and uncertainty, makes them useful in a great variety of applications [8–16].

The most popular area of application is Fuzzy Control (FC), since the appearance, especially in Japan, of industrial applications in domestic appliances, process control, and automotive systems, among many other fields.

In the FC systems, expert knowledge is encoded in the form of fuzzy rules, which describe recommended actions for different classes of situations represented by fuzzy sets.

In fact, any kind of control law can be modeled by the FC methodology, provided that this law is expressible in terms of "if ... then ..." rules, just like in the case of expert systems. However, FL diverges from the standard expert system approach by providing an interpolation mechanism from several rules. In the contents of complex processes, it may turn out to be more practical to get knowledge from an expert operator than to calculate an optimal control, due to modeling costs or because a model is out of reach.

A concept that plays a central role in the application of FL is that of a linguistic variable. The linguistic variables may be viewed as a form of data compression. One linguistic variable may represent many numerical variables. It is suggestive to refer to this form of data compression as granulation.

The same effect can be achieved by conventional quantization, but in the case of quantization, the values are intervals, whereas in the case of granulation the values are overlapping fuzzy sets. The advantages of granulation over quantization are as follows:

- it is more general;
- it mimics the way in which humans interpret linguistic values;
- the transition from one linguistic value to a contiguous linguistic value is gradual rather than abrupt, resulting in continuity and robustness.

FC describes the algorithm for process control as a fuzzy relation between information about the conditions of the process to be controlled, x and y, and the output for the process z. The control algorithm is given in "if ... then ..." expression, such as:

If x is small and y is big, then z is medium;

If x is big and y is medium, then z is big.

These rules are called *FC rules*. The "if" clause of the rules is called the antecedent and the "then" clause is called consequent. In general, variables x and y are called the input and z the output. The "small" and "big" are fuzzy values for x and y, and they are expressed by fuzzy sets.

Fuzzy controllers are constructed of groups of these FC rules, and when an actual input is given, the output is calculated by means of fuzzy inference.

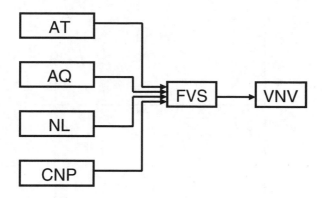

Fig. 2. FVS stucture.

3   Proposed Fuzzy-Based System

The proposed system stucture is show in Fig. 2. We call this system: Fuzzy-based Visiting Spots (FVS) system. In this work, we consider four parameters: Ambient Temperature (AT), Air Quality (AQ), Noise Level (NL) and Current Number of People (CNP) to decide the sightseeing spots Visit or Not Visit (VNV). The AT is the temperature at the sightseeing spots. We use the air pollution data around sightseeing spots to decide the AQ. The NL is the amplitude level of the noise. For CNP, we consider the number of people that have visited these spots. These four parameters are not correlated with each other, for this reason we use fuzzy system. The membership functions for our system are shown in Fig. 3. In Table 1, we show the fuzzy rule base of our proposed system, which consists of 135 rules.

The input parameters for FVS are: AT, AQ, NL and CNP. The output linguistic parameter is VNV. The term sets of AT, AQ, NL and CNP are defined respectively as:

$$
\begin{aligned}
AT &= \{Very\ Cold,\ Cold,\ Normal,\ Hot,\ Very\ Hot\} \\
&= \{VC,\ Co,\ No,\ Ho,\ VH\}; \\
AQ &= \{Good,\ Normal,\ Bad\} \\
&= \{Good,\ Nor,\ Bad\}; \\
NL &= \{Low,\ Middle,\ High\} \\
&= \{Lo,\ Mi,\ Hi\}; \\
CNP &= \{Few,\ Normal,\ Many\} \\
&= \{F,\ N,\ M\}.
\end{aligned}
\tag{1}
$$

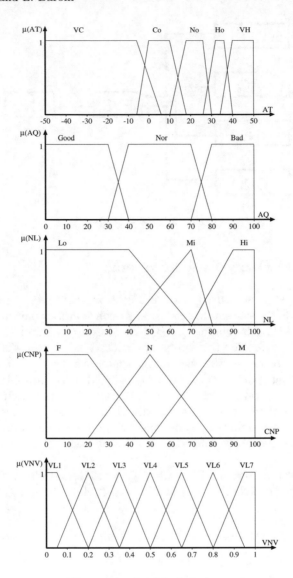

Fig. 3. Membership functions.

Table 1. FRB.

Rule	AT	AQ	NL	CNP	VNV	Rule	AT	AQ	NL	CNP	VNV	Rule	AT	AQ	NL	CNP	VNV
1	VC	Good	Lo	F	VL3	46	C	Bad	Lo	F	VL3	91	H	Nor	Lo	F	VL4
2	VC	Good	Lo	N	VL3	47	C	Bad	Lo	N	VL3	92	H	Nor	Lo	N	VL5
3	VC	Good	Lo	M	VL4	48	C	Bad	Lo	M	VL4	93	H	Nor	Lo	M	VL6
4	VC	Good	Mi	F	VL1	49	C	Bad	Mi	F	VL1	94	H	Nor	Mi	F	VL2
5	VC	Good	Mi	N	VL2	50	C	Bad	Mi	N	VL2	95	H	Nor	Mi	N	VL3
6	VC	Good	Mi	M	VL3	51	C	Bad	Mi	M	VL3	96	H	Nor	Mi	M	VL4
7	VC	Good	Hi	F	VL1	52	C	Bad	Hi	F	VL1	97	H	Nor	Hi	F	VL1
8	VC	Good	Hi	N	VL1	53	C	Bad	Hi	N	VL1	98	H	Nor	Hi	N	VL2
9	VC	Good	Hi	M	VL2	54	C	Bad	Hi	M	VL2	99	H	Nor	Hi	M	VL3
10	VC	Nor	Lo	F	VL2	55	No	Good	Lo	F	VL6	100	H	Bad	Lo	F	VL3
11	VC	Nor	Lo	N	VL2	56	No	Good	Lo	N	VL7	101	H	Bad	Lo	N	VL3
12	VC	Nor	Lo	M	VL3	57	No	Good	Lo	M	VL7	102	H	Bad	Lo	M	VL4
13	VC	Nor	Mi	F	VL1	58	No	Good	Mi	F	VL5	103	H	Bad	Mi	F	VL1
14	VC	Nor	Mi	N	VL1	59	No	Good	Mi	N	VL6	104	H	Bad	Mi	N	VL2
15	VC	Nor	Mi	M	VL2	60	No	Good	Mi	M	VL7	105	H	Bad	Mi	M	VL3
16	VC	Nor	Hi	F	VL1	61	No	Good	Hi	F	VL3	106	H	Bad	Hi	F	VL1
17	VC	Nor	Hi	N	VL1	62	No	Good	Hi	N	VL4	107	H	Bad	Hi	N	VL1
18	VC	Nor	Hi	M	VL1	63	No	Good	Hi	M	VL5	108	H	Bad	Hi	M	VL2
19	VC	Bad	Lo	F	VL1	64	No	Nor	Lo	F	VL5	109	VH	Good	Lo	F	VL3
20	VC	Bad	Lo	N	VL1	65	No	Nor	Lo	N	VL6	110	VH	Good	Lo	N	VL3
21	VC	Bad	Lo	M	VL2	66	No	Nor	Lo	M	VL7	111	VH	Good	Lo	M	VL4
22	VC	Bad	Mi	F	VL1	67	No	Nor	Mi	F	VL4	112	VH	Good	Mi	F	VL1
23	VC	Bad	Mi	N	VL1	68	No	Nor	Mi	N	VL5	113	VH	Good	Mi	N	VL2
24	VC	Bad	Mi	M	VL1	69	No	Nor	Mi	M	VL6	114	VH	Good	Mi	M	VL3
25	VC	Bad	Hi	F	VL1	70	No	Nor	Hi	F	VL2	115	VH	Good	Hi	F	VL1
26	VC	Bad	Hi	N	VL1	71	No	Nor	Hi	N	VL3	116	VH	Good	Hi	N	VL1
27	VC	Bad	Hi	M	VL1	72	No	Nor	Hi	M	VL4	117	VH	Good	Hi	M	VL2
28	C	Good	Lo	F	VL5	73	No	Bad	Lo	F	VL4	118	VH	Nor	Lo	F	VL2
29	C	Good	Lo	N	VL6	74	No	Bad	Lo	N	VL5	119	VH	Nor	Lo	N	VL2
30	C	Good	Lo	M	VL7	75	No	Bad	Lo	M	VL6	120	VH	Nor	Lo	M	VL3
31	C	Good	Mi	F	VL3	76	No	Bad	Mi	F	VL3	121	VH	Nor	Mi	F	VL1
32	C	Good	Mi	N	VL4	77	No	Bad	Mi	N	VL3	122	VH	Nor	Mi	N	VL1
33	C	Good	Mi	M	VL5	78	No	Bad	Mi	M	VL4	123	VH	Nor	Mi	M	VL2
34	C	Good	Hi	F	VL2	79	No	Bad	Hi	F	VL1	124	VH	Nor	Hi	F	VL1
35	C	Good	Hi	N	VL3	80	No	Bad	Hi	N	VL2	125	VH	Nor	Hi	N	VL1
36	C	Good	Hi	M	VL4	81	No	Bad	Hi	M	VL3	126	VH	Nor	Hi	M	VL1
37	C	Nor	Lo	F	VL4	82	H	Good	Lo	F	VL5	127	VH	Bad	Lo	F	VL1
38	C	Nor	Lo	N	VL5	83	H	Good	Lo	N	VL6	128	VH	Bad	Lo	N	VL1
39	C	Nor	Lo	M	VL6	84	H	Good	Lo	M	VL7	129	VH	Bad	Lo	M	VL2
40	C	Nor	Mi	F	VL2	85	H	Good	Mi	F	VL3	130	VH	Bad	Mi	F	VL1
41	C	Nor	Mi	N	VL3	86	H	Good	Mi	N	VL4	131	VH	Bad	Mi	N	VL1
42	C	Nor	Mi	M	VL4	87	H	Good	Mi	M	VL5	132	VH	Bad	Mi	M	VL1
43	C	Nor	Hi	F	VL1	88	H	Good	Hi	F	VL2	133	VH	Bad	Hi	F	VL1
44	C	Nor	Hi	N	VL2	89	H	Good	Hi	N	VL3	134	VH	Bad	Hi	N	VL1
45	C	Nor	Hi	M	VL3	90	H	Good	Hi	M	VL4	135	VH	Bad	Hi	M	VL1

and the term set for the output *VNV* is defined as:

$$VNV = \begin{pmatrix} VisitLevel1 \\ VisitLevel2 \\ VisitLevel3 \\ VisitLevel4 \\ VisitLevel5 \\ VisitLevel6 \\ VisitLevel7 \end{pmatrix} = \begin{pmatrix} VL1 \\ VL2 \\ VL3 \\ VL4 \\ VL5 \\ VL6 \\ VL7 \end{pmatrix}.$$

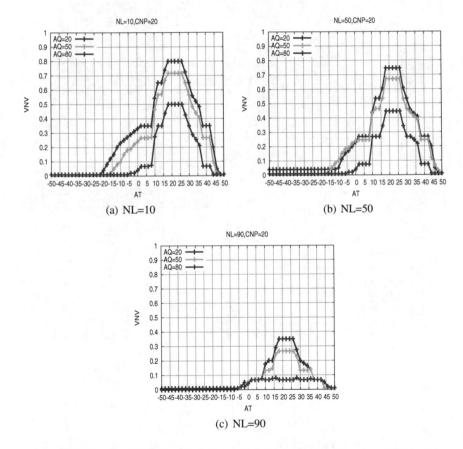

(a) NL=10

(b) NL=50

(c) NL=90

Fig. 4. Relation of VNV with AT and AQ for different NL when CNP = 20.

4   Simulation Results

In this section, we present the simulation results for our proposed fuzzy-based system. In our system, we decided the number of term sets by carrying out many simulations.

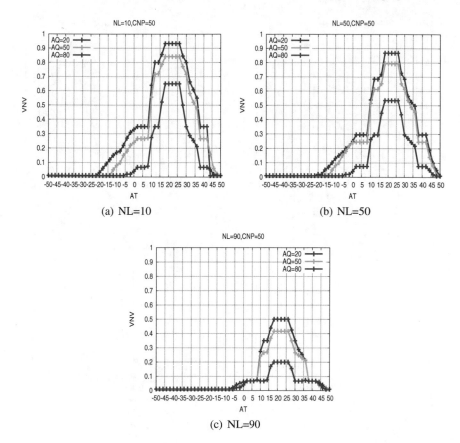

Fig. 5. Relation of VNV with AT and AQ for different NL when CNP = 50.

From Figs. 4, 5 and 6, we show the relation of VNV with AT, AQ, NL and CNP. In these simulations, we consider the NL and CNP as constant parameters. In Fig. 4, we consider CNP value 20 units. We change the AQ value from 20 to 80 units. When the AQ inceases, the VNV is decreased. When AT is around 18 °C−26 °C, the VNV is the best. When NL increases, the VNV is decreased. In Figs. 5 and 6, we increase CNP value to 50 and 80 units, respectively. We see that, when the CNP increases, the VNV is increased.

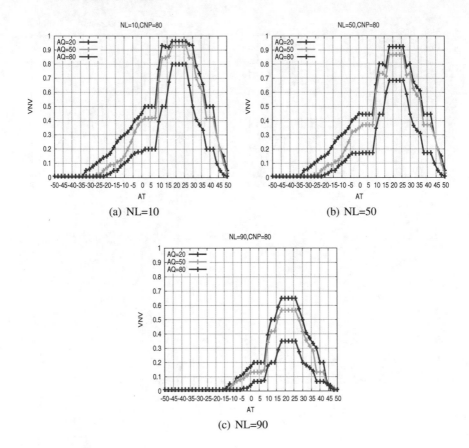

Fig. 6. Relation of VNV with AT and AQ for different NL when CNP = 80.

5   Conclusions and Future Work

In this paper, we proposed a fuzzy-based system to decide the sightseeing spots. We took into consideration four parameters: AT, AQ, NL and CNP. We evaluated the performance of proposed system by computer simulations. From the simulations results, we conclude that when CNP is increased, the VNV is increased. When AQ is increased, the VNV is decreased. When the AT is around $18\,°C-26\,°C$, the VNV is the best. But, by increasing NL, the VNV is decreased.

In the future, we would like to make extensive simulations to evaluate the proposed system and compare the performance of our proposed system with other systems.

References

1. Luo, J., Joshi, D., Yu, J., Gallagher, A.C.: Geotagging in multimedia and computer vision a survey. Multimedia Tools Appl. **51**(1), 187–211 (2011)
2. Chenyi, Z., Qiang, M., Xuefeng, L., Masatoshi, Y.: An obscure sightseeing spots discovering system. In: 2014 IEEE International Conference on Multimedia and Expo (ICME), pp. 1–6. IEEE (2014)
3. Cheng, Z., Ren, J., Shen, J., Miao, H.: Building a large scale test collection for effective benchmarking of mobile landmark search. In: Advances in Multimedia Modeling, pp. 36–46. Springer, Heidelberg (2013)
4. Chen, W., Battestini, A., Gelfand, N., Setlur, V.: Visual summaries of popular landmarks from community photo collections. In: 2009 Conference Record of the Forty-Third Asilomar Conference on Signals, Systems and Computers, pp. 1248–1255. IEEE (2009)
5. Cao, X., Cong, G., Jensen, C.S.: Mining significant semantic locations from GPS data. Proc. VLDB Endow. **3**(1–2), 1009–1020 (2010)
6. Inaba, T., Obukata, R., Sakamoto, S., Oda, T., Ikeda, M., Barolli, L.: Performance evaluation of a QoS-aware fuzzy-based CAC for LAN access. Int. J. Space-Based Situated Comput. **6**(4), 228–238 (2016). https://doi.org/10.1504/IJSSC.2016.082768
7. Terano, T., Asai, K., Sugeno, M.: Fuzzy Systems Theory and Its Applications. Academic Press, Inc. Harcourt Brace Jovanovich, Publishers (1992)
8. Spaho, E., Kulla, E., Xhafa, F., Barolli, L.: P2P solutions to efficient mobile peer collaboration in MANETs. In: Proceedings of 3PGCIC 2012, pp. 379–383, November 2012
9. Kandel, A.: Fuzzy Expert Systems. CRC Press, Boca Raton (1992)
10. Zimmermann, H.J.: Fuzzy Set Theory and Its Applications. Second Revised Edition. Kluwer Academic Publishers, Boston (1991)
11. McNeill, F.M., Thro, E.: Fuzzy Logic. A Practical Approach. Academic Press Inc., Cambridge (1994)
12. Zadeh, L.A., Kacprzyk, J.: Fuzzy Logic for the Management of Uncertainty. Wiley, Hoboken (1992)
13. Procyk, T.J., Mamdani, E.H.: A linguistic self-organizing process controller. Automatica **15**(1), 15–30 (1979)
14. Klir, G.J., Folger, T.A.: Fuzzy Sets, Uncertainty, and Information. Prentice Hall, Englewood Cliffs (1988)
15. Munakata, T., Jani, Y.: Fuzzy systems: an overview. Commun. ACM **37**(3), 69–76 (1994)
16. Yi, L., Kouseke, O., Keita, M., Makoto, I., Leonard, B.: A fuzzy-based approach for improving peer coordination quality in MobilePeerDroid mobile system. In: Proceedings of IMIS 2018, pp. 60–73, July 2018

A Method for Detecting and Alerting the Presence of a Reverse Running Vehicle

Tsukasa Kato[1](✉) and Hiroaki Nishino[2]

[1] Graduate School of Engineering, Oita University, Oita, Japan
v18e3006@oita-u.ac.jp
[2] Faculty of Science and Technology, Oita University, Oita, Japan
hn@oita-u.ac.jp

Abstract. Many researches have recently been paying attention to designing safe driving support systems based on V2X (vehicle to X) technologies. A vehicle can share various information with other vehicles through wireless connections. These days increasing number of accidents caused by reverse running cars become a serious problem. While the accidents occur on both expressways and general roads, some preventing mechanisms against the problem are progressing in expressways. In the case of general road, however, drivers who are on the one-way road have no way to recognize the existence of any reverse running cars until they can visually confirm the existence. If they don't assume the existence, the possibility of accident become high when a reverse running car actually exists. In this paper, we propose a method to solve the problem. It automatically detects a road type (one-way road or not) by analyzing current location, checking the existence of a reverse running car by wirelessly communicating with surrounding cars if the road is one-way, and alerting the driver of the results with text and sound. We design and prototype a system and verify the effectiveness of the proposed method through experiments.

1 Introduction

Recent advances of connected car research field such as V2X (vehicle to X) communication technologies stimulate many research and development activities for traffic safety monitoring and accident prevention. Increasing number of traffic accidents caused by reverse running cars have been identified as serious problems. Because the reverse running cases have higher risk of fatal and injury accidents than in forward running cases, we must take some measures to prevent the accidents. About 200 reverse running cars have been observed on expressways every year and this number increases on general roads. There is a trial to alert drivers of the reverse running cars by installing umbonal objects for giving a light impact only to the reverse running cars on a highway. It also warns the forward running car drivers by displaying the existence of the reverse running car ahead before they encounter on an electronic message board. Therefore, the forward running car drivers can avoid a possible accident on the highway because they respond in a calm manner when actually encounter the reverse running car.

© Springer Nature Switzerland AG 2020
L. Barolli et al. (Eds.): INCoS 2019, AISC 1035, pp. 46–56, 2020.
https://doi.org/10.1007/978-3-030-29035-1_5

On the other hand, most general roads have no electronic message board like highways, the drivers have no means to know the existence of the reverse running car beforehand. They should take actions to avoid the accident after they visually recognize the. To make matters worse, actively alerting the reverse running car drivers is difficult because traffic sign is an only way to notify the drivers on general roads. Uncareful drivers may easily miss the one-way road sign and some obstacles may hide the sign. Elderly drivers tend to drive the wrong way on expressways, but in the case of one-way roads, any drivers who are not familiar with the area may drive the wrong way for the above reasons regardless of their age. Since most drivers don't assume the existence of the reverse running car on a one-way road, they cannot cope with it when they suddenly find it. Such situation may lead to a serious accident.

In this paper, we propose a system for automatically detects a road type (one-way road or not) by analyzing current location, checking the existence of a reverse running car by wirelessly communicating with surrounding cars if the road is one-way, and alerting both of the reverse and forward running car drivers via text message and sound. We prototype a system and verify it through experiments.

2 Related Work

There are many research projects for supporting safety driving based on vehicle-to-vehicle (V2V) communication. Peng et al. proposed a method for acquiring relative trajectories of peripheral cars to detect their traveling lanes using and DSRC (dedicated short range communication) based V2V communication [1]. While obtaining the accurate position of a surrounding car needs multiple expensive sensors and a complex control system, they proposed a low-cost solution by exchanging normal GPS data with an error of several meters between vehicles traveling on a multi-lane freeway. It detects relative positions of surrounding vehicles that are accurate enough for use with merge and lane-change assist functions. They verified the effectiveness of the proposed method via experiments.

Elleuch et al. proposed an intersection collision avoidance system based on V2V communication and a real-time database [2]. They pay attention to the higher probability of accidents at intersections where traffic volume is high, they designed a method to avoid the accidents by cooperatively exchanging some information between neighboring vehicles. In the proposed method, a car approaching to an intersection periodically transmits its position, speed, and traveling direction to surrounding cars via DSRC based broadcast communication. The neighboring cars store the received data in their local real-time database and analyzes if any collision is anticipated in the intersection. It decides the priority of crossing the intersection between vehicles where a possible collision is expected. Then, it makes the lower priority vehicle and its following cars to automatically decelerate or stop to avoid the accident.

These days an increase in collisions caused by reverse driving of elderly drivers became a social problem. Hirano et al. studied the effect of visual attention to make elderly drivers aware of dangerous events during driving [3]. They particularly focus on a fact that elderly drivers tend to lose their cognitive ability and take longer time to response when multiple hazardous events occur. They set up a driving simulator for generating three hazardous events simultaneously such as deceleration of preceding car, wobbling of nearby bicycle, and approaching of oncoming car, and ask both elderly and young drivers to drive with or without visual warning of these events. They reported the elderly drivers' recognition rate of the events became higher with the visual warning and their reaction time were shortened though they did not reach the young driver' improvements.

The increase in bicycle accidents including serious injury and death is another social problem. This is due to the fact that many bicycles tend to travel in reverse direction on the roadway though the bicycles should run on the left side of the road. Yamanaka et al. analyzed the effect of a reverse running warning device placed on a bicycle lane [4]. They made a device for detecting reverse running bicycles and lighting up a warning message on an LED display board. They conducted an experiment to use the device on a real road and compared the case to place a person for watching and raising a warning panel. Although the effectiveness of the human panel warning was higher, a certain amount of effect was confirmed in reducing the number of reverse running bicycles.

In this paper, we propose a method for automatically detecting a road type (one-way or not) and searching any reverse running vehicle by exchanging information with other neighboring cars if own car is traveling on a one-way general road.

3   System Implementation

3.1   System Organization

Figures 1 and 2 show the proposed system organization and its appearance, respectively. As shown in Fig. 2, the system consists of an Android tablet as a main controller node, a Raspberry Pi board as a communication controller node, and a Wi-Fi SUN communication module.

To start the system, a user firstly connects the Raspberry Pi board with the Android tablet via a Bluetooth connection and start the main controller program implemented as an Android application on the tablet. Next, the user starts driving the car after pressing the system start button as shown in Fig. 3. Then, the main controller acquires its current position using a GPS function every five to ten seconds indicated as item 1 in Fig. 1. After that, it detects if the vehicle travels on a one-way road or not by inquiring to the Google Maps Roads API [5] with the acquired position data. If it is on the one-way road, the main controller also derives its traveling direction by calculating the difference between previous and current position coordinate values, and notifies the road type (one-way) and the traveling direction to Raspberry Pi (communication controller) indicated as item 2 in Fig. 1.

The communication controller tries to send the direction data to neighboring cars and acquires their counterpart information via Wi-Sun connections [6] indicated as items 3, 4, 5, and 6 in Fig. 1. If the direction received from a neighboring car is opposite to the own vehicle, it notifies the result to the main controller indicated as item 7 in Fig. 1. Finally, the main controller outputs a warning text message with a sound as shown in Fig. 4. It makes the driver aware of the possible danger due to the existence of the reverse running car on the traveling road. The system iterates these processes without driver's intervention and allows the driver to concentrate on driving. The driver can prepare for the time to meet with the reverse car and quickly takes actions to avoid an accident when he/she visually recognizes the approaching car.

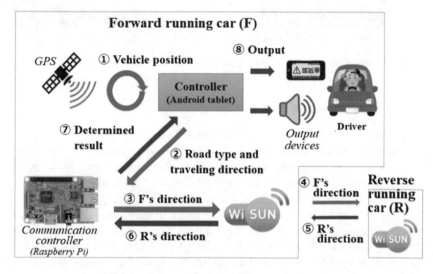

Fig. 1. Organization of the proposed system.

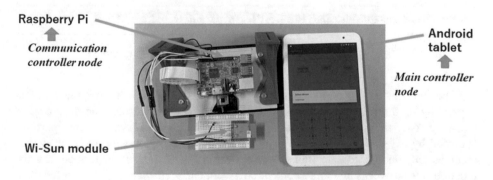

Fig. 2. Appearance of the proposed system

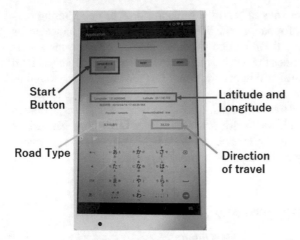

Start Button

Latitude and Longitude

Road Type

Direction of travel

接近車有

Fig. 3. Main controller operation panel.   **Fig. 4.** Warning message panel.

3.2   Road Type Detection

When the system recognizes that the own vehicle is traveling on a one-way road, it repeats the series of processes described in the previous subsection. Therefore, the system firstly needs to automatically detect whether the current road type is one-way or not. We utilize the Google Maps Roads API for detecting the road type [5]. The system inquires the road type to the Google Maps Roads API by rewriting the URL format as shown in Fig. 5. The system sets the current location in the "**parameters**" part as a pair of latitude and longitude acquired by the Android tablet of and the Google Map API key obtained in advance in the "**YOUR _ API _ KEY**" part. This function replies the road information in the vicinity of the inquired location. The system gets an answer among the following three responses: no road, a one-way road, and a two-way road.

---

https://roads.googleapis.com/v1/nearestRoads?parameters&key=YOUR_API_KEY

Fig. 5. URL query format to the Google Maps Roads API.

     The number of location coordinate values included in the response differs depending on the road type (one-way or two-way). If the road near the location is a two-way road, the system receives a pair of input coordinate values (two portions enclosed by red squares) as shown in Fig. 6. In the case of one-way road, the system receives only one coordinate value set as shown in Fig. 7. If there is no road near the location, the answer is an empty content only with curly brackets ({ }). When the main controller running on the Android tablet obtains its location information every five to ten seconds, it sends a query with the location values and analyzes its response to find the road type. If the response indicates the road type is one-way, then the system proceeds to the following processes to check the existence of approaching cars. If the road type is two-way, the controller iterates the acquisition of current location and the inquiry of road type.

```
{
  "snappedPoints": [
    {
      "location": {
        "latitude": 33.184487987039,
        "longitude": 131.61718811540237
      },
      "originalIndex": 0,
      "placeId": "ChIJbYUh6pWYRjURVKxVq4uTZLw"
    },
    {
      "location": {
        "latitude": 33.184487987039,
        "longitude": 131.61718811540237
      },
      "originalIndex": 0,
      "placeId": "ChIJbYUh6pWYRjURVaxVq4uTZLw"
    }
  ]
}
```

Fig. 6. An example response in the case of two-way road.

```
{
  "snappedPoints": [
    {
      "location": {
        "latitude": 33.180184745379229,
        "longitude": 131.61599688138935
      },
      "originalIndex": 0,
      "placeId": "ChIJMS3DX5GYRjURi7xIQOevCbg"
    }
  ]
}
```

Fig. 7. An example response in the case of one-way road.

3.3   Calculation of Traveling Direction

After the system detects the current traveling road type as one-way, it proceeds to the step for checking whether any approaching vehicles exist or not. If such a car is found nearby, that car or own may be traveling in reverse direction and the possibility of an accident is increased. Then, the system tries to inform the driver of the result as warning against an encountering accident. This process needs to calculate own car's traveling direction, exchanging the direction information with other neighboring cars, and comparing the own traveling direction with others. If the other car's direction is opposite to the own car, the system immediately issues the warning to the driver.

While the Google Maps Roads API can determine the road type, it cannot check the direction of a traveling car. We, therefore, implement a method for calculating the own car's direction when it is traveling a one-way road. As shown in Fig. 8, the system calculates the azimuth angle from a pair of position coordinates (latitude and longitude) at two locations t_i and $t_i + \Delta t$ when the positive direction of the vertical axis is oriented toward east. It records the current position coordinates in every step, calculating the azimuth angle θ between the current and past positions recorded at the previous processing step, and using it as the current traveling direction. The system sends its current direction to a neighboring vehicle via a Wi-SUN connection, and simultaneously receives the other car's direction calculated in the same way manner. Then, it compares its own direction with other car's to see if the difference in direction is greater than a predefined threshold value. If the condition is true, the system decide the two cars are moving in opposition directions to each other and finally presents the warning message on a tablet display with an alert sound.

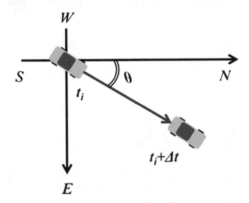

Fig. 8. Azimuth calculation from vehicle locations.

3.4   Inter-vehicle Communication Based on Wi-SUN

The system uses a wireless communication function called Wi-SUN (Wireless-Smart Utility Network) for sharing information with neighboring vehicles. Wi-SUN was originally developed by NICT (National Institute of Information and Communications Technology), Japan, and now is standardized as IEEE 802.15.4g [6]. We adopt the ROHM BP35A1, a Wi-SUN communication module conforming to the standard, as shown in Fig. 9 [7]. Since the module is equipped with an antenna necessary for sending and receiving data and operates for a long time with low power consumption, incorporating the module into the system is easy. It is directly connected to the Raspberry Pi board, the communication controller, via a connector.

Wi-SUN uses 920 MHz communication band which can use freely without license in Japan and have some advantages over 2.4 GHz Wi-Fi communication facility as described in Table 1. Although the communication speed is inferior to Wi-Fi, the communication range to cover is up to 1 km and its diffraction performance is high. Since the system is installed in a moving car to perform data communication, Wi-SUN is considered to be an appropriate solution due to its wide communication range and high reachability of radio waves (diffraction performance). Its low power consumption performance is also a useful feature as an on-vehicle device.

Fig. 9. Wi-SUN communication module ROHM BP35A1.

Table 1. Wi-SUN characteristics over 2.4 GHz Wi-Fi communication band.

	Wi-SUN (920 MHz band)	Wi-Fi (2.4 GHz band)
Communication speed	Up to 200 kbps	10 Mbps and more
Radio interference	Few	Many
Power consumption	Low	High
Communication distance	Up to 1 km	Several hundred meters
Diffraction performance	Good	Bad

3.5 Data Transmission Protocol

Wi-SUN basically defines the lowest two layers (physical and data link layers) in the protocol stack and we can choose upper layer protocols. We select TCP/IP for data transmission protocol. If any packet loss occurs in the data exchange between vehicles facing each other, the system may not able to detect the possible accident caused by the reverse running car. As a result, the reliability of data transmission becomes a first priority for data communication.

The system firstly needs to establish a connection to a nearby vehicle using Wi-SUN to start communication. One car sets a dedicated channel for the communication with its IP address and PAN (personal area network) ID as a master and the other car tries to discover any available communication channel nearby by scanning with the same Pairing ID as a slave. Next, the master only responds to a slave who scans the channel with the same Pairing ID and the slave acquires the channel bandwidth, PAN ID and the master's MAC address to set up the connection in the master's response. Then, the slave converts the MAC address to the master's local IP address and tries to set up the TCP connection through three-way handshake procedure. After that, the master and slave nodes can exchange their traveling directions for analyzing the possibility of accident caused by running cars on a one-way road as described in Sect. 3.4.

The V2V communication we implemented by using Wi-SUN is basically one-to-one communication between a pair of master and the slave as described above. On a congested one-way road, however, many vehicles need to efficiently exchange traveling information. We added a function to prevent from reconnecting with a vehicle previously talked with and exchanged information, and search for another channel set by a different vehicle never talked with. Additionally, efficiently exchanging and sharing the information among many vehicles is difficult if the master and slave relationship is fixed. We, therefore, implement a function to make the slave vehicle once received the traveling information of a nearby car change its role as a master for providing the received information to other vehicles.

4   Preliminary Experiment and Improvement

4.1   Preliminary Experiment

We conducted a preliminary experiment indoors to verify the usefulness of the system.

We employed six university students. Each experimental task requires a pair of subjects. We asked each subject to carry the system with an Android tablet, Raspberry Pi board, and Wi-SUN module as a set, and asked to go straight on a corridor assuming a one-way road. We asked them to start walking from both ends in the L-shaped corridor without knowing each other's presence and verified whether the system could warn and avoid the risky condition before they collided.

We also asked them to perform three trials with different warning output in each task for comparing three different alerts: an image only, a sound only, and both image and sound alerts. After they finished the task, we asked them to answer a questionnaire consisting of the following two questions: "is the system useful as a mean for recognizing the partner before meeting?" in five ranks (5 is "useful" and 1 is "not useful"), and "which warning message is the most effective as an alert?" (select one among the three alerts). Tables 2 and 3 show the results of the usefulness and the best warning message questions, respectively.

Although the number of subjects was not sufficient, everyone highly evaluated the usefulness of the system function. Additionally, five among six subjects supported the alert with image and sound was the best way to notify the possible risk. Multiple subjects mentioned some positive comments like "I was able to proceed with caution due to a prior warning", and "simultaneous visual and auditory stimuli were effective in accurately delivering important information." On the other hand, other subject stated a negative comment like "The warning sound was offensive and annoying." This may due to a sharp sound we used to present in an alert for reliably conveying the imminent danger to the driver. We should reconsider to improve the sound alert.

Table 2. Results on the usefulness question.

Subject	A	B	C	D	E	F	Average
Value	5	4	5	4	4	5	4.5

Table 3. Results on the best alert method question.

Alert method	Image only	Sound only	Image and sound
Number of supporters	0	1	5

4.2   Discussion

As for the warning sound, since the current system repeatedly presents the same sharp sound, the subject who claimed his discomfort might feel the iterated noisy sound unpleasant. Additionally, other subject stated that the sound only alert didn't tell him what was going on in the experimental task. We, therefore, changed the sound to a synthesized human voice indicating the existence of approaching vehicle ahead. We asked subjects to hear the voice and confirmed that they felt it was more natural and less offensive as an alert message. As a result, the driver may be able to grasp the situation only by the voice alert.

Regarding the V2V communication performance, the TCP/IP connection may be too heavy to especially support congested situation. Therefore, we should verify UDP as an alternative protocol to see if it can satisfy with the performance without degrading the reliability requirement.

5   Conclusions and Future Work

In this paper, we proposed a driving safety support system for detecting the existence of reverse running vehicle if he/she is traveling on a one-way road and alert the possible danger by image and sound. It automatically detects the current road type (one-way or not) using the current location and calculates its traveling direction if it is on a one-way road. Then, it exchanges the direction information with other neighboring cars via a Wi-SUN wireless connection and analyzes if any approaching vehicle exists or not. It finally presents the warning message with an image on a tablet screen and a sound output to make the driver pay attention to the situation.

We prototyped the system based on the proposed method and verified the usefulness of the system function through an experiment. Although it was a quasi-experiment asking two persons walking in reverse directions for each other by carrying the system indoors, but we confirmed that the proposed reverse moving object detection and warning functions work effectively.

We designed and implemented the reverse running vehicle detection function as our own method and it cannot detect who is the real reverse running car. We would like to further improve the function to accurately identify the reverse running vehicle and alert neighboring cars to prepare the possible danger as a future work. Conducting outside experiments using real cars is another important issue to verify the practicality of the system as a driving safety support system.

References

1. Peng, Z., Hussain, S., Hayee, M.I., Donath, M.: Acquisition of relative trajectories of surrounding vehicles using GPS and DSRC based V2V communication with lane level resolution. In: Proceedings of the 3rd International Conference on Vehicle Technology and Intelligent Transport Systems, vol. 1, pp. 242–251 (2017)
2. Elleuch, I., Makni, A., Bouaziz, R.: Cooperative intersection collision avoidance persistent system based on V2V communication and real-time databases. In: 2017 IEEE/ACS 14th International Conference on Computer Systems and Applications, pp. 1082–1089 (2017)
3. Hirano, S., Enomoto, M., Sekine, M., Tanaka, K.: Effects of visual alerts on elderly drivers' recognition of multiple, simultaneous hazards. Trans. Hum. Interface Soc. **21**(1), 111–120 (2019). (in Japanese)
4. Yamanaka, S., Takehira, S., Dohko, T., Ikeda, N.: An analysis of the effect of on-street warning device for wrong way cycling on bicycle lane. J. Jpn. Soc. Civil Eng. **73**(5), 711–715 (2017). (in Japanese)
5. https://developers.google.com/maps/documentation/roads/intro. Accessed June 2019
6. Harada, H., Mizutani, K., Fujiwara, J., Mochizuki, K., Obata, K., Okumura, R.: IEEE 802.15.4g based Wi-SUN communication systems. IEICE Trans. Commun. **E100–B**(7), 1032–1043 (2017)
7. http://www.rohm.co.jp/web/japan/products/-/product/BP35A1. Accessed June 2019

A Robot TA System Promoting Students' Active Participation in Lectures

Masashi Kato[1]([✉]), Ryoichi Nagata[2], and Hiroaki Nishino[2]

[1] Graduate School of Engineering, Oita University, Oita, Japan
v18e3007@oita-u.ac.jp
[2] Faculty of Science and Technology, Oita University, Oita, Japan
{nagata-r,hn}@oita-u.ac.jp

Abstract. Q&A session is an important part in face-to-face lectures that teacher explains and students listen. There are, however, some factors for hindering students' self-active questions such as a student may concern about other students and atmosphere in a classroom. As a result, the lecture may progress without obtaining sufficient students' understanding. In this paper, we propose a system allowing the students for asking questions to the teacher via a humanoid robot. The proposed system is developed as a web application, so the students can readily use the system by using their own smartphones or PCs with a standard web browser and a network connection. The students can actively ask questions to the teacher and other students by typing and sending their questions, and at the same time the humanoid robot raises its hand and speaks. The system enables both teacher and students to share the questions in the classroom. We implement the system by combining a web application and a humanoid robot. We elaborate the background and objective of the research, system implementation method, and experimental results in this paper.

1 Introduction

Most university lectures especially in professional classes are face-to-face lectures that a teacher standing on a stage keeps talking and students only listen. In this lecture style, a question-and-answer (QA) session is an important part for improving students' understanding. However, there are few students who raise their own hands and actively ask questions during a lecture. There are some factors for inhibiting the active QA dialogue like the students may concern about other students' appearance and response, their ability to appropriately reveal a problem, and the atmosphere of the classroom [1]. The teacher may prompt the student to ask questions, they hesitate to ask questions while other students don't ask due to the same reasons. As a result, the class may progress without obtaining students' sufficient understanding and satisfaction. Therefore, constructing an environment to promote active QA session during lectures is an urgent issue.

In this paper, we propose a system that a humanoid robot asks questions to the teacher instead of students. The students only describe their questions to their smartphones and then a humanoid robot delivers the questions. Typed questions are anonymously sent to the system and displayed in chat format. It enables the students to

© Springer Nature Switzerland AG 2020
L. Barolli et al. (Eds.): INCoS 2019, AISC 1035, pp. 57–67, 2020.
https://doi.org/10.1007/978-3-030-29035-1_6

actively ask questions without worrying about circumstances when they interrupt by questions. The raised questions are presented to all users (a teacher and students) in real time. Additionally, questioning with audible voice allows to easily share the questions with other students. Furthermore, it is expected for the students to be trained in QA sessions through robot's actions.

The system should prompt the students actively ask questions without putting any burden even if they do not usually ask questions. The system allows the students for asking questions to all users participating in the lecture simply by typing and sending the question contents. Since the students execute this function using their own devices such as smartphones and PCs with a standard web browser and a network connection, they do not feel any burden and easily ask questions. Because the smartphone possession rate of Japanese university students is 98.8%, providing the proposed system as a web application allows them to readily try it in lectures.

In recent years, humanoid robots have been deployed in our lives. In the field of human robot interaction (HRI), the presence of social robots in a child's learning environment achieves positive effects for learning [2]. In addition, humanoid robots' appearances, gestures, and conversation abilities close to human beings make the users feel closer and familiar without discomfort. Therefore, even if the humanoid robot is placed in a lecture, it can perform some expected activities and give positive effects without disturbing the teacher and the students by appropriately controlling its motions and remarks. If the students can see the humanoid robot as a member in the class and the robot actively raises its hand and speaks questions, it lowers the threshold for the students to ask questions and promotes to do the same. Making the students voluntarily raise their hands and ask questions is our final goal.

The rest of the paper is organized as follows. After we describe related research in Sect. 2, we describe the functional overview of the proposed system, we elaborate the implementation method of the system and its GUI design in Sect. 3. After that, we show some experimental results and discussions in the preliminary evaluations in Sect. 4. Finally, we conclude the paper in Sect. 5.

2   Related Work

There are several related research projects for supporting lectures by using ICT. Fujisawa [3] proposed a system to display tweets on slides using twitter to seek student opinion during a lecture. This system displays students' opinions on a slide at regular intervals submitted by students anonymously from PCs. He describes the anonymous submission is an effective way to create an atmosphere that is easy to post the opinions in a classroom. However, there are some problems for delaying the progress of the lecture if the lecturer overlooks the tweets and tries too much feedback.

Atsumi et al. [4] proposed a collaborative system with a teaching assistants (TAs) in which open chat and robots are cooperatively used to support exercise classes. In this system, a robot cooperates with TAs to support exercises by automatically judging and responding to student questions.

Endo [5] studied on the facilitating effect of students' questioning behavior through the Internet. He showed a way to summarize students' comments, questions, and impressions of the class written by using the memo pad in the last 10 min of the class, collecting them into one CSV file, dividing it into six categories, and responding them in the next class. By doing this, he tried to promote student's questioning behaviors and their good understandings. He stated that students become easy to express affinity expressions to the teacher by adopting the note pad as an expression medium and electronic media are useful for reducing resistance to questioning. Additionally, he noted that answering individual questions in the class is an effective way for sharing and relativizing teaching contents and is an advantageous trial in various lectures.

Awedh et al. [6] studied on supporting collaborative learning using a learning tool called "Socrative" with smartphones. They stated exchanging information between teachers and students or between students will improve students' learning ability. Sun et al. [7] proposed a system for making a robot assist teacher in facilitating learning. They developed a humanoid robot as a teaching assistant system that explains two types of ice break games with a teacher. They evaluated the system to investigate the influence of robot TA on learning. They noted that Robot TA is an effective way for supporting the teaching process by teachers and promoting the learning effect of students.

We aim to implement a system for creating an environment where students can actively ask their questions and evaluate the prototype system to verify the effectiveness of the proposed method.

3   Proposed System

Figure 1 shows the system organization. The system consists of a teacher's web application (TW app), a student's web application (SW app), a web server on which the web applications are installed, and a humanoid robot. A teacher and students can access the system by using their own smartphones and PCs with a standard web browser and a network connection. They can directly login to the web applications from their web browsers.

Firstly, the teacher needs to connect his/her device with the robot via a local connection. When a student inputs a question with text data on his/her device, the question is sent to the server. Then, the server sends the received question to all participating devices. This allows all users (a teacher and students) to see the question in real time. The question is displayed with its number and time. If the teacher's browser receives the question from the server and the robot's hand is in a neutral position, the TW app sends a set of commands to the robot for raising its hand and making the robot say "I have a question". This procedure notifies all users that someone sent the question to the system. Both the teacher and the students can freely send questions in the system. Because the system does not display the questioner's name, the students can casually ask their question. The teacher can control the timing for the robot to speak out the question by giving a permission from his/her device. The teacher can also select the vocalized question from the current question list and from the system log. Because the users can download all the questions raised through the system as CSV files, they can confirm the contents even after the lecture is over.

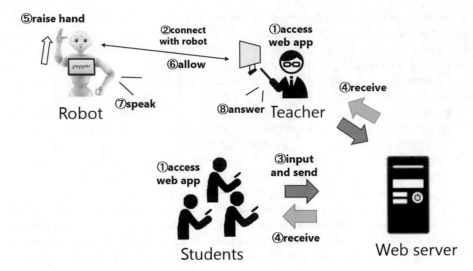

Fig. 1. System organization.

If the teacher can select key questions and make the robot voice them during the lecture, it should be a valuable assistant function. However, when the teacher is giving a lecture, checking questions one by one and deciding an important question to send the robot is too much for the teacher to do and it is not realistic. Therefore, we propose a function to automatically pump up important questions based on the students' evaluation. The system provides a function for the students to rate questions whether they can agree with them or not. Then the system picks up highly rated questions and underline them in the list or display the top five highly rated questions at certain time intervals. This eliminates the need for the teacher to examine all questions. Additionally, it clarifies the questions that students regard as important.

3.1 Humanoid Robot

We adopted a humanoid robot "Pepper" in the system as shown in Fig. 2. Pepper is developed and marketed by Softbank Robotics Corp. Making the robot vocalize a question enables to share the contents with all users in the classroom. At this time, conveying the question contents only by voice without using a robot is also possible. However, our final goal is to make the students raise their own hands and actively ask questions. We, therefore, think that our approach can create a mood for voluntarily asking questions by encouraging the students through making the robot behave like a person to raise its hand and ask question.

The system uses QiMessaging JavaScript for driving Pepper from the teacher's device. It is a library provided by Aldebaran allowing HTML5 applications to set up a connection with the Pepper's OS called NAOqi and issue control commands through API supported by the NAOqi OS. After the TW app acquires the QiMessaging JavaScript installed on the server, the teacher can access and control Pepper via the TW app running with the browser on his/her device. The software components of both TW app

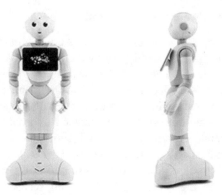

Fig. 2. Humanoid robot "Pepper".

and NAOqi OS should be connected in the same wired or wireless local area network. The teacher needs to enter the robot's IP address to set up a connection and the system displays the message in the teacher's browser console log if the connection successfully completes.

There are three actions taken by the robot such as raising its hand, lowering the hand (neutral position), and speaking. The teacher gives instructions to activate the robot functions using the above-mentioned Qimessaging.js mechanism. Figure 3 shows example robot actions for raising its hand. The system drives a servomotor placed at the right shoulder to raise the hand by making a half turn and lower it by making a reverse half turn.

Hand Raised   **Hand Lowered**

Fig. 3. Robot motions to raise the right hand.

3.2   Web Application System Implementation

We designed and developed the system as a web application system with HTML and JavaScript. It allows the users to readily use by their own smartphones and PCs. There are two types of client side applications for the teacher (TW app) and the students (SW app). Because the teacher's device needs a one-to-one connection with the robot for activating some actions, we divide the modules for the teacher's device from the students'. Both modules are uploaded to the server and provide respective services to the teacher's and the students' devices accessed from web browsers.

We use the communication functions based on Socket.IO to exchange the question contents and display them on all users' devices in real time when a question is raised. We install Socket.IO on the server and write both server and client programs in JavaScript to support asynchronous duplex communications between them. When a client sends a question to the server, the server broadcast the received question to all clients. Then all clients display the question. Since the system automatically performs this process, the users receive the new question without updating the web browser.

Figure 4 shows the system processing flow. When a student inputs his/her question on his/her browser, the SW app running on his/her device sends the question to the server. Next the server broadcasts the question to all clients (teacher's and students' devices). Then, the TW app running on the teacher's device detects the new question arrival and sends a command to the robot for raising its hand and vocalizing "I have a question." All users notice the new question by this robot action. After that, the teacher decides whether he/she permit the question to be vocalized by the robot and sends the question to the robot at his/her convenience. When multiple questions are sent to the server, the robot is only instructed to raise its hand and say "I have a question." The teacher has a control for selecting appropriate questions to be vocalized by the robot and their timing.

The users can download the questions as a CSV file as shown in Fig. 5 so that they can confirm the questions after the lecture. We use JavaScript's blob class to implement the download function. A set of questions is divided into a single question each separated by an enter code, a single question is divided into the number, the time, and its contents each separated by a comma, and then converted it into a CSV file. A function enabling the students to rate questions is not implemented in the current system, so it will be added in the future.

3.3   GUI Design and System Flow

There are two web applications in the system. Figure 6 shows the captured image of the teacher's application (TW app) and Fig. 7 is for the students' application (SW app). The system assumes that a teacher uses the TW app on his/her device and multiple students simultaneously use the SW app on their own terminals as shown in Fig. 8. First of all, the teacher needs to connect his/her device to the robot by entering the robot's IP address and pressing the "Connect button" in the TW app screen. Next, the system allows all the students to freely type questions on their device after connecting to the SW app on the server. If a student wants to ask question, he/she inputs the question in the "Question enter area" and presses the "Send button" on his/her device.

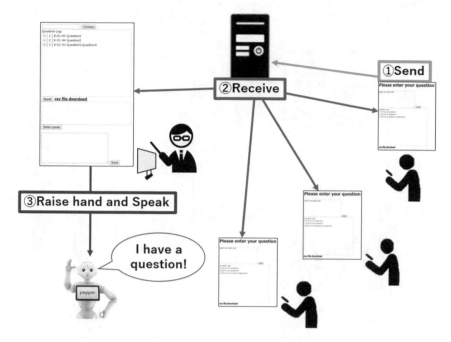

Fig. 4. System processing flow.

Fig. 5. Converting questions to CSV file format.

After the server receives the question and broadcasts to all users, it is displayed in the "Question Log area" of receiver's device with its number and the received time. Because the question number is initialized to one when the web application is started, it may be different between users even for the same question if the timing to start the web application is different.

Since the teacher can also raise a question, the teacher may be better to send a first question after starting the system in a lecture for making students think that someone has already asked a question and try to follow that person. When multiple questions have been sent and displayed on the users' terminals, the teacher presses the "Speak button" to make the robot vocalize the first question displayed in the teacher's "Question Log area" window. Since the robot raises the hand when a new question arrives, pressing the "Speak button" by the teacher instructs the robot to lower the hand

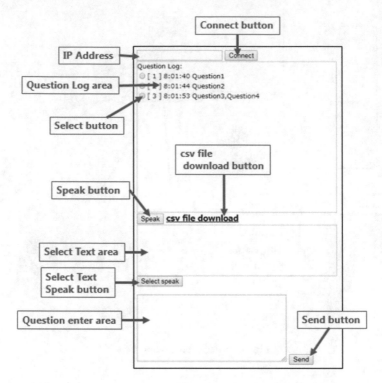

Fig. 6. Captured image of teacher's web application (TW app).

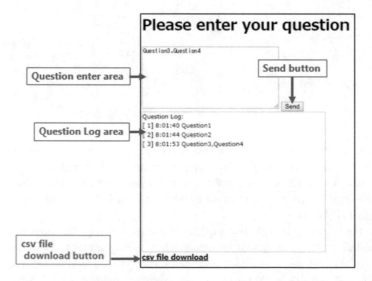

Fig. 7. Captured image of students' web application (SW app).

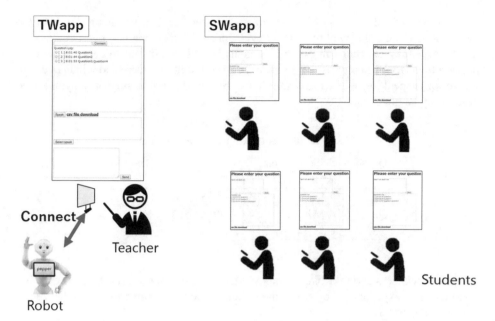

Fig. 8. Example of using web application in lecture.

and vocalize the first question. Pressing the "Speak button" again by the teacher sends the next question to the robot and the robot vocalizes it without raising the hand. If the teacher wants to present specific questions, he/she can press their corresponding "Select button" displayed in the "Question Log area" window. The selected questions are listed in the "Select Text area" and they are sent to the robot when the teacher presses the "Select Text Speak button" for vocalization. The users can download the received questions as a CSV file by pressing the "csv file download button".

4 Preliminary Experiment

We conducted a preliminary experiment to evaluate the functionality and usefulness of the developed system and the impression about the humanoid robot. A total ten university students (six undergraduate and four graduate students) participated in the experiment. We firstly explain how to use the system and make them freely use it to become familiar with the system. After that, we asked them to complete the questionnaire consisting of the following three questions:

Q1: Is it easy to understand how to use the students' web application?
Q2: Is it easy to understand how to use the teacher's web application?
Q3: Do you want to make the robot speak in a lecture?

Table 1 shows the evaluation results. It lists all the scores given by 10 subjects as labeled from A to J and the average scores of the three questions. The value one is bad and five is good for all questions. The average scores of Q1, Q2, and Q3 are 4.2, 3.9, and 3.8, respectively. All questions are rated as better than 3 which is the center of the evaluation, so we confirmed the overall evaluation of the system is good. On the other hand, some subjects stated negative comments like "It is difficult to confirm the connection to the robot in the case of TW app", "It is difficult to understand which question the robot is speaking when many questions are raised" and "Using only synthesized voice may be good enough".

Table 1. Results of preliminary experiment

Question	Subjects' score										Average score
	A	B	C	D	E	F	G	H	I	J	
Q1	4	5	4	5	5	5	5	4	5	3	4.2
Q2	4	5	2	5	5	4	5	4	5	2	3.9
Q3	4	5	4	5	5	3	2	5	5	1	3.8

In the future, we would like to improve the system based on these comments and conduct more experiments by using the system in actual lectures to verify the practicality of the system.

5 Conclusion

In this paper, we proposed the system to solve the problem that students don't raise their hands and actively ask questions during the lecture. The system enables them to type and send questions from their own smartphones or PCs and make the robot raise its hand to notice the existence of questions and vocalize them. We designed and prototyped the system and conducted a preliminary experiment to verify the effectiveness of the system.

As a future work, we would like to improve the system functions referring to the comments obtained in the experiment and conduct more experiments by using the system in real lectures. We also need to implement the function as described in Sect. 3 for allowing students to rate questions and automatically pump up important ones based on the students' evaluation.

References

1. Fuji, T., Yamaguchi, Y.: Question behaviors of undergraduates in classes: why undergraduates don't ask questions in classes? Kyushu Univ. Psychol. Res. **4**, 135–148 (2003). (in Japanese)
2. Koizumi, S., Kanda, T., Miyashita, T.: Collaborative learning experiment with social robot. J. Robot. Soc. Jpn. **29**(10), 902–906 (2011). (in Japanese)
3. Fujisawa, K.: Lecture with Twitter on a large class: showing tweets on slides. Syst. Control Inf. **55**(10), 446–451 (2011). (in Japanese)

4. Atsumi, M., Murata, Y., Yasukawa, A.: A collaborative system with teaching assistants based on linkage of open chat and robots. In: The 31st Annual Conference of the Japanese Society for Artificial Intelligence, pp. 1–4 (2017). (in Japanese)

5. Endo, K.: A pilot study on the facilitating effect of the questioning behavior of students through the internet. J. Educ. Appl. Inf. Commun. Technol. **14**, 11–15 (2011). (in Japanese)

6. Awedh, M., Mueen, A., Zafar, B., Manzoor, U.: Using socrative and smartphones for the support of collaborative learning. Int. J. Integr. Technol. Educ. (IJITE) **3**(4), 17–24 (2014)

7. Sun, Z., Li, Z., Nishimori, T.: Development and assessment of robot teaching assistant in facilitating learning. In: Proceedings of the International Conference on Educational Innovation through Technology (EITT), pp. 165–169 (2017)

User-Density Dependent Autonomous Clustering for MANET Based on the Laplace Equation

Rio Kawasaki[1]([✉]), Chisa Takano[2], and Masaki Aida[1]

[1] Graduate School of Systems Design, Tokyo Metropolitan University,
Tokyo 191-0065, Japan
`kawasaki-rio@ed.tmu.ac.jp, aida@tmu.ac.jp`
[2] Graduate School of Information Sciences, Hiroshima City University,
Hiroshima 371-3194, Japan
`takano@hiroshima-cu.ac.jp`

Abstract. Forming hierarchies is critical for the scalable route management of MANETs, so autonomous clustering of users is essential if we are to autonomously establish hierarchical structures. Particularly with regard to user movement, autonomous clustering should be able to cope with various movement characteristics, including random movement of users and movement accompanied by user density changes such as commuting. Since movement with user-density changes dramatically changes MANET topology, the conventional autonomous clustering approach fails to offer sufficiently stable cluster structures. Recent studies posit that stable cluster structures are possible even if user density varies significantly, by autonomously generating the cluster heads and then replacing them as needed. This approach is effective only when the cost of cluster head replacement is small. In this paper, we propose an autonomous distributed clustering technology based on the Laplace equation that gives stable cluster structures without changing the cluster heads.

1 Introduction

The Mobile ad hoc network (MANET) is a communication scheme in which mobile terminals within radio range of each other can communicate; data transfer to a remote target node is made possible by multi-hop links. Unlike cellular systems, MANET does not require a base station, so it is expected to offer communication during severe disasters. One of the most important technical issues in MANET is lowering the route management cost by raising efficiency and scalability. Since nodes move with users and the topology structure form by wireless links is frequently changed, it is necessary to identify the position of the node of the communication partner, as well as the most appropriate route between nodes; all actions should be performed in an autonomous manner.

The simplest method of routing in MANET is the flooding method [6]. However, in the flooding method, the amount of control packets for route search

© Springer Nature Switzerland AG 2020
L. Barolli et al. (Eds.): INCoS 2019, AISC 1035, pp. 68–79, 2020.
https://doi.org/10.1007/978-3-030-29035-1_7

increases exponentially with network size. This increase of control packet number places excessive loads on the network, and is the chief factor degrading scalability. Hierarchical routing was proposed to solve the scalability problem; the hierarchical structuring of MANET is traditionally achieved by clustering nodes [5].

Clustering for routing in MANET means the grouping of nodes, in hierarchical routing, the group is called a cluster. In hierarchical routing, flooding is limited to within each cluster and routing is divided into intra-cluster routing and inter-cluster routing. This eliminates the significant increase in control packet number as network scale increases and so improves scalability.

In MANET, since no node controls the entire network topology, clustering should be conducted by each node in an autonomous distributed manner. This means each node can behave based on just the local information known to each user. Since the user density in a certain spatial area is not local information, it cannot be used in the procedure of autonomous clustering. Random walk models [4] and random waypoint models [8] are often used as node movement models. These models assume that the user moves around in a relatively random manner, and that changes in the spatial density of nodes are relatively minor. Many studies used the node movement characteristics yielded by those movement models in implementing autonomous clustering. An example of node movement with density variation is a train station at rush hour. In such situations, autonomous clustering does not operate properly. Due to this problem, cluster structures become unstable in such situations.

Inoue et al. [2] showed a solution to this problem by generating cluster heads at random. The method combines two mechanisms: one is that all nodes can become cluster heads randomly with a certain probability, and the other is that the cluster structure managed by existing cluster heads is lost over time. This method introduces the concept of metabolism to clustering as new cluster structures are created continuously. Here, since the expectation value of the number of cluster heads generated autonomously in a certain area is proportional to the node density in that area, many clusters are generated in areas where the user density is relatively high. Therefore, it is possible to attain cluster structure stability even in areas with high node densities.

The method of [2] assumes that any node can become a cluster head. Since the battery consumption of the cluster head is intense, it is an advantageous to prevent the loads of battery consumption from concentrating on specific nodes. The creation of a new cluster means that the new head begins to manage routing control information and node information in the cluster, while cluster head metabolism requires an exchange that replicates control information. Therefore, the effectiveness of the method of [2] is effective only if the cost for changing the cluster head is small. However, if the cost of changing the cluster head is large, even if the load of battery consumption is concentrated on a specific node, it may be better to not change the cluster head. In this paper, we propose an autonomous clustering method for fixed cluster heads that can cope with node density changes.

The paper is organized as follows. Section 2 briefly reviews an existing autonomous clustering method based on the diffusion equation and its extended variant that offers cluster head metabolism proposed in [2]. In Sect. 3, we propose a new autonomous clustering method based on the Laplace equation. The proposed autonomous clustering method can cope with node density changes without changing cluster head nodes. In Sect. 4, we evaluate the characteristics of clusters generated by the proposed method by comparing them with Voronoi division. Finally, we conclude this paper and describe future issues in Sect. 5.

2   Cluster Construction Technology Based on the Diffusion Equation

In this section, we describe an autonomous clustering method based on the diffusion equation [3] that is robust against random mobility of nodes. After that, we describe an extension of the autonomous clustering method that introduces cluster head metabolism as proposed in [2].

2.1   The Diffusion Equation

We briefly explain the diffusion equation that is the basis of the autonomous clustering method, by using one-dimensional space as a simple example.

For density function $p(x,t)$ of a certain state quantity at position x and time t in a one-dimensional space, we assume that the temporal change of $p(x,t)$ occurs only by continuous movement. Then, the temporal evolution equation of $p(x,t)$ is expressed as

$$\frac{\partial p(x,t)}{\partial t} = -\frac{\partial J(x,t)}{\partial x}, \tag{1}$$

where $J(x,t)$ is a vector in the one-dimensional space whose right direction is positive; it represents the amount of movement of $p(x,t)$ per unit time. Assume $J(x,t)$ is given by

$$J(x,t) = -\kappa \frac{\partial p(x,t)}{\partial x}, \tag{2}$$

where $\kappa > 0$ is a constant called the diffusion coefficient. (2) is called Fick's laws of diffusion. Substituting Fick's laws of diffusion (2) into the continuous Eq. (1), yields the temporal evolution equation of $p(x,t)$, which is expressed as

$$\frac{\partial p(x,t)}{\partial t} = \kappa \frac{\partial^2 p(x,t)}{\partial x^2}. \tag{3}$$

This is called the diffusion equation.

2.2   Autonomous Clustering Based on the Diffusion Equation

We explain operation rules of each node that perform the autonomous clustering method based on the diffusion equation. For a certain node, i, let ∂i be the set of nodes adjacent to node i, and Δt be the time interval of the operations. Let $q_i(t_k)$ be the state quantity of node i at time $t_k = 1, 2, \ldots$). Then, the temporal evolution of $q_i(t_k)$ is expressed as

$$q_i(t_{k+1}) = q_i(t_k) - \Delta t \sum_{j \in \partial i} J_{i,j}^{\mathrm{diff}}(t_k), \tag{4}$$

where $J_{i,j}^{\mathrm{diff}}(t_k)$ represents the amount of movement per unit time that the state quantity moves from node i to node j due to the effect of diffusion, and $J_{i,j}^{\mathrm{diff}}(t_k)$ is expressed as

$$J_{i,j}^{\mathrm{diff}}(t_k) = -\frac{\sigma^2}{d_i + 1}(q_j(t_k) - q_i(t_k)), \tag{5}$$

where d_i denotes the nodal degree of node i and σ^2 is a positive parameter that determines the strength of diffusion. In (5), the diffusion coefficient is $\sigma^2/(d_i+1)$, and the reason for making it dependent on the nodal degree is to reduce the variation in diffusion speed over the network.

State quantity q_i of each node i is successively updated by the diffusion effect. Then the cluster head nodes and configuration of clusters are determined based on the state quantity of a node at a certain time. Details are as follows.

Each node i compares the state quantity q_i possessed by itself with that of the adjacent node. If the node has maximum value, the node is a cluster head. Otherwise, the node selects the node having the largest state quantity from among the adjacent nodes, and the selected node further selects the node with the largest state quantity from among the adjacent nodes. When such an operation is repeated, the node whose state quantity is maximal (a cluster head) is reached. Let all the nodes that reach a the same cluster head by node selection belong to the same cluster. Figure 1 shows schematically how this cluster configuration procedure is executed on a one-dimensional network. Here, CH denotes a cluster head, and each cluster is displayed in different colors. The amount of state q_i of each node i is smoothed over time due to the diffusion effect. Therefore, stable clusters can be obtained by stopping the smoothing and maintaining the cluster structure as in Fig. 1. However, if users move such that the density of nodes increases, clusters merge, and a unified large cluster emerges. Inoue et al. [2] tackled this problem by randomly generating cluster heads. The state quantity q_i of each node i is smoothed with time due to the effect of diffusion, and the cluster structure is lost. Simultaneously, to compensate for lost cluster heads, nodes of the cluster head are randomly generated with a certain probability. In this method, cluster heads are generated in proportion to the user density in areas where the user density is high. Therefore, it is possible to prevent excess unification of clusters and ensure that the number of clusters reflects user density even though all actions are performed in an autonomous distributed manner.

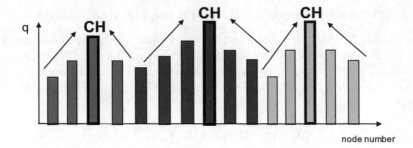

Fig. 1. One line of autonomous clustering

3   Autonomous Clustering Method Based on the Laplace Equation

In the work of [2], described in the previous section, the role of cluster heads is not concentrated on a specific node, since cluster heads are assumed to be changed over time. Therefore, load balancing for battery consumption can be realized. On the other hand, if the cost of changing a cluster head node is large, it is desirable minimize the number of cluster head changes. When specific nodes continue to act as cluster heads (but cluster head nodes can move to the same extent as other nodes), it can be expected that the number of cluster heads is proportional to the user density, as in the method of randomly generating cluster heads. Therefore, it is possible to prevent excess unification of clusters even in areas where user density is high. In order to realize autonomous clustering with fixed cluster heads, it is necessary to generate an appropriate cluster structure around the cluster head regardless of node movement. In this section, we propose a new autonomous clustering method based on the Laplace equation; it fixes specified nodes as cluster heads, in order to avoid the cost of cluster head metabolism.

3.1   The Laplace Equation

The Laplace equation yields a stationary solution for the diffusion equation. The Laplacian $\triangle$ in n dimensional Cartesian coordinate system $(x_1, x_2, \ldots, x_n)$ is expressed as

$$\triangle = \frac{\partial^2}{\partial x_1{}^2} + \frac{\partial^2}{\partial x_2{}^2} + \ldots + \frac{\partial^2}{\partial x_n{}^2}. \tag{6}$$

The Laplace equation is defined as

$$\triangle \phi = 0, \tag{7}$$

where ϕ is a scalar function as the solution of (7); it is called the harmonic function. Harmonic functions have the property that the value of ϕ at each point is equal to the mean value of the values at ϕ around that point.

The relation between the Laplace equation and the diffusion equation is explained by using a simple example in one-dimensional space. The Laplace equation in one-dimensional space is expressed as

$$\triangle\phi(x) = \frac{\partial^2\phi(x)}{\partial x^2} = 0. \tag{8}$$

When $p(x,t)$ of the diffusion Eq. (3) becomes a steady solution, which happens after sufficient time has elapsed, $\phi(x)$ becomes a function independent of t written as

$$\frac{\partial\phi(x)}{\partial t} = \kappa\frac{\partial^2\phi(x)}{\partial x^2} = 0, \tag{9}$$

therefore, the stationary solution of the diffusion Eq. (3) is a harmonic function.

3.2 Autonomous Clustering Based on the Laplace Equation

The solution to the Laplace equation is the stationary solution of the diffusion equation. In order to conduct clustering based on the Laplace equation, we start by setting the initial conditions and the boundary conditions appropriate for the diffusion procedure on the network; this yields a stationary solution with desirable cluster structure. In other words, the technical issue is to maintain cluster structure by giving appropriate initial conditions and boundary conditions.

First, we will explain the rules for particular nodes (hereinafter, specified nodes) that will be cluster heads. The specified node applies an initial condition that gives an initial value q_{ch} higher than other nodes so it becomes the cluster head from the beginning. Furthermore, in order to prevent the initial value given to the specific nodes from decreasing over time due to the effect of diffusion, the state quantity keeps the initial value q_{ch} by compensating for the amount that diffuses to adjacent nodes. So, the temporal evolution of specified node i is given as

$$q_i(t_{k+1}) = q_i(t_k) = ... = q_i(t_0) = q_{\mathrm{ch}} \tag{10}$$

By setting such boundary conditions, the role of cluster head is fixed to the specified nodes.

Next, we will explain the rules of nodes other than the specified nodes. When updating the state quantity of a node according to the rules of diffusion on the network (4), the state quantity of the specified node always maintains the initial value q_{ch}, so the state quantity is continuously supplied to the specified nodes from outside of the network. Therefore, if there is no mechanism for shifting the state quantity of nodes to outside the network, the state quantities of all nodes will approach q_{ch} with time. So, we need a new boundary condition to shift extra state quantity. We add the following rule as a new boundary condition applied to nodes other than specified nodes. Each node decreases its state quantity at a

Table 1. Parameter setting

Parameters	r_{dec}	q_{ch}	σ^2	t
Value	0.99	3.0	1.0	$0 \leq t \leq 2000$

constant rate r_{dec} when updating state quantity. The updating rule of the state quantity for each node i is expressed as

$$q_i(t_{k+1}) = \left(q_i(t_k) - \Delta t \sum_{j \in \partial i} J_{i,j}^{\text{diff}}(t_k) \right) r_{\text{dec}}, \tag{11}$$

By setting the initial conditions and the boundary conditions as described above, autonomous clustering with fixed cluster heads is possible.

3.3   Example of Clustering on One-Dimensional Network

We demonstrate autonomous clustering formed by the proposed method by using a simple one-dimensional network. The reason for using the one-dimensional network is that it provides simple visual understanding of the distribution of state quantities. In the network model, 100 nodes from 0 to 99 are connected in a one-dimensional configuration; nodes 0, 25, 50, and 75 are specified nodes. Other simulation conditions are as shown in Table 1. Here, t is the simulation time.

Figure 2 shows the distributions of the state quantities of the nodes at $t = 1000$ and $t = 2000$ as taken from the simulations results obtained using the above settings. A cluster is formed around each specified node, and it can be seen that the cluster structure does not change even as time passes.

4   Characteristic Evaluation of Autonomous Clustering Based on the Laplace Equation

In this section, experiments are used to investigate the characteristics of autonomous clustering based on the Laplace equation by using a network model. In particular, we clarify the characteristics of the method by comparing the generated clusters with Voronoi division of the network.

4.1   Cluster Structure and Voronoi Division

First, the Voronoi division of a two-dimensional plane is explained. The Voronoi division is a diagram in which multiple points (called seeds) are set on the plane and the area is divided according to which point on the plane is closest to the seeds [1]. Voronoi division is used for area analysis and planning of base station allocation [7]. Here, the correspondence between Voronoi division and

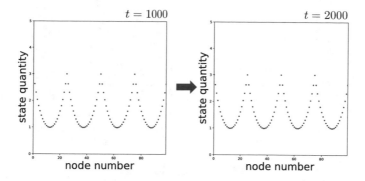

Fig. 2. Clustering experiment result in one-dimensional network

clustering on networks is explained. The Voronoi division on a two-dimensional plane is based on the Euclidean distance of all points from the seeds on the two-dimensional plane. On the other hand, in MANET clustering, the objects to be divided are nodes. Assuming the seeds are cluster heads, the distance between the base station to each node is the number of hops from the cluster head to each node; assuming cluster heads are given in a network, the Voronoi division is determined for all nodes. The nodes belonging to each of the Voronoi divided regions belong to the same cluster. In the following, we compare the cluster structure generated by autonomous clustering based on the Laplace equation with that yielded by Voronoi division of the network, to clarify the characteristics of autonomous clustering based on the Laplace equation. By defining gateway nodes in the cluster structure generated by autonomous clustering based on the Laplace equation and focusing on them, we compare Voronoi division and autonomous clustering based on the Laplace equation. The comparison is as follows.

- Calculate the number of hops from the gateway node to the cluster head of the cluster to which the node belongs in advance, and define the value as HOP.
- Calculate the difference in HOP values between adjacent gateway nodes that belong to different clusters.
- When the difference between the HOP values is less than 1, the gateway nodes are at the edge of a Voronoi division. This means the cluster does not contradict the Voronoi division. On the other hand, if the difference in HOP value is more than 2, the gateway nodes are not at the edge of a Voronoi division.

4.2   Experiment Environment

Here, we explain the network model used in the experimental evaluations of autonomous clustering. A total of 500 nodes are distributed over the region of $[0, 1] \times [0, 1]$ two-dimensional space. We set the four specified nodes that are

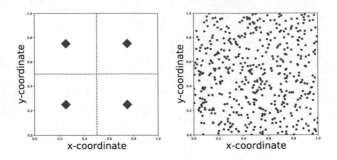

Fig. 3. Node placement

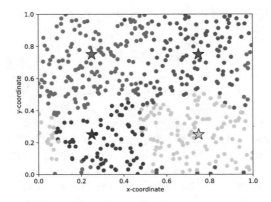

Fig. 4. Result of autonomous clustering

fixed to the position coordinates of (0.25, 0.25), (0.75, 0.25), (0.25, 0.75), and
(0.75, 0.75), as shown on the left of Fig. 3. The other 496 nodes are located at
positions where coordinates of the x axis and the y axis are determined randomly
by uniform random numbers in the interval [0, 1]. The initial positions of these
500 nodes are as shown on the right of Fig. 3. There is no node movement in this
evaluation.

Furthermore, we use a unit disk graph (UDG) as an appropriate network
model for MANET. UDG is a network generated by the rule that two nodes
are connected by a link if and only if the nodes are within a certain distance.
Here, a certain distance corresponds to radio range (communicable distance).
In this experiment, we set the communicable distance to 0.07. Moreover, in
order to exclude the influence of the boundary of two-dimensional space, we use
periodic boundary conditions. This means the left and right sides of the two-
dimensional square space are adjacent, and the and the upper and lower sides
are also adjacent, like a torus.

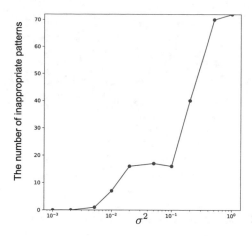

Fig. 5. Parameter σ^2 dependence on the similarity to Voronoi partitioning

4.3   Experimental Evaluations of Cluster Structure

A cluster structure was generated by applying autonomous clustering based on the Laplace equation to the above network model. Here, the values of the parameters used in clustering are the same as those in Table 1. As a result, the state of the cluster at $t = 2000$ is as shown in Fig. 4. Since the cluster structure does not change after $t > 200$, this is the steady state. In Fig. 4, each cluster is distinguished by color, the cluster head is indicated by ★, and the other nodes are indicated by •. Here, it can be seen that the specified nodes are cluster heads, since the ★ symbol is displayed at the coordinates of the plane where the specified nodes are installed. In addition, it can be seen that cluster division corresponds to the cluster heads.

We found 72 patterns in gateway nodes that did not match the Voronoi division (hereinafter, we refers to these as inappropriate patterns). As parameter σ^2 representing the strength of the diffusion is reduced, the number of inappropriate patterns decreases as shown in Fig. 5, and the cluster structure approaches Voronoi division. When the parameter representing the strength of diffusion is $\sigma^2 \geq 0.002$, the number of inappropriate patterns is 0, and it can be seen that the cluster structure matches that of Voronoi division. Figure 6 illustrates the cluster structure corresponding to Voronoi division, given the parameter $\sigma^2 = 0.002$. In this figure, the gateway nodes are indicated by • highlighted by black border.

The following assessments can be drawn from the above evaluation results. In autonomous clustering based on the Laplace equation, the topology structure of the network and the number of hops from the cluster head are factors that determine the state quantity of each node at steady state. In particular, with regard to the number of hops from the cluster heads, the state quantity decreases as the number of hops increases and the distance from the cluster heads. As shown in Fig. 2 in the preliminary evaluation, as the number of hops from the cluster head increases, the gradient of the state quantity distribution decreases

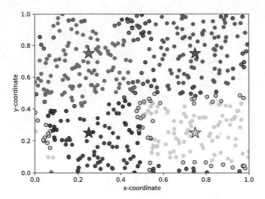

Fig. 6. Cluster structure consistent with Voronoi division

and the flow of the state quantity by diffusion becomes thinner. This phenomenon becomes more effective as the diffusion coefficient decreases due to the decrease of σ^2. The number of hops from the cluster head is measured in the shortest path, but the flow of state quantities by diffusion flows across multiple paths upstream of the cluster head. Therefore, in general, state quantities flow into nodes from multiple nodes, including paths that have different numbers of hops from cluster heads. However, if σ^2 is small, the flow from the minimum number of hops from the cluster head becomes dominant, and a cluster with the same structure as the Voronoi division appears.

5   Conclusions

In this paper, we proposed a new autonomous clustering method for MANETs; it specifies the cluster head nodes in advance by using the Laplace equation, in order to enable autonomous clustering that reflects MANET user density. In the proposed method, the cluster head nodes do not change making it superior to existing solutions when the cost of changing cluster heads is excessive. In addition, the characteristics of the cluster structure were investigated by comparing and evaluating the clusters generated by the proposed method with the those yielded by Voronoi division. As a result, we found that the structure of the cluster was changed by the difference in the strength of diffusion given. By reducing the diffusion coefficient, we can generate a cluster structure based on the minimum number of hops from the cluster head that matches the results of Voronoi division. As future work, in order to confirm the effectiveness of the proposed method, we plan to investigate the characteristics of clustering in an environment where node density changes with node movement.

Acknowledgments. This research was supported by Grant-in-Aid for Scientific Research (B) No. 17H01737 (2017–2019) and (C) No. 18K11271 (2018–2020) from the Japan Society for the Promotion of Science (JSPS).

References

1. Spielman, D.: Spectral graph theory. In: Combinatorial Scientific Computing, chap. 18. Chapman and Hall/CRC, Boca Raton (2012)
2. Inoue, K., Takano, C., Aida, M.: User-density dependent autonomous clustering in MANET. In: The 21st International Symposium on Wireless Personal Multimedia Communications (WPMC 2018), Thailand, pp. 369–374 (2018)
3. Aida, M.: Distributed Control and Hierarchical Structure in Information Networks. Corona Publishing Co. Ltd., Tokyo (2015). (in Japanese)
4. Kumar, M.K.J., Rajesh, R.S.: Performance analysis of MANET routing protocols in different mobility models. IJCSNS Int. J. Comput. Sci. Netw. Secur. $9(2)$ (2009)
5. Sood, M., Kanwar, S.: Clustering in MANET and VANET: a survey. In: Circuits, Systems, Communication and Information Technology Applications (CSCITA), pp. 375–380, April 2014
6. Ramanathan, R., Redi, J.: A brief overview of ad hoc networks: challenges and directions. IEEE Commun. Mag. 40, 20–22 (2002)
7. Yu, S.M., Kim, S.-L.: Downlink capacity and base station density in cellular networks. In: International Symposium on Modeling Optimization in Mobile Ad Hoc Wireless Networks (WiOpt), pp. 119–124, May 2013
8. Camp, T., Boleng, J., Davies, V.: A survey of mobility models for ad hoc network research. Wirel. Commun. Mob. Comput. $2(5)$, 483–502 (2002)

Data and Subprocess Transmission on the Edge Node of TWTBFC Model

Yinzhe Guo[1(✉)], Ryuji Oma[1], Shigenari Nakamura[1], Dilawaer Duolikun[1], Tomoya Enokido[2], and Makoto Takizawa[1]

[1] Hosei University, Tokyo, Japan
{yinzhe.guo.3e,ryuji.oma.6r}@stu.hosei.ac.jp,
nakamura.shigenari@gmail.com, dilewerdolkun@gmail.com,
makoto.takizawa@computer.org
[2] Rissho University, Tokyo, Japan
eno@ris.ac.jp

Abstract. In our previous studies, the TBFC (Tree-Based Fog Computing) model is proposed to reduce the electric energy consumed by fog nodes and servers in the fog computing model. Here, fog nodes are hierarchically structured in a height-balanced tree, where a root node is a cloud of servers and leaf nodes are edge nodes which communicate with devices. Each node receives data from child nodes and sends the processed data to a parent node. In the TWTBFC (Two-Way TBFC) model, is nodes send processed data not only to a parent node but also to each child node. Then, edge nodes make a decision on actions of their child actuators. However, in addition to messages to be delivered to servers, more number of messages are transmitted to edge nodes. In this paper, in order to reduce the number of messages which each fog node sends to the child nodes, each node only at some level of the tree collect output data of the other nodes of the same level. The nodes are referred to as aggregate nodes. Then, each aggregate node sends the collected data to the descendant edge nodes. Each edge node makes a decision or actions and send actions to the child actuators by using the data.

Keywords: Energy-efficient fog computing ·
IoT (Internet of Things) · Tree-based fog computing (TBFC) model ·
Two-way TBFC (TWTBFC) model · Aggregate nod

1 Introduction

The IoT (Internet of Things) [6,8] is composed of not only computers like servers and clients but also millions of devices, i.e. sensors and actuators installed in various things like glasses and cars [11,13]. Compared with traditional information systems like the cloud computing (CC) model [5], the IoT is more scalable and huge amount of data from sensors are transmitted in networks and are processed by application processes servers. In addition, this means, networks and servers are heavily loaded.

© Springer Nature Switzerland AG 2020
L. Barolli et al. (Eds.): INCoS 2019, AISC 1035, pp. 80–90, 2020.
https://doi.org/10.1007/978-3-030-29035-1_8

The fog computing (FC) model [15] is proposed to reduce the network and server traffic of the IoT (Internet of Things). In addition, huge amount of electric energy is consumed by nodes. In order to not only increase the performance but also reduce the electric energy consumption of the IoT, the TBFC (Tree-based Fog Computing) model is proposed in our previous studies [3,11,12,14]. Here, fog nodes are hierarchically structured in a height-balanced tree. A root node is a cloud of servers and leaf nodes are edge nodes which receive sensor data from sensors and send actions to actuators. Each fog node has one parent node and child nodes. Each node receives input data from the child nodes. Then, the fog node processes the input data and sends the output data, i.e. data obtained by processing the input data, to a parent node. A server in a cloud finally receives data processed by fog nodes. Then, the server makes a decision on actions to be done by actuators and delivers the actions to the actuators through networks of fog nodes. While the traffic of sensors at network can be reduced, it takes time to deliver actions to actuators.

In our previous studies [9,10], the TWTBFC (Two-Way TBFC) model is also proposed so that actions of actuators are decided by fog nodes nearer to devices in order to more promptly deliver actions to actuators. Here, each node not only sends output data obtained by processing input data to a parent node in a same way as the TBFC model but also forwards the output data to the child nodes. However, since processed data is transmitted in both upward and downward ways, traffic in the network and processing load of each node increase.

In this paper, we propose a new TWTBFC model to reduce the downward traffic to edge nodes. Here, nodes at some level of the tree are taken as *aggregate* nodes. The aggregate nodes exchange their output data with one another and collect the output data from every other aggregate node. Then, each aggregate node obtains aggregate data which is a collection of output data of all the aggregate nodes. Aggregate nodes send the aggregate data to the child nodes. Then, a collection of the output data is transmitted from each aggregate node down to the descendant edge nodes. Then, edge nodes make a decision on actions and send the actions to child actuators. In the evaluation, we show the number of messages and energy of nodes to obtain the aggregate data and deliver the aggregate data to edge nodes for aggregate level l.

In Sect. 2, we present the TWTBFC model of the IoT. In Sect. 3, we discuss how to deliver the output data to edge nodes. In Sect. 4, we evaluate the TWTBFC model.

2 Two-Way Tree-Based Fog Computing (TWTBFC) Model

2.1 TBFC Model

The fog computing (FC) model [15] to efficiently realize the IoT [11] is composed of sensor and actuator devices, fog nodes, and clouds. Clouds are composed of servers like the cloud computing (CC) model [5]. In the TBFC (tree-based fog computing) model [12,14], fog nodes are hierarchically structured in a height-balanced tree as shown in Fig. 1. Here, the root node f denotes a cloud of servers. Fog nodes at the bottom level are *edge* nodes which communicate with sensors and actuators. In the TBFC model, every edge node is at the same level.

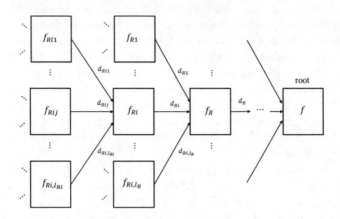

Fig. 1. TBFC model.

Each node f_R has l_R (≥ 0) child nodes $f_R, ..., f_{R,l_R}$. Here, f_{Ri} shows the ith child node of the fog node f_R and in turn f_R is a parent node of the node f_{Ri}. $ch(f_R)$ is a set $\{f_{R1}, ..., f_{R,l_R}\}$ of child nodes of a node f_R. $pt(f_{Ri})$ is a parent node f_R of a node f_{Ri}. For example, the second child node of a root node f is f_2. The second child node of the node f_1 is f_{12} whose second child node is f_{122}. Thus, the label R of a fog node f_R is a sequence of numbers and shows a path from the root node f_0 to the fog node f_R. Let $as(f_R)$ be a set of ancestor nodes of a node f_R. $ds(f_R)$ shows a set of descendant nodes of a node f_R and $sg(f_R)$ is a set of nodes which are at the same level as a node f_R.

In the cloud computing (CC) model, an application process p is performed on servers to process sensor data sent by sensors in networks. In this paper, an application process p is assumed to be a sequence of subprocesses $p_0, p_1, ..., p_{h-1}$. The subprocess p_{h-1} takes input data from sensors, the subprocess p_0 is performed on a root node f, i.e. server. Each subprocess p_i takes data from a subprocess p_{i-1} and gives the processed data to a subprocess p_{i+1}. In the TBFC model of height h, each subprocess p_i is performed on nodes of level i.

Let $p(f_R)$ show a subprocess to be performed in a node f_R.

A node f_R takes input data d_{Ri} from each child node f_{Ri} ($i = 1, ..., l_R$). D_R shows a collection of the input data $d_{R1}, ..., d_{R,l_R}$. The node f_R obtains output data d_R by doing the computation $f(p_R)$ on the input data D_R. Then, the node f_R sends the output data d_R to a parent node $pt(f_R)$.

A notation $|d|$ shows the size [Byte] of data d. The ratio $|d_R|$ / $|D_R|$ is the *output ratio* ρ_R of a node f_R. If a fog node f_R obtains an average value of the input data $d_{R1}, ..., d_{R,l_R}$, the output ration ρ_R is $1/l_R$. Let i_R and o_R be the sizes of $|D_R|$ and $|d_R|$ of input data D_R and output data d_R, respectively. Here, $o_R = \rho_R \cdot i_R$.

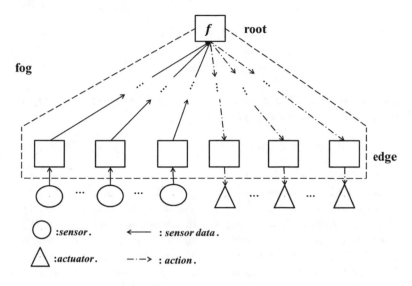

Fig. 2. Data and action transmissions.

In the TBFC model, sensor data is transmitted to a root node by fog nodes and actions are transmitted from a root node to edge nodes as shown in Fig. 2. However, it takes time to deliver data from edge nodes to a root node and actions from the root node to edge nodes.

2.2   Fog Nodes

A node f_R is assumed to be implemented to be a sequence of input (I_R), computation (C_R), and output (O_R) modules. The input module I_R receives input data D_R from the child nodes and the output module O_R sends output data d_R to the parent node $pt(f_R)$. The computation module C_R is a subprocess $p(f_R)$ which generates the output data d_R by processing the input data D_R. In this paper, we assume the I_R, C_R, and O_R modules are sequentially performed in a fog node f_R on receipt of the input data D_R.

It takes time to perform the modules of a node f_R. Let $TI_R(x)$, $TC_R(x)$, and $TO_R(x)$ show the execution time [sec] of the input I_R, computation C_R, and output O_R modules of a node f_R for data of size x, respectively. The execution time $TC_R(x)$ depends on the computation complexity of a subprocess $p(f_R)$. In this paper, the computation complexity of the subprocess $p(f_R)$ is $O(x)$ or $O(x^2)$, $TC_R(x)$ is $ct_R \cdot C_R(x)$ where $C_R(x) = x$ or $C_R(x) = x^2$ and ct_R is a constant. A pair of execution time $TI_R(x)$ and $TO_R(x)$ to receive and send data of size x, respectively, are proportional to the data size x, i.e. $TI_R(x) = rt_R \cdot x$ and $TO_R(x) = st_R \cdot x$, where, st_R and rt_R are constants.

$$TC_R(x) = ct_R \cdot C_R(x). \tag{1}$$

$$TI_R(x) = rt_R \cdot x. \tag{2}$$

$$TO_R(x) = st_R \cdot x. \tag{3}$$

It takes time $TF_R(x)$ [sec] to receive and process input data D_R of size x and send the output data d_R in each node f_R:

$$TF_R(x) = TI_R(x) + TC_R(x) + \delta_R \cdot TO_R(\rho_R \cdot x). \tag{4}$$

Here, if f_R is a root node, $\delta_R = 0$, else $\delta_R = 1$.

We measure the execution time $TI_R(x)$, $TC_R(x)$, and $TO_R(x)$ of the I_R, C_R, and O_R modules of a fog node f_R are measured for data size x as shown in Fig. 3. Here, the node f_R is implemented in a Raspberry Pi 3 model B [2] which is connected with a 10 Gbps network. A pair of fog nodes communicate with each other in UDP [4]. A computation module C_R which just selects one value in the input data of size x is realized in the node f_R. A fog node f_R sends and receives data of size x by *send* and *receive* system calls of UDP [4], respectively. As shown in Fig. 3, the execution time $TI_R(x)$ of the I_R module is five times longer than the execution time $TO_R(x)$ of the O_R, i.e. $rt_R = 5 \cdot st_R$ and $ct_R = rt_R/2$. That is, $ct_R : st_R : rt_R = 1 : 2.5 : 0.5$.

$EI_R(x)$, $EC_R(x)$, and $EO_R(x)$ show the electric energy [J] consumed by the input I_R, computation C_R, and output O_R modules [11] of a node f_R for input data of size x, respectively. In this paper, we assume each node f_R follows the SPC (Simple Power Consumption) model [6–8]. The power consumption of a node f_R to perform the computation module $C_R(= p(f_R))$ is $maxE_R$ [W]. The energy consumption $EC_R(x)$ [J] of the computation module C_R of a node f_R to process input data of size x (> 0) is $EC_R(x) = maxE_R$ [W] $\cdot TC_R(x)$ [sec].

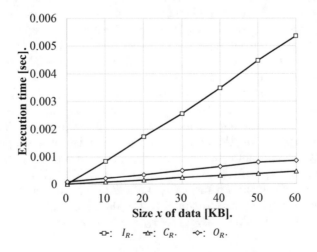

Fig. 3. Execution time of each module.

A pair of the electric power PI_R and PO_R [W] are consumed by the input I_R and output O_R modules, respectively [6–8]. PI_R and PO_R are $re_R \cdot maxE_R$ and $se_R \cdot maxE_R$, respectively, where $0 < se_R \le re_R \le 1$, i.e. $re_R = 0.729$. $se = 0.676$ and $re = 0.729$ in the Raspberry Pi 3 model B node. The energy consumption $EI_R(x)$ and $EO_R(x)$ [J] to receive and send data of size x ($>$ 0) are $EI_R(x) = PI_R[\text{w}] \cdot TI_R(x)[\text{sec}]$ and $EO_R(x) = PO_R[\text{w}] \cdot TO_R(x)[\text{sec}]$, respectively. Each node f_R consumes the energy $EF_R(x)$ to process the input data D_R of size x:

$$EF_R(x) = EI_R(x) + EC_R(x) + \delta_R \cdot EO_R(\rho_R \cdot x)$$
$$= (re_R \cdot TI_R(x) + TC_R(x) + \delta_R \cdot se_R \cdot TO_R(\rho_R x)) \cdot maxE_R$$
$$= (re_R \cdot rt_R \cdot x + ct_R \cdot C_R(x) + \delta_R \cdot se_R \cdot st_R \cdot \rho_R x) \cdot maxE_R. \quad (5)$$

3 Two-Way Transmission of Data

As presented in the preceding section, each fog node f_R of level l receives input data $D_R = \{d_{R1}, ..., d_{R,l_R}\}$ from child nodes $f_R, ..., f_{R,l_R}$ and obtains output data d_R by processing the output data D_R. In addition to sending the output data d_R to the parent node $pt(f_R)$, each node f_R sends the output data d_R to every fog node f_s of the same level l. In turn, each node f_R receives the output data d_s from every other node f_s of the same level l, as shown in Fig. 4. Here, nodes of level l are referred to as *aggregate* nodes and the level l is aggregate level in the tree. Let AN_l be a set of aggregate nodes of level l in the tree. Let O_l be a set of output data obtained by all the aggregate nodes of the level l, i.e. $O_l = \{d_s | f_s \in AN_l\}$ The set O_l of output data of aggregate node, is referred to as *aggregate* data. Each aggregate node f_R obtains the aggregate data O_l by collecting output data d_s form every other aggregate node f_s. Then, each aggregate node f_R sends the aggregate data O_l down to the child nodes $f_{R1}, ..., f_{R,l_R}$ the deliver to the descendant edge nodes.

Each edge node f_R receives the aggregate data O_l from the parent fog node $pt(f_R)$. Then, the edge node f_R makes a decision on actions to be performed on child actuators by processing the aggregate data O_l. The edge node f_R sends the actions to the child actuators.

Let us consider a node f_R which has child fog nodes $f_{R1}, ..., f_{Rl_R}$. Let x_R stand for size of the output data d_R. The size x_R of the output data d_R of a fog node f_R is given as follows: $x_R = \rho_R \cdot (\Sigma_{i=1}^{l_R} x_{Ri})$. Here, ρ_R is the output ratio of the nodes $f_{R1}, ..., f_{R,l_R}$. If f_R is an edge node, each size x_{Ri} shows the size of the output data d_{Ri} from a child sensor. Thus, the size x_{Ri} of the output data d_{Ri} of each node f_{Ri} of level l can be obtained. Each aggregate node f_R of aggregate level l obtains the aggregate data O_l whose size $as_l (= |O_l|)$ is $\Sigma_{f_s \in AN_l} (\rho_s \cdot \Sigma_{i=1}^{l_s} x_{si})$.

Each aggregate node f_R consumes electric energy and time to collect the aggregate data O_l from other aggregate nodes of level l. Each aggregate node f_R of level l sends the output data d_R to and receives the output aggregate data d_s from every other aggregate node f_s. Then, the aggregate node f_R forwards

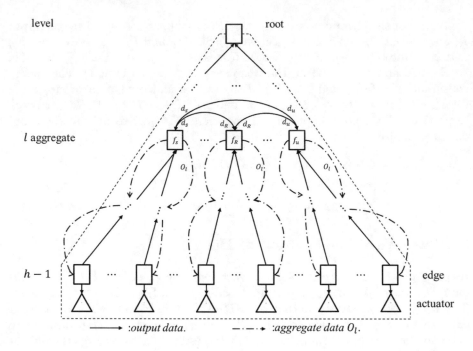

Fig. 4. Two-way transmission.

the aggregate data O_l to the child nodes $f_{R1}, ..., f_{R,l_R}$. O_l is a set $\{d_s | f_s \in AN_l\}$ and $|O_l|$ is $\Sigma_{f_s \in AN_l} |d_s|$. It takes time ATO_R [sec] to send the output data d_R to and receive the output data d_s from every other aggregate node f_s, and to send the aggregate data O_l to the child nodes:

$$ATO_R = TO_R(o_R) \cdot |AN_R| + \Sigma_{f_s \in AN_R} TI_R(o_s) + TO_R(|O_l|). \qquad (6)$$

The node f_R additionally consumes the energy AEO_R as follows:

$$AEO_R = (se_R \cdot TO_R(o_R) \cdot |AN_R| + TO_R(|O_l|) + re_R \cdot \Sigma_{f_s \in AN_R} TI_R(o_s)) \cdot maxE_R. \qquad (7)$$

Each descendant node f_R of the aggregate nodes receives the aggregate data O_l and forwards the aggregate data O_l to the child nodes $f_{R1}, ..., f_{R,l_R}$. Each node f_R of level k ($< l$) consumes energy to forward the aggregate data O_l to the descendant edge nodes.

$$AEO_R = (re_R \cdot TI_R(|O_l|) + se_R \cdot TO_R(|O_l|)) \cdot maxE_R. \qquad (8)$$

The higher the aggregate level l is the smaller size of the aggregate data O_l and the fewer number of messages are exchanged among the aggregate nodes. However, the more number of messages are transmitted to deliver the aggregate data O_l to edge nodes.

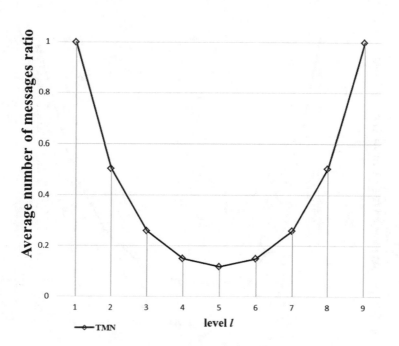

Fig. 5. Average number of messages.

4 Evaluation

We evaluate the TWTBFC model of the IoT in terms of electric energy consumption of fog nodes and number of messages transmitted by fog nodes. The TWTBFC model is composed of fog nodes structured in a tree. In this paper, we consider a height-balanced k-ary tree of fog nodes whose height is h. The output ratio ρ_R of each fog node f_R is assumed to be the same, i.e. $\rho_R = \rho$. We assume a root node is a server f_0 with two Inter Xeon E5-2667 CPUs [1], where the minimum electric power consumption $minE_0$ is 126.1 [W] and maximum electric power consumption $maxE_0$ is 301.3 [W]. Each fog node f_R is realized by a Raspberry Pi Model B [2]. Here, the minimum electric power $minE_R$ is 2.1 [W] and the maximum electric power $maxE_R$ is 3.7 [W]. [3]. The computation rate CR_R of each fog node f_R is $0.879/4.75 = 0.185$, where the computation ration rate of the root node f is 1. There are k^l aggregate nodes at aggregate level l in the tree. As presented in the preceding section, the aggregate nodes exchange the aggregate data O_l with one another. Each aggregate node f_R sends the output d_R to $(k^l - 1)$ aggregate nodes and receives output data from the other $(k^l - 1)$ aggregate nodes. Hence, totally, $k^l \cdot (k^l - 1)$ messages are transmitted. Here, suppose the size of sensor data which each edge node receives from the child sensors is one. A node of level $h - 1$ receives data of total size k from k

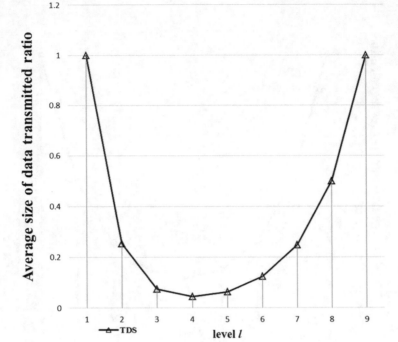

Fig. 6. Average size of data transmitted.

edge nodes and sends the output data d_R of size ρk. Thus, each aggregate node f_R receives the output data D_R of size $(\rho k)^{h-1-l} \cdot k$ and generates the output data of size $(\rho k)^{h-l-1}$. Hence, the total size $k^l \cdot (k^l - 1) \cdot (\rho k)^{h-l-1}$ of data is exchanged among the aggregate node.

Each aggregate node f_R sends the aggregate data O_l whose size is $k \cdot (\rho k)^{h-1-l}$. The aggregate node f_R sends the aggregate data O_l to k child nodes $f_{R1}, ..., f_{Rk}$. Each child node f_{Ri} forwards the aggregate data O_l to k child nodes. Thus, totally $k^{l-1} \cdot (k + k^2 + ... + k^{h-1})$ $(= k^l(1 - k^h)/(1 - k))$ messages are transmitted from aggregate nodes to edge nodes. Each message carries the aggregate data O_l. Hence, the total size $k \cdot (\rho k)^{h-1-l} \cdot k^l \cdot (1 - k^l)/(1 - k) = \rho^{h-l-1} \cdot k^h \cdot (1 - k^l)/(1 - k)$ of data is transmitted. The total number TMN_l of messages and the size TDS_l of data are transmitted.

$$TMN_l = k^l \cdot [(k^l - 1) + (k(1 - k^{h-l-1})/(1 - k))]. \tag{9}$$

Figures 5 and 6 show the average number TMN_l/k^l and the average size TDS_l/k^l for each aggregate node at aggregate level l. As shown in the figures, the number of messages and the size of data transmitted can be minimized for aggregate level $l = 5$.

$$TDS_l = (\rho \cdot k)^{h-l-1} \cdot k^l \cdot (k^l - 1) + (k^{h-l} \cdot \rho^{h-l-1}) \cdot (1 - k^{h-l-1})/(1 - k). \tag{10}$$

5   Concluding Remarks

In this paper, we proposed the modified model to efficiently realize the TWTBFC model. Here, one aggregate level l is selected and fog nodes at the level l are aggregate node, which exchange the output data with one another. Each aggregate node collects the output data to an aggregate data and forward the aggregate data to the descendant edge nodes. Then, each edge node makes a decision on actions to be performed on the child actuators by using the aggregate data and sends the actions to the actuators. In the evaluation, as showed the number of messages and energy consumption of nodes to exchange output data forward aggregate for aggregate level.

References

1. Dl360p gen8. www8.hp.com/h20195/v2/getpdf.aspx/c04128242.pdf?ver=2
2. Raspberry pi 3 model b. https://www.raspberrypi.org/products/raspberry-pi-3-model-b/
3. Chida, R., Guo, Y., Oma, R., Nakamura, S., Duolikun, D., Enokido, T., Takizawa, M.: Implementation of fog nodes in the tree-based fog computing (TBFC) model of the IoT. In: Proceedings of the 7th International Conference on Emerging Internet, Data and Web Technologies (EIDWT-2019), pp. 92–102 (2019)
4. Comer, D.E.: Internetworking with TCP/IP, vol. 1. Prentice Hall, Upper Saddle River (1991)
5. Creeger, M.: Cloud computing: an overview. Queue **7**(5), 3–4 (2009)
6. Enokido, T., Ailixier, A., Takizawa, M.: A model for reducing power consumption in peer-to-peer systems. IEEE Syst. J. **4**, 221–229 (2010)
7. Enokido, T., Ailixier, A., Takizawa, M.: Process allocation algorithms for saving power consumption in peer-to-peer systems. IEEE Trans. Industr. Electron. **58**(6), 2097–2105 (2011)
8. Enokido, T., Ailixier, A., Takizawa, M.: An extended simple power consumption model for selecting a server to perform computation type processes in digital ecosystems. IEEE Trans. Industr. Inf. **10**, 1627–1636 (2014)
9. Guo, Y., Oma, R., Nakamura, S., Duolikun, D., Enokido, T., Takizawa, M.: Evaluation of a two-way tree-based fog computing (TWTBFC) model. In: Proceedings of the 13th International Conference on Innovative Mobile and Internet Services in Ubiquitous Computing (IMIS-2019), pp. 72–81 (2019)
10. Guo, Y., Oma, R., Nakamura, S., Duolikun, D., Enokido, T., Takizawa, M.: A two-way flow model for fog computing. In: Proceedings of the Workshops of the 33rd International Conference on Advanced Information Networking and Applications (WAINA-2019), pp. 612–620 (2019)
11. Oma, R., Nakamura, S., Duolikun, D., Enokido, T., Takizawa, M.: An energy-efficient model for fog computing in the internet of things (IoT). Internet Things **1–2**, 14–26 (2018)
12. Oma, R., Nakamura, S., Duolikun, D., Enokido, T., Takizawa, M.: Evaluation of an energy-efficient tree-based model of fog computing. In: Proceedings of the 21st International Conference on Network-Based Information Systems (NBiS-2018), pp. 99–109 (2018)

13. Oma, R., Nakamura, S., Duolikun, D., Enokido, T., Takizawa, M.: A fault-tolerant tree-based fog computing model (accepted). Int. J. Web Grid Serv. (IJWGS) **15**(3), 219–239 (2019)
14. Oma, R., Nakamura, S., Enokido, T., Takizawa, M.: A tree-based model of energy-efficient fog computing systems in IoT. In: Proceedings of the 12th International Conference on Complex, Intelligent, and Software Intensive Systems (CISIS-2018), pp. 991–1001 (2018)
15. Rahmani, A.M., Liljeberg, P., Preden, J.-S., Jantsch, A.: Fog Computing in the Internet of Things. Springer, Cham (2018)

A Wheelchair Management System Using IoT Sensors and Agile-Kanban

Takeru Kurita[1], Keita Matsuo[2(⊠)], and Leonard Barolli[2]

[1] Graduate School of Engineering, Fukuoka Institute of Technology (FIT),
3-30-1 Wajiro-Higashi, Higashi-Ku, Fukuoka 811-0295, Japan
mgm19103@bene.fit.ac.jp
[2] Department of Information and Communication Engineering,
Fukuoka Institute of Technology (FIT),
3-30-1 Wajiro-Higashi, Higashi-Ku, Fukuoka 811-0295, Japan
{kt-matsuo,barolli}@fit.ac.jp

Abstract. Recently, the handicapped persons require more high quality electric wheelchair to move more flexible in the home or in a hospital. Also, if we use electric wheelchair the maintenance is needed for keeping safety. For this reason, we propose a wheelchair management system that can manage electric wheelchairs using Agile-Kanban. Agile is a technique to develop the software and manage the work efficiency. Kanban is a method to support Agile development. In this work, we design and implement a wheelchair management system using IoT sensors and Agile-Kanban. Our proposed system can manage the wheelchair efficiently. We present the design and implementation of the proposed system and show that it can measure the wheelchair's states and wheelchair's activity.

1 Introduction

Recently, many communication systems have been developed in order to support humans. Especially, IoT technologies, IoT devices and sensors can provide many services for humans. The IoT sensors can be embedded in electronic devices. For instance, if a refrigerator has some IoT sensors inside, it can tell us the ingredients for cooking through the Internet anytime and anywhere. The temperature of the air conditioner in the room can be controlled. Also, television, cooking heater, microwave, bath-unit, toilet-unit, light and so on are going to provide new services.

Currently, the vehicles can be connected to the Internet. Toyota uses the word "Connected Car System". This system can collect the data using mobile network and sensors in the car. These are different kinds of data such as the data taken by the car when is moving, car speed, braking or acceleration, weather condition and traffic jam information. These data are gathered to the datacenter through the network by using Internet technologies. The data can be analyzed and used for solving some traffic problems. There are also other applications using IoT such as the use of IoT technologies for agriculture, factories, offices, schools, hospitals and so on.

© Springer Nature Switzerland AG 2020
L. Barolli et al. (Eds.): INCoS 2019, AISC 1035, pp. 91–100, 2020.
https://doi.org/10.1007/978-3-030-29035-1_9

We consider the use of Agile and Kanban (the meaning of Kanban in Japanese is sign board) for Wheelchair Management system. Agile is software developing method that denotes "the quality of being agile, readiness for motion, nimbleness, activity, dexterity in motion". The software development methods can offer the answer to the business community asking for lighter weight along with faster and nimbler software development processes [1,5]. Kanban is one of methods that can support Agile process [11].

The structure of this paper is as follows. In Sect. 2, we introduce the IoT and Kanban. In Sect. 3, we describe the Agile-Kanban, In Sect. 4, we present our proposed wheelchair management system. In Sect. 5, we show the conclusions and future work.

2   IoT and Kanban

2.1   IoT

Our life style is drastically changing by IoT technologies. The IoT can connect various things to the Internet such as household electronics appliances, vehicles, robots, communication devices, and some application softwares. Thus improving our quality of life. In Fig. 1, we show the image of IoT environment. IoT system has many sensors, which can be used for operating factories, farms, offices, houses, and so on. However given the fact that most of IoT sensors are resource limited and operate on batteries, the consumption power and life time of sensors are important issues for wireless IoT sensor networks design [9,12].

There are some approaches for decreasing the number of packets in the networks. The Opportunistic Networks (OppNets) is one of them. It can provide an alternative way to support the diffusion of information in special locations in a city, particularly in crowded spaces where current wireless technologies can exhibit congestion issues [8].

2.2   Kanban

Kanban is a system produced by Toyota. The system can efficiently produces vehicles. The main goal of Kanban was to produce cars for the same or lower price than the competitors. This Kanban system is called Toyota Production System (TPS). The TPS is also known as Just in Time (JIT) manufacturing and the basic principle is to produce "only what is needed, when it is needed and in the amount needed". Currently, there are many research work using Kanban for developing software [6,7,14]. At the beginning, the Kanban was used in manufacturing processes, however its applications in other areas is continuously growing due to its proven successfulness.

We have shown a schematic illustration of Kanban system in Fig. 2. The system has two Kanbans. The Kanban system has 3 sections: Upstream, Downstream and Store. The Upstream section role is for manufacturing parts by making some blocks of products. Downstream section is used for assembling parts

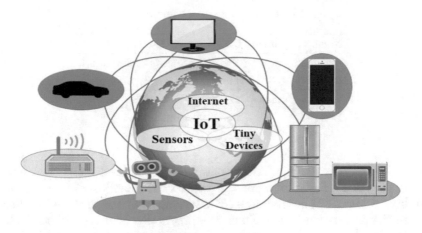

Fig. 1. Image of IoT environment.

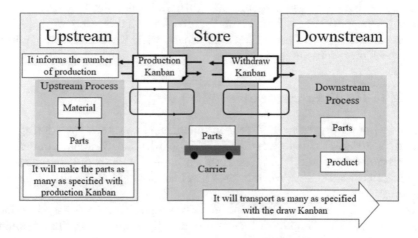

Fig. 2. Structure of Kanban system.

and completing the products. Store section is used for keeping some parts needed for working operation. If the amount of stocking parts increases, the manufacturing efficiency decreases. In order to solve this problem, Kanban system uses Production-Kanban and Withdraw-Kanban.

The upstream section produces some parts and stock them in the store section. While, the downstrean section uses the parts in the store to assemble the products. The Production-Kanban can only move between upstream and store, while the Withdraw-Kanban is moving between store and downstream as shown in Fig. 2.

We describe the moving of Kanbans in Fig. 3. If there is a shortage of parts in downstream section, the Withdraw-Kanban moves from downstrean section to store section. Then, the Withdraw-Kanban informs the number of shortage

Table 1. Description of Kanban names.

Name of Kanban	Description
Backlog	The sensor needs maintenance
Ready	The sensor is ready for using
Work in progress	The sensor is working
Done	The sensor finish the work

parts to Production-Kanban. After that, the Production-Kanban instructs the upstream section to manufacture the number of shortage parts informed by Withdraw-Kanban. The Production-Kanban will put the number of shortage parts in the production queue. Then, the production of parts will starts. Thus, the Kanban system can control the supply and demand of parts, which leads to an efficient manufacturing.

Recently, Kanban system combined with Agile is used for software development. Some papers use Kanban and Agile for collaboration work [3,10]. In another paper, the authors use Agile approach with Kanban for managing the security risk on e-commerce [2].

3   Agile-Kanban System

In this section, we present IoT sensor management system considering Agile-Kanban (see Fig. 4). This system uses Kanboard. The Kanboard requires the web server, database and PHP. The Kanboard offers 4 kinds of Kanbans to users which are Backlog, Ready, Work in progress and Done. Each Kanban description is shown in Table 1.

Kanboard is a free and open source software [4]. Kanboard user interface is shown in Fig. 5. The Kanboard focus on simplicity minimalism. The number of features is limited. By Kanboard can be known the current status of a project because it is visual. It is very easy to understand and there is no need to explain and no training is required.

Kanboard has a number of features as follows.

- Action Visualization,
- Limit work in progress to focus on the goal,
- Drag and drop tasks to manage the project,
- Self-hosted,
- Simple installation.

In Fig. 5, we present a user interface of Kanboard, which shows 4 Kanbans: Backlog, Ready, Work in progress and Done.

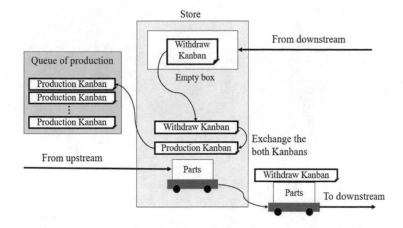

Fig. 3. Moving of Production-Kanban and Withdraw-Kanban.

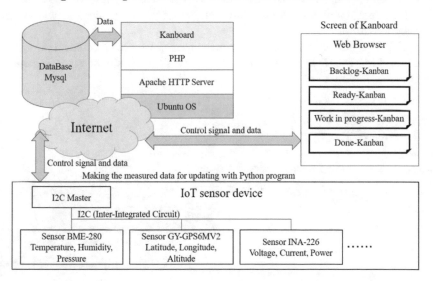

Fig. 4. Agile-Kanban system.

4   Proposed Wheelchair Management System

With the rapid development of science and technology, traditional hand-propelled wheelchairs have gradually evolved into electric wheelchairs that can be operated by individuals with mobility impairments [13]. Especially, the handicapped persons require more high quality electric wheelchair to move more flexible in the home or in a hospital. Also, if we use electric wheelchair the maintenance is needed for keeping safety, changing the battery, checking the state of motor (current, voltage, temperature) and condition of whole electric wheelchair.

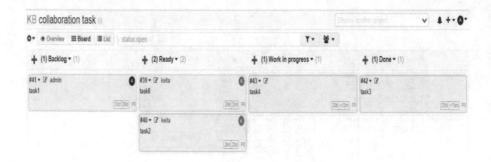

Fig. 5. User interface of Kanboard.

As hospital, nursing or aged facilities need to use many electric wheelchairs. Thus, it will spent a lot of time and manpower. Therefore, we propose a wheelchair management system that can manage many electric wheelchai rs using Agile-Kanban.

In Fig. 6 is shown the user interface of proposed wheelchair management system and in Fig. 7 is shown the schematic illustration of the management system.

Fig. 6. User interface of wheelchair management system.

The Kanbans can move in different states such as Backlog, Ready, Work in progress and Done. One Kanban corresponds to one wheelchair. The Kanban can change some states with drag and drop. For example, when there is a Kanban in the state of Backlog, it means that the wheelchair needs to do maintenance. When Kanban is in the state of Ready, it means that the wheelchair is ready to be used. Work in progress means the wheelchair is working. In this case, the measurement data of wheelchair by sensors will be uploaded to the database. When Kanban is in the state of Done, it means the end of work.

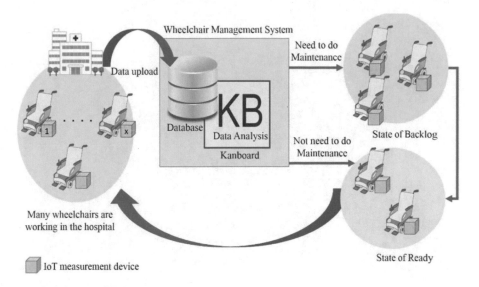

Fig. 7. Schematic illustration of proposed wheelchair management system.

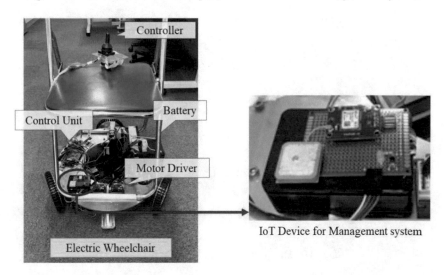

Fig. 8. IoT device for wheelchair management system.

We implemented the IoT device in order to measure the states of wheelchair as shown in Fig. 8. The device can get wheelchair's states such as the battery's voltage, the motor's current so on.

When the Kanban is in the "Work in progress" state, the measured data of voltage, current and temperature on wheelchair are shown in Fig. 9. In Fig. 10 is shown the image of moving Kanban on the wheelchair management system using Kanboard. While, in Fig. 11 is shown the wheelchair status.

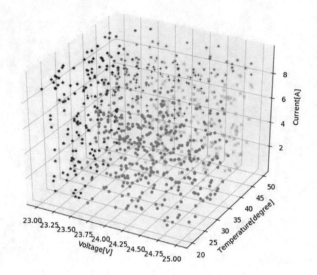

Fig. 9. Measured data on the wheelchair.

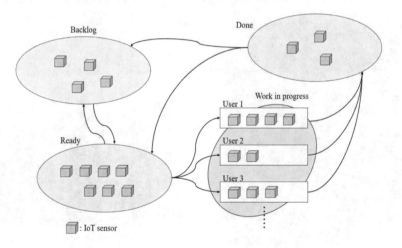

Fig. 10. Different states of Kanban moving.

The benefits of wheelchair management system are as following.

- The system can manage many wheelchairs.
- When the wheelchairs are not used, the management system can cut down the consumption power for saving wheelchair's battery.
- The system is able to analyze the wheelchair status such as wheelchair's working time, frequency of use and wheelchair's activity.
- The system can be installed in other factory machines, vehicles, robots, offices or school facilities, home electrical appliances and so on.

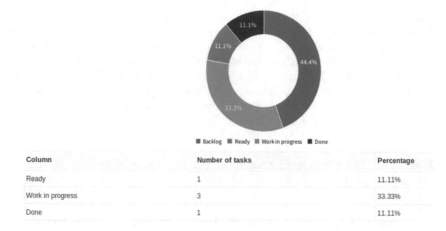

Column	Number of tasks	Percentage
Ready	1	11.11%
Work in progress	3	33.33%
Done	1	11.11%

Fig. 11. Wheelchair status.

5   Conclusions and Future Work

In this paper, we introduced Agile-Kanban. We presented in details the Kanban system and Kanboard. The proposed system can manage many wheelchairs. Also, the proposed system can measure the wheelchair's states and wheelchair's activity. By making an efficient management, the wheelchair battery life time is increased. In addition, we designed and implemented a IoT measurement device using Agil-Kanban.

In the future work, we would like to improve the proposed system and carry out experiments.

References

1. Abrahamsson, P., Salo, O., Ronkainen, J., Warsta, J.: Agile software development methods: review and analysis. arXiv preprint arXiv:1709.08439 (2017)
2. Dorca, V., Munteanu, R., Popescu, S., Chioreanu, A., Peleskei, C.: Agile approach with Kanban in information security risk management. In: 2016 IEEE International Conference on Automation, Quality and Testing, Robotics (AQTR), pp. 1–6. IEEE (2016)
3. Hofmann, C., Lauber, S., Haefner, B., Lanza, G.: Development of an agile development method based on kanban for distributed part-time teams and an introduction framework. Procedia Manuf. **23**, 45–50 (2018)
4. Kanboard: Kanboard. https://kanboard.org/
5. Kiely, G., Kiely, J., Nolan, C.: Scaling agile methods to process improvement projects: a global virtual team case study (2017)
6. Kirovska, N., Koceski, S.: Usage of Kanban methodology at software development teams. J. Appl. Econ. Bus. **3**(3), 25–34 (2015)
7. Maneva, M., Koceska, N., Koceski, S.: Introduction of Kanban methodology and its usage in software development (2016)

8. Miralda, C., Donald, E., Kevin, B., Keita, M., Leonard, B.: A delay-aware fuzzy-based system for selection of IoT devices in opportunistic networks considering number of past encounters. In: Proceedings of the 21st International Conference on Network-Based Information Systems (NBiS-2018), pp. 16–29 (2018)
9. Nair, K., Kulkarni, J., Warde, M., Dave, Z., Rawalgaonkar, V., Gore, G., Joshi, J.: Optimizing power consumption in IoT based wireless sensor networks using bluetooth low energy. In: 2015 International Conference on Green Computing and Internet of Things (ICGCIoT), pp. 589–593. IEEE (2015)
10. Padmanabhan, V.: Functional strategy implementation-experimental study on agile Kanban. Sumedha J. Manag. **7**(2), 6–17 (2018)
11. Petersen, J.: Mean web application development with agile Kanban (2016)
12. Sheng, Z., Mahapatra, C., Zhu, C., Leung, V.: Recent advances in industrial wireless sensor networks towards efficient management in IoT. IEEE Access **3**, 622–637 (2015)
13. Tseng, T.H., Liang-Rui, C., Yu-Jia, Z., Bo-Rui, X., Jin-An, L.: Battery management system for 24-V battery-powered electric wheelchair. Proc. Eng. Technol. Innov. **10**, 29 (2018)
14. Yacoub, M.K., Mostafa, M.A.A., Farid, A.B.: A new approach for distributed software engineering teams based on Kanban method for reducing dependency. JSW **11**(12), 1231–1241 (2016)

A Remote Puppet Control System for Humanoid Communication Robot

Toshiyuki Haramaki[✉] and Hiroaki Nishino

Division of Computer Science and Intelligent Systems,
Faculty of Science and Technology, Oita University, Oita, Japan
{haramaki,hn}@oita-u.ac.jp

Abstract. When a speaker communicates a message to the listener, it can be expected to be transmitted more accurately by attaching an appropriate gesture to message. That theory is valid, whether the speaker and the listener are near or far away. However, for a speaker to deliver messages and gestures to a remote listener synchronously, it is necessary to have a real-time conversation system using video images. Therefore, in this research, we propose a mechanism for presenting gestures synchronized with gestures at minimal cost when sending messages to remote listeners. To realize that, we develop a method to move the robot near the speaker instead of conveying the gesture of the speaker on the video shot. In this proposed method, motion data of the hand of the speaker is acquired using a motion sensor, the data is processing, converting into information for moving the motor corresponding to the joint of the robot, and the information is transmitting to the remote robot, the robot moves with motion according to the intention of the speaker. To realize the proposed method, we prepared a mechanism to convert the hand gesture of the speaker into the motion data of the robot in real-time and manipulate the robot in a faraway place. In our system, we realize a function to transmit information of extreme urgency such as disaster information all at once to robots installed in each household. Based on these requirements, we developed a prototype of an information presentation system. By utilizing this system, we expect the speaker's message to communicate with the multiple listeners effectively.

1 Introduction

Recently, robot technology has been rapidly evolving. It needs to go to enrich people's lives by utilizing robots for society. For example, security robots that protect public security, nursing robots for supporting care recipients. Extreme high safety is the most important thing to put these robots into practical use in society. In addition, people need to feel an affinity with robots. Implementing the function to realize smooth communication between the robot and the human being is a particularly important task. In recent years, communication robots who can talk with people appeared. Those robots to interact with people using voices and motions. These robots are expected in various fields, and they can make people's lives better. We have acquired surrounding ecological data using various sensors, analyzed the data by computer, derived useful information, and effectively present that information to people. The purpose of these

L. Barolli et al. (Eds.): INCoS 2019, AISC 1035, pp. 101–112, 2020.
https://doi.org/10.1007/978-3-030-29035-1_10

studies is that computers help people in diverse situations of human life. And, we aim to build a system that prevents the occurrence of obstacles and accidents by advising people on the risk factors that threaten people's save lives.

In this research project, we have developed a system that makes it easier to transmit and visualize data obtained by measuring intra-building Wi-Fi radio intensity [1]. As another research, by continuing to observe the indoor environment, we realized a system that warns by voice message when the temperature and humidity become high and the risk of heat stroke increases [2]. And, we have constructed a system aiming at sustaining concentration by working on olfaction when the driver of the car is going ahead [3]. Further, we built a system that effectively communicates information through the robot according to the driving situation of the car and the content of information transmission [4, 5]. In these research efforts, the method of moving the robot is a way of predetermining the motion pattern of the robot and calling it when it becomes necessary.

Because of those researches, when conveying a message to a person utilizing a robot, we found that it is nice to send the movement of the robot synchronously with the voice message. However, continuing to steer the remote robot at the same time as the speaker is speaking is a complicated task. For a robot to present a motion simultaneously with a word to a listener at a faraway place, a mechanism is required to enable the speaker to steer the robot with a simple gesture. Also, it is necessary to have a mechanism in which a robot located in a faraway place moves without delay according to the manipulation of the speaker. Another way for humans to operate the robot in real-time is used by the control pad. In this method, the motion of the robot can be easy realized. However, it is not suitable for moving multiple drives on the robot simultaneously. Recently, there is research on how to capture and move the movement of the human body. This research is also conducted in our laboratory as introduced earlier. However, this method requires high-performance computers and complicated processing.

In this paper, we propose a mechanism to acquire the movement of the hand of a person with a sensor and move the robot in real-time according to it. This method can be realized with a mechanism simpler than whole-body motion capture and can realize more complicated motion than the method using the control pad. By realizing this we expect to able to develop a new communication framework using robots. And we define a robot motion design scheme that encompasses various combinations among different sensing technologies and robots. Then, we elaborate an implementation method for a specific case among the combinations. With the improbable message transmission method incorporating the mechanism proposed here, the speaker can transmit the message and the motion to the detached listener through the robot only with a simple operation. We expect that the acceptance level of listeners' messages will increase with the combination of inaccessible language communication and nonverbal communication and communication robots. Our goal goes to construct a robot motion design method that does not depend on specific devices and technologies as much as possible.

2   Related Work

In this research, the humanoid robot expects the speaker to play a role as an interface for correctly communicating what he wants to tell the hearer. Research that transmits information via a robot has been conducting in various ways. Matsui et al. propose that the humanoid robot has some advantage for mapping man movements to robot motions [6]. Patsadu et al. proposed a method to detect the gesture of the whole body of a person by learning about the joint shape of the person photographed by the depth camera of the light coding system [7]. It is a challenging task for appropriately designing and creating motion data so that the robots act like a human. The most promising method goes to directly use human behaviors as is and map them to the series of commands for driving the robots. Jung et al. propose that not only when the robot is speaking, but also when you are not singing voice sends the message appropriately by using of motion of the robot [8].

Even in prior research, they used humanoid robots as interfaces for delivering messages to humans. Williams and Breazeal have developed a robot of AIDA (Affective Intelligent Driving Agent) that sends a message to encourage safe driving [9]. They installed it on the dashboard of the car and conducted experiments and proved that providing the message via the robot is effective. Zeng et al. are working on research to realize a guide dog by a robot for guiding the visually impaired and living support [10]. Peng et al. introduced a rehabilitation support robot that realizes two-way communication to restore the physical function of a stroke patient and showed that it is effective for improving the motor function of patients [11]. Miyachi et al. introduction that both the recreation for health caregivers and the health gymnastics were effective by using the humanoid robot's speaking and action functions together at the care facility [12].

There are numerous studies so far in the field of Telepresence, Telexistence, in a method of presenting information to a remote party. Telepresence is a term indicating a technology that provides a realistic feeling as if they are faced with members in remote areas. As one method of realizing telepresence, Maimone and Fuchs implemented a real-time 3D telepresence environment realized by synchronizing several inexpensive depth cameras [13]. Adalgeirsson and Breazeal developed a robot that enables a speaker to navigate a remote robot to achieve nonverbal communication. Thereby proving that the impression of the listener improved [14]. Wang et al. developed a micro telepresence robot that projects itself to another space remotely, moves around, communicates via video and audio, using the smart device of the user [15]. Telexistence is a term for the technology that enables a human being to get a real-time sense of being a place other than where they exist and being able to interact with the remote environment [16–18].

3 System Organization

The proposed system measures the gesture of the speaker in real-time, turns it into the motion of the robot, transmits its motion data, and drives the robot in the remote place. In the proposed system, one speaker can operate all at once on multiple robots. Figure 1 shows the configuration.

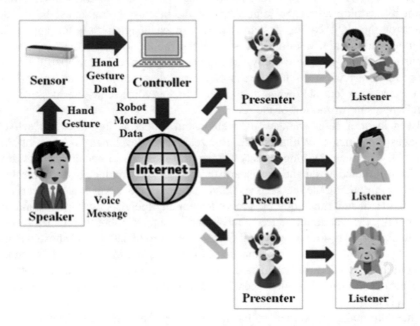

Fig. 1. Dataflow of proposed robot motion broadcasting system.

This system has five objects Speaker, Sensor, Controller, Presenter, Listener. Speaker performs a word message and a hand gesture to operate the robot at the same time for the listener. The sensor acquires the speaker's hand gesture and converts it to data. The controller converts hand gesture data into robot motion data and transmits it via the internet. Presenter receives motion data of robot and voice message and outputs motion with the message. The listener receives information sent from the robot.

Figure 2 shows to generate the movement of the robot from the movements of both hands of the operator. The sensor acquires the movement of the operator's hand as an image with depth information. By analyzing the image, the movement of the operator's palm or finger is detected in real-time. Based on the detection results, the system continues to transmit operator motion data to the robot controller. The controller that receives it converts the data into the robot's motion command and moves the robot's motor. In general, as a method of using the gesture of the speaker for the motion of the robot, it is possible to think of a method in which the speaker moves from the entire body and imitates the movement as it is on the robot. Based on the detection result, data for the robot to move corresponding to the operator's gesture is generated. For moving the robot using that data, the robot moves as intended by the operator.

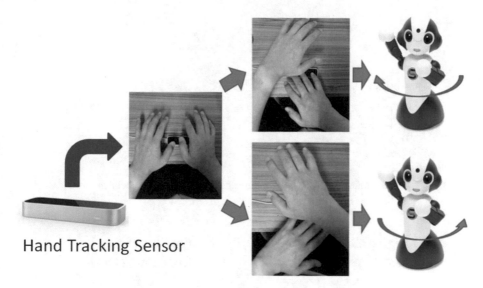

Fig. 2. Move the robot with Hands Gesture.

The reason why we did not contain the method of directly using the speaker's gesture in this research is to make the robot generate motions consciously rather than moving the robot with gestures that the speaker performs unconsciously. Moreover, to recognize the gesture of the whole body, it is a need for high-performance computer and photographic equipment.

In this system implementation, we chose to target only the speaker's hands and fingers as a method of recognizing the speaker's gesture. Therefore, we utilized Leap Motion [19] as photographic equipment to detect the movement of the hands and fingers of the speaker. Figure 3, 4, 5 and 6 explain the mapping between the gesture of the speaker's hand acquired by Leap Motion and the robot movement. We utilized Sota [20] made by Vstone, which is used as a robot on the listener side in this research. The robot has eight servo motors in the body. These motors are one of the left shoulder, left elbow, right shoulder, right elbow, the base body and three at the neck, causing free rotation movement, respectively. By coordinating and moving these motors at the same time, we can create natural motion like human beings.

Figure 3 shows that the operator raises and lowers the sensor's hand to move the left and right shoulder of the robot up and down. Likewise, Fig. 4 shows that the speaker is achieved by opening and gripping the hand on the sensor to bend and extend the elbow on the left and right of the robot. Also, Fig. 5 shows that the speaker pulls one of the left and right hands towards you to move the body of the robot. Finally, Fig. 6 shows that the speaker can freely move the robot's neck by changing the orientation of the palm on the sensor. Between these gestures and the movement of the

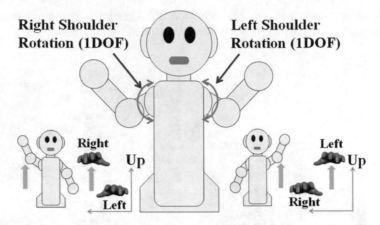

Fig. 3. Mapping from human hand gesture to robot shoulder motion.

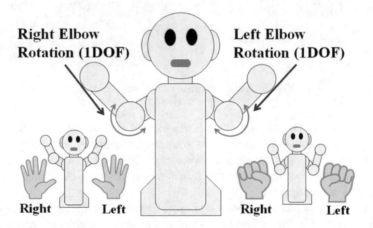

Fig. 4. Mapping from human hand gesture to robot shoulder motion.

robot, there are rules and calculation formulas for conversion respectively. The system converts the information of the captured hand to the corresponding motor angle value of the robot.

In this research, to realize nonverbal communication in conformity with the message by the speaker during the conversation, we detect the movement of the hands and fingers of the speaker and generate the motion of the robot based on the result. It is necessary to perform processing to convert a human gesture into the movement of a robot. We describe the procedure for realizing the proposed system.

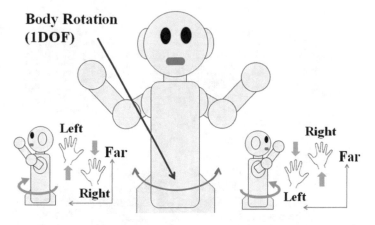

Fig. 5. Mapping from human hand gesture to robot body motion.

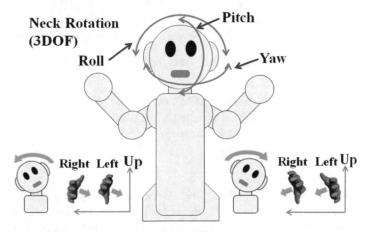

Fig. 6. Mapping from human hand gesture to robot neck motion.

4   System Implementation

Figure 7 shows in the overall configuration in the implementation of this system and the modules installed in each component. This system consists of three components. They are "Controller", "Broadcasting Server", "Communication Robots". In this system, the controller first acquires the user's gesture data continuously using the sensor. The gesture-motion conversion module converts the received gesture data to the angle value of the motor to realize the movement of the robot. The broadcasting server continues to transmit the motor angle value to the message server all the time through the robot motion publisher module of the controller. According to the value, the communication robots continue to drive the motor using the robot motion activation module.

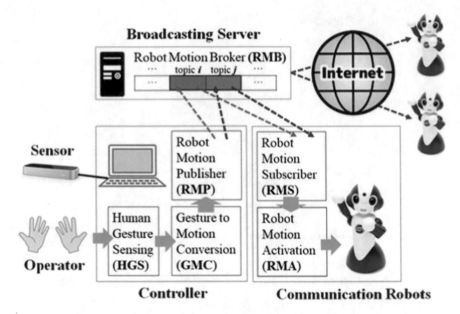

Fig. 7. Dataflow of proposed robot motion broadcasting system.

The data flowing between the parts are a total of eight angle values of the joints for the robot to move and a flag indicating whether the hand of the speaker is recognized. Next, we show in the details of each component.

4.1   Controller

In this system, the controller acquires the gesture of the hand of the speaker, converts the data into motion data of the robot, and is in charge for transmitting the data.

First, the Human Gesture Sensing (HGS) module converts the gesture into data. This work is achieved by using the Leap Motion SDK used as a sensor this time. Next, using Gesture to Motion Conversion (GMC) module, convert gesture data to motion data of the robot. To convert it, it is necessary to prepare calculation formulas after investigating the obtainable range of the gesture data and the operable angle of each joint of the robot in advance. Finally, Robot Motion Publisher (RMP) module puts the motion data of the robot in a transmittable state. In this system, the MQTT [21] Protocol is utilized for data transmission. By choosing this protocol, lightweight data transmission and reception by the Pub-Sub method can be realized.

4.2   Broadcasting Server

In this system, the server functions as a data broker for transmitting and receiving data between the controller and the robot. For the broker, each datum is stored separately in each topic, and these data sent by specifying it as the latest topic from Publisher are rewritten. The subscriber gets the latest data for the specified topic. This data transmission/reception method has an advantage that there is no requirement to establish synchronization between the transmission side and the reception side. By dividing the topic into a plurality of topics, it is possible to send motion simultaneously to robots of different shapes.

4.3 Communication Robots

In this system, the communication robot uses the Robot Motion Subscriber (RMS) module to acquire robot motion data from Server. The acquired data are set as the angle value of each motor by the Robot Motion Activation (RMA) module. By getting it, the robot moves as intended by the speaker. The robot used this time is under a built-in controller with a Linux OS. The RMS module that receives robot motion data and the RMA module that activates the robot motion is established in the development environment such as Eclipse and transferred to the built-in controller. The robot has functions such as the acquisition of a voice message by microphone and speech by the speaker.

5 Preliminary Experiments

We constructed a prototype system built on the design described in the previous section and carried out the preliminary experiment. Figure 8 shows the image of the installed system. The OS of the control PC is Linux (Ubuntu). All modules are running on the environment and are implemented using Python language. Leap Motion connects to the controller PC. The PC converts the gesture of the hand obtained by the sensor into the angle value of the eight motors of the robot. To broadcast these values to multiple robots, we implemented a function using a free MQTT broker Mosquitto [22]. In the robot, all modules implemented in Java language. We utilized the Eclipse Paho library [23] to read robot angle values from the Broker.

Because of the operation experiment of this prototype system, when the controller and the robot are connecting to the same network, the time took for the robot to move after recognizing the hand of the speaker is less than 0.1 s confirmed. Also, the speed of transmitting robot motion from the controller to the robot was over 20 times per second. Moreover, even if the controller and the robot communicate via the 4G mobile phone communication network, we confirmed that the robot moves in real-time depending on the movement of the hand of the speaker. In this system, even if the motion data passes through the mobile phone line, the time to reflect the gesture of the speaker's hand on the motion of the robot is expected to be very rapid.

We implemented this prototype system and conducted numerous experiments to check whether the proposed mechanism works properly. As a result, we confirmed that the system functions properly, and at the same time gained some new knowledge. One of their knowledge is that the operators need to be familiar with the gesture operations defined in this system. Users can operate the robot accurately at remote locations by practicing this system for several tens of minutes in advance. Another knowledge is possible for the behavior of the remote robot may not match the user's maneuvers. The reason is that, because of the user operating the robot. Interference may take place in

Fig. 8. Prototype system implemented for preliminary evaluation.

the movement of the robot's arms and neck. Since the robot only moves the servomotor corresponding to each joint by a specific value, it is necessary to prepare control values so that the robot is not involved in the system.

6   Conclusions

In this study, we developed a prototype of a system that presents information simultaneously via robots to people at remote locations. The movements of those robots were created linked to the movements of human hands. Because of conducting experiments using them, we were able to present gestures and voice messages using multiple robots without delay as designed. In the future, it is assumed that robots with faster control PCs and improved built-in controllers will be used. If this can be realized, it can be called upon to create smoother and more natural robot movements. In addition, for humans to operate robots naturally, it is necessary to recognize the correspondence between human movements and robot movements and the operable range and map those movements appropriately. At present, confirmation and mapping of the range of motion of both human and robot are performed manually. But in the future, we aim to automate these tasks.

We have also conducted experiments and have recognized that it is necessary to make the gesture accustomed to some extent to generate motion in real-time using the gestures of the speaker's hand. In our previous work [5], robot movements used with messages took the form of converting human hand gestures into movements, editing movement data, and calling it. To solve this problem, I would like to consider how to use the gesture to call the robot's motion pattern made by the speaker in advance.

References

1. Shimizu, D., Haramaki, T., Nishino, H.: A mobile wireless network visualizer for assisting administrators. In: Proceedings of the 6th International Conference on Emerging Internet, Data and Web Technologies, pp. 651–662, February 2018
2. Yatsuda, A., Haramaki, T., Nishino, H.: An unsolicited heat stroke alert system for the elderly. In: Proceedings of the 2017 IEEE International Conference on Consumer Electronics - Taiwan, pp. 347–348, June 2017
3. Okazaki, S., Haramaki, T., Nishino, H.: A safe driving support method using olfactory stimuli. In: Proceedings of the 12th International Conference on Complex, Intelligent and Software Intensive Systems, pp. 958–967, June 2018
4. Haramaki, T., Yatsuda, A., Nishino, H.: A robot assistant in an edge-computing-based safe driving support system. In: Proceedings of the 21st International Conference on Network-Based Information Systems, pp. 144–155, September 2018
5. Haramaki, T., Goto, K., Tsutsumi, H., Yatsuda, A., Nishino, H.: A real-time robot motion generation system based on human gesture. In: Proceedings of the 13th International Conference on Broad-Band Wireless Computing, Communication and Applications, pp. 135–146, October 2018
6. Matsui, D., Minato, T., Mac Dorman, K., Ishiguro, H.: Generating natural motion in an android by mapping human motion. In: Proceedings of the 18th IEEE/RSJ International Conference on Intelligent Robots and Systems, pp. 3301–3308, July 2005
7. Patsadu, O., Nukoolkit, C., Watanapa, B.: Human gesture recognition using Kinect camera. In: Proceedings of the 9th International Conference on Computer Science and Software Engineering, pp. 28–32, May 2012
8. Cicirelli, G., Attolico, C., Guaragnella, C., D'Orazio, T.: A kinect-based gesture recognition approach for a natural human robot interface. Int. J. Adv. Robot. Syst. **12**(3), 22 (2015)
9. Williams, K., Breazeal, C.: Reducing driver task load and promoting sociability through an affective intelligent driving agent (AIDA). In: Proceedings of IFIP Conference on Human-Computer Interaction, pp. 619–626, September 2013
10. Zeng, L., Einert, B., Pitkin, A., Weber, G.: HapticRein: design and development of an interactive haptic rein for a guidance robot. In: Proceedings of International Conference on Computers Helping People with Special Needs, pp. 94–101, July 2018
11. Peng, L., Hou, Z., Peng, L., Luo, L., Wang, W.: Robot assisted rehabilitation of the arm after stroke: prototype design and clinical evaluation. Sci. China Inf. Sci. **60**, 073201:1–073201:7 (2017)
12. Miyachi, T., Iga, S., Furuhata, T.: Human robot communication with facilitators for care robot innovation. In: Proceedings of International Conference on Knowledge Based and Intelligent Information and Engineering Systems, pp. 1254–1262, September 2017
13. Maimone, A., Fuchs, H.: Encumbrance-free telepresence system with real-time 3D capture and display using commodity depth cameras. In: Proceedings of the 10th IEEE International Symposium on Mixed and Augmented Reality, pp. 137–146, October 2011

14. Adalgeirsson, S.O., Breazeal, C.: MeBot: a robotic platform for socially embodied telepresence. In: Proceedings of the 5th ACM/IEEE International Conference on Human-Robot Interaction, pp. 15–22, March 2010
15. Wang, J., Tsao, V., Fels, S., Chan, B.: Tippy the telepresence robot. In: Proceedings of International Conference on Entertainment Computing, pp. 358–361, October 2011
16. Tachi, S.: Real-time remote robotics-toward networked telexistence. IEEE Trans. Comput. Graph. Appl. **18**(6), 6–9 (1998)
17. Tachi, S., Minamizawa, K., Furukawa, M., Fernando, C.L.: Telexistence - from 1980 to 2012. In: Proceedings of IEEE/RSJ International Conference on Intelligent Robots and Systems, pp. 5440–5441, October 2012
18. TELEXISTENCE inc. https://tx-inc.com/. Accessed June 2019
19. Leap Motion. https://www.leapmotion.com/. Accessed June 2019
20. Sota. https://sota.vstone.co.jp/home/. Accessed June 2019. (in Japanese)
21. MQTT. http://mqtt.org/. Accessed June 2019
22. Mosquitto. https://mosquitto.org/. Accessed June 2019
23. Eclipse Paho. https://www.eclipse.org/paho/. Accessed June 2019

Network Design Method by Link Protection Considering Probability of Simultaneously Links Failure

Keyaki Uji and Hiroyoshi Miwa[✉]

Graduate School of Science and Technology, Kwansei Gakuin University,
2-1 Gakuen, Sanda-shi, Hyogo, Japan
{keyaki-uji,miwa}@kwansei.ac.jp

Abstract. Information networks are required to be reliable. It is important to design robust networks that are resistant to network failure. For that propose, it is necessary to decrease the failure probability of links by the backup resource and the fast recovery mechanism and so on. However, it costs very much to protect all the links. Therefore, it is necessary to preferentially protect only some highly required links and improve the reliability of the entire network. In this paper, we consider the simultaneously failure of links. A set of links has the probability that all the links in the set simultaneously are broken. When a family of the sets is given, we address the network design maximizing the network reliability (the probability that the entire network is connected) by protecting the limited number of links. We formulate this network design problem and propose a polynomial-time heuristic algorithm. Furthermore, we evaluate the performance by using the topology of some actual information networks, and we show that the proposed algorithm works well.

1 Introduction

It is important to design and operate a robust network that is resistant to network failures in order to realize a highly reliable information network. It is desirable that a communication path exists between nodes even at the event of a failure, but it is unrealistic to completely prevent a link or a node from failing at the event of a disaster, and it is assumed that a failure occurs stochastically.

There is a method of link protection to reduce the link failure probability by the backup resource and the fast recovery mechanism. The link protection improves the network reliability, that is, the probability that the entire network is connected. However, a lot of cost is required, if all links must be protected so that the failure probability is sufficiently small. Therefore, it is a practical solution to preferentially protect only some highly required links so that the reliability of the entire network is large.

In this paper, we consider the simultaneously failure of links. A set of links has the probability that all the links in the set simultaneously are broken. When a family of the sets is given, it is necessary to maximize the network reliability

© Springer Nature Switzerland AG 2020
L. Barolli et al. (Eds.): INCoS 2019, AISC 1035, pp. 113–122, 2020.
https://doi.org/10.1007/978-3-030-29035-1_11

(the probability that the entire network is connected) by protecting the limited number of links. First, we formulate the network design problem and show the NP-hardness of this optimization problem. Next, we design a polynomial-time heuristic algorithm. Moreover, we evaluate the performance by using the topology of some actual communication networks, and we show that the proposed algorithm works well.

2 Related Works

There are some results on link protection [1–4]. The reference [1] deals with the problem of determining a set of protected links so that the diameter of a graph becomes a given value or less. Since the small diameter of a network leads to the small communication delay, an algorithm for this problem gives a network design method. In addition, the references [2,3] deals with the problem of determining a set of protected links that keep the reachability to a specific node, and the reference [3] proposed a polynomial-time approximation algorithm to this problem. The reference [4] deals with the problem of determining protected links so that the reachability is kept within a fixed distance to multiple mirror servers even at the event of a failure. It is known that this problem is NP-hard and that the polynomial-time algorithm for a single-link failure is proposed. The reference [5] deals with the problem of determining protected links so that between the master server and all edge-servers are connected and that the capacity restrictions are satisfied. The reference [6] deals with the problem of determining protected links so that all server nodes are connected regardless of whether they are master or edge-servers and so that the capacity constraint is satisfied. The reference [7] deals with the problem of determining protected links so that the master server and all edge-servers are connected, that the capacity constraint is satisfied, and that the increase ratio of a distance in a network to the distance in a broken network does not exceed a threshold. The reference [8] deals with the problem of determining server placement and link protection simultaneously.

When each edge has the probability that the edge is removed, the network reliability is defined so that the entire network is connected. The network design problems that maximize the reliability under cost constraint or that minimize cost under reliability constraint are extensively studied (ex. [9,10]).

As desctribed above, there are many studies on link protection, link failure probability and connection probability of the entire network, but there are few combinations of them. As a problem similar to this problem, the problem of determining the protection links to maximize the network reliability under the number of protection links is restricted was defined in [11] by the authors. In the reference [11], the NP-hardness is proved, a polynomial-time heuristic algorithm is proposed, and the performance evaluation is conducted for various communication networks. It was assumed that each edge independently and probabilistically fails for the sake of simplicity. However, it is necessary also to consider the simultaneously failure of links. A set of links has the probability that all the links in the set simultaneously are broken. When a family of the sets is given, it is necessary to maximize the network reliability by protecting the limited number of links. In this paper, we deal with such a problem.

3   Link Protection Ploblem Considering Simultaneously Links Failure

Let $G = (V, E)$, where V and E are the vertex set and the edge set of G, respectively, be the graph representing a network structure. When $K = \{F_1, F_2, \ldots, F_k\}$ is the family of edge subsets. Let $H : K \to \mathbb{R}_+$ be a failure function assigning F_i $(i = 1, 2, \ldots, k)$ to the probability that all edges contained in F_i are simultaneously removed. Each set F_i $(i = 1, 2, \ldots, k)$ corresponds to the set of the links that are simultaneously broken at a failure.

For an edge subset E_P, let the set of $\tilde{K} \subseteq K$ that $G_r = (V, E \setminus ((\bigcup_{\forall F \in \tilde{K}} F) \setminus E_P))$ is connected be $R(K, E_P)$. E_P corresponds to the set of the protected links and the probability that the protected link is broken is zero.

The network reliability $Rel(G, E_P)$ is defined as follows.

$$\sum_{\tilde{K} \in R(K, E_P)} \prod_{\forall F (\in \tilde{K})} H(F) \prod_{\forall F' (\in K \setminus \tilde{K})} (1 - H(F'))$$

This means the probability that, even if any simultaneous failures occur, an information network is connected under the condition that the protected links are not broken.

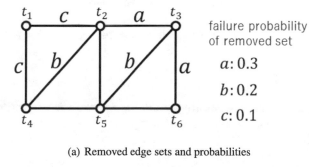

failure probability
of removed set

$a: 0.3$

$b: 0.2$

$c: 0.1$

(a) Removed edge sets and probabilities

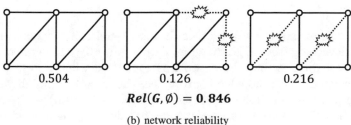

0.504   0.126   0.216

$Rel(G, \emptyset) = 0.846$

(b) network reliability

Fig. 1. Example of network

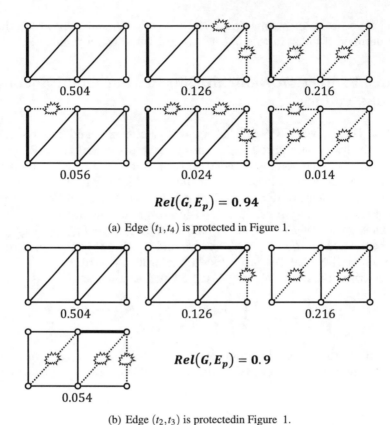

$$Rel(G, E_p) = 0.94$$

(a) Edge (t_1, t_4) is protected in Figure 1.

$$Rel(G, E_p) = 0.9$$

(b) Edge (t_2, t_3) is protectedin Figure 1.

Fig. 2. Example protected edge

Problem PPPSLF

Input: *Network* $G = (V, E)$, *failure probability function* H, *a family of edge sets* $K = \{F_1, F_2, \ldots, F_k\}$, *a positive integer* p.
Constraint: $|E_P| \leq p$
Objective: $Rel(G, E_P)$ *(maximization)*
Output: *Set of protected edges,* $E_P(\subseteq E)$.

We show an example of this problem in Figs. 1 and 2. The failure probability and the removed edge set are given in Fig. 1(a). Figure 1(b) enumerates the events and their probabilities that the network is connected when $p = 0$, that is, no edge is protected. When no edge is protected, the network reliability $Rel(G, \emptyset)$ is the sum of these probabilities, which is 0.846. In Fig. 2(a), we assume that $p = 1$ and that (t_1, t_4) is the protected edge. In this case, the network reliability is 0.94. On the other hand, in Fig. 2(b), we assume that $p = 1$ and that (t_2, t_3) is the protected edge. In this case, the network reliability is 0.9. In both cases, the network reliability is increased by the edge protection, but the network reliability is different according to protected edge. In this network, when the number of the protected edge is restricted to one or less, the protected edge that maximizes the network reliability is shown in Fig. 2(a).

We prove that the problem is NP-hard.

Theorem 1. *PPPSLF is NP-hard.*

Proof. When $K = \{\{e\}|e \in E\}$, that is, each removed edge set consists of one edge, PPPSLF coincides with the problem in [11]. Since the problem in [11] is NP-hard, PPPSLF is also NP-hard.   □.

4   Polynomial-Time Heuristic Algorithm for PPPSLF

Since the problem PPPSLF is generally NP-hard as shown in the previous section, we design a polynomial-time heuristic algorithm.

First, we calculate the probability that each edge is not removed from K and the failure probability function H. We call this probability the edge reliability. Let the edge reliability function be $h : E \to \mathbb{R}_+$. Let N be $N = (G, h)$.

We make the network $N' = (G, -log(1 - h))$ that the cost of each edge e is $-log(h(e))$.

Then, we find the minimum spanning tree $T_{min} = (V, E_{Tmin})$ in N'.

The tree T_{min} consists of edges with small failure probability. Let E_t be the set of the smallest $|E_{Tmin}| - p$ edges from E_{Tmin}. Then, we find the maximum spanning tree $T_{max} = (V, E_{Tmax})$ including E_t in N'.

The tree T_{max} is the spanning tree consisting of E_t and the edges with large failure probability. Let the edge set of $E_{Tmax} \setminus E_t$ be the protected edge set E_P.

The edges in E_P are protected and the edges in E_t have small failure probability; therefore, the probability that T_{max} consisting of E_P and E_t is connected is high. Since T_{max} is a spanning tree, the network reliability of G is high, when E_P is the protected edge set.

The protedted edge set E_P can be determined in the polynomial time, because the above algorithm finds the minimum and the maximum spanning trees. We show the algorithm for PPPSLF in Algorithm 1.

5   Performance Evaluation

In this section, we evaluate the performance of the proposed algorithm in Sect. 4 by using the graph structures of actual ISP backbone networks provided by CAIDA (Center for Applied Internet Data Analysis) [12].

We define a family of removal sets, $K = \{F_1, F_2, \ldots, F_t\}$ as follows. For vertex $v_i(\in V)$, let F_i be the union of the set of the edges incident to v_i and the set of the edges between vertices adjacent to v_i. We show the network with 14 vertices and 26 edges in Fig. 3(a) and an example of a removal set in Fig. 3(b). Let the failure probability of each removal set be 0.2.

Algorithm 1. Algorithm PPPSLF

Input: Network $G = (V, E)$, failure probability function H, a family of edge sets
$\quad\quad K = \{F_1, F_2, \ldots, F_k\}$, a positive integer p.
Output: E_P

1 $E_P \leftarrow \emptyset$
2 **while** *there is an unscanned edge $e \in E$* **do**
3 $\quad$ Let $\{F_1^e, F_2^e, \ldots, F_{b(e)}^e\}(\subseteq K)$ be the set including e
4 $\quad h(e) \leftarrow \prod_{i=1}^{b(e)}(1 - H(F_i^e))$
5 Make Network $N = (G, h)$
6 Find the minimum spanning tree $T_{min} = (V, E_{Tmin})$ of network $N' = (G, -log(h))$
7 Sort $h(e)$ $(e \in E_{Tmin})$ in ascending order.
8 Let E_t be the set of the smallest $|E_{Tmin}| - p$ edges in E_{Tmin}.
9 Find $T_{max} = (V, E_{Tmax})$ which is the maximum spanning tree including E_t of network
$\quad N' = (G, -log(h))$
10 $E_P \leftarrow E_{Tmax} \setminus E_t$
11 **return** E_P

$$Rel(G, \emptyset) = 0.0440$$

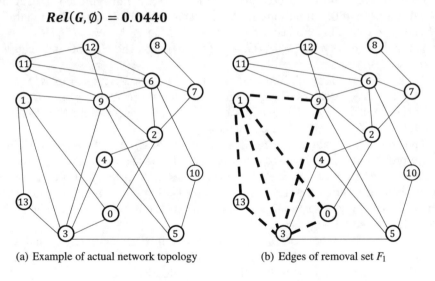

(a) Example of actual network topology   (b) Edges of removal set F_1

Fig. 3. Actual network topology and removal set.

For comparison, we use the algorithm PPPSLF-beta1 that omits the calculation of the highly reliable spanning tree in the algorithm PPPSLF. We show the algorithm PPPSLF-beta1 in Algorithm 2.

When the number of protected edges is 3, 6, and 8 in the network in Fig. 3(a), we show the set of the protected edges by Algorithms 1 and 2 in Figs. 4, 5 and 6, respectively.

The network in Fig. 3(a) has 10 edges included in four or more removal sets. However, the edge $(3, 13)$ is protected in the Fig. 5(a), which is included in three

---

Algorithm 2. Algorithm PPPSLF-beta1

Input: Network $G = (V, E)$, failure probability function H, a family of edge sets
$K = \{F_1, F_2, \ldots, F_k\}$, a positive integer p.

Output: E_P

1 $E_P \leftarrow \emptyset$

2 **while** *there is an unscanned edge* $e \in E$ **do**

3 $\quad$ Let $\{F_1^e, F_2^e, \ldots, F_{b(e)}^e\}(\subseteq K)$ be the set including e

4 $\quad$ $h(e) \leftarrow \prod_{i=1}^{b(e)}(1 - H(F_i^e))$

5 Make Network $N = (G, h)$

6 Find the minimum spanning tree $T_{min} = (V, E_{Tmin})$ of network $N' = (G, -log(h))$

7 Sort $h(e)$ $(e \in E_{Tmin})$ in ascending order.

8 Let E_P be the set of the smallest p edges in E_{Tmin}.

9 **return** E_P

$$P = 3, Rel(G, E_P) = 0.0687 \qquad\qquad P = 3, Rel(G, E_P) = 0.0550$$

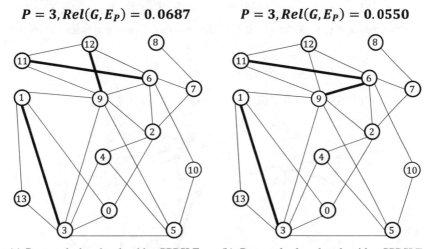

(a) Protected edges by algorithm PPPSLF.   (b) Protected edges by algorithm PPPSLF-beta1.

Fig. 4. Protected edges when $p = 3$

removal sets. This means that the edges with high failure probability are not always protected, but the protected edges depend on the graph structure. Algorithm 2 make a cycle as shown in Fig. 6(b). As a result, although the number of the protected edges are more than the number of the edges protected by Algorithm 1, but the network reliability is smaller.

Next, we show the relationship between the network reliability and the number of protected edges in Table 1. In this numerical experiments, for all removal sets, the failure probability of 0.1 or less is randomly determined for a removal set.

$P = 6, Rel(G, E_P) = 0.1946$

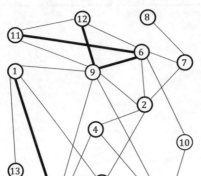

$P = 6, Rel(G, E_P) = 0.1074$

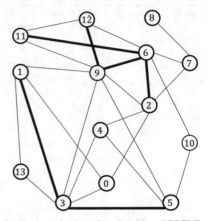

(a) Protected edges by algorithm PPPSLF.

(b) Protected edges by algorithm PPPSLF-beta1.

Fig. 5. Protected edges when $p = 6$

$P = 8, Rel(G, E_P) = 0.2604$

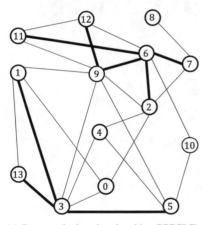

$P = 8, Rel(G, E_P) = 0.1074$

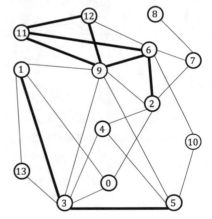

(a) Protected edges by algorithm PPPSLF.

(b) Protected edges by algorithm PPPSLF-beta1.

Fig. 6. Protected edges when $p = 8$.

Algorithm PPPSLF-beta2 determines the protected edges that the network reliability is the maximum among 100 times randomly chosen protected edges.

The results indicate that the network reliability is almost always high by Algorithm PPPSLF regardless of the number of protected edges in any network.

Table 1. Relationship between the network reliability and the number of protected edges

n	m	$Rel(G, \emptyset)$	P	$Rel(G, E_P)$		
				PPPSLF	beta1	beta2
14	26	0.5366	4	0.7521	0.5883	0.5643
			7	0.7917	0.6394	0.5883
			10	0.9395	0.6394	0.6012
16	23	0.4294	4	0.6353	0.5834	0.4689
			7	0.7626	0.6410	0.5038
			10	0.8878	0.7200	0.6011
19	33	0.3915	4	0.5426	0.4282	0.3915
			7	0.6669	0.4282	0.4377
			10	0.7901	0.5389	0.4638
27	35	0.2341	4	0.2729	0.2531	0.2341
			7	0.3658	0.2729	0.2531
			10	0.4676	0.2998	0.2531

6   Conclusions

In this paper, we proposed the network design method based on link protection under the condition that multiple links simultaneously fail at a event of a failure. The link protection, to reduce the link failure probability by the backup resource and the fast recovery mechanism, improves the network reliability, that is, the probability that the entire network is connected. However, a lot of cost is required, if all links must be protected so that the failure probability is sufficiently small. Therefore, it is a practical solution to preferentially protect only some highly required links so that the reliability of the entire network is large.

First, we formulated the network design problem and showed the NP-hardness of this problem. Next, we proposed a polynomial-time heuristic algorithm. Moreover, we evaluated the performance by using the topology of some actual communication networks. The results show that the proposed algorithm works well.

Acknowledgements. This work was partially supported by the Japan Society for the Promotion of Science through Grants-in-Aid for Scientific Research (B) (17H01742) and JST CREST JPMJCR1402.

References

1. Imagawa, K., Fujimura, T., Miwa, H.: Approximation algorithms for finding protected links to keep small diameter against link failures. In: Proceedings of the INCoS2011, Fukuoka, Japan, 30 November–1–2 December 2011 (2011)
2. Imagawa, K., Miwa, H.: Detecting protected links to keep reachability to server against failures. In: Proceedings of the ICOIN 2013, Bangkok, 28–30 January 2013 (2013)
3. Imagawa, K., Miwa, H.: Approximation algorithm for finding protected links to keep small diameter against link failures. In: Proceedings of the INCoS 2011, Fukuoka, Japan, 30 November 2011–2 December 2011, pp. 575–580 (2011)
4. Maeda, N., Miwa, H.: Detecting critical links for keeping shortest distance from clients to servers during failures. In: Proceedings of the HEUNET 2012/SAINT 2012, Turkey, 16–20 July 2012 (2012)
5. Irie, D., Kurimoto, S., Miwa, H.: Detecting critical protected links to keep connectivity to servers against link failures. In: Proceedings of the NTMS 2015, Paris, pp. 1–5, 27–29 July 2015 (2015)
6. Fujimura, T., Miwa, H.: Critical links detection to maintain small diameter against link failures. In: Proceedings of the INCoS 2010, Thessaloniki, pp. 339–343, 24–16 November 2011 (2011)
7. Yamasaki, T., Anan, M., Miwa, H.: Network design method based on link protection taking account of the connectivity and distance between sites. In: Proceedings of the INCoS 2016, Ostrava, pp. 339–343, 7–9 September 2016
8. Irie, D., Anan, M., Miwa, H.: Network design method by finding server placement and protected links to keep connectivity to servers against link failures. In: Proceedings of the INCoS 2016, Ostrava, pp. 439–344, 7–9 September 2016 (2016)
9. Aggarwal, K.K., Chopra, Y.C., Bajwa, J.S.: Topological layout of links for optimising the overall reliability in a computer communication system. Microelectron. Reliab. **22**, 347–351 (1982)
10. Jan, R.-H., Hwang, F.-J., Chen, S.-T.: Topological optimization of a communication network subject to a reliability constraint. IEEE Trans. Reliab. **42**, 63–70 (1993)
11. Uji, K., Miwa, H.: Method for finding protected links to keep robust network against link failure considering failure probability. In: Proceedings of the International Conference on Intelligent Networking and Collaborative System (INCoS 2017), Toronto, Canada, 24–26 August 2017 (2017)
12. CAIDA. http://www.caida.org/

A Novel Bounded-Error Piecewise Linear Approximation Algorithm for Streaming Sensor Data in Edge Computing

Jeng-Wei Lin[1], Shih-wei Liao[2], and Fang-Yie Leu[3(✉)]

[1] Department of Information Management, Tunghai University,
Taichung City, Taiwan
jwlin@thu.edu.tw
[2] Department of Computer Science and Information Engineering,
National Taiwan University, Taipei City, Taiwan
liao@csie.ntu.edu.tw
[3] Department of Computer Science, Tunghai University, Taichung City, Taiwan
leufy@thu.edu.tw

Abstract. Many studies show that many Data compression schemes, like Bounded-Error Piecewise Linear Approximation (BEPLA) methods, have been proposed to lower the length sensor data, aiming to mitigating data transmission energies. When an error bound is given, these data compression schemes consider how to represent original sensor data by using fewer line segments. In this paper, besides BEPLA, we further deal with resolution reduction, which called Swing-RR (Resolution Reduction) sets a new restriction on the position of line segment endpoints. Our simulating results on existing datasets indicate that the length of compressed data is actually lowered.

1 Introduction

Data compression is an effective method to lower data size, particularly for delivering IoT data. In general, compression ratios of lossless compression methods are often not high. Most of the methods follow Huffman coding [1–3]. Also, lossy compression methods do not guarantee the minimum difference between the original data and compressed data. Literature [4–6] indicates that Bounded-error approximation to the original sensor data can retain a certain level of quality of the compressed data. Currently, many Bounded-Error Piecewise Linear Approximation (BEPLA) methods have been released. Swing filter [5], Slide filter [5] and Cont-PLA [4] are typical examples. Most of them focused on how to approximate to the original data with fewer line segments.

In this paper, we further deal with resolution reduction when BEPLA is created. When selecting endpoint for BEPLA, Resolution reduction generates an extra restriction with which fewer bits are required to encode the endpoints of line segments. A variant method of BEPLA with Swing filter [5], named Swing-RR, is proposed. The time and space complexities of Swing-RR are optimal O(1) in processing data. Using existing datasets, our simulation results indicate that even Swing-RR requires longer

© Springer Nature Switzerland AG 2020
L. Barolli et al. (Eds.): INCoS 2019, AISC 1035, pp. 123–132, 2020.
https://doi.org/10.1007/978-3-030-29035-1_12

line segments than some state-of-the-art approaches, the length of sensor data are obviously reduced by resolution reduction, indicating that the mean square errors between the original data and compressed data are also smaller as expected, thus benefiting later analyses of the compressed data.

Definition 1: BEPLA. Given time series $y(t) = (y_1, y_2, \ldots, y_n)$ and preassigned error bound ε, BEPLA $y'(t) = (s1, s2, \ldots, sk)$, $k \leq n$, is a PLA to $y(t)$ and the $l\infty$-error between $y(t)$ and $y'(t)$ is bounded by ε.

In recent years, BEPLA for sensor data have attracted researchers' eyes once again [5–8]. Given time series y(t) and preassigned maximal error ε, we can set upper bound H(t) and lower bound G(t) for $y'(t)$, as listed in Eqs. 1 and 2, respectively.

$$H(t) = y(t) + \varepsilon \tag{1}$$

$$G(t) = y(t) - \varepsilon \tag{2}$$

Although the two εs in both Eqs. 1 and 2 are the same, they may be different in different applications.

In fact, BEPLA $y'(t) = (s_1, s_2, \ldots, s_k)$ is a PLA to $y(t)$ and the l_∞-error between y (t) and $y'(t)$ is bounded by ε if and only if all points in all line segments are between G (t) and $H(t)$. For any line segment s_i, the length of which is m, it is easy to prove that l_1-error and l_2-error between s_i and $y(t)$ are bounded by $m\varepsilon$ and $\sqrt{m}\varepsilon$, respectively. There are two types of BEPLA, joint and disjoint. In joint BEPLA, consecutive line segments share an endpoint, i.e., s_{i+1}.start is actually s_i.stop ($1 \leq i < k$). In addition to the start point of the first line segment (i.e., s_1.start), for each line segment, the stop point is recorded. Specifically, $y'(t)$ is recorded as $(s_1$.start.y, $(s_1$.length, s_1.stop.y), $(s_2$.length, s_2. stop.y), $\ldots$, $(s_k$.length, s_k.stop.y)). Note that s_1.start.x = 1, and s_i.stop.x = s_{i+1}.start. x = s_i.start.x + s_i.length.x − 1 for all is, $1 \leq i < k$. When the value of y is stored in p bits and the length of a line segment is stored in q bits, $y'(t)$ is recorded in $p + (p + q) * k$ bits. Thus, compression ratio for joint BEPLA can be calculated by using Eq. 3.

$$Compression\ Ratio_{joint} = \frac{p(k+1) + qk}{32n} = \frac{p + (p+q)k}{32n} \tag{3}$$

where n is the number of elements in the original time series $y(t)$, and 32 is the number of bits used to represent y_i in y(t), for all is. In disjoint BEPLA, a line segment s_{i+1} does not have to start from the stop point of s_i ($1 \leq i < k$). In practice, s_{i+1} starts from the next time tick after the stop point of s_i, i.e., s_{i+1}.start.x = s_i.stop.x + 1 for all is, $1 \leq i < k$. Since there are no shared endpoints between consecutive line segments, both start and stop points of a line segment need to be recorded. Specifically, $y'(t)$ is recorded as $((s_1$.start.y, s_1.length, s_1.stop.y), $(s_2$.start.y, s_2.length, s_2.stop.y), $\ldots$, $(s_k$.start. y, s_k.length, s_k.stop.y)). Note that s_1.start.x = 1, and s_i.stop.x = s_i.start.x + s_i.length.x − 1 for all is, $1 \leq i \leq k$. Thus, $y'(t)$ is recorded in $(2 * p + q) * k$ bits. Compression ratio for disjoint BEPLA is calculated by using Eq. 4.

$$Compression\,Ratio_{disjoint} = \frac{(2p+q)k}{32n} \tag{4}$$

In fact, there are $k - 1$ hidden line segments whose length is 1 time tick. They are $(s_i.\text{stop}.y, 1, s_{i+1}.\text{start}.y)$ for all is, $1 \leq i < k$, and embedded in the representation of y' (t). In other words, there are actually $2 * k - 1$ line segments. These hidden line segments are not explicitly recorded since they can be derived from the representation of disjoint BEPLA.

Many BEPLA algorithms have been so far proposed [4–10], trying to extend the line segments as long as possible and thus minimizing the number of line segments. In 2009, Elmeleegy et al. [5] reinvented Swing filter and Slide filter for joint and disjoint BEPLA, respectively. Swing filter is the simplest, but not optimal in terms of number of line segments. Its complexities on a data point in both time and space are O (1). In fact, Swing filter was presented by Gritzali and Papakonstantinou in 1983 [9]. An optimal joint BEPLA algorithm, referred to Cont-PLA in this paper, was introduced by Hakimi and Schmeichel in 1991 [4]. Slide filter is optimal in minimizing number of line segments. Actually, an optimal disjoint BEPLA algorithm was proposed by O'Rourke in 1981 [10]. Xie et al. [7] in 2014 improved the running time efficiency of Slide filter. Zhao et al. [8] in 2016 improved the efficiency of Swing filter based on [7]. Using convex hull and similar techniques, the amortized time complexity of both Cont-PLA and Slide filter on a data point are also O (1).

For data compression consideration, each line segment consumes $p + q$ bits in joint BEPLA, while $2 * p + q$ bits in disjoint BEPLA. Since disjoint BEPLA algorithms have higher freedom in selecting the start points of line segments, they usually use fewer line segments. As shown in [7], in terms of bits representing the resultant BEPLA, Cont-PLA and Slide filter mutually outperformed each other on different datasets, and both achieved 15–25% superiority over Swing filer in all datasets. In 2015, Luo et al. [6] introduced Mixed-PLA that uses both joint and disjoint line segments. The authors employed dynamic programming technique and showed that Mixed-PLA were roughly 15% better than Cont-PLA and Slide filter in terms of bits representing the resultant BEPLA.

2 Approaches

Algorithm 1 [13] shows the pseudocode of Swing-RR (). Given error bound ε, maximal delay, and resolution r bits in length, whenever a new data point d is sensed, d is processed by Swing-RR(d) and line segments are generated on the fly.

Similar to Swing filter, Swing-RR maintains a data structure for holding possible line segments. As shown in Fig. 1, two auxiliary lines $s \rightarrow u_s$ and $s \rightarrow l_s$ are maintained. The possible line segments for the current processing window must lie within $s \rightarrow u_s$ and $s \rightarrow l_s$. Both auxiliary lines start from the start point s of the current processing window.

Algorithm 1: Swing-RR(d), given error bound ε, maximal delay *delay*, and resolution r bits in length

---

(1) segment.length++   // segment.length is initialized to −1 for a new window
(2) if (segment.length == 0) then   // the first data point in this window
(3)   segment.start s = picking up a coded data point between $d+\varepsilon$ and $d-\varepsilon$;
(4) else if (segment.length == 1) then   // the second data point in this window
(5) $u_s = d+\varepsilon$; $l_s = d-\varepsilon$;
(6)   check_range();
(7) else if (($d+\varepsilon$ is below $s{\rightarrow}l_s$) or ($d-\varepsilon$ is above $s{\rightarrow}u_s$)
(8)   {close this window; generate a line segment; initialize a new window;}
(9) else
(10)   if ($d+\varepsilon$ is below $s{\rightarrow}u_s$) $u_s = d+\varepsilon$;   // $s{\rightarrow}u_s$ swing down
(11)   if ($d-\varepsilon$ is above $s{\rightarrow}l_s$) $l_s = d-\varepsilon$;   // $s{\rightarrow}l_s$ swing up
(12)   check_range();
(13) end if
(14)
(15) Function check_range()
(16) u = the point at $d.t$ extended from $s{\rightarrow}u_s$;
(17) l = the point at $d.t$ extended from $s{\rightarrow}l_s$;
(18) u^- = the largest coded data point smaller than or equal to u;
(19) l^+ = the smallest coded data point larger than or equal to l;
(20)   if ($l^+ \leq u^-$) then   // there exists at least one coded data point
(21)   segment.stop = ($l^+ + u^-$) / 2;
(22)   if (segment.length $\geq$ *delay*)// Swing-RR is forced to output a line segment
(23)   {close this window; generate a line segment; initialize a new window;}
(24)   end if
(25)   else   // there exist no coded data points
(26)   {close this window; generate a line segment; initialize a new window;}
(27)   end if

When a window is initialized, a coded data point in the bounded range of the first sensed data point is chosen, probably randomly (please refer to lines 2–3 of Algorithm 1 and Fig. 1a). In this experiment, a coded data point nearest to the original data is chosen.

When the second data point d in this window is processed, two support points, $u_s = d + \varepsilon$ and $l_s = d - \varepsilon$, are initialized accordingly (please refer to lines 4–6 of Algorithm 1 and Fig. 1b). The upper support point u_s bounds the maximal slope of possible line segments, i.e., $s \rightarrow u_s$. The lower support point ls bounds the minimal slope of possible line segments, i.e., $s \rightarrow l_s$.

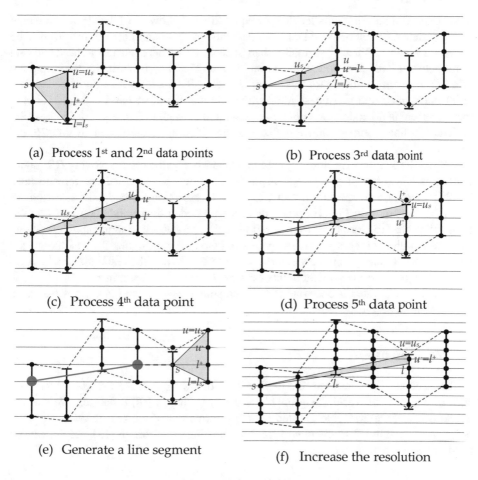

(a) Process 1ˢᵗ and 2ⁿᵈ data points

(b) Process 3ʳᵈ data point

(c) Process 4ᵗʰ data point

(d) Process 5ᵗʰ data point

(e) Generate a line segment

(f) Increase the resolution

Fig. 1. Swing-RR [13].

Similar to Swing filter, whenever a new data point d is sensed, the two support points, u_s and l_s, are maintained according to the positions of $d + \varepsilon$ and $d - \varepsilon$. $s \to u_s$ may swing down, and $s \to l_s$ may swing up (please refer to lines 7–11).

If $d + \varepsilon$ is below $s \to l_s$ and thus $s \to u_s$ will swing down too much, a line segment is then generated. Also, if $d - \varepsilon$ is above $s \to u_s$ and thus $s \to l_s$ will swing up too much, a line segment is generated (please refer to lines 7–8). Otherwise, u_s and l_s are maintained to update the range of possible line segments (please refer to lines 10–11). If $d - \varepsilon$ is above $s \to l_s$, $s \to l_s$ swings up by updating $l_s = d - \varepsilon$, as shown in Figs. 1a and b. If $d + \varepsilon$ is below $s \to u_s$, $s \to u_s$ swings down by updating $u_s = d + \varepsilon$, as shown in Figs. 1c and d.

When resolution reduction is adopted, please see function check_range(), Swing-RR() further checks to see whether there are coded data points between l and u, which are extended from $s \to l_s$ and $s \to u_s$, respectively. Specifically, Swing-RR() calculates l^+ and u^- where l^+ is the smallest coded data point larger than or equal to l, and u^- is the largest coded data point smaller than or equal to u, as shown in Figs. 1b –d. We note that s, l^+, and u^- must be coded data points, while u_s, l_s, u, and l are not restricted to be coded data points. When l^+ is smaller than or equal to u-, there must be at least one data point between l and u. A coded data point between l^+ and u^- is chosen as the stop point of a line segment candidate for this window. Swing-RR() adopts the middle coded data point between l^+, and u^- (please refer to line 21). On the other hand, when l^+ is larger than u^-, there is no coded data point between l and u, as shown in Fig. 1d. Swing-RR() generates a line segment, and initializes a new window (please refer to line 26 and Fig. 1e).

When the length of a line segment candidate is equal to *delay*, Swing-RR() generates the line segment candidate and initializes a new window (please refer to lines 22–24). Adoption of resolution reduction puts a restriction on endpoint selection for BEPLA. As shown in Fig. 1d, when there are no coded data points between l and u, Swing-RR() has to close the current window and generates a line segment, while Swing filter can further process new data points. Obviously, the higher the resolution, the more the coded data points between $d - \varepsilon$ and $d + \varepsilon$. When the resolution r increases, there might be more coded data points between l and u. As shown in Fig. 1f, when r is increased by one, there is only one coded data point between l and u. Swing-RR() does not have to close the window.

3   Experiments

In this section, we investigate the performance of Swing-RR. An archive which consists of several real world datasets [11] is used. As that in [6], 8 datasets are chosen: Cricket_X and wafer. Table 1 listed related information of the two datasets. All data points are stored in IEEE 754 single precision floating point format [12], i.e., 32 bits are used to store a data point.

Table 1. Dataset description[a]

Dataset	Length	Minimal	Maximal
Cricket_X	117,000	−4.766200	11.494000
wafer	152,000	−3.054000	11.787000

[a]The precision of data points is six digits after the decimal point.

Note that x-coordinates of line segment endpoints in BEPLA produced by Swing filter and Swing-RR are aligned with time ticks. As a result, the lengths of line segments are all the same in the following experiment.

3.1   Experimental Setup

In addition to typical error bounds ranging from 0.5% to 5% of the whole range of possible data points, in this experiment, we also examine scenarios of a small error bound, ranging between 0.1% and 0.4%. Note that in this case, a higher resolution is needed to ensure the existence of some coded data points, the values of which are between the upper and lower bounds. As well, when the error bound is small, the space for BEPLA follows. Consequently, the expected lengths of line segments of BEPLA are also short, thus further shortening the maximal delay so that fewer bits are required to record the lengths of these line segments. Table 2 shows the resolution and maximal delays (in time ticks) employed in this experiment.

Table 2. Scenario examined in the experiment.

Error bound (ε)	Resolution	Maximal delay
0.5%	7, 8, 9, 10	63, 127, 255, 511
1%	6, 7, 8, 9	63, 127, 255, 511
2%	5, 6, 7, 8	63, 127, 255, 511
3%	5, 6, 7, 8	63, 127, 255, 511
4%	4, 5, 6, 7	63, 127, 255, 511
5%	4, 5, 6, 7	63, 127, 255, 511

Maximal delays are set usually based on the applications of sensor networks. The morc in real time requirements, the smaller the maximal delays. However, the lengths of line segments are more restricted by the given error bound. When the maximal delays are too short, long line segments, if exist, are forced to be cut. When the maximal delays are longer than the lengths of most line segments, bits usage in recording the lengths of line segments is inefficient.

Given an error bound ε, for all datasets, Swing-RR utilizes different resolutions, particularly from the minimal resolution to higher. When minimal resolution is employed, there is at least one coded data point, the value of which is between $y - \varepsilon$ to $y + \varepsilon$ for a data point y. When one more bit is used for the resolution, the number of coded data points will be doubled.

We compare the performance of Swing-RR and Swing filter [5]. Three criteria are investigated, including compression ratio, lengths of line segments and their distribution, and mean square error (MSE). MSE is calculated by Eq. 5.

$$MSE = \sqrt{\frac{\sum_{i=1}^{n} |y_i - y_i'|^2}{n}} \tag{5}$$

Previous methods focused on how to approximate to the original data by using fewer line segments. With resolution reduction, we further examine the compression ratios for different resolutions and maximal delays. Investigation on the lengths of line segments and their distribution helps us understand the tradeoff regarding the selection of maximal delays. BEPLA generated by Swing-RR and Swing filter are all bounded by ε. MSE related with l2-error provides additional information about these BEPLA. In general, a BEPLA with a smaller MSE fits better to the original data than those with larger MSEs do.

3.2   Compression Ratio

Figure 2 shows the sizes of the BEPLA generated by Swing filter and Swing-RR, given different resolutions to the Cricket_X dataset. Scenarios on maximal delays of 63 and 127 are shown. As described above, Swing-RR outperforms Swing filter significantly. Swing-RR with the minimal resolution achieves a much better compression than Swing filter does.

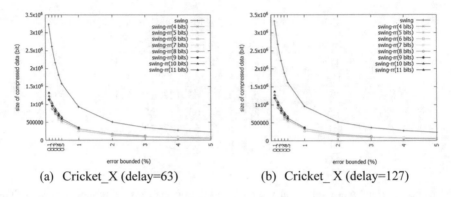

(a) Cricket_X (delay=63)   (b) Cricket_ X (delay=127)

Fig. 2. Performance of Swing-RR on different resolutions.

Figure 3 shows the size of the BEPLA generated by Swing filter and Swing-RR, given different maximal delays to the wafer dataset. To more clearly show the differences, the value of Swing filter is not depicted. For a typical error bound, i.e., from 0.5% to 5%, and resolutions, i.e., 7 and 8 bits, Swing-RR with a maximal delay of 127 or 255 compresses the wafer dataset better than Swing-RR with a maximal delay of 63 or 511 does. When the maximal delay is too short, as mentioned before, long line segments will be cut. When the maximal delay is longer than the lengths of most line segments, some bits allocated to record the length will not be used, thus bit efficiency is reduced.

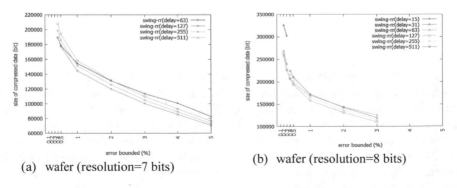

(a) wafer (resolution=7 bits)   (b) wafer (resolution=8 bits)

Fig. 3. Performance of Swing-RR on different maximal delays.

As described above, Slide filter and Cont-PLA mutually outperform each other given different datasets, while Mixed-PLA outperforms Swing-filter, Slide filter, and Cont-PLA on the 8 real world datasets [6, 7]. The size of BEPLA generated by Mixed-PLA is about 50%–60% of that produced by Swing filter. These methods focus on how to minimize the number of line segments. Furthermore, when using minimal resolution, Swing-RR requires only 20%–25% of bits.

4 Conclusions

In this paper, Swing-RR is proposed to generate disjoint BEPLA with Resolution Reduction for sensor data compression. Two real world datasets [12] are utilized to evaluate performance of Swing-RR. Simulation results illustrate that Swing-RR is better than Swing filter. For error bounds between 0.5% and 5%, Swing-RR with minimal resolution has better compression ratios. Swing-RR uses 7 (for 0.5%) to 4 (for 5%) bits, rather than 32 bits, to store the approximated data point. Compared to the number of bits needed by Swing filter to represent its BEPLA, Swing-RR uses only 20%–25% of bits, while state-of-the-art approaches employ 50%–60% of bits. Consequently, fewer bits are delivered via network and smaller disk space are consumed to keep the sensor data in its data center. Generally, the power consumption is hugely lowered. Further, the MSE of BEPLA produced by Swing-RR is smaller than the MSE generated by Swing filter.

In this study, Swing-RR utilizing the minimal resolution produces higher compression ratios than employing a higher resolution does. Most line segments are short. Only a small portion is long. Thus, it is worth to utilize fewer bits to encode the lengths of segments.

References

1. Kawahara, M., Chiu, Y.J., Berger, T.: High-speed software implementation of Huffman coding. In: Proceedings of 1998 Data Compression Conference, p. 553 (1998)
2. Marcelloni, F., Vecchio, M.: A simple algorithm for data compression in wireless sensor networks. IEEE Commun. Lett. 12(6), 411–413 (2008). https://doi.org/10.1109/LCOMM. 2008.080300
3. Javed, M.Y., Nadeem, A.: Data compression through adaptive Huffman coding schemes. In: Proceedings of IEEE TENCON 2000, pp. 187–190 (2000)
4. Hakimi, S.L., Schmeichel, E.F.: Fitting polygonal functions to a set of points in the plane. CVGIP: Graph. Models Image Process. 53(2), 132–136 (1991)
5. Elmeleegy, H., Elmagarmid, A.K., Cecchet, E., Aref, W.G., Zwaenepoel, W.: Online piecewise linear approximation of numerical streams with precision guarantees. Proc. VLDB Endow. 2(1), 145–156 (2009)
6. Luo, G., Yi, K., Cheng, S.W., Li, Z., Fan, W., He, C., Mu, Y.: Piecewise linear approximation of streaming time series data with max-error guarantees. In: 2015 IEEE 31st International Conference on Data Engineering, pp. 173–184 (2015)

7. Xie, Q., Pang, C., Zhou, X., Zhang, X., Deng, K.: Maximum error-bounded piecewise linear representation for online stream approximation. VLDB J. **23**(6), 915–937 (2014)
8. Zhao, H., Dong, Z., Li, T., Wang, X., Pang, C.: Segmenting time series with connected lines under maximum error bound. Inf. Sci. **345**, 1–8 (2016)
9. Gritzali, F., Papakonstantinou, G.: A fast piecewise linear approximation algorithm. Sig. Process. **5**(3), 221–227 (1983)
10. O'Rourke, J.: An on-line algorithm for fitting straight lines between data ranges. Commun. ACM **24**(9), 574–578 (1981)
11. Chen, Y., Keogh, E., Hu, B., Begum, N., Bagnall, A., Mueen, A., Batista, G.: The UCR time series classification archive 2015. http://www.cs.ucr.edu/~eamonn/time_series_data/. Accessed 7 July 2018
12. IEEE Computer Society. IEEE Standard for Floating-Point Arithmetic. IEEE Standard 754-2008
13. Lin, J.W., Liao, S.W., Leu, F.Y.: Sensor data compression using bounded error piecewise linear approximation with resolution reduction. Energies **12**, 1–20 (2019)

A Message Relaying Method with a Dynamic Timer Considering Non-signal Duration from Neighboring Nodes for Vehicular DTN

Shogo Nakasaki[1], Makoto Ikeda[2(✉)], and Leonard Barolli[2]

[1] Graduate School of Engineering, Fukuoka Institute of Technology,
3-30-1 Wajiro-higashi, Higashi-ku, Fukuoka 811-0295, Japan
tshogonakasakit@gmail.com
[2] Department of Information and Communication Engineering,
Fukuoka Institute of Technology,
3-30-1 Wajiro-higashi, Higashi-ku, Fukuoka 811-0295, Japan
makoto.ikd@acm.org, barolli@fit.ac.jp

Abstract. For End-to-End (E2E) communication in a sparse vehicular network is needed a flexible message delivery method that can be applied for large transmission delay and link disconnection. In this paper, we propose a message relaying method with dynamic timer considering non-signal duration from neighboring vehicles for Vehicular Delay/Disruption/Disconnection Tolerant Networking (DTN). From the simulation results, we found that our proposed method can provide a high delivery rate by using the dynamic timer in Vehicular DTN.

Keywords: Message relaying method · Vehicular DTN · Dynamic timer

1 Introduction

In recent years, many companies have been actively developing a 3D laser scanner and connected cars to provide safe autonomous driving in urban areas. In general, vehicles have been used for shopping, travel, and logistics, but now vehicles are going to be connected to the network and becoming mobile nodes [4,7,11,12, 15,16,31]. Vehicles will connect to the next generation of the wireless network that will become not only intermediate nodes but also source and destination and will be used for various applications [9,10,17,24,26,30]. The vehicles will be equipped with On-Board-Unit (OBU) by a plug-in and utilize actuator and sensor information via Controller Area Network (CAN).

Delay/Disruption/Disconnection Tolerant Networking (DTN) is a message relaying method in Vehicular to Vehicular (V2V) communication [13,22,23,25]. In Vehicular DTN, consumption of network resources and storage usage of each

© Springer Nature Switzerland AG 2020
L. Barolli et al. (Eds.): INCoS 2019, AISC 1035, pp. 133–142, 2020.
https://doi.org/10.1007/978-3-030-29035-1_13

node becomes a critical problem due to the DTN nodes duplicate messages to others.

In our previous work [14], we have proposed a recovery method considering the different thresholds for reducing delay and increasing the delivery ratio in Vehicular DTN. We have shown that the proposed recovery method with thresholds can decrease storage usage, but the approach decreased the delay time and needs adaptive threshold to consider different network conditions. Also, we proposed a threshold-based adaptive method [29] for message suppression in Vehicular DTNs. The proposed adaptive method decreased the replicated bundle messages compared with the conventional method, but the non-signal duration was not considered.

In this paper, we propose a message relaying method with a dynamic timer considering non-signal duration from neighboring vehicles for Vehicular DTN. We evaluate the proposed method by simulations and consider delivery ratio, delay, overhead, and storage usage as evaluation metrics.

The structure of the paper is as follows. In Sect. 2, we give the related work. In Sect. 3 is described the message relaying method considering the dynamic timer. In Sect. 4, we provide the description of the simulation system and the evaluation results. Finally, conclusions and future work are given in Sect. 5.

2   Related Work

DTN can provide a reliable internet-working for space tasks [1,3,5,8,19,28]. The space networks have possibly long delay, frequent link disconnection and frequent disruption. In DTN, the messages are stored and forwarded by nodes. When the nodes receive messages, they store the messages in their storage. After that, the nodes duplicate the messages to other nodes when it is transmitted. This technique is called message switching. The architecture is specified in RFC 4838 [6].

Epidemic routing is well-known routing protocol for DTN [18,20,27]. Epidemic routing uses two control messages to duplicate messages. Nodes periodically broadcast the Summary Vector (SV) message in the network. The SV contains a list of stored messages of each node. When the nodes receive the SV, they compare received SV to their SV. The nodes send the REQUEST message if received SV contains unknown messages.

In Epidemic routing, consumption of network resources and storage usage becomes a critical problem. Because the nodes duplicate messages to neighbors in their communication range. Moreover, received messages remain in the storage, because the messages are continuously duplicated even if the destination node receives the messages. Therefore, recovery schemes such as anti-packet or timer are needed to limit the duplicate messages. In the case of anti-packet, the destination node broadcasts the anti-packet, which contains the list of messages that are received by the destination node. Nodes delete the messages according to the anti-packet. Then, the nodes duplicate the anti-packet to other nodes. However, network resources are consumed by anti-packet. In the case of the timer, messages have a lifetime. The messages are punctually deleted when the lifetime of the messages is expired. However, the setting of a suitable lifetime is difficult. In this paper, we propose a message relaying method considering a dynamic timer to limit the duplicate messages in Vehicular DTN.

3 Message Relaying Method Considering Dynamic Timer

In this section, we explain in detail of proposed message relaying method considering dynamic timer.

3.1 Overview of Dynamic Timer

In Vehicular DTN, we propose a method using a dynamic timer to control the bundle drop deadline. In the conventional method, a fixed value of lifetime will be settled at the time of message generation. If the lifetime is expired, the vehicle deletes the bundle message in their storage. However, the conventional method does not consider network conditions around each vehicle such as vehicle density, distance to end-point, and so on. Therefore, we propose a dynamic timer method considering non-signal duration from neighboring vehicles. The non-signal duration means that there is no vehicle around. In this method, each vehicle sets a dynamic timer for messages, and the management of data is independent. Thus, it is possible to set an appropriate lifetime without decreasing the message delivery ratio.

3.2 Timer Setting

We present the flowchart of the proposed method in Fig. 1. Each node periodically checks the number of received SV. The proposed method measures a non-signal time (NT) from neighboring vehicles. If the current NT is greater than maximum NT (NT_{max}), NT_{max} will be updated. The formula of Dynamic Timer (DTimer) is:

$$DTimer = NT_{max} + Interval, \qquad (1)$$

where *Interval* indicates the check interval, which uses for checking the number of received SVs.

When the number of received SVs is 0, our method resets the timer. In this way, our method can keep the bundle message and prevent a drop in packet arrival rate even if the vehicles are disconnected in the network due to the vehicle movement. In general, if the DTimer is set in the message, the lifetime is not reset. But, in our proposed method, the DTimer reset is allowed even if the lifetime is reset. We consider the *Interval* to keep the message in the storage for checking the next SV. Even if the vehicle suddenly leaves the neighboring vehicles and there is no signal time, our method keeps the message.

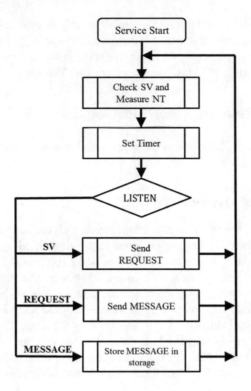

Fig. 1. Flowchart of our proposed method.

4   Modeling and Simulation Results

In this paper, we evaluate the proposed method for different vehicle densities. We implemented the proposed dynamic timer method on the Scenargie [21] network simulator.

4.1   Scenario Setting

We consider a grid scenario with a maximum density of 250 vehicles/km². We show a road model in Fig. 2. Table 1 shows the simulation parameters used on the Scenargie network simulator. Message start-point and end-point are static, and other vehicles move on the road based on the random way-point mobility model.

The start-point sends bundle messages to end-point considering ITU-R P.1411 propagation model [2]. When the vehicle receives bundle messages, they store the bundle messages in their storage. After that, the vehicles duplicate the bundles to other vehicles. Therefore, we consider the interference of obstacles on 5.9 GHz band. We use four types of routing protocols as follows:

1. **Timer(Average)**: Epidemic activated timer method with the duration set to average delay,

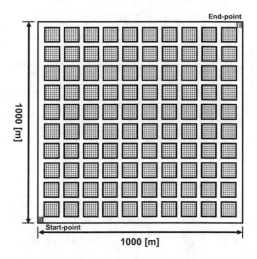

Fig. 2. Road model.

Table 1. Simulation parameters.

Parameter	Value
Simulation time ($T_{\max}$)	600 [s]
Area dimensions	$1,000$ [m] $\times$ $1,000$ [m]
Density of vehicles	60, 80, 120, 250 [vehicles/km^2]
Minimum speed ($V_{\min}$)	8.333 [m/s]
Maximum speed ($V_{\max}$)	16.666 [m/s]
Message start and end time	10–400 [sec]
Message generation interval	10 [sec]
Message size	1000 [bytes]
PHY model	IEEE 802.11p
Propagation model	ITU-R P.1411
Antenna model	Omni-directional

2. **Timer(Average*2)**: Epidemic activated timer method with the duration set to double average delay,
3. **Timer(Max)**: Epidemic activated timer method with the duration set to maximum delay,
4. **Proposed Method**: Epidemic with proposed dynamic timer considering the non-signal duration.

Before the simulation, we evaluated the delay performance using conventional Epidemic in this road model. We consider these results for setting the average delay and maximum delay. In the case of conventional Epidemic with Timer(Max), all bundles will be reach to the end-point.

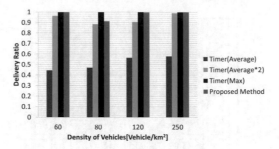

Fig. 3. Delivery ratio

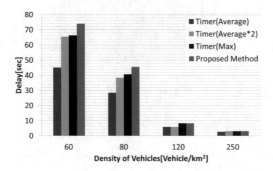

Fig. 4. Delay

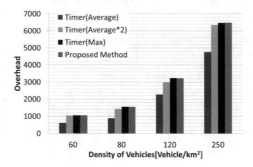

Fig. 5. Overhead

For simulations, we consider four evaluation parameters: delay, delivery ratio, storage usage, and overhead. The delay indicates the transmission delay of the message to reach the end-point. The delivery ratio indicates the ratio of bundles successfully received by end-point to the total number of bundles sent by the start-point. The storage usage indicates the average of the storage usage of each vehicle. The overhead indicates the number of times for sending duplicate messages. Each timer function will be activated after the simulation time is 60 s.

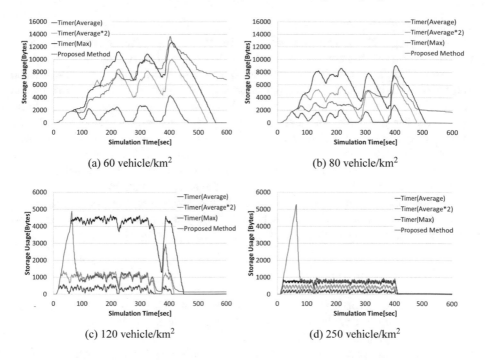

Fig. 6. Storage usage for different vehicle densities.

4.2 Simulation Results

We present the simulation results of the delivery ratio for different timers in Fig. 3. In the case when a timer is set to Timer(Average) and Timer(Average*2), all messages could not reach the end-point. When timer is set to the Timer(Average), half of the sent messages could not reach the end-point. When the timer is set to Timer(Max), the results of the delivery ratio for different vehicles reached 100%. When the density of vehicles is 60 and 250 vehicles/km^2, our proposed method delivered all messages to end-point. In other cases, we observed that the performance was close to the Timer(Max).

We show the simulation results of delay for different timers in Fig. 4. The delay results were calculated from the messages that reached the end-point. It should be noted that the number of received messages is different. The results of the delay is decreased by increasing the density of vehicles. We found that our proposed method has a large delay. In the case of dense network, Timer(Max) is the same as the proposed method.

The results of the overhead are shown in Fig. 5. The overhead results of the proposed method and Timer(Max) are higher than Timer(Average) and Timer(Average*2). We found that there is a trade-off between the delivery ratio and overhead.

We show the simulation results of average storage usage in Fig. 6. For 80 vehicles/km^2, the proposed method used less storage, but the storage usage was increased after 260 s. For 120 vehicles/km^2, the proposed method and Timer(Average*2) have similar results after 100 s. The delivery ratio of the proposed method at this time is higher than that of Timer(Average*2).

From 300 to 500 s in 60 vehicles/km^2, and from 100 to 400 s in 250 vehicles/km^2, the results of storage usage of the proposed method are the same as that of Timer(Max). Similar to the relationship between overhead and delivery ratio, storage usage and delivery ratio had a trade-off relationship. For 60 and 80 vehicles/km^2, we observed that the storage usage of the proposed method at 600 s was higher compared with the other methods. This is because a large number of vehicles can not receive the signals from neighboring vehicles in case of the sparse network. As a result, many vehicles reset the timer.

In addition, in the case of a conventional Epidemic that does not use a timer, the average storage usage at 600 s is 39 kbytes, thus the proposed method significantly reduces the storage usage.

5   Conclusions

In this paper, we proposed a message relaying method with dynamic timer considering non-signal duration from neighboring vehicles for Vehicular DTN. We evaluated the proposed method considering delivery ratio, delay, overhead and storage usage as evaluation metrics. From the simulation results, we found that the proposed method shows a high delivery rate by using the dynamic timer.

In future work, we would like to consider a new selection method for NT_{max}.

References

1. Delay- and disruption-tolerant networks (DTNs) tutorial. NASA/JPL's Interplanetary Internet (IPN) Project (2012). http://www.warthman.com/images/DTN_Tutorial_v2.0.pdf
2. Rec. ITU-R P.1411-7: Propagation data and prediction methods for the planning of short-range outdoor radiocommunication systems and radio local area networks in the frequency range 300 MHz to 100 GHz. ITU (2013)
3. Araniti, G., Bezirgiannidis, N., Birrane, E., Bisio, I., Burleigh, S., Caini, C., Feldmann, M., Marchese, M., Segui, J., Suzuki, K.: Contact graph routing in DTN space networks: overview, enhancements and performance. IEEE Commun. Mag. **53**(3), 38–46 (2015)
4. Araniti, G., Campolo, C., Condoluci, M., Iera, A., Molinaro, A.: Lte for vehicular networking: a survey. IEEE Commun. Mag. **21**(5), 148–157 (2013)
5. Caini, C., Cruickshank, H., Farrell, S., Marchese, M.: Delay- and disruption-tolerant networking (DTN): an alternative solution for future satellite networking applications. Proc. IEEE **99**(11), 1980–1997 (2011)
6. Cerf, V., Burleigh, S., Hooke, A., Torgerson, L., Durst, R., Scott, K., Fall, K., Weiss, H.: Delay-tolerant networking architecture. IETF RFC 4838 (Informational), April 2007
7. Dias, J.A.F.F., Rodrigues, J.J.P.C., Xia, F., Mavromoustakis, C.X.: A cooperative watchdog system to detect misbehavior nodes in vehicular delay-tolerant networks. IEEE Trans. Industr. Electron. **62**(12), 7929–7937 (2015)

8. Fall, K.: A delay-tolerant network architecture for challenged Internets. In: Proceedings of the International Conference on Applications, Technologies, Architectures, and Protocols for Computer Communications, SIGCOMM 2003, pp. 27–34 (2003)

9. Grassi, G., Pesavento, D., Pau, G., Vuyyuru, R., Wakikawa, R., Zhang, L.: VANET via named data networking. In: Proceedings of the IEEE Conference on Computer Communications Workshops (INFOCOM WKSHPS 2014), pp. 410–415, April 2014

10. Hou, X., Li, Y., Chen, M., Wu, D., Jin, D., Chen, S.: Vehicular fog computing: a viewpoint of vehicles as the infrastructures. IEEE Trans. Veh. Technol. **65**(6), 3860–3873 (2016)

11. Kenney, J.B.: Dedicated short-range communications (DSRC) standards in the united states. Proc. IEEE **99**, 1162–1182 (2011)

12. Lin, D., Kang, J., Squicciarini, A., Wu, Y., Gurung, S., Tonguz, O.: MoZo: a moving zone based routing protocol using pure V2V communication in VANETs. IEEE Trans. Mob. Comput. **16**(5), 1357–1370 (2017)

13. Mahmoud, A., Noureldin, A., Hassanein, H.S.: VANETs positioning in urban environments: a novel cooperative approach. In: Proceedings of the IEEE 82nd Vehicular Technology Conference (VTC-2015 Fall), pp. 1–7, September 2015

14. Nakasaki, S., Yoshino, Y., Ikeda, M., Barolli, L.: A recovery method for reducing storage usage considering different thresholds in VANETs. In: Proceedings of the 21st International Conference on Network-Based Information Systems (NBiS-2018), pp. 793–802, September 2018

15. Ning, Z., Hu, X., Chen, Z., Zhou, M., Hu, B., Cheng, J., Obaidat, M.S.: A cooperative quality-aware service access system for social internet of vehicles. IEEE Internet Things J. **5**(4), 2506–2517 (2018)

16. Ohn-Bar, E., Trivedi, M.M.: Learning to detect vehicles by clustering appearance patterns. IEEE Trans. Intell. Transp. Syst. **16**(5), 2511–2521 (2015)

17. Radenkovic, M., Walker, A.: CognitiveCharge: disconnection tolerant adaptive collaborative and predictive vehicular charging. In: Proceedings of the 4th ACM MobiHoc Workshop on Experiences with the Design and Implementation of Smart Objects (SMARTOBJECTS-2018), June 2018

18. Ramanathan, R., Hansen, R., Basu, P., Hain, R.R., Krishnan, R.: Prioritized epidemic routing for opportunistic networks. In: Proceedings of the 1st International MobiSys Workshop on Mobile Opportunistic Networking (MobiOpp 2007), pp. 62–66 (2007)

19. Rsch, S., Schrmann, D., Kapitza, R., Wolf, L.: Forward secure delay-tolerant networking. In: Proceedings of the 12th Workshop on Challenged Networks (CHANTS-2017), pp. 7–12, October 2017

20. Rohrer, J.P., Mauldin, A.N.: Implementation of epidemic routing with IP convergence layer in ns-3. In: Proceedings of the 10th Workshop on ns-3 (WNS3-2018), pp. 69–76, June 2018

21. Scenargie: Space-time engineering, LLC. http://www.spacetime-eng.com/

22. Stute, M., Maass, M., Schons, T., Hollick, M.: Reverse engineering human mobility in large-scale natural disasters. In: Proceedings of the 20th ACM International Conference on Modelling, Analysis and Simulation of Wireless and Mobile Systems (MSWiM-2017), pp. 219–226, November 2017

23. Theodoropoulos, T., Damousis, Y., Amditis, A.: A load balancing control algorithm for EV static and dynamic wireless charging. In: Proceedings of the IEEE 81st Vehicular Technology Conference (VTC-2015 Spring), pp. 1–5, May 2015

24. Tornell, S.M., Calafate, C.T., Cano, J.C., Manzoni, P.: DTN protocols for vehicular networks: an application oriented overview. IEEE Commun. Surv. Tutor. **17**(2), 868–887 (2015)
25. Uchida, N., Ishida, T., Shibata, Y.: Delay tolerant networks-based vehicle-to-vehicle wireless networks for road surveillance systems in local areas. Int. J. Space-Based Situated Comput. **6**(1), 12–20 (2016)
26. Urquiza-Aguiar, L., Igartua, M.A., Tripp-Barba, C., Caldern-Hinojosa, X.: 2hGAR: 2-hops geographical anycast routing protocol for vehicle-to-infrastructure communications. In: Proceedings of the 15th ACM International Symposium on Mobility Management and Wireless Access (MobiWac-2017), pp. 145–152, November 2017
27. Vahdat, A., Becker, D.: Epidemic routing for partially-connected ad hoc networks. Technical report, Duke University (2000)
28. Wyatt, J., Burleigh, S., Jones, R., Torgerson, L., Wissler, S.: Disruption tolerant networking flight validation experiment on NASA's EPOXI mission. In: Proceedings of the 1st International Conference on Advances in Satellite and Space Communications (SPACOMM-2009), pp. 187–196, July 2009
29. Yoshino, Y., Nakasaki, S., Ikeda, M., Barolli, L.: A threshold-based adaptive method for message suppression controller in vehicular DTNs. In: Proceedings of the 13th International Conference on Broad-Band Wireless Computing, Communication and Applications (BWCCA-2018), pp. 517–524, October 2018
30. Zhang, W., Jiang, S., Zhu, X., Wang, Y.: Cooperative downloading with privacy preservation and access control for value-added services in vanets. Int. J. Grid Util. Comput. **7**(1), 50–60 (2016)
31. Zhou, H., Wang, H., Li, X., Leung, V.C.M.: A survey on mobile data offloading technologies. IEEE Access **6**, 5101–5111 (2018)

A 2-Dimensional Technology and Real-World Interaction View Approach

Toshihiko Yamakami$^{(\boxtimes)}$

ACCESS, Tokyo, Japan
`Toshihiko.Yamakami@access-company.com`

Abstract. The rise of computing technology has increased its influence on societies. As computing technology penetrates into everyday life, we witness more interactions and integration issues with technology and societies In this paper, the author coins two dimensional models which reflects today's technological landscape with the reality. Penetration to physical world and social aspects of human life creates increased border interaction for computing technologies. The author presents implications from these dimensional models.

1 Introduction

Advances of Information and communication technologies have increases their visibility in our daily lives. It provides a fundamental challenge for human beings to develop a balanced view on technology to assess its influence and significance.

In the past, many technology deployments encountered borders of applicability. The barriers of real world were stronger these days. Today's technology provides penetration to everyday life. It enables border-crossing deployment with advanced communication technology and small intelligence-embedded chips.

As the interaction between technologies and everyday lives increases, it is important to capture the time-dimensional and space-dimensional border crossing characteristics of technology deployment in the real world.

In this paper, the author describes 2-dimensional technology-related views. Then, the author presents the implications of similarities of these 2-dimensional views.

2 Background

2.1 Purpose of Research

The aim of this research is to describe interaction patterns between technologies and social aspects of societies.

© Springer Nature Switzerland AG 2020
L. Barolli et al. (Eds.): INCoS 2019, AISC 1035, pp. 143–151, 2020.
https://doi.org/10.1007/978-3-030-29035-1_14

2.2   Related Work

Research of techno-social interaction consists of (a) open data and societies, (b) technological substitution of social components, and, (c) emerging technological UI.

First, in regards to open data and societies, Hunnius et al. discussed open data from techno-political viewpoint [3]. Porwol et al. discussed VR-based participatory democracy [5]. Cordasco et al. discussed lack of engagement in social open data [2].

Second, in regard to technological substitution of social components, Richter et al. discussed social interaction in a smart robotic environment [6]. Yamazaki et al. discussed techno-sociological UI for a robot guide [9]. Shapiro et al. discussed non-scripted social interaction with virtual characters [7]. Su et al. discussed an authentic life with technology [8].

Third, in regard to emerging technological UI, Paine discussed a techno-somatic dimension [4]. Britton discussed the role of body in interacting with cyborgs [1].

The originality of this paper lies in its identification of common aspects of 2-dimensional view models.

3   Method

The author performs the following steps:

- Observe information and communication technologies,
- Present 2-dimensional models,
- Describe the universal similarities among models.

4   Observation

4.1   General Technology Model

The author presents a 2-dimensional model with economic/non-economic axes and technical/non-technical axes in 1990s, as depicted in Fig. 1.

The author first considered this model when he came to think about R&D positioning in the corporate world. At this stage, this was just a concept model to view "engineering" with contrasts with other entities. There are two dimensions, technology one and economical one. The four quadrants are derived from technical/non-technical aspects and economical/non-economical aspects. Engineering is usually driven by economical merits. When it does not pursue economical merits, it is called as Science. Even it does not pursue economical merits, science and engineering are interrelated in multiple aspects and influence each other. Economy usually pursues economical merits. However, in order to healthily harness economy, there should be some non-economic aspects considerations such as regulations, fairesss, accountability, ethical responsibilities. In order to cover these aspects, politics plays a role here. These quadrants denote close relationships of these four aspects and identify borders among them.

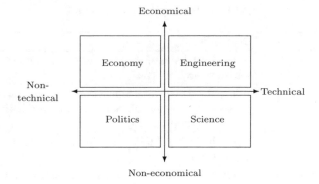

Fig. 1. A 2-dimensional model of technology/society interaction is depicted.

4.2   Computer Technology Models

Then, the author was involved in international standardization of the application layer in communication engineering during 1980–2005. After the experience of international standardization in W3C and WAP, the author experienced a mobile business model development in feature phones and smartphones. This experience brought the author a 4-stage transition model of value composition in communication software as depicted in Fig. 2. For an ordinary engineer, this was eye-opening. When there was little knowledge to generate some products, it was critically important to master the manufacturing. When the technology becomes mature, the focus shifts from manufacturing to non-technological aspects. It describes a general trend that the focus of technology will shift to non-technological aspects when the technology becomes mature.

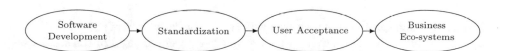

Fig. 2. A 4-stage model of communication software

The description of each stage is depicted in Table 1.

The author was involved in international standardization in 2000s, however, during that time, it was increasingly visible that the standardization yielded to user adoption. For example, a multimedia version of SMS, MMS, was widely installed in most of the feature phones. The use of MMS was very low worldwide due to the expensive charge. And when iPhone emerged, the AppStore market business model changed the mobile Internet business landscape.

The author turns the stage model to a 2-dimensional model as depicted in Fig. 3. This resembles a spiral evolution as the end point will act as a starting point of a new spiral. The two dimensions are intra-aspect/inter-aspect and

Table 1. Description of each stage

Stage	Description
Software development	In the initial stage, the value creation is done at software development
International standardization	When the global interconnection is enabled, the value creation is done at international standardization, which ensures interoperability
User acceptance	When interoperability is enabled, the value creation depends on user acceptance
Business ecosystem	When a mass user base is established, the value creation is done at building third-party ecosystems, such as application stores in smartphones

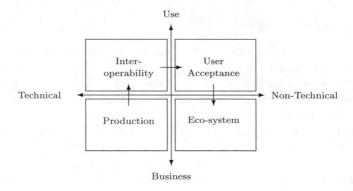

Fig. 3. A 2-dimensional model of software development and real world interaction is depicted.

production/non-production technical aspects. In the early days of software development, only production-oriented technical aspects were focused in the software development industry. As the networking increases importance, interoperability and user-acceptance increased their importance. Finally, business ecosystems are key in interconnected open business systems in the Internet era. This dimensional view fits the spiral evolution of standardized communication technologies.

The author describes a similar 2-dimensional model for software deployment as depicted in Fig. 4.

In this model, two dimensions are coding/real-world and user/system. It also denotes a spiral form of software-lifecycle. It represents the complexity of software deployment today. As the number of components and interactions of software increases. The focus of software development departs from simple coding to lifecycle management. This view model represents such a trend.

The author coins a 2-dimensional model for open source software (OSS) lifecycle as depicted in Fig. 5. The word OSS was coined in 1990s. It was a new paradigm to deal with complexity of software. In this model, two axes are inter-

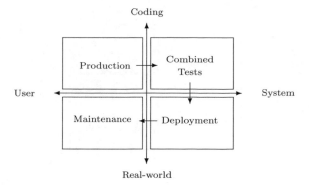

Fig. 4. A 2-dimensional model of software deployment is depicted.

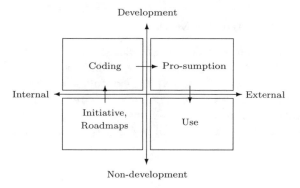

Fig. 5. A 2-dimensional model of OSS lifecycle

nal/external and development/no-development. This represents the interaction of coding and use.

Then, the author describes an IoT transition. IoT (the Internet of Things) transitions from IoT data access to the final digital transformation are depicted in Fig. 6.

Digital transformation is a new paradigm that was born in 2010s. IoT, data analytics, machine learning, communication technologies and cloud computing drive it. It starts simple data access with embedded sensors. Then, it enables manipulation with actuators. When the coordination among a massive number of sensors and actuators are enabled, it create IoT ecosystems. When it matures, digital transformation of business is deployed. This will impact the industrial landscapes.

This dimensional view can be used as a litmus test for digital transformation of IoT.

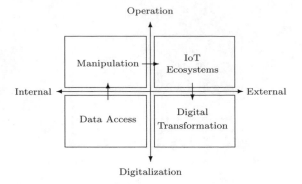

Fig. 6. A 2-dimensional model of digital transformation is depicted.

4.3   Knowledge Awareness Model

A similar dimensional model exists at knowledge awareness as depicted in Fig. 7. This model is useful for identifying one's own blind spots in knowledge management in relation to analysis of internal and external knowledge. This model does not relate to computing technologies. However, it relates to the awareness and underlying realities, which has some similarities with other models.

This can be used for exploration of technology-empowered knowledge transfer.

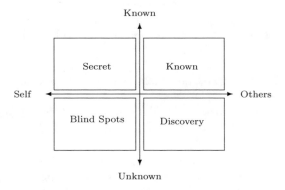

Fig. 7. A 2-dimensional model of knowledge awareness

5   Discussion

5.1   Advantages of the Proposed Method

The proposed models demonstrate that increasing real-world integration and conflicts with heterogeneous entities can captured in 2-dimensional views. It is useful to capture the emerging design patterns of (a) conflicts, (b) migration, and (c) life-cycle management.

The similarity in a wide range of different artifacts lies in the following two factors described in Table 2.

Table 2. Multiple factors that drive similarity in a wide range of different models

Factor	Description
Real-world integration	The external world, such as social communities or real-world deployment, has bidirectional influence on technologies
Border crossing	Technologies need to accommodate corder crossing capabilities
Dynamism	Penetration of the real-world leads to the needs to deal with a wide range of dynamism in the real world

The reasons of these universal similarities of dimensional transition models are depicted in Table 3.

Table 3. There are multiple reasons that drive the similarities in dimensional models.

Reasons	Description
Conflict	Explicit border crossing creates conflicts and tensions
Migration	Explicit border crossing requires special skills and operation as migration
Feedback-loop	Technology is not a standalone entity, but an integral part of feedback loop with the real world
Integration	In order to enable smooth transition over a border, it requires integration
Life-cycle	In order to deal human, social and political aspects, it requires iterative process of dealing with requirements from these heterogeneous artifacts

Technological penetration of IoT, artificial intelligence and softwarization of process and operation increases the emerging design patterns to deal with (a) conflicts, (b) migration, and (c) life-cycle management. The proposed models demonstrate that increasing real-world integration and conflicts with heterogeneous entities can captured in 2-dimensional frameworks.

The described models are categorized as depicted in Table 4.

These types are used to understand integration models to have interaction with heterogeneous entities. IoT and artificial intelligence drive injection of technological components in everyday lives and businesses. The proposed dimensional perspective is useful to provide a framework to deal with conflicts and migration.

Table 4. Categories of dimensional models

Type	Description	Examples
Segmentation	Dimensions identifies segmentation	Technology, Awareness
Transition	It describes transitions and life-cycle management	Software development, technology adoption
Border interaction	It identifies border interactions	OSS

5.2 Limitations

This research is descriptive and qualitative. The dimensional models are exploratory and lack quantitative verification.

Each dimension is exploratory and lacks theoretical backgrounds. The weight or influence of each dimension are not described.

Quantitative measures to identify dimensions and positioning in the dimensional views are missing. Reproducible methods to position items in the dimensional views are not provided.

Detailed analysis of case studies remain future studies.

Quantitative comparisons with other models are not presented in this paper.

6   Conclusion

Advances in ICT (Information and Communication Technologies) enables border-less computing. IoT is one of such examples. It blurs the once-rigid boundaries. Digital transformation also challenges the fixed boundary between a real world and a virtual world.

The author presented multiple dimensional view models and similarities among them.

Technology increasingly needs to deal with dynamisms, such as migration, integration, and evolution.

Computer technologies has increasing opportunities to penetrate into the real world and create interaction with heterogeneous entities, such as human, social and physical artifacts. This increases challenges to deal with border-crossing transitions and life-cycle management.

The proposed approach of 2-dimensional models are useful in analyzing such heterogeneous interaction. It can provide a stepping stone for a wide range of heterogeneous interaction in the era of digital transformation.

It is important to understand the emerging interaction of technological components and other entities. The 2-dimensional view model approach is promising to identify, highlight and contrast border-crossing process of today's technologies.

References

1. Britton, L.M.: Manifesting the cyborg via techno body modification: from human computer interaction to integration. In: Companion of the 2017 ACM Conference on Computer Supported Cooperative Work and Social Computing, CSCW 2017, Companion. ACM, New York (2017)
2. Cordasco, G., De Donato, R., Malandrino, D., Palmieri, G., Petta, A., Pirozzi, D., Santangelo, G., Scarano, V., Serra, L., Spagnuolo, C., Vicidomini, L.: Engaging citizens with a social platform for open data. In: Proceedings of the 18th Annual International Conference on Digital Government Research, DG.O 2017, pp. 242–249. ACM, New York (2017)
3. Hunnius, S., Krieger, B.: The social shaping of open data through administrative processes. In: Proceedings of The International Symposium on Open Collaboration, OpenSym 2014, pp. 16:1–16:5. ACM, New York (2014)
4. Paine, G.: Designing the techno-somatic. In: Proceedings of the 2nd International Workshop on Movement and Computing, MOCO 2015, pp. 48–51. ACM, New York (2015)
5. Porwol, L., Ojo, A.: Through vr-participation to more trusted digital participatory democracy. In: Proceedings of the 19th Annual International Conference on Digital Government Research: Governance in the Data Age, DG.O 2018, pp. 95:1–95:2. ACM, New York (2018)
6. Richter, V., Kummert, F.: Towards addressee recognition in smart robotic environments: an evidence based approach. In: Proceedings of the 1st Workshop on Embodied Interaction with Smart Environments, EISE 2016, pp. 2:1–2:6. ACM, New York (2016)
7. Shapiro, D., Tanenbaum, K., McCoy, J., LeBron, L., Reynolds, C., Stern, A., Mateas, M., Ferguson, B., Diller, D., Moffitt, K., Coon, W., Roberts, B.: Composing social interactions via social games. In: Proceedings of the 2015 International Conference on Autonomous Agents and Multiagent Systems, AAMAS 2015, pp. 573–580. International Foundation for Autonomous Agents and Multiagent Systems, Richland (2015)
8. Su, N.M., Stolterman, E.: A design approach for authenticity and technology. In: Proceedings of the 2016 ACM Conference on Designing Interactive Systems, DIS 2016, pp. 643–655. ACM, New York (2016)
9. Yamazaki, A., Yamazaki, K., Ohyama, T., Kobayashi, Y., Kuno, Y.: A techno-sociological solution for designing a museum guide robot: regarding choosing an appropriate visitor. In: Proceedings of the Seventh Annual ACM/IEEE International Conference on Human-Robot Interaction, HRI 2012, pp. 309–316. ACM, New York (2012)

Event Management Service System

Vladyslav Berezhetskyi[1], Artur Tomczak[1], Daniel Soliwoda[1],
and Emiljana Hoti[2(✉)]

[1] Lodz University of Technology, Lodz, Poland
[2] Faculty of Management, Comenius University in Bratislava,
831 04 Bratislava, Slovakia
emiljanahoti@fm.uniba.sk

Abstract. The purpose of the study is to estimate and investigate the advantage of using microservices capability, to develop the system with new functionalities without a relatively large interference in the logic of the individual modules. Similarly using another method as alternative as monolithic architecture which is created as independent on development of an application as a set of small independent services. Creating architecture using microservices brings advantages as freedom in choice of technology and focusing on one service and disadvantages such as delays with remote connections or difficulties related to maintaining transaction security. With the example of Event Watcher system presented it is possible to use completely different technologies in the production of microservices. In the results it was found that the application process of mircorservices were easy to manage and REST communication with service easy to be substituted, testers didn't had to wait for runnable project, they could work with those micro-services which already they had.

1 Introduction

Creating architecture using micro services is a relatively new approach. The micro service style is based on application development as a set of smaller services that operate as separate processes. In most cases they communicate by the HTTP resources, API is the most popular method. The micro-services centralization is as low as possible, which allows programming of each of them in a different language [5]. Another method that can be considered as the alternative to microservice is monolithic architecture. Which is designed as independent; the components are interdependent. In order for the code to be com-piled and executed, all components must be implemented. [Mario Villamizar]. Monolithic applications can be difficult to understand and modify. This is tough especially when the application is getting bigger. Considering the complexity of the additional code, it is difficult to add new developers to the project or replace existing ones [11]. The microservice architecture eliminates some of the problems associated with monolithic applications. According to Lianpig Chen, the benefits of developing applications in the micro services architecture include: Deployment Independency – the implementation of changes in the project was more efficient because, unlike monolithic applications, there was no need for merge the change into the master branch by all development teams. Simpler Deployment

L. Barolli et al. (Eds.): INCoS 2019, AISC 1035, pp. 152–161, 2020.
https://doi.org/10.1007/978-3-030-29035-1_15

Procedures – it could be noticed inter alia, in enabling the team to set their own procedures without the need to overly adapt them to all development teams [12]. Faster decision-making - due to the decentralization of applications and the independence of development teams, there was no need to consult minor changes in the functionalities of individual services [2].

1.1 Microservice Architecture

It is possible to distinguish the basic layers characterizing micro services: database, logical, micro services and user view. Information in micro-websites are usually stored in separate databases, in addition, data is only updated via the service API. Logical and view layers are responsible for handling and displaying data to the user. The microservice layer is focused on individual services [5].

Microservices architecture brings advantages but also disadvantages. Pros can include: freedom in the choice of technology and focusing on one service, in addition, due to services dispersion, parallel operations are possible. Microservices architecture also contribute to frequent releases. The disadvantages, however, include delays associated with remote connections, difficulties related to maintaining transaction security, and it is almost impossible to the transfer of code between sites [5]. This architecture additionally creates delays and difficulties with configuration.

There are two types of communication in the microservice architecture [11]:

– Synchronous – user expects a response from the service, in this case the most common is REST with the HTTP protocol
– Asynchronous – client does not expect a response from the website, in such case the most frequently used protocols are AMQP, STOMP and MQTT.

1.2 Literature Review

The micro-services approach is a relatively new term in software architecture patterns. The micro-service architecture is an approach to developing an application as a set of small independent services. Each of the services is running in its own independent process. Services can communicate with some lightweight mechanisms (usually it is something around HTTP). Such services could be deployed independently. In addition, the centralized management of these services is a completely separate service too. It may be written in different programming languages, use own data models, etc. [5, 13].

The main component of any such system are microservices. A microservice is described as a (small) application in its own right, able to evolve independently and choose its own architecture, technology, and platform, and can be managed, deployed and scaled independently with its own release lifecycle and development methodology [7]. Accordingly, a microservice-based architecture is defined as an architecture pattern for distributed applications where the application is comprised of a number of smaller "independent" components that are small application in themselves [1]. For example, envisage an online ordering application. This application performs multiple functions such as providing address validation, a product catalogue, customer credit check, etc. With the microservice architecture pattern, individual applications are created for address validation, customer credit check and online ordering; these application are then grouped together to create the ordering application (which now is a composite application). This approach to application development challenges the traditional "monolithic" application and services. Hence, a microservice is built around the scope of "business-bounded context and its concerns will apply to that context [6]. As in the example above, a microservice can then be mapped to enterprise systems such as customer relationship management, payroll, etc. or to smaller functions within these systems [15].

2   Main Principles of Microservices

When creating a micro-service one should be guided by certain rules. This allows you to optimize the operation of selected micro services. This significantly influences the communication between the layers with which the services communicate. In the course of works on microservices, the focus should be on ensuring the greatest possible autonomy for individual microservices. The key principles to be followed in the construction of architecture and implementation are presented in Fig. 1.

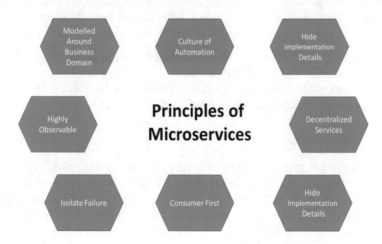

Fig. 1. Principles of microservices (Source: http://digitalcto.com/microservices/)

It is worth discussing the individual principles of creating micro-services. This will allow a broader understanding of their meaning and functioning:

- Adopt a Culture of Automation.
- Hide Internal Implementation Details.
- Independently Deployable.
- Decentralize All the Things.
- Highly Observable.
- Isolate Failure.
- Model around Business Concepts [14].

3   EventWatcher – Admin Panel Microservice

3.1   Overview

Due to the improved architecture based on micro-services, system design becomes more and more important. With the technological development and popularization of fast internet networks, the idea was born to add another layer responsible for the service to the standard three-layer architecture. In the case of a micro-service, acting as an administrator panel, apart from the three standard layers, an additional layer responsible for communication with other services should be distinguished. For this purpose, the . Net WebAPI tools were used. Web Application Programming Interfaces using the power of HTTP protocol communicates with other services in the system. This communication allows moving part of the application logic to another service and thus allows isolating the client [4] from the server. For communication purposes, one of the two popular types of services was selected.

Due to the universality of the JSON format, the choice was made for the REST API service. Another service that could be used is a SOAP service based on XML.

The choice of the REST service was supported by such aspects as [10]:

- A unified interface
- Staturelessness
- Client-server separation
- Interlayer
- Code on request

The most important of them is discussed below. The key features of the uniform interface include the fact that the server sends only a part of the information and not the entire resource. The client, if entitled, can delete or modify the data. Additionally, you can use HTTP protocol features: header, body, and query. In queries aimed at authentication of the user logging into the Event Watcher Admin Panel system, is used to send a password encrypted with the SHA256 method [9] after sending the password, it is verified on the side of another service and then in the feedback, the permission or lack of permission is received [6, 8].

Another important feature is the separation of the client from the server. This separation works both ways. The client does not know how the data is stored and the

Fig. 2. Event watcher - login window

server does not know about the state of the client. An interesting advantage is the ability to have a cache when using the REST API. Its existence positively influences the scalability of the application (Fig. 2).

After logging in to the admin panel, the user has access to the list of all events. In addition, it is possible to add and remove selected events from the list. Placing an event wizard on the side of this service eliminates the problem of development of other services included in the system. The logic and responsibility for proper functioning of particular system functionalities have been separated. The picture below presents a diagram of usage cases, which describes the possible actions that may occur during the use of the system.

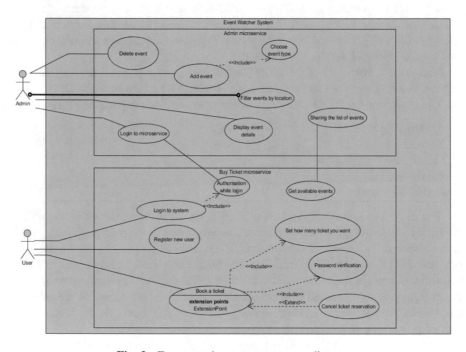

Fig. 3. Event watcher system use case diagram

Figure 3 shows which functionalities of the system its individual users use. The boundaries between micro-services are clearly visible. Microservices in Fig. 3 are treated as subsystems of the Event Watcher system. This procedure is purposeful and is intended to show the division of responsibility between the services. Communication links between micro-services have also been taken into account. It can be read that the admin panel makes the list of events available to the service responsible for ticket sales. After having accessed the API for creating events, the microsite is able to communicate with the authorization module, which is located outside its area.

3.2   Tests

Jmeter [3] was used to test the service responsible for sharing the list of events, which allowed the performance of the API to be determined under load due to the large number of queries. 50 queries were simulated to determine the maximum and minimum API query time. The analysis shows that the longest time was 2905 ms and the shortest time was 336 ms.

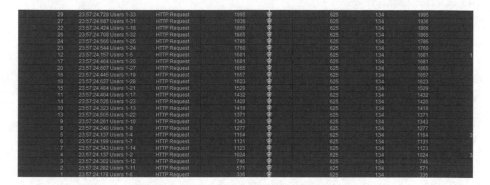

Fig. 4. Event watcher event share API tests

The fourth column of the table in Fig. 4 shows the time spent on the request. There can also notice that the queries in the tests have green marks [column 5] that indicates the correct operation of the Web API.

4   EventWatcher - Reservation Microservice

4.1   Overview

Nowadays one of the most important future, which characterize good website, is simply user-friendly frontend. Part responsible for reservations was created in such way to let end user to manage their reservations. User can to authorize himself by login and via message, book and cancel reservations. Logical layer of the microservice was made in Django framework, user interface is based on technologies like HTML, Java Scrips and CSS. Server communicates with user interface by Representational State Transfer (REST) that provides simplicity in data transfer. To send messages GSM module was used.

User authentication has been done in two parts. To enter the reservation panel user has to log in (or create an account first). The second part is message authorization. To buy tickets user has to authorize himself by validation code send on mobile number inputted while reservation process. Website sends request to server with is using GSM module to send message (Fig. 5).

GSM module specification:

– Chip: SIM800L
– Card slot: MicroSIM

#	Location	Event type	Seats left	Price [$]	Date	Details	Buy Ticket
1	Dyblin	Opera	246	120,00	2019-01-10 21:00	Details	Buy a ticket
2	Warsaw	Concert	250	230,00	2019-01-10 21:00	Details	Buy a ticket

Fig. 5. User interface reservation panel table.

- Handle: SMS/GPRS
- Frequencies: QUAD BAND 850, 900, 1800, 1900 MHz
- Interface: UART (RX, TX, GND)
- UART voltage: 2,85 V–5 V
- Antenna connector: IPX
- Antenna: 2dBi[46]

Events data is collected via REST communication from admin microservice. Reservation however are stored in booking microservice, which provides simplicirty in user data management. Details button show all of the extra information provided in admin panel.

Figure 6 shows buying tickets process. In the first step user have to choose how many tickets he want to buy (maximum is 6). Second step shows authorization part in which user has to enter key send via GSM module at mobile number inputted during creating account process. To resolve problem connected with relatively long time of message management multithreading was used. If authorization key is valid user gets information that reservation process has finished successfully.

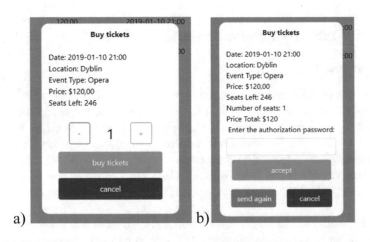

Fig. 6. Buy tickets process. (a) Tickets amount (b) Enter authorization key

User is able to see its reservations in profile panel. As Fig. 7 shows user can also cancel its reservation. Request will be send to reservation microservice and tickets can be bought again.

Your tickets:

#	Location	Event type	Number of seats	Price total[$]	Date	Cancel reservation
1	Dyblin	Opera	3	360	2019-01-10 21:00	Cancel reservation

Fig. 7. User reservations panel table.

Microservice is able to manage tickets reservations and solves problem of user validation. Token authentication provides additional user security against unauthorized operations. It also solves problem of message authorization by multithreading implementation. Separation between services in reservation storage ensures that user will be able to see its reservation even if admin microservice will be unavailable.

4.2 Tests

The main task of the tests is to check the availability of the application for new users' registration and the login integrity. To solve this problem, an application for testing was created which has two functions: the function of checking the login and the function of registering new users (Fig. 8).

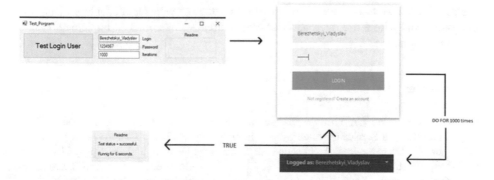

Fig. 8. User login test

To test the first functionality, the tester should enter the username and password of an existing user and the number of iterations to check. The program will start automatically logging in and counting the execution time. At the end of the test, if the result is successful, you can see the execution time by which to calculate the speed of access to the account. The time of one visit very much depends on the speed of connection to the server and other parameters but the main task of this test is to check the health of the site.

5 Conclusions

The advantage of using microservices is the ability to expand the system with new functionalities without a relatively large interference in the logic of individual modules. Additionally, each microservice can be delivered independently from other services, including various publishing cycles. Based on the example of the Event Watcher system presented in the article, it is possible to use completely different technologies in the production of microservices. Moreover, each microservice acts as a separate application, process or instance of a virtual machine, so its "problems" do not affect the operation of other services [16–18].

It is worth to pay attention at the design process itself. Due to its small size, designing micro-services is easier. Looking at the drawbacks that exist in systems based on microservices, it is worth paying attention to the problems associated with maintenance and integration of all elements. Transactional activities related to data processing require reliability of individual nodes and elimination of disturbances in the operation of microservices, which is often difficult to maintain [19, 20].

Implementation process showed that microservices are easy to manage. Nobody has to wait for others service implementation. REST communication with service can be easily replaced by example data at the simple endpoint. Such way can be useful because it does not stop any team implementations. In microservices architecture tests are relatively easy to implement, testers doesn't have to wait for runnable project, they can work only with those microservices which already have finished implementation of some functionalities.

References

1. Xu, C., Zhu, H.: CAOPLE: a programming language for mircroservices SaaS. In: IEEE Symposium on Service-Oriented System Engineering (SOSE), pp. 34–43 (2016)
2. Chen, L.: Microservices: Architecting for Continuous Delivery and DevOps (2018)
3. Halili, E.H.: A Practical Beginner's Guide to Automated Testing and Performance Measurement for Your Websites (2008)
4. Molnár, E., Molnár, R.: Web intelligence in practice. J. Service Sci. Res. 6(1), 149–172 (2014)
5. Thönes, J.: I.–1, January 2015
6. Kryvinska, N.: Building consistent formal specification for the service enterprise agility foundation. J. Service Sci. Res. 4(2), 235–269 (2012)
7. Limpiyakorn, R.W.: Development of IT Helpdesk with microservices. In: International Conference on Electronics Information and Emergency Communication (ICEIEC), pp. 31–34 (2018)
8. Gregus, M., Kryvinska, N.: Service Orientation of Enterprises - Aspects, Dimensions, Technologies. Comenius University in Bratislava, Bratislava (2015). ISBN 9788022339780
9. Yung, M., Liu, P.: Information Security and Cryptology: Fourth International Conference, Inscrypt 2008, Beijing, China, 14–17 December 2008. Springer, Heidelberg (2009)
10. Masse, M.: REST API Design Rulebook: Designing Consistent RESTful Web Service Interfaces. O'Reilly Media, Inc., Sebastopol (2011)

11. Dmitry, N., Manfred, S.-S.: On micro-services architecture. Int. J. Open Inf. Technol. **2**(9), 24–27 (2014)
12. Kryvinska, N., Gregus, M.: SOA and its Business Value in Requirements, Features, Practices and Methodologies. Comenius University in Bratislava, Bratislava (2014). ISBN 9788022337649
13. Kryvinska, N., Kaczor, S.: It is all about services - fundamentals, drivers, and business models. J. Serv. Sci. Res. **5**(2), 125–154 (2013)
14. Newman, S.: Building Microservices: Designing Fine-Grained Systems. O'Reilly Media Inc., Sebastopol (2015)
15. Xiao, Z., Wijegunaratne, I.: Reflections on SOA and microservices. In: 4rth International Conference on Enterprise Systems, pp. 60–67 (2017)
16. Poniszewska-Marańda, A.: Modeling and design of role engineering in development of access control for dynamic information systems. Bull. Polish Acad. Sci. **61**(3), 569–580 (2013)
17. Majchrzycka, A., Poniszewska-Marańda, A.: Secure development model for mobile applications. Bull. Polish Acad. Sci. **64**(3), 495–503 (2016)
18. Poniszewska-Marańda, A., Rutkowska, R.: Access control approach in public software as a service cloud. In: Zamojski, W., et al. (eds.) Theory and Engineering of Complex Systems and Dependability. Advances in Intelligent and Soft Computing, vol. 365, pp. 381–390. Springer, Heidelberg (2015). ISSN 2194-5357, ISBN 978-3-319-19215-4
19. Tkachenko, R., Izonin, I.: Model and principles for the implementation of neural-like structures based on geometric data transformations. In: Hu, Z., Petoukhov, S., Dychka, I., He, M. (eds.) Advances in Computer Science for Engineering and Education. ICCSEEA 2018. Advances in Intelligent Systems and Computing, vol 754, pp. 578–587. Springer, Cham (2019)
20. Poniszewska-Marańda, A.: Conception approach of access control in heterogeneous information systems using UML. J. Telecommun. Syst. **45**(2–3), 177–190 (2010)

Method for Extracting Positions
of Players from Video of Lacrosse Game

Miki Takagi and Hiroyoshi Miwa(✉)

Graduate School of Science and Technology, Kwansei Gakuin University,
2-1 Gakuen, Sanda-shi, Hyogo, Japan
{miki.t,miwa}@kwansei.ac.jp

Abstract. It has always been important to analyze games in the sports field. Especially, the motion tracking of a player is useful for analysis of strategy; therefore, it must be recorded. The outline of the movement path of a player has been recorded manually based on video so far. Recently, the movement path of a player can be recorded automatically from video; however, since a dedicated system using many high-performance cameras under an environment where various conditions are satisfied is necessary, it can only be used in some professional sports such as the soccer. On the other hand, there is a large demand for low-cost and simple systems so that they can be used easily in amateur sports. In this paper, we propose a method for extracting spatial position of players from video of lacrosse game. The method can detect players, even if video includes shakes. We implement the proposed method and evaluate the performance based on actual video of lacrosse game.

1 Introduction

Analysis of games in the sports field is indispensable for the ability analysis of the player, improvement guidance, strategy planning, and so on for not only professional sports but also amateur sports. Especially, the motion tracking of a player is useful; therefore, a lot of data has been recorded. The outline of the movement path of a player has been recorded manually based on video so far, that is, many members for game analysis recorded the movement path of a player for all players of not only same team but also the opponent team manually by watching video. This work takes much time and effort, although it is very important fot game analysis.

Recently, the movement path of a player can be recorded automatically from video. For example, a method using GPS is investigated. However, in the method using GPS, the position error can not be sufficiently suppressed at present, and it is difficult to distinguish approaching players. In addition, the device for motion tracking is obstacle for players, since the device is not sufficiently small. Therefore, there are still many problems in practical use. On the other hand, a method using video image is developed; however, since a dedicated system using many high-performance cameras under an environment where various conditions

© Springer Nature Switzerland AG 2020
L. Barolli et al. (Eds.): INCoS 2019, AISC 1035, pp. 162–171, 2020.
https://doi.org/10.1007/978-3-030-29035-1_16

are satisfied is necessary, it can only be used in some professional sports. Since there is a large demand for low-cost and simple systems so that they can be used easily in amateur sports, a method that work even with low-cost and simple system is desirable.

In this paper, we propose a method for extracting spatial position of players from video of lacrosse game. The method can detect players, even if video includes shakes. We implement the proposed method and evaluate the performance based on actual video of lacrosse game.

2   Related Works

There are some studies on the extraction of the moving routes of players in the sports field. For example, Baysal et al. [1] proposed a method detecting a soccer field using characteristic points and a method detecting a player using SVM (support vector machine) from one video image. Xing et al. [2] proposed a method detecting a field based on the colors and a method detecting the moving route of a player using Bayesian estimation for a video of games of the soccer and the basketball.

Pers et al. [3] studies a real-time player tracking method using a single fixed point camera from above the field in a room.

Utsumi et al. [4] studies a method based on a template matching by color for tracking moving objects when objects interfer each other.

There are some methods based on particle filters. For example, Ericson et al. [5] implemented an automatic particle filter tracking system using Gaussian from four fixed point cameras for a futsal game. Zhu et al. [6] studies a method detecting and tracking multiple objects using SVM and particle filters. Mathner et al. [7] studies a method tracking players using particle filters from the video of a fixed point camera for beach volleyball.

It is possible to apply the above method, if the lines of a field are clearly recognized and there is no shake by wind. However, in an actual game of amateur sports, these conditions are not often satisfied. Therefore, a method detecting a filed and players even such a condition is necessary.

3   Method for Extracting Position of Players from Video of Lacrosse Game

In this section, we propose a method for extracting position of players from video of lacrosse game.

We assume that videos are taken from some high positions to get the whole view of a field. Since the view is from diagonally above, it is impossible to use person identification technology using deep learning as it is. The recognition ratio is actually low, because it is not images from the angle used for learning in these existing technologies. In addition, due to camera shake by wind and so on, an error with the actual position frequently occurs, and due to the change of the brightness of sunlight, the color of the field and players also frequently changes. If we can use a dedicated system using high-performance cameras under the condition of a clear-color maintained field and the clear-color uniform of players, most of these problems can be solved. However, if we use a standard PC, some household videos, rough field, clothes that are not necessarily uniform in practice games, we need a method that can solve these problems.

We describe the basic idea of the proposed method.

First, the positions of the corners of a field are detected by extracting the color of the red cones located at the corners. If the line of the field is clearly visible, the red cones are not necessary. In general, since the field is rough, it is difficult to detect the lines. In such a case, the marks to identify the field are necessary.

Next, the connected regions corresponding to the players are extracted by color extraction as well. The video is converted to the gray scale video, and the connected regions whose color is black are extracted. Similarly, the connected regions whose color is white in order to extract only the player's white shirt after the video is converted to the HSV video. Then, the union of these parts are extracted, and each area is labeled. However, this includes a lot of noise, and even a single player may be separated into two parts. Therefore, we execute the following procedures. When a region is too small or too large, the label of the region is deleted; Some close regions are merged to an area; missing regions among the successive video frames are interpolated. Thus, we can extract all players from the video.

We show the overview of the proposed method in Fig. 1.

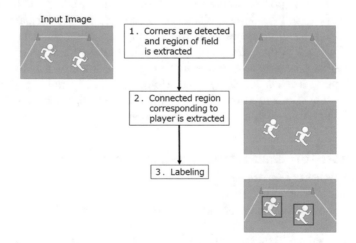

Fig. 1. Overview of proposed method

We describe the method in detail in the following sections.

3.1 Detection of Field

The positions of the corners of a field are detected by extracting the color of the red cones located at the corners (Fig. 2). The corners of a field can be detected by this method, because the extraction of the color of red is easy and relatively robust against the change of the sunlight. Therefore, even if there are camera shakes and oscillation, correction is possible from tracking the positions of the corners. If the line of the field is clearly visible, the red cones are not necessary.

Fig. 2. Detection of positions of corners

3.2 Detection of Players

The players are extracted by color extraction. The video is converted to the gray-scale video. A gray-scale image is an image in which the brightness is represented by white, black, and gray 254 gray scales. The connected regions whose color is black are extracted by a threshold value according to the brightness of the lower body in the video. Similarly, the video is converted to the HSV video. The connected regions whose color is white are extracted by a threshold value according to the brightness of the upper body. This is because the clothes of the upper body and the lower body of a player have often different colors. Thus, we extract the connected regions of the upper and the lower body of a player, respectively. The values of the thresholds are manually adjusted in advance.

Furthermore, we apply the contraction and expansion processing to the image. The contraction processing is a process of removing the pixels at the boundary between the object and the background by one pixel and is a technique used for removing noise such as small points; the expansion processing is a process of increasing the pixels of the boundary by one pixel and is a technique used for filling small holes and small irregularities. Even if the parts of the upper body and the lower body are not connected, since the divided objects are not far apart, it can be repaired by the expansion processing. The noise can be removed by the contraction processing. Since these functions are generally implemented in an image processing library such as OpenCV, we can easily use the functions. Thus, we repair the image and extract the region of a player for all players. We show an example in Fig. 3.

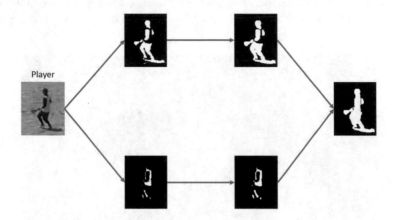

Fig. 3. Method of extracting player

3.3   Labeling

There still remains the connected regions that is not a player. Therefore, we execute the following procedures.

First, too small or too large regions in size, as well as the regions outside the field, are removed (Fig. 4). Next, sufficiently close different connected regions are unified to a connected region (Fig. 5). Then, compared to the previous frame, the connected area missing in the next frame is interpolated (Fig. 6).

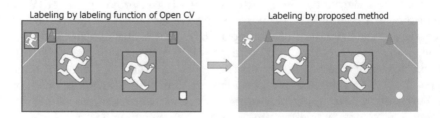

Fig. 4. Deletion procedure

Thus, each connected region indicates a player and its label. Consequently, we can extract all players from video.

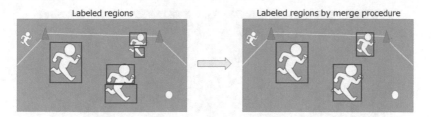

Fig. 5. Merge procedure

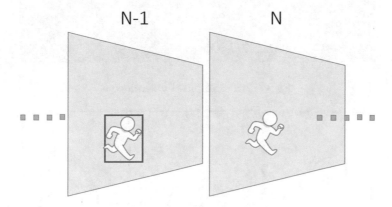

Fig. 6. Interpolation procedure

4   Performance Evaluation of Proposed Method

We implemented the proposed method described in Sect. 3 and we developed a system extracting position of players from video of lacrosse game. In this section, we evaluate the performance of the method by applying the system to the video of actual lacrosse games.

We describe the specification of the video used for the evaluation. The video used in this paper is the video of a game of lacrosse taken from a height of 3 m using a monopod. Figure 7 is an image from the video taken from the back of the goal. This is a 7 to 7 match in the area surrounded by the red line in Fig. 8. The length of the video is about 50 s between the interruptions of a game. The image size is 1080×1920.

4.1   Performance Evaluation of Function of Extracting Region of Field

We evaluate the performance of the function extracting the region of the field.

First, we evaluated the extraction of a field. Figure 9 is an example of the image after the projection conversion is performed on the original image. In order to extract the region of the field, four points at the corners are necessary; however, only two points are included in the video. Even the method described

Fig. 7. An example of video image

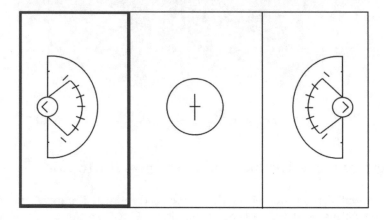

Fig. 8. Target area

in Sect. 3 cannot extract the region of the field completely; however, we evaluate the performance of the method under the condition of the lack of the points.

We added manually fixed two points at the first video frame so that we can extract the shape of the field in the following frames, and we apply the method to the video.

The extraction of red cones was successful through all frames. On the other hand, as for the region of the field, 706 frames of 1505 frames (47%) were correctly extracted. The reason that 53% could not be extracted successfully is the large change of the positions of manually added two points by wind. However, the proposed method can absorb the small shake and detect the region of the field. Since the red cones are completely detected, if three or four points are included in the video, the region of the field can be extracted completely.

Fig. 9. Extraction of region of field

4.2   Performance Evaluation of Function of Extracting Players

We evaluate the performance of the proposed method extracting players. Figure 10 shows an example of the connected regions extracted by the method in Sect. 3.2.

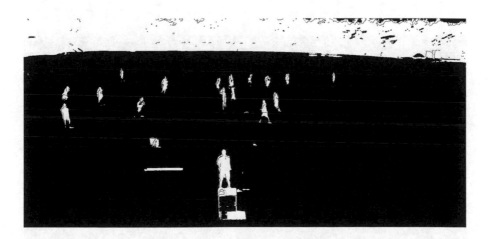

Fig. 10. Extracted connected region

We show an example of the players labeled by the function of OpenCV in Fig. 11 and an example of the players labeled by the method in Sect. 3.3 in Fig. 12. We can see the reduction of noise by the proposed method from these figures. The lack of the players occurred in the frames of 5.3% of the total frames; the failure of labels occurred in 0.4% of the total frames; the failures that one player is recognized as some different players occurred in the frames of 0.06% of the total frames. That is, the 5.76% of the total frames fails in the extraction of players. Conversely, the proposed method can extract players by the recognition ratio of 94.24%.

Fig. 11. Labeling by OpenCV

Fig. 12. Labeling by proposed method

5   Conclusion

In this paper, we proposed a method for extracting spatial position of players from video of lacrosse game. The method can detect players, even if video includes shakes. We implemented the proposed method and evaluated the performance based on actual video of lacrosse game. As a result, the system by the proposed method can extract players by high recognition ratio.

Acknowledgements. This work was partially supported by the Japan Society for the Promotion of Science through Grants-in-Aid for Scientific Research (B) (17H01742) and JST CREST JPMJCR1402.

References

1. Baysal, S., Duygulu, P.: A soccer player tracking system using model field particles. IEEE Trans. Circ. Syst. Video Technol. **26**(7), 1350–1362 (2016)
2. Xing, J., Ai, H., Liu, L., Lao, S.: Multiple player tracking in sports video: a dual-mode two-way Bayesian inference approach with progressive observation modeling. IEEE Trans. Image Process. **20**(6), 1652–1667 (2011)
3. Pers, J., Vuckovic, G., Kovacic, S., Dezman, B.: A low-cost real-time tracker of live sport events. In: Proceedings of 2nd International Symposium on Image and Signal Processing and Analysis, pp. 362–365 (2001)
4. Utsumi, O., Miura, K., Ide, I., Sakai, S., Tanaka, H.: An object detection method for describing soccer games from video. In: Proceedings of IEEE Conference on Multimedia and Expo ICME02, vol. 1, pp. 45–48 (2002)
5. Morais, E., Ferreira, A., Cunha, S., Barros, R., Rocha, A., Goldenstein, S.: A multiple camera methodology for automatic localization and tracking of futsal players. Pattern Recogn. Lett. **39**, 21–30 (2014)
6. Zhu, G., Xu, C., Huang, Q., Gao, W.: Automatic multi-player detection and tracking in broadcast sports video using support vector machine and particle filter. In: Proceedings of International Conference Multimedia & Expo, pp. 1629–1632 (2006)
7. Mauthner, T., Koch, C., Tilp, M., Bischof, H.: Visual tracking of athletes in beach volleyball using a single camera. Int. J. Comput. Sci. Sport **6**, 21–34 (2007)

Digital University Admission Application System with Study Documents Using Smart Contracts on Blockchain

Kosuke Mori and Hiroyoshi Miwa[(✉)]

Graduate School of Science and Technology, Kwansei Gakuin University,
2-1 Gakuen, Sanda-shi, Hyogo, Japan
{mrkosuke,miwa}@kwansei.ac.jp

Abstract. In order to promote data distribution and utilization of personal data, the concept of data portability attracts attention. As an example of such personal data, there is a study document of a high school student used for entrance examination etc. Since study documents are still paper medium in Japan, computerization of study documents is planned. However, since a study document contains a lot of confidential information, the information must not be leaked even if erroneous transmission or mistake occurs. In addition, it is necessary to prevent falsification of information. The format of a study document is now planned to be updated by Ministry of Education, Culture, Sports, Science and Technology, and, in a updated study document, not only teachers input information but also student him/herself also conducts research activities, extra-curricular activities, qualifications/certification etc. Consequently, the authority of input/viewing becomes extremely complicated. In this paper, we propose a digital university admission application system with study documents and e-portfolio using smart contracts on blockchain. Furthermore, we implement the proposed system using Ethereum platform. Due to the property of the blockchain, the falsification of information is extremely difficult, information leakage is prevented by combining with encryption, and it is effective from the viewpoint of data portability.

1 Introduction

The concept of data portability attracts attention to promote data distribution and utilization of personal data.

As an example of such personal data, there are university admission application and study document of a high school student. In Japan, document for university admission application is still written on paper, and signatures and seals on the documents are used as the guarantee of the documents. On the other hand, the Ministry of Education, Culture, Sports, Science and Technology (MEXT) in Japan promotes the reformation for the purpose of utilization of the document for university admission application in conjunction with the contents of study documents stored in an e-Portfolio system such as Japan e-Portfolio [1]. However, if paper is used as the documents, it is unrealistic from the viewpoint

L. Barolli et al. (Eds.): INCoS 2019, AISC 1035, pp. 172–180, 2020.
https://doi.org/10.1007/978-3-030-29035-1_17

of cost, human operation cost, and time cost. Therefore, the computerization of the document for university admission application is indispensable. MEXT is planning to realize the computerization in 2021 and the group of Kwansei Gakuin University in Japan undertakes the project of MEXT for the investigation research and the proving test of a system now.

A system for utilization of the document for university admission application in conjunction with the study document in e-Portfolio must satisfy various requirements such as data portability, high security level such as information leakage prevention, cooperation with the system in school which recorded data such as achievement results of students, cooperation with the e-portfolio system, cooperation with the university application system, and so on.

In the world, there are some systems for university admission application such as Common Application [2], Coalition [3], Universal College Application [4], UCAS [5]. However, the rules of university admissions in Japan are largely different from other countries; therefore, it is impossible to use directly a system of another country. It is necessary to make a system of university admission application which is appropriate to the situation in Japan.

In this paper, we propose a digital university admission application system with study documents and e-portfolio using smart contracts on blockchain. It is proven that the falsification is very difficult due to the property of the blockchain, and it is effective to utilize digitized official documents among some different organizations. Furthermore, we develop a prototype of the proposed system using a platform of Ethereum.

2 Digital University Admission Application System Using Smart Contracts on Blockchain

2.1 Digital University Admission Application System

In this section, we describe a digital university admission application system with study documents and e-portfolio, and then we propose the system architecture using smart contracts on blockchain.

First, we describe a digital university admission application system with study documents. This system includes not only the information specified as what should be written in study documents defined by MEXT but also the information of e-portfolio such as Japan e-Portfolio (JeP). A study document of a student includes the scores of all the subjects, other academic records, curricular activities, extra-curricular activities, and so on. The information of e-portfolio of a student includes the description of the process of learning for each activity, which is written by the student. Study documents and e-portfolio have duplicate and mutual complementary information. For example, the information of qualification is included in both, but in the present study documents, the correctness of the information of qualification is guaranteed only by school principal's seal. On the other hand, the e-portfolio have accumulated the evidence directly from certification organizations; therefore, the confirmation work in high school can be omitted and reliable evidence can be obtained. Thus, the information of study documents and e-portfolio, which is personal information of students, is handled by this system in an integrated fashion. In the rest of the paper, we assume that the system handles the set of articles, each of which is the different information of a student included in the study document and e-portfolio. Each article has the property of the permissions of Input, View, Send, Receive, and the combinations of them, and the permission group are defined.

The users of this system are defined as students, high schools, universities, organizations for certifications of qualifications, and the education management center. The education management center develops, manages and operates the whole system. Furthermore, the center manages the registration of users to the system. An organization such as the education management center is needed, because the identification and the authentication of a user is necessary and they need the procedures other than the system.

We describe the property of the permissions of articles in detail. First, the articles of "Study Document Information (SDI)" of a student can be input, viewed, and sent only by the high school to which the student belongs. Neither students nor certification organizations can input and view it. The universities to which the student applies and the education management center can receive and view it. Next, the articles of "e-Portfolio Information (EPI)" of the student can be input, viewed, and sent by the student, but certification organizations cannot input, view, send, and receive it. The high school to which the student belongs, the universities to which the student applies, and the education management center can receive and view it. The articles of "Qualification Information (QI)" can be input, viewed, and sent by the organization that certified the qualification of the student. The student can receive, view, and send it. The high school to which the student belongs can view it. The universities to which the student applies and the education management center can receive and view it. The articles of SDI are input by the high school to which the student belongs and sent to the student. The articles of EPI are input by the students, and the articles of QI are input by the certification organizations. The articles of SDI, EPI, and QI are sent to the education management center by the student, and the center receives them and transmits the information to the universities to which the student applies. All articles are encrypted, and only the operation according to each access authority is possible.

Figure 1 shows the overview of the system.

2.2   Architecture of System Using Smart Contracts on Blockchain

We describe the system architecture of the digital university admission application system with study documents and e-portfolio using smart contracts [6,7] on blockchain [8].

We use Ethereum [9], one of blockchain technology in this system. Ethereum is a platform for building Decentralized Applications (Dapps) and smart contracts. Execution history of smart contracts is recorded in the blockchain on a P2P network called Ethereum network. Ethereum has a programming language

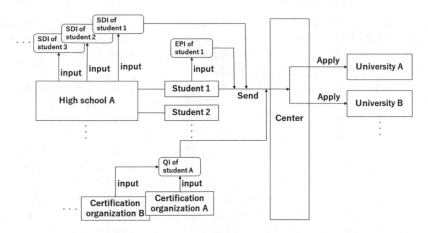

Fig. 1. Overview of digital university admission application system with study documents

to describe smart contracts and a program execution environment called EVM (Ethereum Virtual Machine). The participants to the network can write and execute smart contract programs in the blockchain on the network. Generally, Ethereum is a public blockchain, but it can also be used to build a private blockchain.

We extend ERC721 [10], which is one of the Ethereum's smart contract standards. ERC 721 is an implementation of "non-fungible tokens" that can record the ownership and the transaction history of a token and that a token cannot be split. ERC721 is extended so that a token holds not only the owner's address but also the issuer's address. We use the smart contracts by the extended ERC721 in the rest of the paper. In the proposed system, a token is the set of the articles of SDI, EPI, or QI, respectively. We call these tokens, SDI token, EPI token, and QI token, respectively. Each student is the issuer of the tokens, that is, the set of the tokens whose issuer is the same is all the information of a student. The access authority to the tokens differs depending on the properties of the permissions. This can be realized by using a public key cryptosystem and common key sharing.

The system uses a private blockchain managed by the education management center. The public blockchain approves and records transactions by unspecified number of nodes, which is a lot of network participants. This method prevents the falsification of information by malevolent users and ensures transparency of information. However, the public blockchain has some demerits. It takes much time to approve transactions, and participants have to pay a fee to the node that is responsible approval work. On the other hand, in a private blockchain that a single administrator manages, participants need to be approved by the administrator in order to participate in network. There is no need to pay a fee for the approval of a transaction, and a large amount of approvals can be transacted in a short time. As a demerit of a private blockchain, there is a possibility of falsification by the administrator. Therefore, the organization corresponding to the administrator must be public and reliable. In the proposed system, we assume that the education management center plays an important role as a public organization supported by MEXT.

The proposed system consists of the registration function, the issuing function, the transmitting function, the input function, and the inquiry function. These are described as follows.

The registration function registers the addresses of high schools, students, universities and certification organizations in the smart contract. Only the education management center, which is the organization issuing the smart contract for the system, has a privilege to use this function. Therefore, the users who can use this system can be limited to only the Ethereum address approved by the center, and the reliability of the system can be guaranteed. The registration function not only registers the users but also links the addresses of high schools and students, by registering the hash value of the pair of the high school address and the student address in the smart contract from the student directory submitted by the high school.

The issuing function issues tokens. Only students have a privilege to use this function. First, a student requests the smart contract to issue tokens. If the tokens have not been issued, the smart contract requests the token contract to issue the token, and if it has been issued, the contract process is aborted. By a student becoming token's issuer, the data portability on the information of the student can be ensured.

The transmitting function sends tokens. The address of the owner of a token is set to be the same as the address of the issuer when the token is issued, and changing the address of the owner is regarded as sending the token. If the token is owned by an address other than the issuer of the token, the token can only be sent to the issuer to prevent it from being owned by other third parties. The user who owns the token can input or view it, depending on the property of the token. However, based on the association of high schools and students registered in the smart contract, a high school can view only the token of the high school students, even if it is not the token owner. Other users must be the token owner to view.

Due to the property of blockchain, it is not possible to execute a contract at a specified time. For this reason, it is possible that the transmitted token is not returned. Therefore, the issuer of the token can only change the address of an owner to the address of issuer for own token.

The input function inputs information into tokens. Only users with input privileges, determined by a combination of token property and token owner, can perform this function. When a student sends an SDI token to a high school, the high school inputs the articles of SDI into the token. In an EPI token, a student inputs the articles of EPI into the token. A QI token is input by a certification organization, when it is sent from a student to a certification organization.

The inquiry function is used to view the articles in a token. As for an EPI token and a QI tokens, only when a permitted user becomes the token owner, the articles in the tokens are viewed by decoding encrypted articles with the common key shared between the token issuer and the token owner. An SDI token is not viewed by the issuer of the token. When a high school or the education management center becomes the token owner, the articles in the tokens are viewed by decoding encrypted the articles with the common key shared between the high school and the center.

We describe the flow of the university admission application using the system. A student sends the SDI token to the high school to which the student belongs and receives the SDI token in which the articles are input by the high school. The EPI token is input by the student. The student sends the QI token to a certification organization and receives the QI token in which the certification is input. When the student applies to the universities, the student sends these tokens and the list of the universities to which the student applies to the education management center. The center sends the information of the student to each university. It is possible to send and receive the information between the center and universities by the smart contract or by the usual communication.

In general, students often apply to more than one university. Since a token cannot be split or copied, if a student sends a token to a university, the student cannot send it to another university simultaneously. Therefore, in the proposed system, students send tokens directly not to university but to the center and the center send the information to the university.

Figure 2 shows the flow of the tokens in the proposed system.

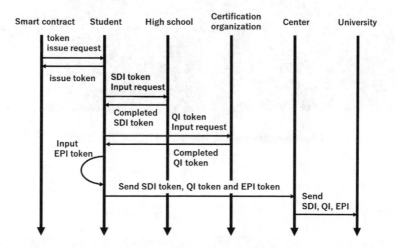

Fig. 2. Flow of tokens in the proposed system

Since this system is based on the smart contracts on blockchain, it is resistant to falsification. In addition, since the identity verification of an input person is also possible, it is more reliable than a conventional system such as signatures and seals. It is also superior from the viewpoint of data portability because students can aggregate their own data as a token issuer. Since this system is designed to include the information of e-portfolio, the cooperation with e-portfolio can be realized naturally. In addition, the information is encrypted, and only authorized users can access it. Therefore, information leakage is prevented.

2.3   Implementation of Prototype of Proposed System

We developed the system proposed in the previous section as a decentralized application on Ethereum using the contract-oriented language called Solidity [11]. We also use Ganache [12], an Ethereum blockchain for Dapps (decentralized applications) development.

We developed a private blockchain, and we confirmed that the smart contract, which registers addresses, issues tokens, and transmit, input, and inquiry for three types of tokens, functions correctly.

Figure 3 shows an example of the execution of inquiry and input functions; Fig. 4 shows an example of the execution of transmitting and inquiry functions.

Fig. 3. Example of inquiry and input functions

3   Conclusion

In this paper, we proposed the digital university admission application system with study documents and e-portfolio using smart contracts on blockchain, and we developed a prototype of the proposed system using a platform of Ethereum.

Since this system is based on the smart contracts on blockchain, it is resistant to falsification. In addition, since the identity verification of an input person is also possible, it is more reliable than a conventional system such as signatures and seals. The data portability is also realized, because students can aggregate their own data as a token issuer. The information is encrypted, and only authorized

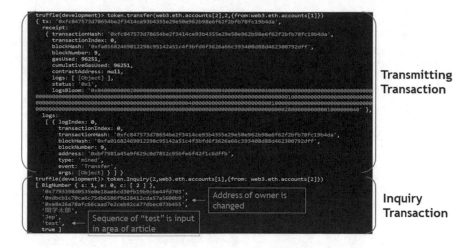

Fig. 4. Example of transmitting and inquiry functions

users can access it. Therefore, information leakage is prevented. Above all, since paper is not used, high schools can significantly reduce the costs of issuing study document and mailing costs. In universities, since all information is electronic, universities greatly reduce time and cost of manually checking and inputting data.

As a future work, operation verification and demonstration experiments in the realistic scale by large number of users in the scale of actual data item number are remained.

Acknowledgements. This work was partially supported by the Japan Society for the Promotion of Science through Grants-in-Aid for Scientific Research (B) (17H01742) and JST CREST JPMJCR1402.

References

1. Japan e-Portfolio. https://jep.jp/
2. Common Application. https://www.commonapp.org/
3. Coalition for Access, Affordability, and Success. http://www.coalitionforcollegeaccess.org/
4. Universal College Application. https://www.universalcollegeapp.com/
5. UCAS. https://www.ucas.com/
6. Szabo, N.: Smart contracts (1994). http://www.fon.hum.uva.nl/rob/Courses/InformationInSpeech/CDROM/Literature/LOTwinterschool2006/szabo.best.vwh.net/smart.contracts.html
7. Szabo, N.: The idea of smart contracts (1994). http://www.fon.hum.uva.nl/rob/Courses/InformationInSpeech/CDROM/Literature/LOTwinterschool2006/szabo.best.vwh.net/idea.html
8. Nakamoto, S.: Bitcoin: A Peer-to-Peer Electronic Cash System. https://bitcoin.org/bitcoin.pdf

 9. Ethereum. https://www.ethereum.org/
10. EIPs. https://github.com/ethereum/EIPs/blob/master/EIPS/eip-721.md
11. Solidity. https://github.com/ethereum/solidity
12. Ganache. https://truffleframework.com/ganache

Algorithm Using Deep Learning for Recognition of Japanese Historical Characters in Photo Image of Historical Book

Liao Sichao and Hiroyoshi Miwa[✉]

Graduate School of Science and Technology, Kwansei Gakuin University,
2-1 Gakuen, Sanda-shi, Hyogo, Japan
{liaosichao,miwa}@kwansei.ac.jp

Abstract. In Japan, there are vast amount of classical books written in cursive Japanese that cannot be read by modern people. It is difficult to recognize each cursive Japanese separately, because they are written connected. Furthermore, there are many types of shape of characters. Therefore, an efficient method to convert them into modern characters automatically is required. Some methods recognizing a block of a few characters using deep learning have been studied so far. However, every page in a Japanese historical book is stored by a photo image; therefore, it is desirable to recognize all characters in a photo image at once. In this paper, we propose a method using deep learning to recognize cursive Japanese in a photo image without separating a block of characters manually. Furthermore, we evaluate the performance of the proposed algorithm using photo images of an actual book.

1 Introduction

Historical documents includes important information sources over a wide range of field. There are a lot of historical documents also in Japan. Especially, in the Edo period (1603–1868), the level of education rose and the number of people who could read and write increased. In the same era, a large number of books made by woodblock printing were on the market. According to the General Catalog of National Books [1], there are 1.7 million books in total published in Japan prior to 1867. In addition, since many documents are not registered in the catalog, the number of documents is expected to be much larger.

However, since most of historical documents are written on papers, the documents face the risk of loss due to disasters and aging degradation. Therefore, the historical documents must be urgently recorded by media other than paper. At present, documents are recorded by photo image page by page.

Cursive Japanese characters used for historical documents in Japan are called Kuzushiji. The characteristics of the layout are that plural characters are connected, that the size of characters varies, that characters are not always arranged

© Springer Nature Switzerland AG 2020
L. Barolli et al. (Eds.): INCoS 2019, AISC 1035, pp. 181–189, 2020.
https://doi.org/10.1007/978-3-030-29035-1_18

in the vertical direction, and that some characters are omitted. Furthermore, there are some kinds of characters in the Japanese language; Kanji (Chinese characters), Hiragana (a kind of Japanese characters), and so on. One character of Hiragana has some different types of shape, because a type of shape of one Hiragana character was made from one Chinese character and some different Chinese characters were assigned to one Hiragana character. Since the Japanese old language has these characteristics, the special knowledge is necessary to read Kuzushiji and it is very difficult for modern Japanese to read Kuzushiji. Therefore, for modern people in order to get information and knowledge from historical documents, Kuzushiji must be converted to modern Japanese characters, that is, they must be reprinted. However, since it is almost impossible to convert a enormous large number of historical documents by only experts, a system to convert automatically is needed.

Some methods have been developed to convert Kuzushiji to modern characters so far. As a conventional automatic recognition technique of Kuzushiji [2], there is a method in which each character is manually extracted from image data, and a character is automatically recognized by using a similar image retrieval based on feature quantity from a glyph image database. However, the special knowledge is still necessary, since it is necessary to extract every character manually by experts. A method for automatically cutting out all characters from a page is not known yet. Recently, character recognition using deep learning has been successful. The open data set of Kuzushiji is published [3], and some methods for recognition of Kuzushiji have been also proposed [4,5]. However, these methods deal with the recognition of one character or successive a few characters extracted manually beforehand. Since every page in a Japanese historical book stored by a photo image, it is desirable to recognize all characters in a photo image at once.

In this paper, we propose a method using a combination of algorithms automatically extracting characters and deep learning to recognize cursive Japanese characters in the photo image of a page in historical documents without manually cutting out characters. Furthermore, we evaluate the performance of the proposed algorithm using photo images of an actual book provided in [3] and reprint sentences provided by National Institute of Japanese Literature [6].

2   Related Works

There is a method automatically cutting out a character vertical column from the photo image of a page [7]. Since documents in Japanese historical documents are written vertically, the vertical columns must be extracted above all. The outline of the method is described as follows. The image data is binarized so as to reduce the influence of background noise and dirt, then, the number of pixels darker than a threshold is counted in the vertical direction, and the set of the vertical lines including more darker pixels is extracted as a vertical column. There is also a method cutting out a block of Kuzushiji from a vertical column [7]. Since documents in Japanese historical documents are written with a brush, the line connecting characters tends to be thin. A block of characters is extracted using this property.

As a conventional automatic recognition method of Kuzushiji, there is a method in which each character is manually cut out from image data, and the character is recognized by using similar image retrieval based on feature quantity from a glyph image database [2]. Recently, there is the research on the recognition of one character of Kuzushiji using deep learning [8]. The learning data of Kuzushiji was published by the Center for Open Data in the Humanities (CODH) [3]. The data set includes 70,000 images of the top 10 characters of Hiragana (Kuzushiji-MNIST), 270,912 images of 49 characters of Hiragana, and 140,426 images of 3,832 characters of Kanji. The recognition rate of a single character of Hiragana based on the deep learning model proposed by [4] is 98.9% for the top 10 characters of Hiragana and 97.33% for all characters of Hiragana. The recognition rate of characters of Kanji is not known yet. In 2019, the data set is expanded and it includes 684,165 images of 4,645 characters of Hiragana and Kanji. As for the recognition of multiple characters of Hiragana, that is, to recognize each characters from an image of a block of successive characters, there are some results [5,9]. The recognition rate of a block including a few characters is 74% [9] by the CNN using the data set of blocks not open to the public at present. Since cost and time are needed to make the new data set of blocks in addition to the already made data set of single characters, it is desirable to design a method based on only the data set of single characters.

3   Method of Automatic Reprinting of Historical Books from Photo Images

In this section, we describe the proposed method using a combination of algorithms automatically extracting characters and deep learning to recognize Kuzushiji in the photo image of a page in historical documents without manually cutting out characters.

We show an example of the photo image of a page in Fig. 1.

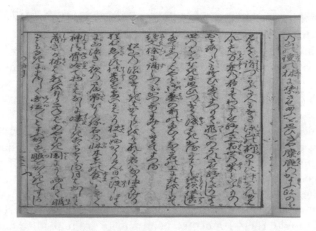

Fig. 1. Example of photo image of page

The outline of the proposed method is described as follows. First, the outline boxes in a page is removed, then the vertical columns are extracted. Next, a vertical columns are divided to a set of blocks including some characters by the decomposition to the connected components and the procedure connecting the components appropriately. The block is divided to some pieces and these pieces are recognized by a single character recognizer by deep learning. This division is iterated by changing the separator lines, and the separator lines are determined so that the probability of recognition is large. As a result, it is expected that each piece corresponds to a character. Consequently, the recognition of Kuzushiji in the photo images is achieved.

We describe the above method in detail as follows.

A provided photo image includes characters and outline boxes. First, a photo image is binarized, the maximum connected component including black pixels is extracted (Fig. 2), and the component is removed, because the connected component whose area is the maximum is the outline box. Thus, the outline box is removed.

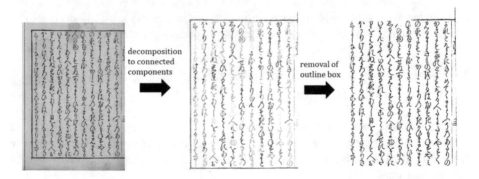

Fig. 2. Removal of outline box

Next, the vertical columns are extracted based on the method in [7]. Since Japanese historical documents are written vertically, the color of a vertical column on which characters are written is close to black and the color of the gap between two vertical columns is close to white. Therefore, the number of pixels darker than a threshold is counted in the vertical direction, and the set of the vertical lines including more darker pixels is extracted as a vertical column (Fig. 3).

Next, we describe the procedure that a vertical column is divided to blocks, each of which contains a few characters. The connected component decomposition of a figure containing darker pixels is executed (Fig. 4). If the values of the vertical coordinate of the centers of two connected components are close each other, these connected components are unified to one component (Fig. 5). If the minimum rectangle including a connected component overlaps considerably other rectangle, these connected components are unified to one component (Fig. 6).

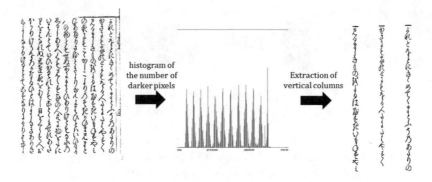

Fig. 3. Extraction of vertical columns

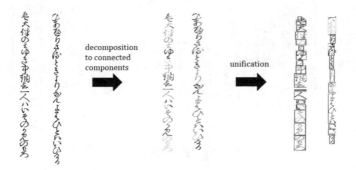

Fig. 4. Division to blocks from vertical column

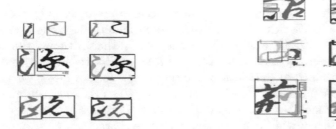

Fig. 5. Unification of connected components by closeness of vertical coordinate

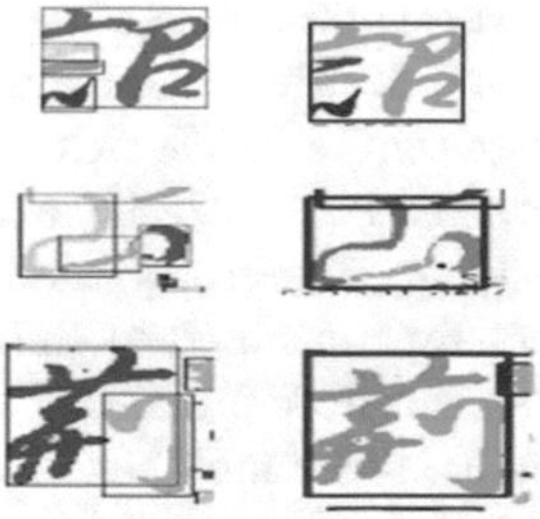

Fig. 6. Unification of connected components by overlap

Next, we make the single character recognizer by deep learning. We use the data set [3] including 684,165 images (581,540 images for learning data and 102,625 images for test data) of 4645 characters of Hiragana and Kanji. After an image is binarized based on a threshold value and dirt and noise are removed, the image is converted to the image of 64 pixels × 64 pixels by padding (Fig. 7).

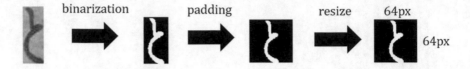

Fig. 7. Preprocessing of data set of Kuzushij

We use VGG16 [10] as the model of deep learning (Fig. 8), which is a CNN (Convolutional Neural Network) having 16 layers. The size of the input layer is 64×64, and the size of the middle layers is $4 \times 4 \times 512$.

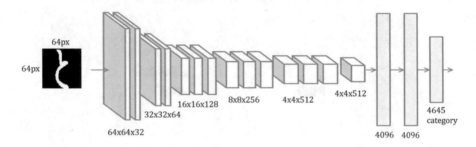

Fig. 8. VGG16

The recognition rate of this single character recognizer is 96% for 4645 characters of Hiragana and Kanji.

A block is divided to some pieces and these pieces are recognized by the single character recognizer. We describe the procedure in detail. First, the number of characters in a block is estimated based on the ratio of the width and the height of a block. In this case, when the ratio is 1.8 or more, we assume that two characters are contained in the block; when the ratio is four or more, we assume that three characters are contained in the block. When a block contains two characters, we assume that the height of the first character is 35% to 65% of the height of the block; when a block contains three characters, we assume that the height of the first character is 20% to 40% of the height of the block and that the sum of the heights of the first and the second characters is 60% to 80% of the height of the block. This division is iterated by changing the separator lines, and the separator lines are determined so that the sum of the probability of recognition by the single character recognizer is the maximum (Fig. 9).

Thus, Kuzushiji, the characters of Hiragana and Kanji, in the photo image of a page, are recognized by the above method.

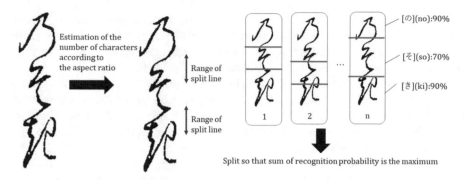

Fig. 9. Recognition of multiple characters in a block

4   Performance Evaluation

In this section, first, we evaluate the performance of the single character recognizer, and then we evaluate the performance of the proposed method.

The data used for the performance evaluation are the photo images and the reprinted characters of pages 3 to 7 in "Ugetsu-Monogatari" (1776) (Fig. 1) provided by CODH. The total number of characters on pages is 1474.

4.1   Recognition Rate of One Character

We use the data set of one characters manually cut out from the photo images of "Ugetsu-Monogatari." The data set is provided by CODH.

We evaluate the performance of the single character recognizer which learned the data set [3] including 684,165 images (581,540 images for learning data and 102,625 images for test data) of 4645 characters of Hiragana and Kanji.

As a result, 1438 characters out of total 1474 characters in "Ugetsu-Monogatari" are correctly recognized by the single character recognizer and the recognition rate is about 97%. Since the recognition rate is 96% (94% for test data) for the data set [3], the accuracy is always sufficiently high.

The recognition rate in [2] is about 80%, although the method of [2] is based on the similarity in the data set that includes the target characters. Our single character recognizer achieves higher recognition rate under the same condition. Therefore, the recognizer based on deep learning is effective also for Kuzushiji characters.

4.2   Recognition Rate of a Page

We apply the proposed method to the photo images of "Ugetsu-Monogatari." It is difficult to extract the Kuzushiji characters one by one exactly from a photo image. To evaluate the performance of the extraction and the recognition totally, we use the edit distance (Levenstein distance) as the performance measure, because the number of characters recognized by the proposed method is not always the same with that reprinted by experts. The edit distance between two words is defined as the minimum number of single character edits (insertions, deletions or substitutions) required to change one sequence into the other. In this evaluation, we used the reprinted sentences in [6].

The correct number of the characters is 1474, on the other hand, the number recognized by the proposed method is 1703, and 580 characters are correctly recognized (the recognition rate is about 39%). The edit distance is 894 (insertion of 105 characters, deletion of 334 characters, and substitution of 455 characters).

The recognition rate is about 39% while the recognition rate of the single character recognizer is about 96%; therefore, this gap is caused by the mistakes of the extraction. The edit distance of 894 in the text containing 1474 characters is large. There are three possible reasons. First, there are the duplication of vertical columns between the successive two pages as we can see in Fig. 1. This increases the number of the deletions. Therefore, the preprocessing to remove the duplication is necessary; however, it is not easy, because the duplication must be recognized automatically. Second, a character that consists of multiple components cannot be recognized, when the connected components are mutually far apart, because these components are not unified (Fig. 10). Third, the number of the characters in a block is estimated erroneously, when a vertically long or short character is included in the block.

Fig. 10. Example of failure of unification

The performance of the single character recognizer in the proposed method is sufficiently high. On the other hand, as for the recognition of the photo image of a page, the recognition rate of 39% and the edit distance of 894 in the text containing 1474 characters are insufficient to understand as sentences. Although the system based on the method can help to read Kuzushiji, it is necessary to improve the performance more.

5   Conclusions

In this paper, we proposed the method using a combination of algorithms automatically extracting characters and deep learning to recognize cursive Japanese characters, Kuzushiji, in the photo image of a page in historical documents without manually cutting out characters.

Since the previous methods that convert Kuzushiji to modern characters deal with the recognition of one character or successive a few characters, the blocks including the characters must be manually cut out to apply the methods to a historical document. Since every page in a Japanese historical document stored by a photo image, it is desirable to recognize all characters in a photo image at once.

The proposed method removes the outline boxes in a page, extracts the vertical columns, divides a vertical column to a set of blocks, and divides a block to some pieces and these pieces are recognized by the single character recognizer by deep learning.

As a result of the evaluation using a historical document "Ugetsu-Monogatari," the performance of the single character recognizer in the proposed method is sufficiently high. On the other hand, as for the recognition of the photo image of a page, the recognition rate of 39% and the edit distance of 894 in the text containing 1474 characters are insufficient to understand as sentences. The system based on the method can help to read Kuzushiji, but the performance is not sufficient as a method to reprint automatically. As the future work, the recognition rate and the edit distance must be improved.

Acknowledgements. This work was partially supported by the Japan Society for the Promotion of Science through Grants-in-Aid for Scientific Research (B) (17H01742) and JST CREST JPMJCR1402.

References

1. Shoten, I.: General Catalog of National Books. Iwanami Shoten, Chiyoda (2002)
2. Yamamoto, S., Osawa, T.: Labor saving for reprinting Japanese rare classical books: the development of the new method for OCR technology including Kana and Kanji characters in cursive style. Inf. Manag. **58**(11), 819–826 (2016)
3. The Dataset of Kuzushiji (Open Data Center for Humanities). http://codh.rois.ac.jp/pmjt
4. Tarin, C., Mikel, B., Asanobu, K., Alex, L., Kazuaki, Y., David, H.: Deep learning for classical Japanese literature. In: Proceedings of 2018 Workshop on Machine Learning for Creativity and Design (Thirty-second Conference on Neural Information Processing Systems), 3 December 2018
5. 21st PRUM Algorithm Contest. https://sites.google.com/view/alcon2017/prmu
6. National Institute of Japanese Literature. http://base1.nijl.ac.jp/infolib/metapub/CsvSearch.cg
7. Shibayama, M., et al.: Research on Higher Accuracy Document Character Recognition Systems (in Japanese). Grants-in-Aid for Scientific Research (B)(1) report on research results, pp. 33–49 (2005)
8. Hayasaka, T., Ohno, W., Kato, Y., Yamamoto, K.: Recognition of Hentaigana by deep learning and trial production of WWW application (in Japanese). In: Proceedings of IPSJ Symposium of Humanities and Computer Symposium, pp. 7–12 (2016)
9. Nguyen, H.T., Ly, N.T., Nguyen, K.C., Nguyen, C.T., Nakagawa, M.: Attempts to recognize anomalously deformed Kana in Japanese historical documents. In: Proceedings of 4th International Workshop on Historical Document Imaging and Processing, pp. 31–36, 10–11 November 2017
10. Simonyan, K., Zisserman, A.: Very deep convolutional networks for large-scale image recognition. In: Proceedings of International Conference on Learning Representations, San Diego, 7–9 May 2015

Proposing a Blockchain-Based Open Data Platform and Its Decentralized Oracle

Akihiro Fujihara[(✉)]

Chiba Institute of Technology, 2-17-1 Tsudanuma, Narashino, Chiba 275-0016, Japan
`akihiro.fujihara@p.chibakoudai.jp`

Abstract. Bitcoin has been attracted attention to issue digital currency and also to manage transactions in a decentralized and trustless way without assuming any reliable third-party organization. To enhance the potential of Bitcoin, blockchain technologies have been studied to find more applications to smart contract, identification, and certificate which is valid even after the original issuer disappears. On the other hand, mobile crowdsensing is known as a method to collectively gather and share local data with the help of many and unspecified participants having their portable devices with sensors like smartphones to extract useful information from them. Expectation-Maximization (EM) algorithm is considered to estimate true information from task results from many independent workers and to evaluate workers' scores. By the combination of blockchain and mobile crowdsensing technologies, a open data platform to save true information into the blockchain and issue currency to workers based on their reliability scores. In this research, we explain how to create this open data platform to generate Big and Long-time Open Data (BaLOD) to certify events that occurred in reality and everyone can validate the events.

1 Introduction

More than ten years have passed since a peer-to-peer electronic cash system known as Bitcoin came along on November 1, 2008 (Japan Standard Time) [1]. Before Bitcoin, it was considered for a long time that placing a credible third-party organization as a centralized reliability base, such as governments or banks, is necessary for the system to control, for example, issuing currency and certifying transactions. After Bitcoin, however, it has been widely recognized that the system can be managed by its anonymous users in a decentralized and trustless way without relying on any credible third-party organization, meaning that many and unspecified nodes on a P2P network can manage and maintain its own open database consistently by recording and sharing information about issuing currency and certifying transactions according to a rule. This claim has been proved by the fact that Bitcoin works as a electronic cash system in reality.

Nowadays, many researchers have been seeking other applications or use cases to extend the technology used in Bitcoin to solve other problems by managing open and reliable database in a decentralized and trustless way. Studies on public

© Springer Nature Switzerland AG 2020
L. Barolli et al. (Eds.): INCoS 2019, AISC 1035, pp. 190–201, 2020.
https://doi.org/10.1007/978-3-030-29035-1_19

blockchain technologies are related to them. The word "Blockchain" was coined from its data structure: Block and chain. In a block, data, such as transactions, are saved. Every block in a blockchain is linked to its previous one to form a structure like a chain by including the digest of the previous block calculated by cryptographic hash function into the block. In general, blockchain have some mechanism to protect data from tampering. One of famous mechanisms is Proof of Work (PoW), which is a high-energy-consuming calculation done by a node called a miner to find a nonce to make the digest of block is less than a target. The result of PoW can easily be validated by other nodes, thus it is difficult for malicious nodes to tamper data in the blockchain. The blockchain constantly takes in a new block created by a miner with passive and emergent consensus between all the other nodes on the P2P network, which is totally different from the sense of consensus in distributed computing. As a result, every block including data is arranged in chronological order. Therefore, blockchain is considered to be a timestamp system to record legitimate history of a wide range of data. In this context, some applications using blockchain technologies are expected, such as automatic execution of contracts called smart contract [2], identification and certificate that work even if their issuers disappeared.

However, most blockchain applications that have been considered are not successful to become in practical use because there are many problems in blockchain technologies. The biggest reason to narrow the range of applications is that blockchain usually requires high publicity to the recorded data. In Bitcoin, for example, all the transactions approved to be saved into the blockchain are open to the public as open data, thus anyone can check to validate the transactions. If an application does not require this high publicity, it is meaningless to utilize blockchain technologies. This also indicates that if we deal with data that privacy matters, we should not use blockchain technologies, but it is enough to use existing distributed database technologies. Furthermore, when we consider to use blockchain for business, we must give up centralization which is the source of huge profits to the centralized company based on its reliability to gather and use data by itself.

Another problem in blockchain technologies is that the system does not have any method to judge truth or falsity of data taken into the blockchain. To remove false information, it is necessary to introduce the source of true information called *Oracle*. To do this, some reliable third-party organization is usually assumed, thus it is difficult to save only true information into the blockchain in a decentralized way in general.

On the other hand, crowdsourcing is known as a method to offer jobs or tasks to many and unspecified persons through the Internet. This method works well when you offer tasks that is easily done by humans but it is not by computers. It also works when you request a survey to distant places where you have difficulty to come by yourself. The requester of jobs and tasks usually pay money to the workers as rewards. In these contexts, crowdsourcing is widely used in practice. To evaluate the quality of workers, the system prepares some scoring mechanism based on the history of jobs and tasks done by workers in the past.

Mobile crowdsensing is a crowdsourcing using mobile devices with sensors like smartphones to gather and share data. There are multiple mobile crowdsensing projects to gather traffic information [3–5]. If gathered data are saved in the blockchain, the traffic information becomes existence of various kinds of vehicles and also it is possible to work as a part of certificates of delay in public transportation systems. This means that traffic information that is provided by public transportation companies at a cost in the past can also be provided by general citizens using blockchain technologies. This might possibly become a platform to save highly public data as Big and Long Open Data (BaLOD) to certify true information that occurs in reality.

However, there are three problems to realize this platform. (1) The capacity of auxiliary storage device grows rapidly to write a large amount of data into the blockchain. (2) Because blockchains take a long time to fix a legitimate history of data, they cannot save data in real time. (3) There is no mechanism to eliminate false information. About the problem (3), especially, it is vulnerable to attack by intentional spreading of false information to confuse users of the platform.

In this paper, we propose a method for nodes on P2P network to manage blockchains of each geographically close-range area called domain. Because the blockchain is created in every domain, information related to its domain is distributed to each domain. Therefore, it is possible to suppress the rapid increase of data storage capacity. It depends on the application, but it also avoid rapid increase of the capacity by forgetting data in a distant past from the blockchain. Also, since nodes geographically close to each other, the delay of transferring data and blocks can be reduced, which improves to save data into the blockchain much faster. In the field of crowdsourcing, Expectation-Maximization (EM) algorithm to estimate the correct information and to score workers' reliability using their reports is known. If we include the calculation by this algorithm is included in the PoW process, it automatically eliminate false information under a certain condition that workers do tasks independently and most of them submit the correct information.

2   Related Works

Crowdsourcing and crowdsensing using blockchain technologies have been proposed [6,7]. In these proposals, it is considered how to write data in the block and make payments to workers using cryptocurrency. In order to eliminate false information, a role called validator is introduced. There is a method to exchange gathered data by workers and rewards for the workers automatically using smart contract.

Decker and Wattenhofer investigated how long it takes to spread mined blocks across a wide area on Bitcoin's P2P network [9–11]. According to their report, the proportional relation between the block size b [kB] and the latency time that blocks are spread to 50% of the total nodes $D_{50\%}$ [sec.] is satisfied.

$$D_{50\%}(b) = 1.80 + 0.066 \times b. \tag{1}$$

Since the upper bound of Bitcoin's block size is about 1 MB, the latency time is 67.8 [sec.] at maximum.

Sompolinsky *et al.* consider a method that blocks are efficiently included into the blockchain [10,11]. The more the difficulty of PoW decreases, the faster blocks can be created. But, the fork of blockchain occurs more frequently. Therefore, many useless blocks that are not taken in the blockchain are created, which degrades the efficiency of creating blocks. To solve this problem, Ethereum blockchain is allowed to take in uncle blocks in addition to the parent block. This rule increases the number of transactions taken in the blockchain, and also the block creation time is shorten to 15 min [8].

Applications of Blockchain technologies to Internet of Things (IoT) are also considered [12]. By this background, a system for traffic information gathering and sharing using blockchain technologies has been proposed [13,14]. In this system, multiple beacons devices are deployed along the road side in each road segments and the devices detect traffic information, such as congestion and road closure, using Wi-Fi radio wave emitted from vehicles going through the segment to record the data in blocks. They also manage blockchain by taking in the blocks. Therefore, the system enables to record traffic information into the blockchain in quasi-real time. Furthermore, the system also issue coinbase to all the nodes joining to create the data, such as vehicles and road side beacon devices as their incentive. When traffic congestion and road closure occur, the blockchain is forked in general. For this reason, the incentive mechanism can also contribute to a behavior change that vehicles reroute to avoid congested roads and road closures to earn more cryptocurrency.

Disaster evacuation guidance using opportunistic networking has been proposed as a disaster evacuation method for disaster victims evacuating to nearby refuges collaboratively with the help of human-carried mobile devices [15–19]. Evacuees start evacuation along a shortest-path-based evacuation route guided by their devices. If a evacuee encounters a road closure by fire, building collapse, congestion, and others, the evacuee records the road status manually or automatically using sensors in the device to recalculates an alternative route. When a couple of evacuees happens to meet with each other, their devices share the disaster information using opportunistic networking to change the path using shared information. Average and maximum evacuation times are evaluated by numerical simulations. As a result, it shows that the guidance contributes to decrease the evacuation times. This method is a kind of crowdsourcing of disaster information between evacuees. However, it is known that at the time of disaster, misinformation or so-called fake news can be easily spread between evacuees. The proposed system never guarantees that gathered disaster information is really true, and when the information is outdated. This problem leads to serious confusion of evacuees to avoids disaster evacuation.

3   Blockchain-Based Open Data Platform

3.1   About Block Generation Time

Because creating a block by PoW is governed by a stochastic process that a miner explore a correct nonce by trying different nonce repeatedly. Therefore, the latency time that a block is created by a miner is described by a stochastic variable T'. The latency time that the next block is created and fixed in the blockchain is designed to takes $\lambda^{-1} = (h/d)^{-1} = 10 + \tau_d$ [min.] on average, where h [min.$^{-1}$] is the maximum hash power of nodes, d is the difficulty of PoW, and τ_d is the average time that blocks are spread to most nodes on P2P network. Therefore, the probability distribution function for the latency time $T' = t$ obeys the exponential distribution with the average λ^{-1}.

$$f_1(t) = \lambda \exp(-\lambda t) \qquad (t \geq 0). \tag{2}$$

Moreover, the latency time that the latest k–blocks are created and fixed in the blockchain is given by $T = \sum_{i=1}^{k} T_i$, where T_i $(i = 1, \cdots, k)$ is the latency time that i−th block in k blocks is created and fixed in the blockchain. When we assume that each T_i is independent of others, the latency time $T = t$ is described by k–convolutions of the exponential distribution, meaning that it obeys the Erlang distribution with the shape parameter k.

$$f_k(t) = \frac{\lambda^k t^{k-1}}{(k-1)!} \exp(-\lambda t). \tag{3}$$

These probability distribution functions are supported by real data of block fixation times in Bitcoin blockchain. The period of gathering the real data is about two month from May to June in 2019, and they are compared with the above theoretical results, which are shown in Fig. 1. The agreement between real data and theory can be confirmed when the fitting parameter is $\lambda^{-1} = 10.29$ [min.], which indicates that the block spread time can be estimated as $\tau_d = 0.29$ [min.]. In addition, the average and standard deviation of the Erlang distribution with the shape parameter k are k/λ and $\sqrt{k}/\lambda$, respectively. Therefore the coefficient of variance which is defined by the standard deviation divided by the average is $1/\sqrt{k}$. This result indicates that the more the block height in the blockchain increases, the lower the variance of the block fixation time becomes. Thus, in the long-time period, the block height grows almost constantly with elapsed time, which is why blockchains works as the database with electric time stamp.

3.2   Domains and Proof of Work at Proximity (PoWaP)

In the proposal, a geographically independent area in which nodes manage its own blockchain is introduced. The area is called a *domain*. In each domain, core nodes are miners to create blocks and they also manages the blockchain. A central core node is also introduced as the open entrance of the domain. When a

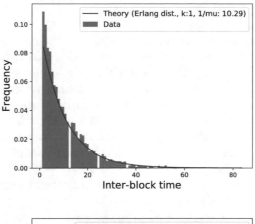

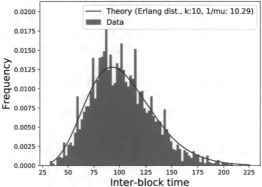

Fig. 1. Histograms of block fixation times (Blue) and probability distribution functions predicted by theoretical results (red). The latency time of $k = 1$ block fixation obeys the exponential distribution (top) and that of $k = 10$ block fixation almost obeys the Erlang distribution (bottom).

core node want to participate in the domain, it firstly connect to the central core node to download and transfer some blocks within a given short time period (for instance, within two seconds). In this manner, Each domain has its rule to check if a new core node can transfer data as fast as nodes in the domain. Therefore, core nodes geographically close with each other naturally construct the domain. If core nodes in the domain has their geographical coordinates, they share the coordinates to have an approximate coordinates of the domain by averaging the coordinates.

Decker and Wattenhofer investigated the latency time that a block spread across the P2P network. As a result, they showed that the latency time is approximately 12.6 s on average [9]. Note that this latency time is close to $\tau_d = 0.29 * 60 = 17.4$ [sec].

On the other hand, a measure to estimate how efficient the system can manage the blockchain is defined [8]. This efficiency E is defined as follows.

$$E = V/(V + I), \tag{4}$$

where V is the number of blocks taken in the blockchain and I is the number of blocks that finally are not taken in the blockchain. It is known that V is proportional to the mining time by PoW and I is proportional to the block transfer time. In our real data from Bitcoin blockchain, for example, the efficiency becomes $E = 10/10.29 \approx 0.97$. This result indicates that by speeding up block propagation time, the efficiency E keeps high and the mining time can shorten to create blocks rapidly. For instance, when the block propagation time is two seconds on average and the efficiency keeps high at $E = 0.97$, then block creation time must be $V = EI/(1 - E) = (0.97 * 2.0/(1 - 0.97))/60 \approx 1.08$ min. If conventional six-block confirmation is needed to fix the blockchain, it is estimated that about six and a half minutes is enough.

Workers of mobile crowdsensing is joined as a client node (or edge node) of the core nodes in the domain. Client nodes access core nodes using close-range wireless communication, such as Wi-Fi or Bluetooth, to transfer gathered data. Core nodes share the data with each other on the P2P network. By saving the Merkle root of the digests of Gathered data in a block header, the blockchain managed in the domain validate the existence of the data. The proposed method of efficient and rapid PoW process to create blocks to manage the blockchain is called Proof of Work at Proximity (PoWaP) [13,14].

3.3   Decentralized Oracle (DO) to Openly Estimate True Information

In the field of crowdsourcing, quality control techniques have been investigated. Unless the assumption that a large majority of workers do tasks independently to report the correct results satisfies, it is possible to enhance recording true information into the blockchain using Expectation-Maximization (EM) algorithm. This algorithm enables not only to avoid false information, but also to evaluate reliability scores to each worker.

Here, for simplicity, we consider the case that workers' tasks are reported as binary values (Yes or No). It is known that the following method we are going to explain is easily extended to multi-value answers. We assume that N tasks are prepared by a task requester and J workers to report the tasks are found. The set of tasks don by worker j is described by $I_j \subseteq \{1, \ldots, N\}$, and the set of workers who do task i is described by $J_i \subseteq \{1, \ldots, J\}$. Also, the correct answer of task i is represented by $t_i \in \{0, 1\}$ and $y_{ij} \in \{0, 1\}$ is the answer of task i by worker j. Here, we consider the following two types of conditional probabilities.

$$\alpha_1^{(j)} = p(y_{ij} = 1 | t_i = 1), \tag{5}$$

$$\alpha_0^{(j)} = p(y_{ij} = 1 | t_i = 0), \tag{6}$$

What we are going to estimate is these conditional probabilities and the expected value of the correct answer $E_i(= p(t_i = 1))$ for each task i. For the probability that the correct answer for task i is 1, $p(t_i = 1) = q_i$ $(1 \leq i \leq N)$, and the above conditional probabilities, $\alpha_1^{(j)}$ and $\alpha_0^{(j)}$ $(1 \leq j \leq J)$, these probabilities are initially estimated as follows.

$$q_i = \sum_{j=1}^{J} y_{ij}/J, \tag{7}$$

$$\alpha_1^{(j)} = \alpha_0^{(j)} = \sum_{i=1}^{N} y_{ij}/N. \tag{8}$$

Under these initial conditions, EM algorithm is applied to calculate values shown in the following two steps called E step and M step one after the other.

E step

We assume that workers independently answer $\{y_{ij}\}$ for $j \in J$. Then, using Bayes' theorem, E_i is calculated for all the tasks $1 \leq i \leq N$ as follows.

$$E_i = \prod_{j \in J} p(t_i = 1|y_{ij}) = \frac{q_i \prod_{j \in J} p(y_{ij}|t_i = 1)}{Z_i}, \tag{9}$$

$$Z_i = \prod_{j \in J} p(y_{ij}) = q_i \prod_{j \in J} p(y_{ij}|t_i = 1) + (1 - q_i) \prod_{j \in J} p(y_{ij}|t_i = 0),$$

where

$$p(y_{ij}|t_i = 1) = (\alpha_1^{(j)})^{y_{ij}} (1 - \alpha_1^{(j)})^{1-y_{ij}}, \tag{10}$$

$$p(y_{ij}|t_i = 0) = (\alpha_0^{(j)})^{y_{ij}} (1 - \alpha_0^{(j)})^{1-y_{ij}}. \tag{11}$$

M step

Using the results E_i calculated in E step, the probability that the correct answer of task i is 1 can be estimated as follows.

$$\hat{q} = \sum_{i=1}^{N} E_i/N, \tag{12}$$

for $1 \leq i \leq N$. Also, the conditional probabilities $\alpha_1^{(j)}$ and $\alpha_0^{(j)}$ are also estimated.

$$\hat{\alpha}_1^{(j)} = \frac{\sum_{i=1}^{N} E_i y_{ij}}{\sum_{i=1}^{N} E_i}, \tag{13}$$

$$\hat{\alpha}_0^{(j)} = \frac{\sum_{i=1}^{N} (1 - E_i) y_{ij}}{\sum_{i=1}^{N} (1 - E_i)}, \tag{14}$$

for $1 \leq j \leq J$.

By repeating E step and M step one after the other, estimated values $\hat{q}(=\hat{E}_i)$, $\hat{\alpha}_1^{(j)}$, and $\hat{\alpha}_0^{(j)}$ gradually converge to fixed values. The converged value $\hat{E}_i \geq$ 0.5, the algorithm decides $t_i = 1$ (the correct answer of task i is 1), otherwise $t_i = 0$. By this machine learning algorithm, true information is determined without assuming any reliable authority and it is recorded in the blockchain automatically. This method with open data plus open algorithm that anyone can validate the whole process of decision to decide why data is true information is essential to realize the open data platform. We call this way of information processing *Decentralized Oracle* (DO).

3.4   Estimation of Users' Reliability and Incentive Design

Using the converged values of conditional probabilities $\hat{\alpha}_1^{(j)}$ and $\hat{\alpha}_0^{(j)}$ by EM algorithm, workers' reliability scores can be calculated as follows.

$$\tilde{S}_j = (\alpha_1^{(j)} - \alpha_0^{(j)})^2. \tag{15}$$

Using these scores, we also consider the incentive design of the proposed system. As a naive incentive, worker j obtains his or her reward R_j which is proportional to the score $\tilde{S}_j$, for example,

$$R_j = B \frac{\tilde{S}_j}{\sum_{j=1}^{J} \tilde{S}_j}. \tag{16}$$

Here, the miner who create a block and all the workers who contribute to gather data taken in the block are rewarded using block's coinbase. The total amount of coinbase is issued as $2B + B_{min} = B_1 + B_2$ and $B \propto N$. $B_1(= B_{min} + B)$ is the reward for the miner, where B_{min} is the minimum amount of coinbase that are paid to the miner. $B_2(= B)$ is the total reward paid to all the workers proportional to their reliability scores. In this system, not only the miner get the reward, but also the workers who contribute to gather data get the reward. Therefore, it is expected that the workers will gather and share true information that most honest client nodes can possibly obtain. In Bitcoin blockchain, some miners create blocks with no transaction to obtain only coinbase reward, but they do not expect charges from users who create transactions. This is a self-ish behavior to have trouble to avoid executing transactions. By the proposed incentive design, the miner get rewarded more if transactions are taken in the block more, which avoid the trouble well.

There is a method to attack blockchains called Sybil attack. In this attack, multiple nodes behave differently but one person controls all the nodes to distort the assumption that all the workers report tasks independently to estimate true information by EM algorithm. If the controlled nodes gain the majority, they can easily attack with spreading false information. We must consider some method to avoid this for future work.

3.5   Security

There is an empirical rule known as *blockchain's trilemma*. This is the claim that it is impossible to satisfy all the following three properties about blockchains: (1) Decentralization, (2) Scalability, and (3) Security. In this viewpoint, our proposal is to weaken the conditions (1) and (3) to strengthen the condition (2). Because the number of nodes for each domain decreases in general, the security of tamper tolerance is degraded.

There is a scenario to attack blockchains called *selfish mining* [20]. This attack is done by a miner who does not instantly propagate mined blocks to other nodes, but propagate them after their cryptocurrency is used for double spending. This attack is possible if the miner has a vast amount of computational resources and it is observed in Monacoin and Ethereum Classic [21].

The proposed system do not deal with transactions to save them in the blockchain. But, it deals only with coinbase as the accumulation of reliability scores for miners and workers. Therefore, miners does not have to do double spending attack, but they just create the next block to gain reliability more. Therefore, node's reliability is the asset in this platform.

Blockchain systems commonly have the problem to save blockchain data which constantly grow. To realize the open data platform using blockchain, the problem becomes harder because blocks possibly contains a vast amount of open data in the blockchain to save all of them into auxiliary storage. To avoid this problem, we propose to remove sufficiently old blocks. For example, when we consider a blockchain that grows with one block with size $2\,MB$ on average to create within one minutes, the block height to contain data obtained in one year is about $60*24*365 = 525600$ and the storage size is about $2*10^{-6}*60*24*365 \approx 1.05$ TB. It depends on applications, but the necessary storage size can be easily estimated. Thus, blockchains as the platform for saving open data is possible to properly estimate the storage size.

3.6   Privacy

In some data, such as location information, privacy matters when the data is linked to some attributes of human and it collects data of human mobility traces [22]. It is reported that human mobility patterns are highly predictable when the trace data is used. To avoid these privacy problems, ID of trace data is changed within one day. Data recorded in the blockchain will be open even if it is encrypted because every cryptographic technology will be vulnerable soon or later when the computer evolves to calculate faster enough to break encryption. For this reason, it is very important to consider how to separate public data to save in the blockchain from personal data to avoid saving for privacy protection.

4   Summary

In this paper, a blockchain-based open data platform and its decentralized oracle are proposed. Geographically close areas called domain is defined and core nodes in the domain do PoW to create blocks, which is called PoWaP. Client nodes do mobile crowdsensing to report tasks as workers to gather reported data to be taken in the blockchain. EM algorithm is used to automatically estimate true information from workers' reports and workers' reliability scores as DO. As the incentive design, each block issues coinbase to distribute rewards as cryptocurrency that is proportional to their reliability scores estimated by EM algorithm to both miners and workers. To avoid double spending attack, the system avoids dealing with transactions to exchange cryptocurrency between users, but they accumulate coinbase rewards as the accumulated reliability scores. This opens up the possibility to realize a blockchain-based open data platform, for example, the application to the certification issuing system for delays in public transportation systems.

For future work, it is important to implement the programs that run as core and client nodes to collaboratively form a P2P network as the open data platform to create blockchains. In addition, false information or fake news is a hot topic for SNS users to confuse which information is true or not. This problem is just a matter of information literacy skills for each user. But, if there exists a platform that data and algorithms are open in blockchains and they are distributed around the world for a fingerprint of facts with temper tolerance, we can validate suspicious information and eliminate false information by themselves.

Acknowledgement. This work was partially supported by the Japan Society for the Promotion of Science (JSPS) through KAKENHI (Grants-in-Aid for Scientific Research) Grant Numbers 17K00141 and 17H01742. The author thanks Dr. Shigeichiro Yamasaki and Dr. Hitoshi Okada for useful discussion and comments on the proposal.

References

1. Nakamoto, S.: Bitcoin: A Peer-to-Peer Electronic Cash System (2008). https://satoshi.nakamotoinstitute.org/emails/cryptography/1/#selection-19.14-63.19, https://bitcoin.org/bitcoin.pdf
2. A Next-Generation Smart Contract and Decentralized Application Platform (White paper). https://github.com/ethereum/wiki/wiki/White-Paper
3. Waze, Free Community-based GPS, Maps & Traffic Navigation App. https://www.waze.com/
4. Flight Tracker—Flightradar24—Track Planes in Real-time. https://www.flightradar24.com/
5. Fix My Street. https://www.fixmystreet.jp/
6. Wang, J., et al.: A blockchain based privacy-preserving incentive mechanism in crowdsensing applications. IEEE Access **6**, 17545–17556 (2017)
7. Li, M., et al.: CrowdBC: a blockchain-based decentralized framework for crowdsourcing. IEEE Trans. Parallel Distrib. Syst. 1–15 (2018). https://ieeexplore.ieee.org/document/8540048
8. Buterin, V.: Toward a 12-second Block Time. Ethereum Blog (2014). https://blog.ethereum.org/2014/07/11/toward-a-12-second-block-time/

9. Decker, C., Wattenhofer, R.: Information propagation in the Bitcoin network. In: IEEE P2P 2013 Proceedings (2013). https://ieeexplore.ieee.org/abstract/document/6688704

10. Sompolinsky, Y., Zohar, A.: Accelerating Bitcoin's Transaction Processing. Fast Money Grows on Trees, Not Chains (2013). https://pdfs.semanticscholar.org/4016/80ef12c04c247c50737b9114c169c660aab9.pdf

11. Sompolinsky, Y., Zohar, A.: Secure high-rate transaction processing in bitcoin. In: Financial Cryptography and Data Security, International Conference on Financial Cryptography and Data Security, FC 2015, pp. 507–527 (2015). https://eprint.iacr.org/2013/881.pdf

12. Fernández-Caramés, T.M., Fraga-Lamas, P.: A review on the use of blockchain for the internet of things. IEEE Access **6**, 32979–33001 (2018)

13. Fujihara, A.: Proposing a system for collaborative traffic information gathering and sharing incentivized by blockchain technology. In: The 10th International Conference on Intelligent Networking and Collaborative Systems (INCoS-2018), pp. 170–182 (2018). https://link.springer.com/chapter/10.1007/978-3-319-98557-2_16

14. Fujihara, A.: PoWaP: Proof of Work at Proximity for a crowdsensing system for collaborative traffic information gathering. Internet of Things, Elsever (2019)

15. Fujihara, A., Miwa, H.: Real-time disaster evacuation guidance using opportunistic communication. In: IEEE/IPSJ-SAINT 2012, pp. 326–331 (2012)

16. Fujihara, A., Miwa, H.: Effect of traffic volume in real-time disaster evacuation guidance using opportunistic communications. In: IEEE-INCoS 2012, pp. 457–462 (2012)

17. Fujihara, A., Miwa, H.: On the use of congestion information for rerouting in the disaster evacuation guidance using opportunistic communication. In: IEEE 37th Annual Computer Software and Applications Conference Workshop, ADMNET 2013, pp. 563–568 (2013)

18. Fujihara, A., Miwa, H.: Disaster evacuation guidance using opportunistic communication: the potential for opportunity-based service. In: Big Data and Internet of Things: A Roadmap for Smart Environments, Studies in Computational Intelligence, vol. 546, pp. 425–446 (2014)

19. Fujihara, A., Miwa, H.: Necessary condition for self-organized follow-me evacuation guidance using opportunistic networking. In: Intelligent Networking and Collaborative Systems (INCoS), pp. 213–220 (2014)

20. Eyal, I., Sirer, E.G.: Majority is not enough: bitcoin mining is vulnerable. Commun. ACM **61**(7), 95–102 (2018). https://dl.acm.org/citation.cfm?id=3212998

21. Peck, M.: Let's destroy Bitcoin. MIT Technol. Rev. (2018). https://www.technologyreview.com/s/610809/lets-destroy-bitcoin/

22. Divanis, A.G., Bettini, C. (eds.) Handbook of Mobile Data Privacy, Springer, Heidelberg (2018)

Visualization of Users Raking in Online Dating Service

Jana Nowaková[1(✉)], Martin Hasal[2], and Václav Snášel[1]

[1] Department of Computer Science, VŠB – Technical University of Ostrava,
Ostrava, Poruba, Czech Republic
{jana.nowakova,vaclav.snasel}@vsb.cz
[2] IT4Innovations, VŠB – Technical University of Ostrava,
Ostrava, Poruba, Czech Republic
martin.hasal@vsb.cz

Abstract. Data visualization represents an important tool in data analysis. In this paper, we focus on the visualization of data from online dating service - *Libimseti*. *Libimseti* is a rarity among online dating services because it offers a user's raking system. It means that every user can evaluate other users on the scale from one to ten.

We present a modification of a circular plot to represent general data from the raking system. The introduced visualization provides both aesthetically pleasing and readable graph. It shows a general overview of people's behavior, opinion, and personal preferences in the given service without deeper understating of the data.

It helps to discover users who can suffer from other users negativity or users who provoke others and decrease the popularity of the web itself. Moreover, the presented graph can help to find influencers, who can popularize the website among its users.

Keywords: Data visualization · Online dating service · Rating · Anomaly detection

1 General Introduction

Social networks are the phenomena of the 21st century. Social Networking Sites (SNSs) like Facebook, Twitter, LinkedIn, etc. grow rapidly in the past time. Also, a quickly growing social platform represents an online dating service. Many users appreciate that they are bringing together people according to their preferences, their hobbies, their jobs, or even their problems. Since seeker knows his/her expectations, he/she can seek out a soulmate over a lot of profiles. But many dating services come with a condition which increases the number of users - if you want to look at other people profile, you must have your profile too.

Here come the issues. The more information (like real name, hobbies, photos, etc.) members provide the higher chance is to find an ideal partner. But it also allows the other people to evaluate your profile. Users can lie in their profile, but

© Springer Nature Switzerland AG 2020
L. Barolli et al. (Eds.): INCoS 2019, AISC 1035, pp. 202–211, 2020.
https://doi.org/10.1007/978-3-030-29035-1_20

in the end, they face the true on real meeting with counterpart. Overall, people lie in their profiles, naturally because their self-rating is higher due to living in self-deceptive; or intentionally in the hopes of looking better to potential partners.

In this paper, we introduce a new visualization technique for visualizing of the data with the rating system. We apply this technique on the real dataset from a real online dating service – *Libimseti*. All the data is anonymized, and cannot be connected with the real-life users. *Libimseti* is the classic platform providing profile creation, chat groups, games, videos, etc. Moreover, it has a rank (from one-the worst to ten-the best) system by which users can evaluate the attractiveness of other users. This makes this web unique among other online dating service web pages [11].

We suggest a new circular type visualization for the raking system, which gives an overview of all users of any platform. It visualizes average marks (from one to ten) and a total number of evaluations in one graph. Presented type of visualization gives an immediate image of the whole dataset, i.e., users behavior, distribution of people evaluations, and density of raking over the raking interval. The secondary outcome of the presented method is critical anomaly detection, i.e., people under Cyber-Bullying, people presenting offensive content, or people who provoke others (destroying the experience of other users). The last two groups are a threat only for web owner, but the users from the first group can be under a lot of stress from the negativity of other individuals.

The remainder of this paper is organized as follows. Section 2 summarizes related work in the presented topic of ranking visualization. Section 3 outlines the proposed approach to visual data exploration by the special circular drawing of raking distributions. Section 4 describes the dataset from the real online dating service. Section 5 illustrates the ability of circular drawing to produce an overview of dataset and the work is concluded in Sect. 6.

2   Related Work

Data visualization is a broad topic; for a general overview, we recommend books by Chen et al. [4,5]. Despite the popularity of online dating, the area of visualization of data from online dating service is quite limited, probably due to the keeping of secrets about user behavior by web owners. So from limited sources of related work, we would like to mention the following ones. The visual representations of people on 39 dating sites intended for the older population data from online dating services based on age are presented by Gewirtz-Meydan and Ayalon in [7]. There are also a patented method by Buyukkokten [3] and Leonard [10] for rating and matching associated members in a social network.

Study about users self-presentation in online dating profiles is produced by Toma et al. in [13]. In this work information about participants' physical attributes was collected and compared with their online profile. The results shows that deviations tended to be ubiquitous but small in magnitude. It was discovered, that participants' self-ratings were intentional different than self-deceptive.

The same self-improving of user's profile can be expected in many profiles from our study.

The topic of threats on social networks is presented by Al Hasib in [1]; some of the threats can be discovered by presented visualization. Methods from Rege [12] and Yang [15] shows the importance of finding dating sites validate of online services, where users can be moreless safe. Another approach is to use some statistical methods, for more details see Fiore [6]. Even patented methods can be found in the area of raking in online dating services [8].

3   Methodology of Visualization

There are many methods which measure rating and ranking, for more details see [9]. Mostly they produce some sorted list of methodologically rated elements, or numbers representing the weights of the rated elements. Such approaches are perfect for an educated audience in the given field, which can further operate with produced results. On the other hand, the complexity of results makes it difficult to read for the general public.

Visualization of the data, which is aesthetically pleasing, easy-to-read, and concerning all necessary information, is often more acceptable for the general public than tables with numbers. We suggest visualizing big data at once in the pie chart (circle) graph type. The circle is divided into ten sectors by to the rank system from one (the worst) to ten (the best), see Fig. 1. All sectors are divided into several segments by the distance from the circle center. This division is adjusted by the difference between the minimal and maximal number of ranked users.

Every user represents one point in the circular plot, where angle (θ) gives the average raking of given user and the radial distance from the center (r) shows the total number of votes for given user. The position of every user in the circle plot is calculated as

$$\theta = \frac{2}{10}\pi r_{ave}, \tag{1}$$

$$r = \log_{10}(r_{multi}), \tag{2}$$

where r_{ave} is the average mark from all evaluation of given user, and r_{multi} is total number of all evaluations. Figure 1 shows the visualization of the first ten evaluated users from the *Libimseti* dataset, for more details see Sect. 4.

It is evident that there is a big difference between a total number of raking among different people. This difference can be solved either by adding more scaling circles and increase radius for readability, but it can lead to large graphs with many radial axes, or using the logarithmic scale with the base of ten (this approach is used in this paper, see Eq. (2)).

The advantage of the presented visualization shows the distribution of users raking; if the appropriate area of points is used. In this article, the surface of all points grows with (2) multiplied by 0.8 quotient. This fitted parameter ensures the visibility of points and distribution. Smaller parameter decreases the size of

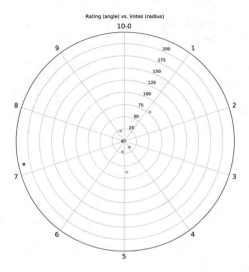

Fig. 1. Visualization of the first ten user's rank from online dating service platform. Every point represents one user. Radial coordinate (distance from a center of a circle) shows a total number of rankings. Position in circle segment (angular coordinate) represents the average of all rankings.

all points as well as readability. Higher fitted parameter causes the overlapping of many nodes, which is also a not desired effect.

The advantage of the presented visualization is in anomaly detection. A secondary goal is to seek the user with low rank and many rankings. Such users can be found in segments 1–3 and close to the outer boundary of the graph. Such users can suffer from the negativity of other users, and it can affect their personal life. Mindful web owner should provide either professional help or increase of raking by any means.

Sometimes users, with very low raking, provoke by their opinions and presented photos. In this case, the negativity of other voters points out to nonstandard behavior and web provider should check the account. Or web owner risks the credibility and the migration of users to the concurrent web. Hence it is quite simple to plot e.g. Figure 1 with marked points by ID of all users to identify such users.

4   Dataset

The dataset was taken from [2]. It contains a real data from czech online dating service - *Libimseti.cz*. Dataset provides an information about

- UserID - user who provided rating,
- ProfileID - user who has been rated (not every profile has been rated),
- Rating - 1–10 scale where one is the worst, ten is the best (integer ratings only),

- Gender - information whether the user is Female, Man, or Unknown (user who did not choose the gender during the registration process).

Libimseti service is a platform where any user can create a profile containing photographs and general information. Moreover, it has chat groups, blogs, games, etc. It means that users can meet each other, e.g., in chat groups and then evaluate the sympathy to others by the raking system. Or the user can slice over users and rank them by photo and info. Basic dataset characteristic is listed in Table 1.

Table 1. *Libimseti* data sets overview.

Gender	Users ID	Evaluated ID	Provided ratings	Average rate - given
Men	76 441	57 606	4 852 458	5.0313
Female	61 365	43 510	10 804 043	6.3581
Unknown	83 164	67 675	1 702 845	5.86014
Total	220 970	168 791	17 359 346	5.7498

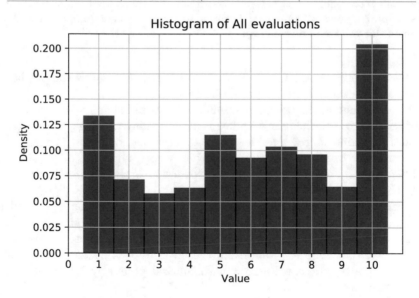

Fig. 2. Histogram of all provided evaluations, where all provided ratings are normalized to form a probability density.

Histogram of all provided evaluations is shown in Fig. 2. More than 20% of all evaluation has rank ten, it demonstrates positivity and willingness of users to evaluate with the highest rank. On the other hand, the second most given rank (around 14%) is one. The rest (two to nine) basically copies normal distribution.

Although, Fig. 2 indicates higher number of the most negative rank one, it has less significant response in histogram of average rating, see Fig. 3. This lower density of low raking (one and two) is decreased by averaging. i.e. one higher ranking decreases the influence of more lower rankings. Average ranking in Fig. 3 has a shape of normal distribution except the rank ten. Comparison of Figs. 3 and 2 can suggest, that a few users take the highest rankings from all users, but this it not clear and evident from histograms. On the other hand the visualization presented in this article (see Sect. 3) discovered such dependencies.

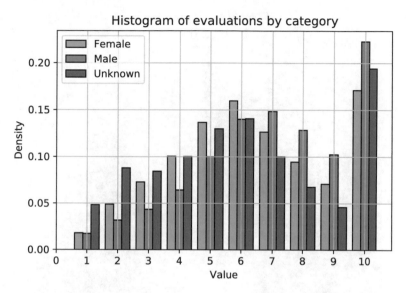

Fig. 3. Histogram of average rating of all users who were evaluated. User are divided into categories by Table 1. All provided ratings are normalized to form a probability density.

5   Results

Implementation for data processing is written in Python 3.7 using Numpy library for computing and Matplotlib library for visualization. Benchmarks were run on PC with Intel(R) Core(TM) i7-4810MQ CPU @ 2.80 GHz, 32 GB RAM, Windows 10 operation system.

The distribution of users rank concerning their number of votes is depicted in Fig. 4, where people with no ranking were excluded in the visualization in Fig. 4. It can be seen that there is a significant difference between users in the total number of rankings. Better optimization of point scattering gives the usage of log 10 scale together with color-scale in agreement with radial distance. The mean of total numbers of ranking is 102.84, and the standard deviation is 413.71. Figure 4 shows a higher density of evaluation in the interval from 3 to 7. Figure 4 also indicates a cluster of people with high rank and many rankings close to the line representing rank ten.

From anomaly detection point of view, it is clear that there are fewer people with lover rank (1 to 3) and over 1 000 votes than in other segments. But still, there are some profiles which should be checked. If a web provider finds users, also known as influencers, they can be chosen from users with more than 10 000 rankings; the majority of such people is very active on chats. The most rated user has 33 389 ratings, and average rank is 9.99.

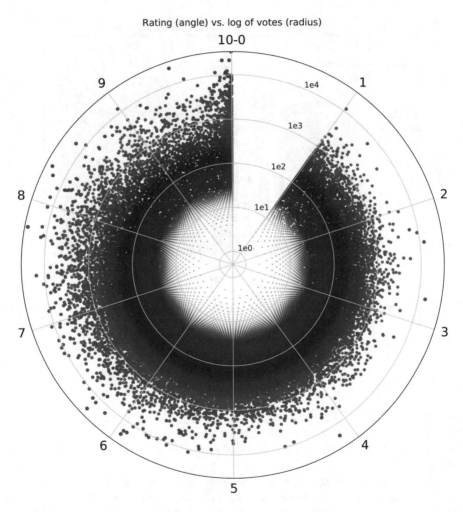

Fig. 4. Visualization of user's rank from online dating service platform. Every point represents one user. Radial coordinate (distance from a center of a circle) shows total number of rankings in log 10 scale. Position in circle segment (angular coordinate) represents average of all rankings.

The distribution of female, male, and unknown rating is shown in Fig. 5. Men receive the higher rank than other groups. Figure 5 can be compared with

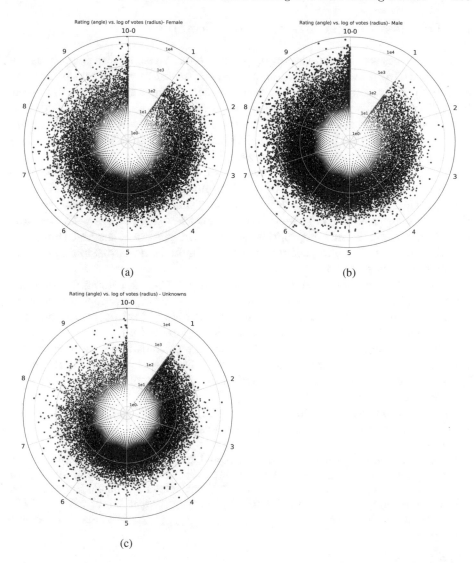

Fig. 5. Comparison of rating distribution of female (a), male (b), and unknowns (c) users. Radial coordinate (distance from a center of a circle) shows total number of rankings in $\log 10$ scale. Position in circle segment (angular coordinate) represents average of all rankings.

Fig. 4, because the same scale was used. The comparison of (a), (b), and (c) in Fig. 5 shows that all groups are evaluated mostly in interval from 3 to 7.5 in average. The density of users increases close to the axis ten, i.e. representing the highest rating; it shows that among all users there are highly valuable individuals. Positive characterization is the lower occupancy of users in the low-rank area with many ratings.

6   Discussion over Results and Conclusion

This work presents a modification of circular plot to visualize ratings. It uses a real dataset from web online dating service - *Libimseti*. On this platform, users evaluate the attractivity of other user on the scale from one-the worst to ten-the best. It makes this web page unique, and processing of the data provides an overview of the general behavior of the people on pages with a raking system. We demonstrated that presented visualization is capable to produce aesthetically pleasing graph together with all necessary information.

Our visualization gives the evaluation of the general behavior of people's opinion as well as the detection of extra popular or unpopular users. It shows the approach of finding and (possibly) helping the users with low rank because they can suffer from other's negativity. Often, the users profiles with low rank have bad quality, old or inappropriate photo – the picture is the first thing that a person will see on online dating sites [14]. Maybe some professional help can be offered to these persons.

Users with a low rank can be persons, who provoke the community by inappropriate photos and opinions, and web owner should take proper attention to it. On the other hand, circular visualization can help to find people as known as influencers. Such people can be used for web popularization.

Mainly presented visualization is aesthetically pleasing and easy-to-read, and it can make it be a popular method in data visualization of raking data. It quickly gives basic information about the distribution of the people's opinion without a deep understanding of the data.

In the future, we would like to compare the actual dataset with an updated version to see the evolution of the people's evaluation.

Acknowledgements. This work was supported by the European Regional Development Fund under the project AI&Reasoning (reg. no. CZ.02.1.01/0.0/0.0/15_003/ 0000466), by the Technology Agency of the Czech Republic under the grant no. TN01000024, and by the project SP2019/135 of the Student Grant System, VŠB-Technical University of Ostrava.

References

1. Al Hasib, A.: Threats of online social networks. IJCSNS Int. J. Comput. Sci. Netw. Secur. **9**(11), 288–93 (2009)
2. Brozovsky, L., Petricek, V.: Recommender system for online dating service. In: Proceedings of Conference Znalosti 2007, VSB, Ostrava (2007). http://www. occamslab.com/petricek/papers/dating/brozovsky07recommender.pdf
3. Buyukkokten, O., Smith, A.D.: Methods and systems for rating associated members in a social network (2011). US Patent 8,010,459
4. Chen, C.H., Hrdle, W., Unwin, A., Chen, C.H., Hrdle, W., Unwin, A.: Handbook of Data Visualization (Springer Handbooks of Computational Statistics), 1 edn. Springer, Heidelberg (2008)
5. Chen, I.X., Yang, C.Z.: Visualization of Social Networks, pp. 585–610. Springer, Boston (2010). https://doi.org/10.1007/978-1-4419-7142-5_27

6. Fiore, A.R.T.: Romantic regressions: an analysis of behavior in online dating systems. Ph.D. thesis, Massachusetts Institute of Technology (2004)
7. Gewirtz-Meydan, A., Ayalon, L.: Forever young: visual representations of gender and age in online dating sites for older adults. J. Women Aging **30**(6), 484–502 (2018)
8. Ji, J.J.: Method and system for online collaborative ranking and reviewing of classified goods or services (2008). US Patent App. 11/952,562
9. Langville, A.N., Meyer, C.D.: Who's #1?: The Science of Rating and Ranking. Princeton University Press, Princeton (2012)
10. Leonard, M.: Matching social network users (2011). US Patent 8,060,573
11. Merkle, E.R., Richardson, R.A.: Digital dating and virtual relating: conceptualizing computer mediated romantic relationships. Family Relat. **49**(2), 187–192 (2000)
12. Rege, A.: What's love got to do with it? Exploring online dating scams and identity fraud. Int. J. Cyber Criminol. **3**(2), 494–512 (2009)
13. Toma, C.L., Hancock, J.T., Ellison, N.B.: Separating fact from fiction: an examination of deceptive self-presentation in online dating profiles. Pers. Soc. Psychol. Bull. **34**(8), 1023–1036 (2008). https://doi.org/10.1177/0146167208318067. PMID: 18593866
14. Weisbuch, M., Ivcevic, Z., Ambady, N.: On being liked on the web and in the "real world": consistency in first impressions across personal webpages and spontaneous behavior. J. Exp. Soc. Psychol. **45**(3), 573–576 (2009)
15. Yang, Y.: Method and apparatus for evaluating trust and transitivity of trust of online services (2007). US Patent 7,249,380

Performance Evaluation of V2V and V2R Communication Based on 2-Wavelength Cognitive Wireless Network on Road State Information GIS Platform

Akira Sakuraba[1], Yoshitaka Shibata[1(✉)], Goshi Sato[2], and Noriki Uchida[3]

[1] Iwate Prefectural University, Takizawa, Japan
{a_saku, shibata}@iwate-pu.ac.jp
[2] Resilient ICT Research Center, Sendai, Japan
sato_g@nict.go.jp
[3] Fukuoka Institute Technology, Fukuoka, Japan
n-uchida@fit.ac.jp

Abstract. In this paper, performance evaluation of V2V and V2I communication introduces cognitive wireless communication method which is designed for exchanging of road sensing information between the vehicle and vehicle and roadside server system for gathering and sharing road state information. N-wavelength wireless cognitive network method is applied to improve the network performance over the conventional wireless network. In our system, as basic control link, 920 MHz and 2.4 GHz and 5.6 GHz bands as data link are used. By combining those wireless links, both long distance and highspeed data transmission are attained at the same time. The prototype system of road state information system is constructed to evaluate and compare the network performance with the conventional single wireless network.

1 Introduction

Mobility is one of the most important means for economic activity to safely and reliably carry persons, loads, feeds and other materials around world. So far, huge number of manual operated cars are produced and their quality and cost performance are improved year and year. Then, self-driving cars have been emerged and running on highways and major public streets on which are well maintained with ideal conditions, such as wide, flat road surface and clear visible center lines in urban area.

On the other hand, in winter season of cold weather countries such as Northern Japan, Northern America and Europe, most of the road surfaces are occupied with heavy snow and iced surface and many slip accidents occurred even though the vehicles attach anti-slip tires. In fact almost more 90% of traffic accidents in northern part of Japan is caused from slipping car on snowy or iced roads.

This could be challenge for realizing high-level autonomous vehicle such as SAE 5 [1] and even for today's manual driving vehicle.

© Springer Nature Switzerland AG 2020
L. Barolli et al. (Eds.): INCoS 2019, AISC 1035, pp. 212–222, 2020.
https://doi.org/10.1007/978-3-030-29035-1_21

In those cases, traffic accidents are rapidly increased. Therefore, safer and more reliable road monitoring and warning system which can quickly transmit the road condition information to drivers and new self-driving method before passing through the dangerous road area is indispensable. Furthermore, the information and communication environment in local or mountain areas, is not well developed and their mobile and wireless communication facilities are unstable along the roads compared with urban area. Thus, once a traffic accident or disaster occurred, information collection, transmission and sharing are delayed or even cannot be made. Eventually the resident's lives and reliabilities cannot be maintained. More robust and resilient information infrastructure and proper and quick information services with road environmental conditions are indispensable.

In order to resolve those problems, various road sensing methods are possible to understand forward road surface condition with various environmental sensors. For example, dynamics sensor such as accelerometer is a popular equipment to determine road surface condition in civil engineering. This type of method collects oscillation of the probe vehicle body with onboard dynamics sensors. However, these sensing methods is designed for offline processing to analyze and determine the road state condition after the probe vehicle has completed the recording of road sensing data. Therefore, more realtime based sensor data processing and sharing with other vehicles is indispensable.

In this work, we propose a the system to deliver road state information and onboard sensor data between vehicles which is conducted by vehicle-to-vehicle (V2V) network, and between vehicle and roadside unit (RSU) which is delivered over vehicle-to-road (V2R) communication. Both wireless communications can be configured without any public wireless network. Our communication system consists of cognitive wireless network by organized multiple wavelength wireless networks, so called N-wavelength wireless cognitive network and can select the best wave-length link to deliver data depending on the network condition and types of data.

We focus to describe V2V and V2R wireless communication methods and their performance evaluation on field experiment in actual public road and vehicle when for the speed of vehicle is varied. Though the field experiment, we could conduct the result that our proposed method has reasonable performance to exchange road state information in actual road environment.

2 Related Works

As advent of road state sensing and mobility network technologies, autonomous as well as connected vehicles have been developed toward the realization of safer and more reliable automotive world. The sensor data from various sensor devices installed on vehicle are collected on own vehicle and analyzed to determine the road state qualitatively or quantitatively.

There are several methods which have been introduced in field of pavement engineering with non-contact onboard sensor to collect pavement condition while probing vehicle is running. In those methods, dynamics sensors especially acceleration sensor is one of typical sensors to obtain road condition such as pothole or bump of pavement. Du et al. proposed a measurement method to estimate the International Roughness Index (IRI) using Z-axis acceleration sensor [2]. Casselgren et al. introduced a road surface condition determination method using near infrared (NIR) sensor [3]. This method uses three different wavelengths of NIR laser sensors to determine the paved road condition into dry, wet, icy, and snowy states.

Floating car data (FCD) is a vehicular telemetry method which provides to collect status of the remote place such as road traffic flow. This system uses public wireless network such 3G/3.5G/4G cellular network for data delivering. On the other hand, CarTel [4] realizes V2R communication on private wireless network which is short-range vehicular adhoc network (VANET) based on 802.11b WLAN. Using this wireless network, about 30 KiB/s as end-to-end throughput from a vehicle to Internet can be attained.

There are also other VANET based on Dedicated Short Range Communication (DSRC) or 802.11p, and 802.11a/b/g/n for the popular wireless standard for V2V and V2R communication. WAVE system [5] is a prototype of DSRC operated on 5.8 GHz band and designed to notify traffic signs for drivers with V2R communication. Road Side Units (RSUs) which is equipped as the improved DSRC based wireless device are now in operation at actual highway of Japan.

V2V applications especially require shorter latency and less packet loss ratio on in addition to high efficiency data delivering on wireless communication. There are several implementations of cognitive radio for V2X communication which consists of multiple wireless links and switches depending on communication requirement. Ito et al. developed 2-wavelength wireless communication consists of 920 MHz band LoRa and 2.4 GHz band with 802.11b/g/n WLAN device to reduce latency which is caused by authentication and association processes before data delivering on WLAN band. They confirmed that the system allowed to increase amount of data delivering between the vehicle and the RSU [6].

In our research, end-to-end performance in terms of communication distance and total data delivery rate between the V2V and V2R are investigated.

3   V2V and V2R Communication Server System

In Fig. 1 shows V2R communication method between the Smart Mobile Box (SMB) which is a communication server system of a vehicle and the Smart Relay Station (SRS) of road side server. First, one of the wireless networks with the longest communication distance can first make connection link between SMB and SRS using SDN function. Through the this connection link, the communication control data of other wireless networks such as UUID, security key, password, authentication, IP address, TCP port number, socket No. are exchanged [7–9]. As approaching each other, the second wireless network among the cognitive network can be connected in a short time and actual data transmission can be immediately started. This transmission process can be repeated during crossing each other as long as the longest communication link is connected. This communication process between SMB and SRS is the same as the communication between SMB to other SMB except for using adhoc mode.

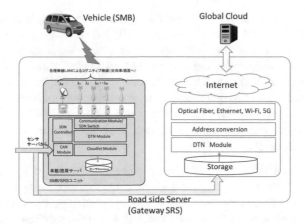

Fig. 1. V2R communication method between the SMBS and SRS

On the other hand, in the V2x communication between vehicle and the global cloud server on Internet, the sensor data from SMB is transmitted to the SRS in road side server. Then those data are sent to the gateway function unit and the address of those data from local to global address and sent to the global cloud server through Internet. Thus, using the proposed V2X communication protocol, not only Intranet communication among the vehicle network, but also Intranet and Internet communication can be realized.

4 Cognitive Wireless Network

Our proposed system has cognitive wireless network in order to perform data delivering with the best network throughput. The system consists of multiple network devices which have different wavelength wireless links. In implementation at this time, we install three different wireless standard based devices which is composed of 2.4/5.6 GHz band WLAN and 920 MHz band LPWA. We expect that system will be able to provide better throughput for application by link selection which corresponds to type of data/information.

2.4 GHz Band WLAN. 802.11b/g/n is the most popular wireless standard and used in many field today, as well as VANET platform. 2.4 GHz band is a license-free band both indoor and outdoor, and also allows to provide adhoc networking easily. However, as this wireless band would be encountered with terrible interference due to carriers of other traffics, therefore 2.4 GHz band has concerned of system could not expect much performance especially in urban area. In current implementation, we designed the system to deliver road state information over 2.4 GHz WLAN which could be composed of less amount of data.

5.6 GHz Band WLAN. We assumed that 5.6 GHz band is based on 802.11ac wave 2 standard, as primary data transmission link for exchange road surface states and raw sensor data. 802.11ac wave 2 standard has capability for higher throughput compared with existing 2.4 or 5 GHz WLAN standard, by multi-user multiple-input and multiple-output (MU-MIMO) and channel bonding. MU-MIMO can deal multiple vehicle communication at once without network performance degradation. Another characteristics is beamforming technology, it can extend longer distance wireless communication with high throughput. Both characteristics are could be expected to provide higher throughput even in longer range, it helps reliable and massive data transmission while vehicle is moving faster. Our system uses 5.6 GHz band WLAN as the primary link which delivers massive amount of data such raw sensor value or images of onboard RGB sensor etc.

920 MHz Band LPWA. 920 MHz band wireless network, as known as LoRa or LPWA, is one of license-free long distance radio communication technology. Network performance of this type wireless technology is designed to perform several tens of kilo bps throughput with several kilometers range distance, it is difficult to transmit large amount of data as primary data transmission link but it is suitable to utilize it as controlling message link. We designed our system which has common control channel among SRSs and SMBs on LPWA long distance wireless network. This is a reason that typical authentication of WLAN such Wi-Fi Protected Access II (WPA2) requires to connect in several to several dozen seconds, vehicle easily moves to out of wireless LAN coverage area while SMB is attempting to authenticate. This method could increase amount of transmittable data than conventional single wavelength WLAN configuration.

5 Data Transmission Flow in 2-Wavelength Network

Figure 2 describes data transmission while V2R and V2 V communication using 2-wavelength network devices from initiation to be connected to WLAN.

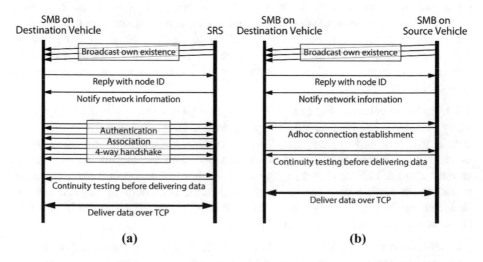

Fig. 2. (a) Message & data flow in V2R (b) Message & data flow in V2V

Messaging flow of V2R scenario is illustrated as Fig. 2a. Initially, SRS broadcasts own existence information via 920 MHz band, which corresponds to beacon frame of access point in 802.11 based WLAN. This information includes traffic direction which SRS collects and provides road states. When running vehicle receives broadcast message from SRS which is installed in the same traffic direction as the vehicle is moving, SMB responses to SRS with own node ID. Responding to that, SRS assigns an IP address for WLAN on SMB and then deliver WLAN network information for SMB. These information is composed of assigned IP address, netmask, default gateway address, ESSID of access point, WPA passphrase etc. due to SRS requests SMB to connect Infrastructure Mode in V2R communication. SMB activates WLAN device after connection information has been received and waits for vehicle is in range of WLAN of SRS. When the vehicle entered into WLAN coverage area, SMB attempts to take 4-way handshake for WLAN connection. If WPA authentication has been succeeded, it also checks for continuity by ICMP message exchanging. At this point WLAN connection has been established completely, thus both vehicle and SRS are able to exchange road state information over TCP each other.

Figure 2b denotes message flow which describes V2V communication from initiation to be established. Message and data flow is very similar procedures, firstly destination vehicle which requires road state information from other vehicle, waits broadcasts own existence information from source vehicle which has already road state information, via 920 MHz band. If SMB on destination vehicle received broadcast message, sends node ID in response. In V2V communication, source vehicle answers IP address related information, ESSID, and wireless-key to destination vehicle. After destination vehicle received these information, it activates WLAN adapter and attempts to connect in adhoc mode on WLAN until both vehicles in range of it. After WLAN connection has been established, destination vehicle measures for continuity, finally road condition exchange will be taken over TCP in adhoc mode between both vehicles if continuity test has been passed.

6   Prototype and Performance Evaluation

6.1   Prototype System

In order to verify the effects of our proposed system, we built prototype system based on three-wavelength wireless communication device which uses Buffalo UI-U2-300D for 802.11b/g/n WLAN USB device, Planex communication GW-900D USB WLAN adapter for 802.11ac WLAN, and Oi Electric OiNET-923 for 920 MHz LWPA wireless network as shown in Fig. 3. Those devices are connected to organize a SMB. SRS also has the similar configuration except for Ruckus Wireless Zoneflex T300 access point for 802.11ac WLAN capable device instead of GW-900D. Controlling both SMB and SRS is based on AMD64 architecture desktop system. We implemented them on Intel NUC Kit NUC5i5RYH barebone PC kit with Ubuntu 16.04 Linux system. Application programs on both SRS and SMB are written by C, Ruby, and Bash Script.

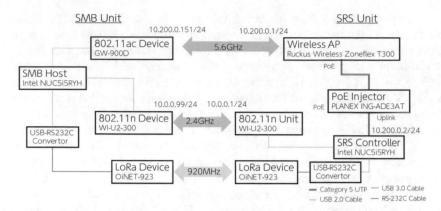

Fig. 3. Device connection and configuration of experimental setting on SRS and SMB

We have experimented a prototype system to evaluate performance our method on actual vehicle on V2R and V2V communications. First, we evaluated V2R communication method between the SMB on vehicle and the SRS on road side server as shown in Fig. 4. For each case, we set two scenarios to measure performance difference which is caused by the moving direction of vehicle. The first scenario requires vehicle to move to eastbound or westbound direction as be described in Fig. 4. Vehicle moves at 20–50 km/h constant speed on either direction of the road. We installed both 2.4 GHz and 5.6 GHz band WLAN unit on the top of left side view mirror on the vehicle and 920 MHz band LoRa at the roof of the vehicle.

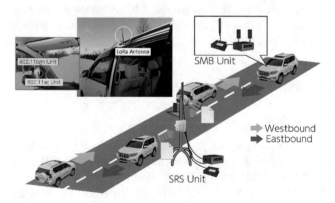

Fig. 4. Overview of V2R communication

On the other hand, Fig. 4 illustrates outline of evaluation for V2V communication which consists of the same 920 MHz band LPWA and 2.4 GHz band WLAN as the V2R communication in actual road environment. In both communications, we observed performance using TCP based networking tool on WLAN conducted by onboard SMBs which are placed on two vehicles moving to another direction (Fig. 5).

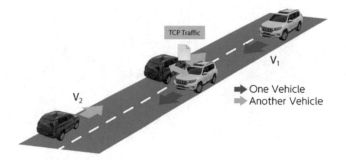

Fig. 5. Overview of both vehicle direction trials and installation of SMB wireless units

As measurement data, several types of network states which includes received signal strength Indicator (RSSI) on each wireless links, the communication completion time to connect WLAN from system initiation, effective throughput and communication distance and total amount of delivered data. In the experiment, we ran at least 5 trials for each vehicle speed from 20–50 km/h and averaged the results to reduce the signal noise and interference on the measurement.

6.2 Performance Results on V2R Communication

Figures 6, 7, 8 and 9 indicates averaged RSSI and averaged throughput of WLAN link which is both plotted in every second. The result recorded as of the time after the link established. The result is separately plotted in each running speeds. We have observed that succeeded rate of TCP connection establishing was almost 100% of whole scenarios.

These graphs denote that both the peak value of RSSI and throughput appeared at earlier elapsed time as the running speed of vehicle increases. Generically, since RSSI has a strong correlation with throughput, this result is reasonable.

Throughput on TCP traffic was affected by the running speed of vehicle. The result shows that Data transmission was available in 80 s after connection establishing when the vehicle was moving at 20 km/h. As increasing running speed of the vehicle from 20 km/h, the data transmission period decreased by 40 s at 50 km/h. The peak of throughput was not drastically changed by varying vehicle's velocity.

6.3 Performance Results on V2V Communication

Figures 10, 11, 12 and 13 indicate the averaged RSSI and averaged throughput of WLAN link which are both plotted in every second after the after the link was established. The result is separately plotted for each running speed. We have observed that succeeded rate of TCP connection establishing was 92.5% of whole scenarios.

These graphs denote that both the peak value of RSSI and throughput appeared at earlier elapsed time as the running speed increased. Generically, RSSI has a strong correlation with throughput, we could confirm that this result was quite reasonable.

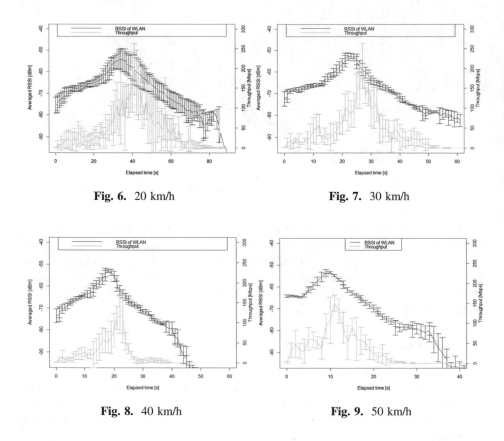

Fig. 6. 20 km/h

Fig. 7. 30 km/h

Fig. 8. 40 km/h

Fig. 9. 50 km/h

Throughput on TCP traffic was affected by running speed. The result shows that possible data transmission period was 38 s after connection establishing when the vehicle was moving at 20 km/h and decreased by 15 s at 50 km/h. The peak value of throughput was not drastically changed even though varying vehicle's running speed.

7 Discussions

Surprisingly the result shows that the system has some capability for realistic vehicle velocity up to 50 km/h which is possible in actual road traffic. However, total delivered data amount was varied even the same condition by environmental factors, such as terrain, existing of oncoming or overtaking vehicle, or distance of the following vehicle, etc. Especially, terrain of the road influents line-of-sight (LOS) or difference height of Tx/Rx antenna on the vehicle which is occurred by terrain, therefore these factors could make significant effects to RSSI. Thus we should consider the position of WLAN antenna on the vehicle.

In higher vehicle velocity scenario, our prototype recorded higher RSSI after connection establishing. We can analyze this result is caused with distance between vehicle was closer while WLAN adapter is establishing with 4-way handshake, the

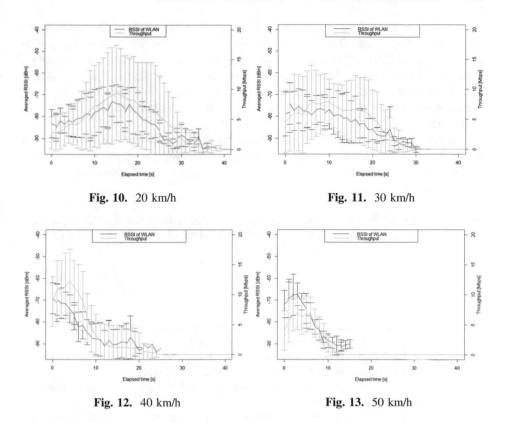

Fig. 10. 20 km/h   Fig. 11. 30 km/h

Fig. 12. 40 km/h   Fig. 13. 50 km/h

distance when it completed to establishing would be shorter than slower velocity scenario. To obtain much amount data with exchanging over V2V communication, we should implement a new WLAN connection procedure which conducts shorter time compared with conventional method, for instance Fast Initial Link Setup (FILS) which is standardized as IEEE 802.11ai [8], it could realize as one of solution for it.

8   Conclusion

In this paper, we proposed a V2V and V2R Communication method based on 2-wavelength cognitive wireless network for exchanging the road state information. This system has three of fundamental functions; observing of physical state of road surface by multiple sensors on the vehicle, analyzing those sensor output for decision of road surface conditions, and transmitting the calculated road condition to roadside device or other vehicle to share the road condition information on which the oncoming vehicle travelled until now. Design of our communication system utilizes cognitive wireless network which consists of WLAN for data delivering and LPWA as the control message link.

We have evaluated our proposed cognitive wireless system for V2V and V2R on actual road environment. The evaluation results indicated that the system has enough capability for delivering road state information over V2V and V2R communication even vehicles are moving in realistic speed on the actual public road. We could also confirm that our system can work in the challenged network environment where public communication networks such as 3G and LTE are not available in rural or mountainous area roads.

As our future work, we will implement the actual data delivering application which is used in the inter-urban highway. We are also planning to carry out the social experiment by collaborating with bus operation company for road state sharing system using improved prototype.

Acknowledgments. This research was supported by Strategic Information and Communications R&D Promotion Program (SCOPE) No. 181502003, Ministry of Internal Affairs and Communications, Japan.

References

1. Taxonomy and Definitions for Terms Related to Driving Automation Systems for On-Road Motor Vehicles J3016_201806, SAE International (2018)
2. Du, Y., Liu, C., Wu, D., Jiang, S.: Measurement of international roughness index by using Z-axis accelerometers and GPS. Math. Probl. Eng. **2014**, 1–10 (2014)
3. Casselgren, J., Rosendahl, S., Eliasson, J.: Road surface information system. In: Proceedings of the 16th SIRWEC Conference (2013)
4. Hull, B., et al.: CarTel: a distributed mobile sensor computing system. In: Proceedings of the 4th Conference on Embedded Networked Sensor Systems (2006)
5. Tsuboi, T., Yamada, J., Yamauchi, N., Hayashi, M.: Dual receiver communication system for DSRC. In: Proceedings of 2008 2nd International Conference on Future Generation Communication and Networking, pp. 459–464 (2008)
6. Ito, K., Hashimoto, K., Shibata, Y.: V2X communication system for sharing road alert information using cognitive network. In: Proceedings of 8th International Conference on Awareness Science and Technology, pp. 533–538 (2017)
7. Ito, K., Hirakawa, G., Shibata, Y.: Experimentation of V2X communication in real environment for road alart information sharing system. In: IEEE AINA 2015, pp. 711–716, March 2015
8. Ito, K., Hirakawa, G., Shibata, Y.: Estimation of communication range using Wi-Fi for V2X communication environment. In: The 10th International Conference on Complex, Intelligent, and Software Intensive Systems, (CISIS 2016), pp. 278–283. Institute of Technology, Fukuoka, July 2016
9. Ito, K., Hashimoto, K., Shibata, Y.: V2X communication system for sharing road alert information using cognitive network. In: The 8th International Conference on Awareness Science and Technology, (iCAST 2017), CD-ROM (2017)

A Machine Learning Approach to Fake News Detection Using Knowledge Verification and Natural Language Processing

Marina Danchovsky Ibrishimova and Kin Fun Li[✉]

Department of Electrical and Computer Engineering, University of Victoria,
Victoria, BC V8P5C2, Canada
{marinaibrishimova, kinli}@uvic.ca

Abstract. The term "fake news" gained international popularity as a result of the 2016 US presidential election campaign. It is related to the practice of spreading false and/or misleading information in order to influence popular opinion. This practice is known as disinformation. It is one of the main weapons used in information warfare, which is listed as an emerging cybersecurity threat. In this paper, we explore "fake news" as a disinformation tool. We survey previous efforts in defining and automating the detection process of "fake news". We establish a new fluid definition of "fake news" in terms of relative bias and factual accuracy. We devise a novel framework for fake news detection, based on our proposed definition and using a machine learning model.

1 Introduction

What is fake news? One of the aims of this paper is to define what constitutes fake news in the context of information warfare, and to propose an automated method for fake news detection based on this definition. Symantec 2019 Internet Threat Report lists information warfare as an emerging cybersecurity threat [1]. Information warfare includes the generation and the spread of fabricated claims for the purpose of manipulation.

Information warfare has existed for as long as warfare has existed. As Sun Tzu wrote in 5th Century BC, "All warfare is based on deception," in his book "The Art of War" [2]. Warfare and politics are deeply intertwined [3]. In 2016, the former President of Bulgaria Rosen Plevenliev warned that the Russian government is trying to influence the outcome of Bulgarian elections after losing the elections [8]. In 2017 Parkinson and Kantchev claimed in a Wall Street Journal article that a Bulgarian security agency allegedly obtained a document from a Russian spy outlining a Russian campaign to interfere in the Bulgarian elections. "The document offered advice on how to burnish the candidate's image by planting stories with Moscow-friendly news outlets. The stories were to be closely coordinated, publishing first in fringe blogs before entering mainstream media en masse to create maximum impact and ultimately become election talking points for the party." [9] Bulgarian security agency neither confirmed nor denied detecting Russian interference [10]. Russia denied all allegations, insinuating that they are fake news [11, 12].

© Springer Nature Switzerland AG 2020
L. Barolli et al. (Eds.): INCoS 2019, AISC 1035, pp. 223–234, 2020.
https://doi.org/10.1007/978-3-030-29035-1_22

If a piece of information is not supported by concrete evidence, then its factual accuracy cannot be established. What is "breaking news" to some becomes fake news to others. A universal definition is therefore required before an adequate computational solution can be created. In Sect. 2 of this paper we study various definitions of fake news and propose a new definition based on absolute factual accuracy and relative reliability of the source. In Sect. 3 we introduce previous work in automating the process of fake news detection. In Sect. 4 we propose a novel fake news detection framework, which utilizes both manual and automated knowledge verification and stylistic features. We discuss our results in Sect. 5. In Sect. 6 we discuss future work, and we draw conclusions in Sect. 7.

2 Defining Fake News

Defining fake news is problematic. People tend to regard any news as "fake" if it does not align with their views or agenda [13]. Edson et al. provided a typology of fake news definitions. They studied 34 different papers on fake news published between 2003 and 2017 and constructed a framework for the different types of fake news based on their definitions [14]. The different types, which include propaganda and advertising/public relations, can all be used in information warfare to influence public opinion on a particular topic [6, 7]. In this paper we focus primarily on "fake news" as a disinformation tool. Lazar et al. define fake news as "fabricated information that mimics news media content in form but not in organizational process or intent" [14]. However, the research of Horne and Adali suggests that there are notable differences in form especially when it comes to the titles of fake news [15]. By the definition provided by Lazar et al., the Wall Street Journal article that depicts Russian interference in Bulgarian elections is likely to be "fake news" because the main source that can verify the factual accuracy of the claims in the article, namely the Bulgarian secret service, is refusing to do so [10]. This makes the story appear fabricated.

So what is the likelihood that this story is in fact fabricated? Since its factual accuracy can be neither confirmed nor denied, then it is equally likely that it is fabricated and factually accurate. The Wall Street Journal has a good reputation, so this likelihood grows slightly larger but probably not by much. After all, even reputable news outlets have produced factually inaccurate news in the past [16, 17].

In 2002 the Associated Press fired one of their reporters after it was discovered that he had been fabricating facts and sources in his news reports for at least 2 years [17]. Just because a piece of information appears in one news outlet, it does not make it factually accurate even if the outlet is a respectable news agency with rigorous organizational processes in place. Nevertheless, a generally reliable source does slightly increase the likelihood that a piece of information is factually correct especially if the opposite cannot be proven. This increase is relative to the overall reliability of the source.

What other factors can increase the likelihood? Are there similar reports that offer more concrete evidence? In the case of the Wall Street Journal article on Russian interference in foreign elections there are. Although the Russians have repeatedly denied interfering in the political affairs of foreign countries [11, 12], a Czech secret service agency pointed at evidence to the contrary as early as 2008 [5]. The Czech Security Information Service unequivocally stated that "operations of intelligence services of the Russian Federation … are by far the most active ones in our territory". In addition, the Czech Security Information Service warned that Russia readopted and repurposed the tactics from Soviet times known as "active measures" [5]. Obviously, this information still does not prove the claim in the Wall Street Journal article about Russian election interference in Bulgaria. However, it does slightly increase the likelihood that it is factually accurate considering that Bulgaria and the Czech Republic, which was formerly a part of Czechoslovakia, are similar in that they were satellite states of the Soviet Union.

In cases where it is difficult to establish the factual accuracy of a piece of information, the line between a fabricated report and a factually accurate report becomes blurry. This is why the fact-checking website Politifact uses a Truth-o-Meter to score political facts [22]. In this paper we introduce "the fake news spectrum" [18]. It takes into account reports that can neither be factually verified nor disputed, even by professional fact-checkers, and even if they come from reliable sources.

In particular, if the accuracy of a report, or a claim can neither be confirmed nor denied, then there is a 50% chance that it is fabricated. If in addition the source is generally reliable or if a similar claim appears in a generally reliable source, then that percentage becomes slightly lower, relative to the overall reliability of the source(s). We can also take into account whether similar claims can be found in various other sources that follow rigorous organizational processes. In addition, we can also take into account whether the claim appears to be based on facts or opinions. In Sect. 4 we describe a machine learning model based on our previous work in incident classification and these observations. In the next section we discuss the existing methods for fake news detection.

3 Automating Fake News Detection

Establishing the factual accuracy of a claim is crucial in determining whether it is "fake news" by most definitions of "fake news". Several manually generated tools for identifying the factual accuracy of a given claim exist. However, such approaches rely on humans who may or may not be objective [18–20]. Additionally, evidence to support the factual accuracy of the claim might not be available as in the case of the Wall Street Journal article on Russian interference in Bulgarian elections. Various automated fact-checking methods have also been proposed. Thorne and Vlachos provide a comprehensive survey of existing automated fact-checking methods. One method uses Recognizing Textual Entailment (RTE) where "RTE-based models assume that the textual evidence to fact check a claim is given" as part of the claim [21]. Another method relies on checking a claim against a knowledge database of proven facts. Yet another method attempts to verify claims by profiling their source and implementing "credit history" of individual sources [21]. Thorne and Vlachos identify issues with all of these methods. Namely, RTE-based methods fail when there is no evidence to support the claim, the "database of proven facts" methods fail when presented with novel claims, and the "profiling the source" methods fail when the source is new. There is an even greater issue associated with fact-checking political news. As Coleman suggests, "Political truth is never neutral, objective or absolute" [4]. Even computational giants such as Google could not tackle this issue and had to shut down their fact-checking tool out of concerns over inaccuracy [22]. Although individually these methods all have weakness, it is worth studying different combinations of them.

In addition to the factual accuracy of a claim, researchers also studied extensively whether its stylistic form can reveal if it is fabricated [23–25]. Oshikawa et al. provide a comprehensive survey on methods using Natural Language Processing (NLP) [26]. Another survey by Groendahl and Asokan asserts that "while certain linguistic features have been indicative of deception in certain corpora, they fail to generalize across divergent semantic domains" [27]. However, they do admit that "some results have been replicated in multiple studies" [27].

Groendahl and Asokan focus primarily on fake news detection methods at the document level as opposed to at the level of the news title. Their survey does not include the work of Horne and Adali who show that the title of a news article is often sufficient to detect if it is fake news [15]. An ordinary news title is typically written in a way to entice the reader to read the entire article. Political disinformation campaigns' main purpose is to spread their narratives to as many people as possible, including to people who do not like to read much. A fake news title as a tool for political disin-formation is typically a summary of the entire article [15]. Horne and Adali explore a wide range of syntactic, psychologic, and stylistic features for machine learning models using several different datasets of political news and come to the conclusion that fake news titles are generally longer, have "significantly fewer stop-words and nouns, while using significantly more proper nouns and verb phrases". Of the many different features they test, Horne and Adali identify the top 4 features for classifying fake news' titles: "the percent of stopwords, number of nouns, average word length, and FKE read-ability." They achieve an accuracy of about 70% [15].

Researchers have also studied hybrid frameworks that employ natural language features and verification features. Conroy et al. survey the various different fake news detection technologies and outline a hybrid framework, which uses content cues (natural language processing tools to detect deceptive language) as well as information about the network, and source verification [28]. However, as Tschiatschek et al. point out, "[it is] difficult to design methods based on estimating source reliability and network structure as the number of users who act as sources is diverse and gigantic (e.g., over one billion users on Facebook); and the sources of fake news could be normal users who unintentionally share a news story without realizing that the news is fake," [29]. Tschiatschek et al. propose a system that relies on crowd-sourcing fake news detection using trusted users to flag potentially deceptive content, which is then forwarded to professional fact-checkers for further investigation. However, there are ethical implications to be considered when profiling users. Zhang and Ghourbani provide a comprehensive survey on fake news detection, "an exhaustive set of hand-crafted features, and the existing datasets for training supervised models" and also acknowledge the importance of having a clear definition of fake news [30].

4 Our Proposed Framework

We propose a hybrid framework for fake news detection, which repurposes the machine learning model for incident classification we previously described [31, 32]. Our incident classification model consists of 5 NLP features combined with 3 knowledge verification features in the form of questions related to the scope, the spread, and the reliability of the source. The ternary answers to these questions are obtained from the user submitting a textual incident report and can be verified independently by the system.

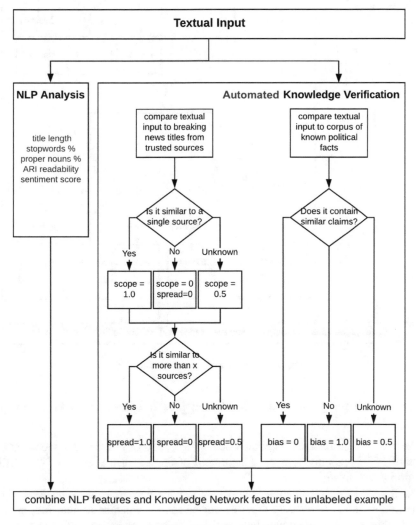

Fig. 1. Automated knowledge verification

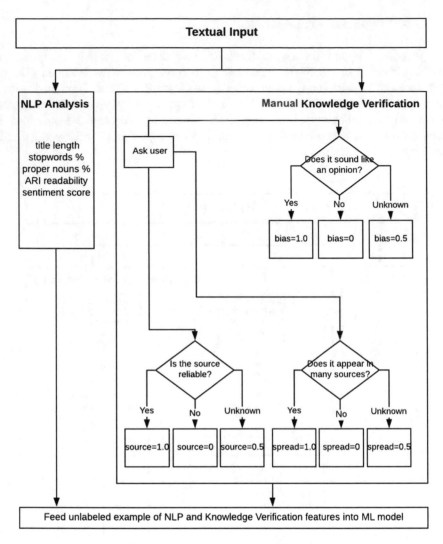

Fig. 2. Manual knowledge verification

We propose using the same general model for the detection of fake news titles where:

1. The 5 NLP features are: stopwords percentage; ratio of proper nouns to nouns; title length; ARI readability; overall sentiment of the text using Google NLP API for sentiment analysis.
2. The 3 knowledge verification features are ternary answers to the questions whether the title is similar to a recent title in a trusted source, whether similar titles appear in more than x sources, and whether the title appears to be based on facts or opinions.

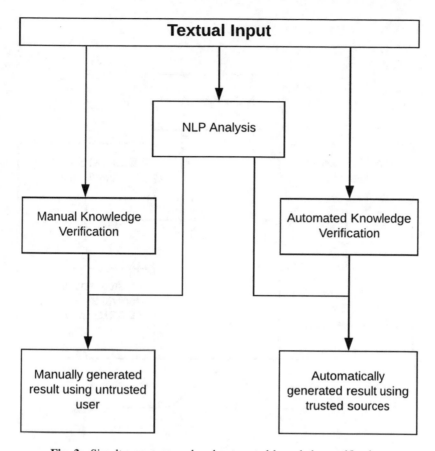

Fig. 3. Simultaneous manual and automated knowledge verification

Our model is different than Horne and Adali in that it uses knowledge verification features in addition to NLP features. Also, our model includes 1 more feature related to sentiment analysis, and we use the automated readability index (ARI), which is a readability test for English texts slightly different than the one used in [15]. Horne and Adali also observed that fake news use more proper nouns but less nouns overall [15], therefore we decided to use the proper nouns (entities) to nouns ratio instead of the number of nouns.

Our system is similar to the one outlined by Conroy et al. in that it is a hybrid framework. However, our proposed verification features are different in that the claim is checked for similarities with recent claims made by a trusted source. The spread of the claim over a variety of trusted sources is also measured. Finally the likelihood that the claim is fact-based or opinion based is established.

Our system can be used in various ways. Namely the verification process can be automated as described in Fig. 1 using trusted sources and datasets of known political facts such as FEVER: a large-scale dataset for Fact Extraction and VERification

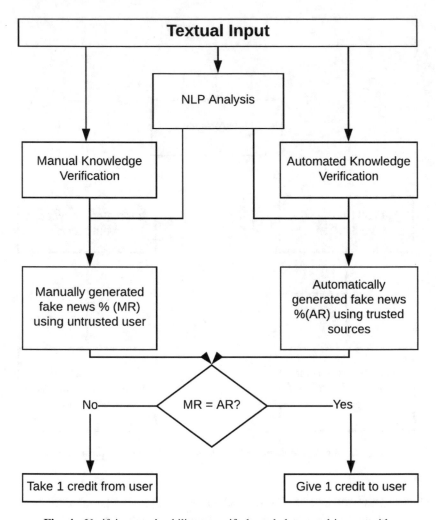

Fig. 4. Verifying user's ability to verify knowledge stated in news title

introduced by Thorne et al. [33] or ClaimBuster by Hassan et al. [34]. Figure 2 shows how the verification process can be manually generated by asking a user to answer questions related to the reliability of the source, the spread of the information, and its bias from the perspective of the user. We propose allowing users to take part in the investigative process by answering the verification questions described in Fig. 2. This way, a manually generated fake news percentage result (MR) is obtained. Behind the scenes an automatically generated fake news percentage result (AR) on the same trained model is also obtained as described in Fig. 1. The system then compares MR and AR as shown in Fig. 3. If the two results are very different or they both suggest the claim is fake, the claim is sent to a professional fact-checker for further investigation. Alternatively, the system can be used to test the fake news detection aptitude of users in a social network. Figure 4 describes one such application. This user-verification system is different from the DETECTIVE system described by Tschiatschek et al. [29] in that it does not rely on users to directly flag news they perceive as fake. Instead, it outsources the knowledge verification to the user and simultaneously performs an automated knowledge verification on its own. It creates two different labeled examples to be fed to a trained machine learning model, and finally it compares the 2 results returned by the model.

5 Result Analysis

We implemented the system described in Fig. 1 using Google NLP API [40] for the NLP analysis portion and News API [41] for the knowledge verification portion. We used the logistic regression model described in our paper on incident classification [31, 32] as our machine learning model for fake news detection. We trained it and validated it on Kaggle Fake News Dataset [42]. We generated a separate test dataset of actual news titles as reported by several verified news outlets with rigorous journalistic practices and we ran each example in our set through our implemented system. Table 1 shows the probability that these news titles are fake as determined by our system.

Table 1. The probabilities that real news titles are fake as determined by our system

Real news	Probability of being fake news
US expected to allow lawsuits against foreign companies doing business in Cuba	10%
Centre-right opposition wins Estonia election as far-right populists make inroads	30%
Vatican to open secret archives of wartime Pope Pius XII	12%
Netherlands summons ambassador to Iran amid diplomatic spat	39%
Residents evacuate as flash floods hit southern Afghanistan	39%
Rep. Dingell on what Cohen's testimony means for future investigations of Trump	45%
OxyContin drug maker mulls bankruptcy due to myriad lawsuits	37%
Fox News confirms exit of Eboni K. Williams	26%
AG William Barr not recusing himself from Russia probe, official says	5%

We generated a separate dataset of fake news by negating crucial words of news titles in the first dataset and we ran each example in our set through our implemented system. Table 2 shows the probability that these fake news titles are fake as determined by our system.

Table 2. The probability that fake news titles are determined as fake

Fake news	Probability of being fake news
US expected to disallow lawsuits against foreign companies doing business in Cuba	80%
Centre-right opposition doesn't win Estonia election as far-right populists make	95%
Vatican won't open secret archives of wartime Pope Pius XII	71%
Netherlands doesn't summon ambassador to Iran amid diplomatic spat	68%
Residents don't evacuate as flash floods hit southern Afghanistan	80%
Rep. Dingell on what Cohen's testimony doesn't mean for future investigations of Trump	84%
OxyContin drug maker doesn't mull bankruptcy due to myriad lawsuits	71%
Fox News denies exit of Eboni K. Williams	74%
AG William Barr recusing himself from Russia probe, official says	40%

6   Future Work

The automated knowledge verification proposed in this paper relies heavily on the notion of similarity. In particular, it relies on establishing semantic similarity. Several state-of-the-art algorithms designed for this purpose have been introduced over the last 20 years, namely latent semantic analysis [35], latent relational analysis [36], explicit semantic analysis [37], temporal semantic analysis [38], distributed semantic analysis [39]. Future work in automating fake news detection using our proposed framework would involve evaluating these algorithms and deciding on one that best fits our purpose.

7   Conclusion

In this paper we define fake news in the context of information warfare. We briefly study the socio-political implications of fake news and we investigate previous efforts in automating fake news detection. We find that the most promising framework for fake news detection uses a combination of source and fact verification and NLP analysis, and we propose a hybrid framework based on our previous work in automating incident classification.

References

1. Symantec Internet Security Threat Report 2019. https://www.symantec.com/en/sg/security-center/threat-report. Accessed 07 Mar 2019
2. Tzu, S.: The Art of War. China. 5th Century BC
3. Kaiser, D.: Politics and War: European Conflict from Philip I1 to Hitler, pp. 149 and 172. Harvard University Press, Cambridge (1990)

4. Coleman, S.: The elusiveness of political truth: from the conceit of objectivity to intersubjective judgement. Eur. J. Commun. **33**(2), 157–171 (2018). https://doi.org/10.1177/0267323118760319
5. Abrams, S.: Beyond propaganda: soviet active measures in Putin's Russia. Connections **15**(1), 5–31 (2016). http://www.jstor.org/stable/26326426
6. Wall, T.: U.S. Psychological Warfare and Civilian Targeting. Peace Rev. **22**(3), 288–294 (2010). SocINDEX with Full Text. Web. 20 Feb. 2015
7. Fried, D., Polyakova, A.: Democratic defence against disinformation. https://www.atlanticcouncil.org/images/publications/Democratic_Defense_Against_Disinformation_FINAL.pdf. Accessed 04 Apr 2019
8. Correra, G.: Bulgaria warns of Russian attempts to divide Europe. BBC (2016). https://www.bbc.com/news/world-europe-37867591. Accessed 19 Feb 2019
9. Parkinson, J., Kantchev, G.: Wall Street J. (2017). https://www.wsj.com/articles/how-does-russia-meddle-in-elections-look-at-bulgaria-1490282352. Accessed 19 Feb 2019
10. Vassilev, I.: Russia's elections manual for BSP and the role played by Bulgarian sociological agencies. Bulgaria Analytica (2017). http://bulgariaanalytica.org/en/2017/03/31/
11. Sputnik News: While West Claims Moscow is Meddling in Bulgarian Elections, Sofia Blames Turkey (2017). https://sputniknews.com/europe/201703251051961027-bulgarian-election-meddling-accusations/, Retrieved 04.04.2019
12. Blyskov, P.: Dmitry Medvedev's interview with Bulgarian newspaper Trud. http://government.ru/en/news/35903/. Accessed 10 Apr 2019
13. Edson, C., Tandoc, Jr., Lim, Z.W., Ling, R.: Defining fake news. Digit. J. **6**(2), 137–153 (2018). https://doi.org/10.1080/21670811.2017.1360143
14. Lazer, D., et al.: The science of fake news. Science **359**(6380), 1094–1096 (2018)
15. Horne, B.D., Adali, S.: This just in: fake news packs a lot in title, uses simpler, repetitive content in text body, more similar to satire than real news. arXiv preprint arXiv:1703.09398 (2017)
16. Wang, P., Angarita, R., Renna, I.: Is this the era of misinformation yet: combining social bots and fake news to deceive the masses. In: Companion Proceedings of the Web Conference 2018 (WWW 2018). International World Wide Web Conferences Steering Committee, Republic and Canton of Geneva, Switzerland, pp. 1557–1561 (2018)
17. https://www.nytimes.com/2002/10/22/us/ap-says-it-couldn-t-find-45-of-fired-writer-s-sources.html. Accessed 10 Apr 2019
18. Factcheck. https://www.factcheck.org/about/our-mission/. Accessed 02 Apr 2019
19. Snopes. https://www.snopes.com/about-snopes/. Accessed 02 Apr 2019
20. PolitiFact. https://www.politifact.com/truth-o-meter/article/2018/feb/12/principles-truth-o-meter-politifacts-methodology-i/. Accessed 02 Apr 2019
21. Thorne, J., Vlachos, A.: Automated fact checking: task formulations, methods and future directions. In: Proceedings of the 27th International Conference on Computational Linguistics (COLING 2018) (2018)
22. https://www.poynter.org/fact-checking/2018/google-suspends-fact-checking-feature-over-quality-concerns/. Accessed 10 Apr 2019
23. O'Brien, N., Latessa, S., Evangelopoulos, G., Boix, X.: The language of fake news: opening the black-box of deep learning based detectors. In: Workshop on AI for Social Good, NIPS 2018 (2018). http://hdl.handle.net/1721.1/120056
24. Rashkin, H., Choi, E., Jang, J.Y., Volkova, S., Choi, Y.: Truth of varying shades: analyzing language in fake news and political fact-checking. In: Proceedings of the 2017 Conference on Empirical Methods in Natural Language Processing, pp. 2931–2937 (2017)
25. Long, Y., Lu, Q., Xiang, R., Li, M., Huang, C.R.: Fake news detection through multi-perspective speaker profiles. In: Proceedings of the Eighth International Joint Conference on Natural Language Processing (Volume 2: Short Papers) (2017)

26. Oshikawa, R., Qian, J., Wang, W.Y.: A Survey on Natural Language Processing for Fake News Detection. Computation and Language (2018). arXiv:1811.00770v1 [cs.CL]
27. Gröndahl, T., Asokan, N.: Text Analysis in Adversarial Settings: Does Deception Leave a Stylistic Trace? Computation and Language (2019). arXiv:1902.08939v2 [cs.CL]
28. Conroy, N.J., Chen, Y., Rubin, V.L.: Automatic deception detection: methods for finding fake news. In: The Proceedings of the Association for Information Science and Technology Annual Meeting (ASIST 2015), 6–10 November, St. Louis (2015)
29. Tschiatschek, S., Singla, A., Gomez Rodriguez, M., Merchant, A., Krause, A.: Fake news detection in social networks via crowd signals. In: Companion Proceedings of the The Web Conference 2018 (WWW 2018). International World Wide Web Conferences Steering Committee, Republic and Canton of Geneva, Switzerland, pp. 517–524 (2018). https://doi.org/10.1145/3184558.3188722
30. Zhang, X., Ghorbani, A.: An overview of online fake news: characterization, detection, and discussion. Inf. Process. Manag. (2019). ISSN 0306-4573, https://doi.org/10.1016/j.ipm.2019.03.004
31. Ibrishimova, M.D.: Cyber incident classification: issues and challenges. In: Xhafa, F., Leu, F.Y., Ficco, M., Yang, C.T. (eds.) Advances on P2P, Parallel, Grid, Cloud and Internet Computing. 3PGCIC 2018. Lecture Notes on Data Engineering and Communications Technologies, vol. 24 (2018)
32. Ibrishimova, M.D., Li, K.F.: Automating incident classification using sentiment analysis and machine learning. In: Traore, I., Woungang, I., Ahmed, S., Malik, Y. (eds.) Intelligent, Secure, and Dependable Systems in Distributed and Cloud Environments, ISDDC 2018. Lecture Notes in Computer Science, vol. 11317 (2018)
33. Thorne, J., Vlachos, A., Christodoulopoulos, C., Mittal, A.: FEVER: a large-scale dataset for fact extraction and verification. In: NAACL-HLT (2018)
34. Hassan, N., Arslan, F., Li, C., Tremayne, M.: Toward automated fact-checking: detecting check-worthy factual claims by claimbuster. In: Proceedings of the 23rd ACM York, NY, USA, pp. 1803–1812 (2017). https://doi.org/10.1145/3097983.3098131
35. Landauer, T.K., Foltz, P.W., Laham, D.: An introduction to latent semantic analysis. Discourse Process. 25(2–3), 259–284 (1998). https://doi.org/10.1080/01638539809545028
36. Turney, P.D.: Measuring semantic similarity by latent relational analysis. In: Proceedings of the Nineteenth International Joint Conference on Artificial Intelligence, IJCAI 2005, Edinburgh, Scotland, pp. 1136–1141 (2005)
37. Gabrilovich, E., Markovitch, S.: Computing semantic relatedness using Wikipedia-based explicit semantic analysis. In: Sangal, R., Mehta, H., Bagga, R.K. (eds.) Proceedings of the 20th International Joint Conference on Artifical Intelligence (IJCAI 2007), pp. 1606–1611. Morgan Kaufmann Publishers Inc., San Francisco (2007)
38. Radinsky, K., Agichtein, E., Gabrilovich, E., Markovitch, S.: A word at a time: computing word relatedness using temporal semantic analysis. In: Proceedings of the 20th International Conference on World Wide Web (WWW 2011), pp. 337–346. ACM, New York (2011). http://dx.doi.org/10.1145/1963405.1963455
39. Nguyen, T.H., Di Francesco, M., Ylä-Jääski, A.: Extracting knowledge from Wikipedia articles through distributed semantic analysis. In: Lindstaedt, S., Granitzer, M. (eds.) Proceedings of the 13th International Conference on Knowledge Management and Knowledge Technologies (i-Know 2013), Article 6, 8 p. ACM, New York (2013). https://doi.org/10.1145/2494188.2494195
40. Google NLP API. https://cloud.google.com/natural-language/. Accessed 02 Apr 2019
41. News API. https://newsapi.org. Accessed 02 Apr 2019
42. Fake News Dataset. https://www.kaggle.com/c/fake-news/data. Accessed 02 Apr 2019

Short-Term Solar Power Forecasting Using SVR on Hybrid PV Power Plant in Indonesia

Prasetyo Aji[1,2(✉)], Kazumasa Wakamori[1], and Hiroshi Mineno[1]

[1] Graduate School of Integrated Science and Technology, Shizuoka University,
Hamamatsu 432-8011, Japan
prasetyo.aji.17@shizuoka.ac.jp
[2] National Laboratory for Energy Conversion Technology,
Agency for the Assessment and Application of Technology (BPPT),
Puspiptek, Tangerang Selatan, Banten 15314, Indonesia
prasetyo.aji@bppt.go.id

Abstract. Considering the environmental issues, the use of renewable energy sources is a far more sustainable solution to meeting the energy demand than fossil fuels. However, the limited availability of renewable energy is a growing problem to be solved. Solar energy has become a popular renewable energy source in several countries such as Indonesia because of their equatorial locations. In this study, limited meteorological measurement has been applied with the aim of forecasting solar power generation for planning photovoltaic (PV) power plants, especially in rural areas, which have limited access to fossil energy. We used limited measurements such as temperature, humidity, and solar radiation. The use of support vector regression (SVR) was applied to improve denoising capabilities and simplify computation. SVR has been evaluated using statistical metrics such as mean absolute percentage error (MAPE), relative root means square error (NRMSE), and coefficient of determination (R^2). The results showed the MAPE value obtained 18.56% from the RBF_SVR. NRMSE value performed excellently with 8.02% from the SW-SVR method. R2 also indicated good forecasting with 0.99. The results showed that promising short-term solar power generation forecasting can be applied to estimate the availability of solar power, plan for an extension, and assess the performance of hybrid power plants in Indonesia.

1 Introduction

Renewable energy is an important issue that is addressed in several international treaties such as the Kyoto Protocol and the Paris Agreement. Several points have been raised, such as the need for reducing global temperatures, increasing renewable energy use, reduction of greenhouse gas emissions, and financial support to implement these programs. The use of fossil fuels has increased the amount of harmful gases being released into the air; by increasing the use of renewable energy, this can be reduced. Solar energy is one of the most effective forms of renewable energy, with advantages such as low maintenance, high efficiency, and abundant natural availability. In addition,

L. Barolli et al. (Eds.): INCoS 2019, AISC 1035, pp. 235–246, 2020.
https://doi.org/10.1007/978-3-030-29035-1_23

the power potential of solar radiation in Indonesia is nearly 207.9 MW, and in the equatorial area, the availability of sunlight is almost constant throughout each year.

Solar power generation data is important in order to estimate and evaluate the performance of solar energy resources. However, there are several challenges faced such as limited measurement of meteorological data and long installation times. Artificial intelligence predictions can be used to derive solar power generation data; there are several artificial intelligence methods that have successfully predicted solar power, one of which is SVR. SVR was chosen because this method has been widely used for classification and regression analysis. Meteorological data such as temperature, humidity, duration of solar radiation, and wind speed are used as input. Zeng and Qiao [1] proposed a least-square (LS) support vector machine (SVM)-based model for short-term solar power prediction (SPP) in the USA from atmospheric transmissivity, sky cover, relative humidity, and wind speed data. This study aims to predict global solar power generation using SVR with meteorological data.

Li et al. [2] has applied the regression object based on the complexity of variables that affect photovoltaic (PV) systems, such as weather systems and electrical installations in systems; regression with several input variables is used to predict solar energy. Some of the variables involved are related to the predictors. Kanwal et al. [3] predicted the availability of power generation such that the generated power would be dispatched to the area requiring it. Energy can be stabilized by coordinating between the independent power generated at the plant and the main power generated by the system, such as the utility grid. Hassan et al. [4] compared multiple regression based on SVR methods such as linear SVR, SVR-RBF-Kernel with linear regression to analyze the solar radiation prediction and yield feasible results for short-term solar power prediction.

In this study, we used the physical measurements of the meteorological station as explanatory variables. The measurements were taken by the weather station system of the hybrid power plant in Baron Technopark, Yogyakarta, Indonesia. Support vector regression (SVR) was assessed by limited meteorological data measurement. The intermittent nature of solar radiation is a common problem in Indonesia; this forecasting method attempts to solve the heavy cloud and clear sky conditions. Implementation in rural areas, which have limited electrical connectivity, caused us to explore the possibility of integrating the forecasting method with the utility power management system; the model will be applied to open data for public services. Additionally, the purpose of this research is also to plan the extension area of the PV power plant and assess the quality of the system.

2 Methodology

2.1 Data Analysis

We proposed a method to process the training and testing data, monitoring data of the hybrid power plant in Baron Technopark. The system consists of a 5 kW wind energy generator, 10 kW wind energy generator, 36 kW photo-voltaic plant, 20 kW diesel engine generator, and a 20 kW lead acid battery system. The system monitoring was

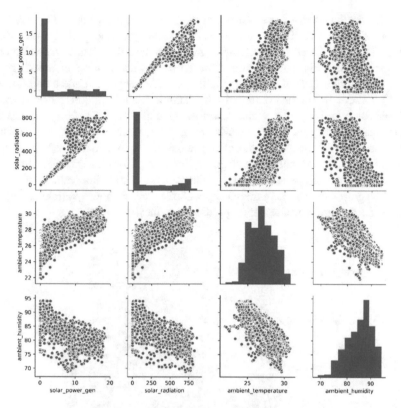

Fig. 1. Scatter plot of explanatory variable and response variable for solar power prediction

also included in the plant. From the monitoring data, it obtained the physical values from sensors on each power generator. Measured data for a 1 month period (June 2017) was collected from the plant, with an average of 10–11 h of bright sunlight per day. The dataset consisted of ambient temperature, humidity, and wind speed, with a total of 30 days of summer data which is divided into 27 days of training data, 3 days of validation data, and 1 day of testing data.

A 15-min resolution time was applied in the system to prevent time shifting and unreasonable data in this period [5]. It was provided after averaging the data from each variable. Figure 1 has indicated a correlation between solar power and other predictors. The solar radiation, ambient temperature and ambient humidity are the parameters which show close relationships to the solar power, with solar radiation being the solar variable showing the closest relationship. From the scatter, the plot looks at the linear relationship between solar power and solar radiation, ambient temperature, and humidity environment. The three explanatory variables showed a significant linear correlation with solar power. From the graph presented, there are several data outliers between the variables. The outliers can be attributed to several factors such as errors in sensor readings, lost data, and loss of power in the sensor. Data distribution on the histogram also shows that solar power has a centralized distribution of data at 0 and

19 kW. This indicates that there is no solar radiation at night, while the 19 kW value is the value of solar radiation in the maximum radiation time period from 10.00 to 14.00 h. The solar power generation histogram also describes data distribution in solar power generation. This data illustrates the 75% data distribution in the range of 0–10 kW. Meanwhile, 25% of the data is in the >10 kW range. The median training data is at 5 kW. There are no data outliers in the dataset. The matrix correlation is shown in Fig. 2. The correlation matrix graph describes the correlation between parameters with a square matrix using Pearson's coefficient relation, which connects the explanatory variables and the response variable in the range of −1 to 1. Solar radiation shows a significant correlation to solar power generation, with an index of 0.98. Other significant variables are temperature and humidity, with correlation values of 0.78 and −0.72. This graph shows that explanatory variables and response variables are closely related to the production of accurate predictive values.

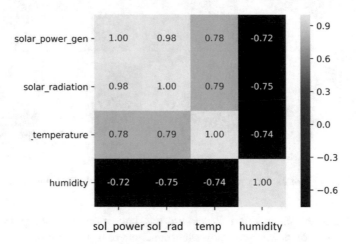

Fig. 2. Matrix correlation between variables

2.2 Support Vector Regression

SVR finds a function of the predicted value and actual value. Then, the function with the highest deviation from the target value is identified. Making the line was regarded as a hyperplane considering the linearity and separation between the actual and predicted values. The graph did not take errors into account as long as they were less than the error boundary on the hyperplane. The fundamental working principle of SVR is to perform data mapping in certain spaces through non-linear mapping and perform direct calculation in the peculiarity space. On the off-chance that a method for registering the internal item in a feature space is accessible specifically as an issue to the first includes focuses, it is conceivable to construct a non-direct learning machine, which is known as an issue processing technique of a kernel function, denoted by K. The flexibility of SVR is attributed to the kernel that represents the information in a higher-dimensional peculiarity space. A linear solution in the feature space corresponds to a non-linear solution in the original input space.

There are methods that employ non-linear kernels to regression problems and correspondingly apply SVR. One such kernel function is the radial basis function (RBF). The main advantage of the RBF is that it is computationally more efficient than the ordinary SVR method because the RBF needs only a solution of linear equations rather than the computationally demanding quadratic programming requirement in standard SVR. The RBF is a more compressed, supported kernel than other kernel functions. In this study, the parameter σ is adapted for the RBF, which is defined as

$$K(x, y) = \exp\left(-\frac{1}{\sigma^2}||x - y||^2\right) \tag{1}$$

Where $K(x, y)$ is kernel function, and x and y are vectors of features computed from training or test samples [6]. In this study, there are several SVR methods has been applied for the dataset to predict the solar power generation. The following are the equations of some of these SVR methods:

$$\text{Linear SVR} = K(x, y) = x \times y \tag{2}$$

$$\text{Polynomial SVR} = K(x, y) = ((x \times y) + c)^d \tag{3}$$

$$\text{RBF SVR} = K(x, y) = \exp\left(-\gamma ||x - y||^2\right) \tag{4}$$

Where x and y are vectors of features computed from training or test samples, and c is a constant intended to balance influence of higher-order versus lower-order terms in the polynomial. γ and σ are the kernel function parameters of the RBF kernel [7].

2.3   Sliding Window-Based Support Vector Regression

The sliding window-based support vector regression (SW-SVR) was used to effectively predict solar radiation [8]. The SW-SVR resolves the computational complexity of the biased data, which is a result of errors, noise, and incomplete datapoints. As it is based on SVR, the above-mentioned regression model effectively handles the multidimensionality problem of the dataset. The model finds a function of the predicted value and actual value. Subsequently, the function with the largest deviation from the desired value is considered. After creating training data based on the matrix correlation. SW-SVR extracts effective training data depends on the movement r meaning the change of a specialized object during prediction horizons. Movements of training data can be calculated by referring to the time when each training data is observed. The estimated movement r_t is given as follows:

$$r_t = ||G_t - G_t'|| \approx \frac{\sum_{i=1}^{N} w_i ||x_i - x_i'||}{\sum_{i=1}^{N} w_i} \, where \, w_i = \frac{1}{||G_t - x_i||^p} \tag{5}$$

N is the number of training data, w is weight vector, and p is a weighted parameter. Subsequently, we obtained the extracted training data, represented by S_t. In the equation below, x_i is the explanatory variable, and y_i is response variable.

$$S_t = \left\{ (x_i, y_i) \middle| \, ||G_t - x_i|| < ||G_t - G_t'|| \right\} \tag{6}$$

The number of weak learners is adjusted, and the weight parameters were determined from relation between the dataset as specialized data. Thereafter, the fit kernel trick and partial least square method were utilized to resolve the multidimensionality problem of the dataset. The deviations in the predicted values were also calculated by extracting the deviations in the specialized data, as shown in the equation below. The deviations occurred when the training data was obtained.

$$H(P) = \frac{\sum_{t=1}^{N} w_i H_t(P)}{\sum_{t=1}^{N} w_t} \, \text{where } w_t = \frac{1}{||G_t - P||^q} \tag{7}$$

Here, G is specialized data before prediction, and G' is specialized data after prediction. $H(P)$ is a hypothesis of each model, and q is a weighted parameter. Despite characteristic variation with time in the test data, SW-SVR always gives priority to specialized models that are more suitable for predicting test data. Owing to the results being comparable, we built hypothesis as trained by linear SVR and thereafter applied the other SVR method for comparison. Finally, the predicted values were obtained.

2.4   Evaluation

SVR methods are commonly evaluated using statistical metrics such as the mean absolute percentage error (MAPE) and relative root mean square error (NRMSE) [9]. Additionally, the coefficient of determination, R^2, is also utilized for evaluation by some researchers. All the above-mentioned evaluation methods are used to derive the correlation between the output and input. Explanatory data has some dataset which multidimensional data. It predicted response variable. R^2 is calculated by subtracting the residual sum of squares from 1, and then dividing the result by the total sum of squares.

MAPE is used as the index of prediction error, and the building time is calculated based on the CPU clock time as the index of computational complexity. It is calculated using y_i, $\hat{y}_i$ and n as shown in the equation below.

$$MAPE = \frac{1}{n} \sum_{i=1}^{n} \frac{|y_i - \hat{y}_i|}{y_i} \times 100\% \tag{8}$$

Where y_i and $\hat{y}_i$ are the actual and predicted values, respectively. n is the number of test data. The absolute value obtained by dividing the difference between y_i and $\hat{y}_i$ by the actual value y_i is summed for every predicted point in time and divided by the number of fitted points n. Multiplying the resulting value by 100 gives us the percentage error.

R^2 is often used in statistics for estimating model performances. It provides the fraction of the calculated values that are the closest to the measurement data. While ideal values of all other statistical indicators used in this study are 0, the R^2 values are close to 1, as shown in the equation below:

$$R^2 = 1 - \frac{SS_{RES}}{SS_{TOT}} = \frac{\sum_i (y_i - \hat{y}_i)^2}{\sum_i (y_i - \bar{y}_i)^2} \tag{9}$$

Where SS_{RES} represents the residual sum of squares; SS_{TOT} (proportional to the variance of the data) represents the total sum of squares; y_i is the actual value, and; $\hat{y}_i$ is the predicted value.

NRMSE is the percentage value of the type of statistical metric, i.e., the RMSE.

$$NRMSE = \frac{\sqrt{\frac{1}{N}\sum_{i=1}^{N} (y_i - \hat{y}_i)^2}}{\frac{1}{N}\sum_{i=1}^{N} y_i} * 100 \tag{10}$$

RMSE, which is also a method of evaluating metrics, gives the standard deviation of residuals or prediction errors. A residual is the difference between the predicted value and actual value. NRMSE is obtained by dividing the RMSE value by the average measurement value, and then multiplying the result by 100. In our study, we used a notebook computer having the following specifications: Intel i5 7200U CPU, 16 GB RAM, 500 GB HDD, and Intel HD Graphics 620, and the scikit-learn module with Python 2.7 version for running the SW-SVR model, and subsequently, performance evaluation was conducted, and the statistical metrics were obtained [10].

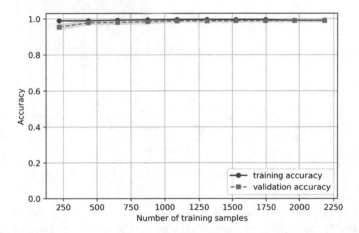

Fig. 3. Learning curve of training dataset

3   Results and Discussion

For building the training and testing dataset for solar power prediction, data were recorded over a period of 30 days. We divided the collected data into training and testing datasets for short-term prediction. Before performing hyperparameter tuning, the training dataset was analyzed using the learning curve of the dataset and validation curve of the parameters.

From Fig. 3, it can be seen that for the training dataset, the accuracy values for training and validation are in good agreement with one another. The model was verified after analyzing its learning curve. If in the learning curve plot, training accuracy and validation accuracy curves lie close to one another in the exterior of the desired accuracy region, it means underfitting has occurred. On the other hand, if there is a gap between the curves of training accuracy and validation accuracy in the region desired accuracy, it means overfitting has occurred.

From the learning curve plot, it can be seen that although the training accuracy and validation accuracy curves were close to each other, the desired accuracy region was still obtained. The model showed reasonable performance on validation and training accuracies. The gap between the training and validation accuracy curves is insignificant, indicating that sufficient data were acquired for accurate parameter selection. From the Fig. 3 also showed a good spot above 1250 training data. It was found that training dataset with less than 500 datapoints led to underfitting, while, the one with more than 2250 datapoints led to overfitting.

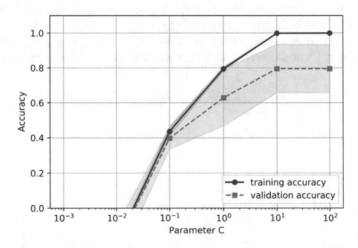

Fig. 4. Validation curve for parameter C of the training dataset

The learning curve was also evaluated using stratified k-fold cross-validation. The data were divided to 9 using the cross-validation parameter, proportional of the training and testing data. From the validation curve, shown in Fig. 4, it is observed that for parameter C, a value of 0.1 gives a low accuracy even though training and validation values are nearly identical. Therefore, there is a possibility of an underfitting situation.

Meanwhile, if the value of parameter C reaches 0.8 or 80%, then the problem of overfitting may occur. Therefore, the selection of parameter C also needs to be considered during hyperparameter tuning to produce accurate predictions.

The SVR utilized hyperparameter tuning in which dependent parameters or kernel functions, such as Epsilon, C and RBF Gamma were optimized [11]. Effects of appropriate hyperparameter tuning reflects in the form of accurate predictions and results.

Table 1. Comparison results of SW-SVR, linear_SVR, RBF_SVR, and Poly_SVR

Method	NRMSE		R^2		MAPE	
	Validation	Test	Validation	Test	Validation	Test
SW-SVR	9.98	8.02	0.98	0.99	21.26	24.04
linear_SVR	9.57	9.37	0.97	0.97	42.78	74.02
RBF_SVR	10.24	9.42	0.97	0.97	23.70	18.56
Poly_SVR	12.19	17.35	0.96	0.91	19.47	24.16

All the comparison results are listed in Table 1. Based on these results, the predicted and real value distributions were obtained. The predicted values were evaluated using NRMSE and MAPE, and the evaluations showed different characteristics. In NRMSE, the data move the squares so that the presence of outliers to be larger if no error distribution outlier would be ideal if at the root squared. Therefore, MAPE can be considered to be more robust as it is less sensitive to outliers, although this assumption cannot be generalized for every dataset.

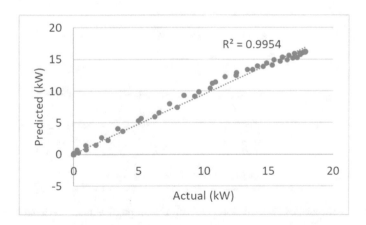

Fig. 5. Linear correlation between actual (kW) and predicted (kW)

Short-term prediction at testing period can be done by considering a minimum MAPE value of 18.56%. The interpretation of MAPE (%) values was explained in terms of forecasting by Lin and Pai [12] as mentioned in the following: less than 10% indicates excellent forecasting, 10%–20% indicates good forecasting, 20%–50%

indicates fair forecasting, and 50% or more indicates poor forecasting. According to this, MAPE value of RBF-SVR falls under good forecasting. There are several variables that affect the solar power generation, such as electrical system, instrument system, and wiring. NRMSE results are shown in the table for short-term prediction at testing period. Minimum NRMSE value is found to be 8.02%, which is obtained from SW-SVR. The interpretation of NRMSE (%) values was suggested by Mohammadi et al. (2015) as mentioned in the following: less than 10% indicates excellent forecasting, 10%–20% indicates good forecasting, 20%–30% indicates fair forecasting, and 30% or more indicates poor forecasting. The short-term prediction at testing period for SW-SVR shows an excellent forecasting result as the NRMSE value is less than 10%.

Table 1 shows the short-term prediction of solar power generation for validation period. In the validation period, MAPE scores of SW-SVR, linear_SVR, RBF_SVR, and Poly_SVR are 21.26%, 42.78%, 23.7%, and 19.47%, respectively. The results vary in the range between 19%–40% approximately, that is between good and fair forecasting. A good forecasting is obtained from Poly_SVR and a fair forecasting from linear_SVR. In addition, the model has been evaluated by using coefficient of determination (R^2). R^2 scores of SW-SVR, linear_SVR, RBF_SVR, and Poly_SVR are 0.98, 0.97, 0.97, and 0.96, respectively. Figure 6. and the R^2 scores indicate linearity between predicted and actual values for all methods. The value of SW-SVR was close to 1 despite the weather conditions where dry or summer season data were recorded along with several rainy cloudy days. NRMSE values in validation period for SW-SVR, linear_SVR, RBF_SVR, and Poly_SVR are 9.98%, 9.57%, 10.24%, and 12.19%, respectively.

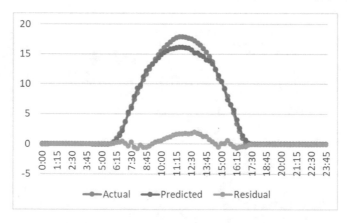

Fig. 6. Solar power prediction plot with actual, predicted, and residual point

In addition, Table 1 shows the short term prediction of solar power generation for testing period. In the testing period, MAPE scores of SW-SVR, linear_SVR, RBF_SVR, and Poly_SVR are 24.04%, 74.02%, 18.56%, and 24.16%, respectively. The results vary in the range between 19%–74% approximately, that is between good and poor forecasting. A good forecasting is obtained from RBF_SVR and a poor forecasting from linear_SVR. In addition, the model has been evaluated by using R^2. In testing period, the R^2 scores of SW-SVR, linear_SVR, RBF_SVR, and Poly_SVR are 0.99, 0.97, 0.97, and 0.91, respectively. These results indicate linearity between predicted and actual values. The value of SW-SVR was close to 1 similar to that of the validation period. NRMSE values in testing period for SW-SVR, linear_SVR, RBF_SVR, and Poly_SVR are 8.02%, 9.37%, 9.42%, and 17.35%, respectively. SW-SVR is the best method according to the NRMSE results. The hyperparameter tuning was calculated until the best results were obtained. The parameters C, gamma, and epsilon obtained contribute to the minimization of NRMSE results of the SW-SVR method. The testing error of SW-SWR for C = 32, gamma = 0.01, epsilon = 0.01, and intercept = 128 is obtained for NRMSE of 8.02%. The parameter C was calculated until a steady-state accuracy of training and validation as shown in Fig. 4. was achieved. The parameter steady state in range of 10 until 100 appeared in case of the results of hyperparameter tuning was in this range.

In addition to Fig. 5, another method that can help to compare and display the quality of predictions is, the residual plot [3]. Figure 6 shows the residual plot between predicted, actual and residual values in the test period of June. From the figure, it is evident that the residual values are centered along the x-axis. Further, outlier values are not found in the image, which implies that the predicted values show good results. In addition, the figure explains the time zone of forecasting for testing data, SW-SVR is closely predicting the solar power, the sky condition as training data made the forecasting matched for time zone before maximum solar power value at noon. In the afternoon, the predicted value missed the maximum data of solar power generation because the results were affected by the weather conditions.

The response variable predicts the solar power generation successfully under all weather conditions by using meteorological measurement. In the future work, essential meteorological data such as clearness index, rainfall, and sunshine duration will be included because it is required to increase the accuracy. Furthermore, by maximizing the classification between sunny, cloudy, and rainy days, we could decrease the significant error.

4 Conclusion and Future Works

In this research, we assessed the performance of SVR methods, such as SW-SVR, linear_SVR, RBF_SVR, and Poly_SVR to predict solar power generation of hybrid power plant in Indonesia. We used ambient temperature, ambient humidity, and solar radiation as explanatory variables. Evaluation of the results was done by statistical metrics such as R^2, NRMSE, and MAPE. SW-SVR predicted solar power generation by using R^2 and obtained a value of 0.99 that is close to 1. Further, the NRMSE score of SW-SVR method is 8.02%, which implies excellent forecasting. Small errors in the result could be due to noisy data, uncompleted data, and missing data. Another evaluation metric MAPE, showed good results for the SVR method with a value of 18.5%.

These results infer that SW-SVR method could be a promising one and can be applied to hybrid power plant in Indonesia. In future works, the algorithm of SW-SVR will be improved by decreasing the error value of the explanatory variables. An alternative to this is to add dataset measurement data such as precipitation, clearness index, and sunshine hours, and consider data from global meteorological measurement and forecast.

Acknowledgments. Prasetyo Aji was supported by Mineno Laboratory of Shizuoka University and by Research and Innovation in Science and Technology Project (RISET-PRO) World Bank Loan No. 8245-ID, Ministry of Research, Technology, and Higher Education of Indonesia. Any opinions, findings, and conclusions expressed in this material are those of the authors, and do not necessarily reflect the views of the funding agencies. Authors also would like to gratitude anonymous reviewers for their very helpful and constructive comments, which improved this manuscript from the original.

References

1. Zeng, J., Qiao, W.: Short-term solar power prediction using a support vector machine. Renew. Energy **52**, 118–127 (2013)
2. Li, Y., He, Y., Su, Y., Shu, L.: Forecasting the daily power output of a grid-connected photovoltaic system based on multivariate adaptive regression splines. Appl. Energy **180**, 392–401 (2016)
3. Kanwal, S., Khan, B., Ali, S.M., Mehmood, C.A., Rauf, M.Q.: Support vector machine and gaussian process regression based modeling for photovoltaic power prediction. In: 2018 International Conference on Frontiers of Information Technology (FIT) (2018). https://doi.org/10.1109/fit.2018.00028
4. Hassan, M.Z., Ali, K.M.E., Ali, A.S., Kumar, J.: Forecasting day-ahead solar radiation using machine learning approach. In: 2017 4th Asia-Pacific World Congress on Computer Science and Engineering (2017). https://doi.org/10.1109/apwconcse.2017.00050
5. Wolff, B., Kühnert, J., Lorenz, E., Kramer, O., Heinemann, D.: Comparing support vector regression for PV power forecasting to a physical modeling approach using measurement, numerical weather prediction, and cloud motion data. Sol. Energy **135**, 197–208 (2016)
6. Mohammadi, K., Shamshirband, S., Anisi, M.H., Alam, K.A., Petkovic, D.: Support vector regression-based prediction of global solar radiation on a horizontal surface. Energy Convers. Manag. **91**, 433–441 (2015)
7. Hassan, M.A., Khalil, A., Kaseb, S., Kassem, M.A.: Potential of four different machine-learning algorithms in modeling daily global solar radiation. Renew. Energy **111**, 52–62 (2017)
8. Aji, P., Wakamori, K., Mineno, H.: Highly accurate daily solar radiation forecasting using SW-SVR for hybrid power plant in Indonesia. In: 2018 4th International Conference on Nano Electronics Research and Education (ICNERE) (2018). https://doi.org/10.1109/icnere.2018.8642593
9. Belaid, S., Mellit, A.: Prediction of daily and mean monthly global solar radiation using support vector machine in an arid climate. Energy Convers. Manag. **118**, 105–118 (2016)
10. Hackeling, G.: Mastering Machine Learning with Scikit-learn. Packt Publishing, Birmingham (2014)
11. Ahmad, M.W., Mourshed, M., Rezgui, Y.: Tree-based ensemble methods for predicting PV power generation and their comparison with support vector regression. Energy **164**, 465–474 (2018)
12. Lin, K.-P., Pai, P.-F.: Solar power output forecasting using evolutionary seasonal decomposition least-square support vector regression. J. Cleaner Prod. **134**, 456–462 (2016)

Optical Axis Estimation Method Using Binocular Free Space Optics

Kouhei Yamamoto, Rintaro Simogawa, Kiyotaka Izumi,
and Takeshi Tsujimura[✉]

Saga University, Saga, Japan
18575034@edu.cc.saga-u.ac.jp,
tujimura@me.saga-u.ac.jp

Abstract. Based on active free space optics, we designed a binocular device with independent receiver and transmitter using positioning photodiode, quadrant photodiode, and voice coil motor. We proposed the alignment method using five photodiodes for the method of beam alignment. The estimation error of alignment was 30.1 mm. Basic communication experiments were conducted using two binocular devices. It is proved that free space optics communication can be performed by the designed active free space optics device.

1 Introduction

Multiple drones and swarm robots need an intercommunication system to transmit their plan of movement during their own operation [1–3]. Radio waves are widely used as data transfer technology. Radio waves propagate omni directionally in air and can be received anywhere within its effective transmission range. However, encryption is required to prevent wiretapping.

Free space optics (FSO) is a communication technology to send a laser beam in the air instead of radio waves [4–6]. It has the same data transfer capabilities as fiber optic networks. The FSO system discharges a collimated laser beam. Therefore, the transmission signal does not spread and the communication is not intercepted. FSO communication is performed by a laser beam traveling straight through the air between a transmitter and a receiver. Traditional FSOs are designed for fixed point-to-point communication [4–17]. One of the disadvantages of conventional FSOs is their weakness in the obstacles to airborne laser beam transmission. Communication is interrupted immediately if problems occur that interfere with the laser beam connection.

When the FSO system is applied to swarm robots or fleets, a laser beam is used to establish an optical mesh network. An essential issue for applying FSO to such unstable situations is the real time correction of laser beam alignment. The directivity of the laser beam is necessary to accurately steer its radiation direction even while the device is moving.

We studied active free space optics system [18–28] to realize ubiquitous broadband communication in user networks. The active free space optics system uses a laser beam controller including a transmitter and a receiver to steer the laser beam. The laser beam

© Springer Nature Switzerland AG 2020
L. Barolli et al. (Eds.): INCoS 2019, AISC 1035, pp. 247–256, 2020.
https://doi.org/10.1007/978-3-030-29035-1_24

is bi-directionally transmitted between a pair of devices. The relative position of devices is not always stationary, and may shift by inches. One of the salient features of the active free space optics system is mobile terminal tracking technology. Laser beam alignment is essential to complete communication between remote transmission devices. And the quality of broadband communication depends on the accuracy of alignment.

This paper created a binocular active free space optics device using positioning photodiode, quadrant photodiode (QPD), and voice coil motor (VCM). We also propose laser beam alignment technology using five photodiodes (PD). Communication accuracy was confirmed by transmission experiments.

2 Device System

Figure 1 is an overview of the designed free space optics system. This is a binocular type in which a receiver and a transmitter are combined. The receiving side consists of positioning photodiode, VCM, beam splitter, and QPD. The transmission unit involves a VCM that controls the radiation direction of the laser beam. Laser beam is transmitted between the two devices in the atmosphere. The laser beam is incident on a receiving lens in the receiving unit of the facing device. The laser beam is split into two directions by a beam splitter after passing through a lens attached to the VCM. One of the laser beams goes straight and enters the receiving side single mode fiber (SMF). The other is irradiated to the QPD for feedback control of the VCM. The feedback system makes the laser beam accurately incident on the SMF. The optical axis position is estimated according to the laser beam intensity detected by the photodiodes. Thereafter, the information is sent to the transmitter.

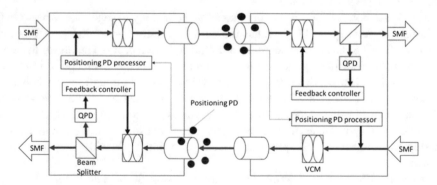

Fig. 1. Free space optics system

Figure 2 shows the transmitter system. The communication laser beam is emitted from 1.0×10^{-2} mm SMF core in diameter, then collimated to 10 mm, and transmitted in the atmosphere. The radiation direction of the laser beam is controlled by operating a VCM.

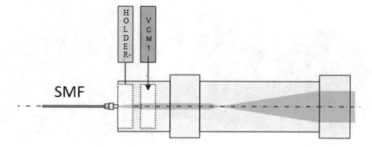

Fig. 2. Transmitter of prototype FSO

Figure 3 shows the receiver system. The collimated laser beam incident from the receiving lens is reduced to 2 mm, and then the VCM is operated to guide it to an 1.0×10^{-2} mm SMF core. At this time, the laser beam split by the beam splitter strikes the QPD to generate a voltage. The position of the laser beam on the QPD is specified by analyzing the output voltage of the QPD. If the VCM adjusts the split laser beam on the center of the QPD, the incident laser beam is accurately guided to the SMF.

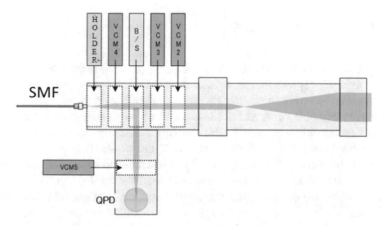

Fig. 3. Receiver of prototype FSO

The positioning PD placed around the receiving lens detect the intensity of the laser beam. This makes a rough estimation of the position of the optical axis. The 1550 nm laser beam used for communication follows Gaussian distribution. Therefore, the optical axis position of the laser beam can be estimated based on Gaussian beam optics. As a result, it is possible to guide laser beam into the receiving lens.

If the laser beam enters the receiving lens, it is split by the beam splitter and emitted to the QPD. The exact position of the laser beam can be found.

Based on the above investigation, the authors made a prototype of binocular FSO device as shown Fig. 4. It is equipped with the transmitter and receiver lens barrels. They discharge and catch a 1550 nm laser beam of 10 mm in diameter.

Fig. 4. Prototype FSO

3 Laser Beam Alignment

We have proposed coarse adjustment of the optical axis of beam alignment. It is assumed that a laser beam is parallel to the z-axis to be illuminated in the x-y plane. The research uses a Gaussian beam optics, and the intensity distribution of the Gaussian beam follows Eq. (1).

$$E_{xy} = E_0 \exp\left\{ -\frac{(x-a)^2 + (y-b)^2}{w^2} \right\}. \tag{1}$$

where, E_{xy} is the voltage at the coordinates of (x, y), (a, b) represents the coordinates of the optical axis, E_0 represents the maximum value of the voltage in the laser beam optical axis, and w^2 represents the dispersion of the laser beam. The number of unknowns parameters is four, and the estimation of the optical axis is performed by four PDs. However, when an optical mesh network is actually assumed, the optical axis of the laser beam and the irradiation surface of the positioning photodiode are not necessarily perpendicular. When the laser beam is irradiated from other than the front, the intensity distribution of the laser beam obtained by the PD becomes a distorted circle. At that time, the correct laser intensity distribution does not form concentrically. So, we assume the laser beam intensity follows the modified normal distribution below.

Assuming that the laser beam has a dispersion of intensity distribution in each of the x-axis and y-axis, the voltage at any position is

$$E_{xy} = E_0 \exp\left\{ -\frac{(x-a)^2}{n^2} - \frac{(y-b)^2}{m^2} \right\}. \tag{2}$$

where, n^2 represents the x-axis dispersion, and m^2 represents the y-axis dispersion.

In this Eq. (2), five unknowns exist. Therefore, five independent equations are required. Five photodiodes are used to estimate the optical axis of the laser beam. We designed a positioning photodiode set made of five photodiodes, whose arrangement is as shown in Fig. 5.

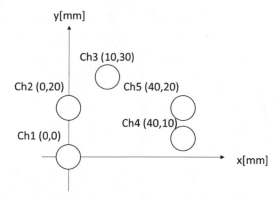

Fig. 5. Arrange of photodiode

The positioning photodiodes give five output voltages e1, e2, e3, e4 and e5 depending on the position of the laser beam spot. We obtain five data sets as (x, y, e1), (x, y + 20, e2), (x + 10, y + 30, e3), (x + 40, y + 10, e4), (x + 40, y + 20, e5). Here, the first and second elements contain PD's x and y coordinates, and the third element shows PD output which represents the laser beam intensity. By giving these data sets to Eq. (2), five simultaneous equations are obtained. The coordinates of the optical axis can be determined by solving this fifth-order simultaneous equation. Here, the following solution is obtained regarding the photodiode arrangement in Fig. 5.

$$a = \frac{245log(E_5) - 240log(E_4) - 80log(E_3) - 5log(E_2) + 80log(E_1)}{13log(E_5) - 12log(E_4) - 4log(E_3) - log(E_2) + 4log(E_1)}. \tag{3}$$

$$b = \frac{20log\left(\frac{E_1}{E_5}\right) - 20log\left(\frac{E_1}{E_4}\right) - 15log\left(\frac{E_1}{E_2}\right)}{2log\left(\frac{E_1}{E_5}\right) - 2log\left(\frac{E_1}{E_4}\right) - log\left(\frac{E_1}{E_2}\right)}. \tag{4}$$

The photodiode set was moved within the range of 80 mm by 80 mm. The laser strikes the photodiode set to obtain a voltage. The optical axis position can be estimated by substituting the voltage into the Eqs. (3) and (4). Figure 6 shows the relationship between the estimated position of the optical axis and the actual laser irradiation position. The distance between the start point of the laser beam and the irradiation surface of the device was 1.5 m.

The black dot in Fig. 6 represent the correct arrival position of the laser beam. The white circle represents the laser beam position estimated by the output obtained from the PD for each measuring position. The figure displays that the measured data concentrate around (20, 10). It suggests the estimation shifts 30.1 mm in average.

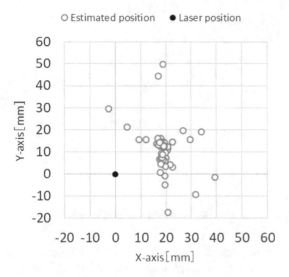

Fig. 6. Result of estimation

4   Basic Communication Experiment

As a basic communication experiment, bi-directional measurement is performed for the eye pattern and bit error rate at a distance between devices 1.5 m, 5.0 m, 35 m, communication time 100 s, communication wavelength 1550 nm, and communication speed 1 Gbps. The eye pattern is a superposition of many sampled transitions of signal waveforms. The eye pattern is a visual representation of whether communication is possible.

An experiment was conducted by changing the distance between the opposing devices as shown in Fig. 7. The bit error rate test set (BERTS), which measures eye patterns and bit error rates, is connected to two FSO devices using SMF as shown in Fig. 8. One of the SMFs was connected to the FSO transmitter and the other to the opposite FSO receiver. The laser from the measuring device flows to the FSO device transmitter. The laser is transmitted from the FSO device transmitter to the opposite FSO device receiver in the air. The laser that has reached the FSO device receiver passes through the SMF and returns to the measuring device. After the measurement is completed, the connection is switched to reverse the direction of the laser between the opposing FSO devices, and the measurement is performed again.

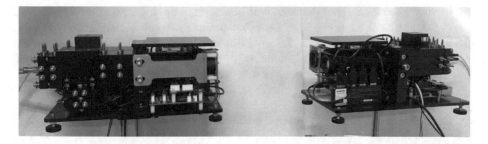

Fig. 7. Communication experiment model

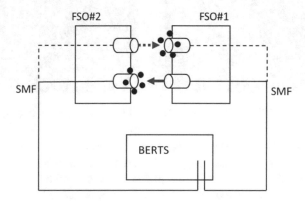

Fig. 8. Communication experiment

The results of basic communication experiments at 1.5 m is shown in Figs. 9 and 10. Figures 9 and 10 indicate the eye patterns open wide enough to find digital signals. They suggest 1.5 m bidirectional transmission is successful between FSO devices 1 and 2.

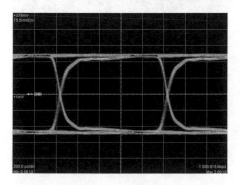

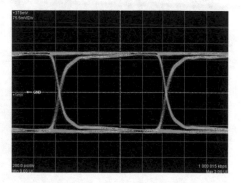

Fig. 9. Eye pattern (1.5 m FSO1 to FSO2)   **Fig. 10.** Eye pattern (1.5 m FSO2 to FSO1)

Figures 11 and 12 show the results of basic communication experiments at 5.0 m.

Figures 11 and 12 show that the eye pattern is clearly visible even when the distance between the devices is 5.0 m. In bidirectional communication between the active free space optics device 1 and the active free space optics device 2, it can be said that communication is possible even at 5.0 m.

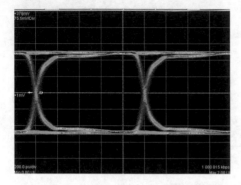

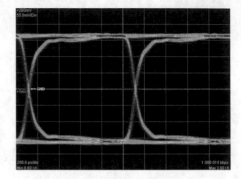

Fig. 11. Eye pattern (5.0 m FSO1 to FSO2)   **Fig. 12.** Eye pattern (5.0 m FSO2 to FSO1)

Figures 13 and 14 show the results of basic communication experiments at 35 m.

From Figs. 13 and 14, the eye pattern can be firmly confirmed even when the distance between the devices is 35 m. In the two-way communication between the active free space optics device 1 and the active free space optics device 2, communication is possible even at 35 m.

The bit error rate was also measured. The bit error rate showed a smaller value than 1.0×10^{-11} even in the 35 m bidirectional communication. Therefore, the communication state in this device is good as shown in Table 1.

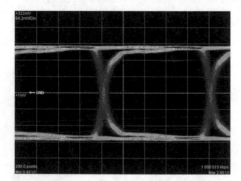

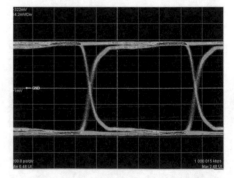

Fig. 13. Eye pattern (35 m FSO1 to FSO2)   **Fig. 14.** Eye pattern (35 m FSO2 to FSO1)

Table 1. Bit error rate result.

Communication distance [m]		1.5	5.0	35
Bit error rate	FSO1 to FSO2	$<1.0 \times 10^{-11}$	$<1.0 \times 10^{-11}$	$<1.0 \times 10^{-11}$
	FSO2 to FSO1	$<1.0 \times 10^{-11}$	$<1.0 \times 10^{-11}$	$<1.0 \times 10^{-11}$

5   Conclusion

We designed a binocular device with independent transmitter and receiver using PPD, QPD, and VCM based on active free space optics system. Based on Gaussian beam optics, we proposed a beam search method using five PDs so that the optical axis could be searched even if the beam intensity distribution is not concentric. Basic communication experiments suggest that the position of the optical axis can be estimated in an accuracy of 30 mm from the measured position. We conducted basic communication experiments at 1.5 m, 5.0 m and 35 m using two devices. The eye pattern could be confirmed at all distances, and the bit error rate was also smaller than 1.0×10^{-11}. From this result, it is proved that free space optics communication is possible with the designed active free space optics device. In the future, investigations are planned to reduce estimation errors and to automatically tune the laser of the device.

Acknowledgments. This work is supported by Strategic Information and Communications R&D Promotion Program (SCOPE) of Ministry of Internal Affairs and Communications, Japan.

References

1. Kantaros, Y., Zavlanos, M.M.: Distributed intermittent connectivity control of mobile robot networks. IEEE Trans. Autom. Control **62**(7), 3109–3121 (2017)
2. Saulnier, K., Saldaña, D., Prorok, A., Pappas, G.J., Kumar, V.: Resilient flocking for mobile robot teams. IEEE Robot. Autom. Lett. **2**(2), 1039–1046 (2017)
3. Wu, C., Chu, X., Wei, Y., Cui, X.: Regional targeting based millimeter-wave beamforming for robot communication in 5G scenes. In: International Conference on Artificial Intelligence, Automation and Control Technologies, Article No. 14 (2017)
4. Pratt, W.K.: Laser Communication Systems, p. 196. Wiley, Hoboken (1969)
5. Ueno, Y., Nagata, R.: An optical communication system using envelope modulation. IEEE Trans. Commun. **20**(4), 813 (1972)
6. Willebrand, H., Ghuman, B.S.: Free-Space Optics: Enabling Optical Connectivity in Today's Networks. Sams Publishing, Indianapolis (1999)
7. Nykolak, G., et al.: Update on 4x2.5 Gb/s, 4.4 km free-space optical communications link: availability and scintillation performance. In: Optical Wireless Communications II, Proceedings of SPIE, vol. 3850, pp. 11–19 (1999)
8. Dodley, J.P., et al.: Free space optical technology and distribution architecture for broadband metro and local services. In: Optical Wireless Communications III, Proceedings of SPIE, vol. 4214, pp. 72–85 (2000)
9. Wang, J., Kahn, J.M.: Acquisition in short-range free-space optical communication. In: Optical Wireless Communications V, Proceedings of SPIE, vol. 4873, pp. 121–132 (2002)

10. O'Brien, D.C., et al.: Integrated transceivers for optical wireless communications. IEEE J. Sel. Topics Quantum Electron. **11**(1), 173–183 (2005)
11. Minch, J.R., et al.: Adaptive transceivers for mobile free-space optical communications. In: IEEE Military Communications Conference, pp. 1–5 (2006)
12. Ghimire, R., Mohan, S.: Auto tracking system for free space optical communications. In: 13th International Conference on Transparent Optical Networks, pp. 1–3 (2011)
13. Yamashita, T., et al.: The new tracking control system for Free-Space Optical Communications. In: International Conference on Space Optical Systems and Applications, pp. 122–131 (2011)
14. Vitasek, J., et al.: Misalignment loss of free space optic link. In: 16th International Conference on Transparent Optical Networks, pp. 1–5 (2014)
15. Dubey, S., Kumar, S., Mishra, R.: Simulation and performance evaluation of free space optic transmission system. In: International Conference on Computing for Sustainable Global Development, pp. 850–855 (2014)
16. Wang, Q., Nguyen, T., Wang, A.X.: Channel capacity optimization for an integrated Wi-Fi and free-space optic communication system. In: 17th ACM International Conference on Modeling, Analysis and Simulation of Wireless and Mobile Systems, pp. 327–330 (2014)
17. Kaur, P., Jain, V.K., Kar, S.: Capacity of free space optical links with spatial diversity and aperture averaging. In: 27th Biennial Symposium on Communications, pp. 14–18 (2014)
18. Tsujimura, T., Yoshida, K.: Active free space optics systems for ubiquitous user networks. In: Proceedings of Conference on Optoelectronic and Microelectronic Materials and Devices (2004)
19. Tsujimura, T., Yoshida, K., Shiraki, K., Sankawa, I.: 1310/ 1550 nm SMF-FSO-SMF no-repeater transmission technique with semi-active FSO Nodes. In: 33st European Conference and Exhibition on Optical Communication, pp. 189–190 (2007)
20. Tanaka, K., Tsujimura, T., Yoshida, K., Katayama, K., Azuma, Y.: Frame-loss-free line switching method for in-service optical access network using interferometry line length measurement. In: Optical Fiber Communication Conference, postdeadline PDPD6 (2009)
21. Tanaka, K., Tsujimura, T., Yoshida, K., Katayama, K., Azuma, Y.: Frame-loss-free optical line switching system for in-service optical network. J. Lightwave Technol. **28**, 539–546 (2009)
22. Tsujimura, T., Tanaka, K., Yoshida, K., Katayama, K., Azuma, Y.: Infallible layer-one protection switching technique for optical fiber network. In: 14th European Conference on Networks and Optical Communications (2009)
23. Tsujimura, T., Yoshida, K., Tanaka, K.: Length measurement for optical transmission line using interferometry. Interferometry. InTech (2012). ISBN 978-953-308-459-6
24. Yoshida, K., Tanaka, K., Tsujimura, T., Azuma, Y.: Assisted focus adjustment for free space optics system coupling single-mode optical fibers. IEEE Trans. Ind. Electron. **60**, 5306–5314 (2013)
25. Tsujimura, T., Muta, S., Masaki, Y., Izumi, K.: Initial alignment scheme and tracking control technique of free space optics laser beam. In: OPICS 2014 (2014)
26. Tsujimura, T., Izumi, K., Yoshida, K.: Optical axis adjustment of laser beam transmission system. In: Fifth International Conference on Digital Information Processing and Communications, pp. 13–18 (2015)
27. Tsujimura, T., Suito, Y., Yamamoto, K., Izumi, K.: Spacial laser beam control system for optical robot intercommunication. In: 2018 IEEE International Conference on Systems, Man, and Cybernetics (2018)
28. Tsujimura, T., Izumi1, K., Yoshida, K.: Collaborative all-optical alignment system for free space optics communication. In: INCoS 2018, LNDECT 23, pp. 146–157 (2019)

Enhancing Security of Cellular IoT with Identity Federation

Bernardo Santos[1](✉), Bruno Dzogovic[1], Boning Feng[1],
Van Thuan Do[1,2], Niels Jacot[2], and Thanh Van Do[1,3]

[1] Oslo Metropolitan Univeristy, Pilestredet 35, 0167 Oslo, Norway
{bersan,bruno.dzogovic,boning.feng}@oslomet.no,
thanh-van.do@telenor.com
[2] Wolffia AS, Haugerudvn. 40, 0673 Oslo, Norway
{vt.do,n.jacot}@wolffia.net
[3] Telenor ASA, Snarøyveien 30, 1331 Fornebu, Norway

Abstract. This paper presents a Cellular Identity Federation solution which both strengthens and simplifies the authentication of Internet of Things (IoT) devices and applications by providing single sign-on between the network layer and IoT applications. They are hence relieved of the burden of authentication and identity management, which could be both technically and economically challenging. The paper aims at clarifying how IoT authentication can be skipped without compromising security. The proposed solution is described thoroughly, and the authentication process is depicted step by step. Last but not least is the comprehensive description of the proof-of-concept which shows the feasibility of the Cellular Identity Federation.

Keywords: Cellular IoT · NB-IoT · Internet of Things · M2M · 5G · Lower Power wide area network

1 Introduction

Cellular Internet of Things is a rather new concept which emerged due to the demands for mobility and extended coverage of the Internet of Things. Briefly, the concept means that the mobile network is used as network layer for **IoT** applications. In fact, the cellular mobile network is originally intended for mobile phones which are on always online and require higher data rates compared to **IoT** devices which quite often have limited power and only communicate occasionally using a few bytes. Consequently, to use the 3G/4G mobile networks for **IoT** are neither cost efficient nor sustainable. To accommodate **IoT** devices, the mobile community has proposed a few wireless access technologies such as Extended Coverage GSM for Internet of Things (EC-GSM-IoT), Long Term Evolution Machine Type Communications Category M1 (LTE MTC Cat M1, also referred to as LTE-M) and Narrowband IoT (NB-IoT) [1]. Most importantly, the next mobile generation mobile network, **5G** is aiming at supporting a variety of **IoT** applications with diversified requirements in terms of mobility, bandwidth, latency and reliability by making use of the concept of network slicing [2].

© Springer Nature Switzerland AG 2020
L. Barolli et al. (Eds.): INCoS 2019, AISC 1035, pp. 257–268, 2020.
https://doi.org/10.1007/978-3-030-29035-1_25

Unfortunately, although better security is offered to **IoT** applications due to the inherent stronger security at the network level compared to other wireless technologies, it is not sufficient and **IoT** applications are left to themselves to ensure appropriate security.

This paper introduces an innovative solution which aims at providing higher level of security at the same time as relieving the administrative burden of the **IoT** applications. The solution makes use of the concept of identity federation which is used in the world wide web to provide single sign-on i.e. the user can just sign in once at one web site and move around to other ones without having to sign in again. In our case the device authentication at the network level is re-used on the **IoT** application to achieve single sign-on and higher level of security. The paper starts with a concise description of related works followed by a short introduction to identity federation. The main objective of the paper is to clarify how identity federation can both enhance and simplify security of cellular IoT applications. We choose to use a rather formal description using Unified Modelling Language (**UML**) [3] diagrams, being the central part of the paper the description of the proposed Cellular Identity Federation solution. The proof of concept implementation at the Secure 5G4IoT Lab[1] is also presented in a comprehensive way. The paper concludes with some suggestions for future works.

2   Related Works

As mentioned in [4] there are currently no known activities aiming at simplifying authentication of IoT devices by providing single sign-on with the cellular network authentication. However, there are a few works focusing on extending the usage of the SIM (Subscriber Identity Module) authentication for other applications on smart phones such as Internet browsing, Web mail, Social networks, financial services, etc.

The Generic Bootstrapping Architecture (**GBA**) [5, 6] is a standard specified by the 3rd Generation Partnership Project (**3GPP**), which achieves the mentioned objective by introducing in the mobile network a new network element called Bootstrapping Server Function (**BSF**), responsible for retrieving authentication vector from the Home Subscriber Server (**HSS**) and carrying out a mutual authentication of the mobile phone aka User Equipment (**UE**). The **BSF** provides the mobile Internet application aka Network Application Function (**NAF**) with encryption key Ks_NAF for the session between the **NAF** and the **UE**. The most serious limitation of this solution lies on the fact that **GBA** requires the presence of the **GBA** client on the mobile phone, which is quite difficult because handset manufacturers do not have the incentive to implement it.

To avoid the need for the BSF the Eureka Mobicome project has been proposing some solutions called Subscriber Identity Module (**SIM**) strong authentication that provides strong authentication from a regular browser on a regular mobile phone carrying a **SIM/USIM** [7, 8]. However, these solutions do not address IoT, in which devices are communicating without the intervention of human beings. ETSI did

---

[1] The Secure 5G4IoT lab results from the collaboration between OsloMet, Telenor and Wolffia within the scope of H2020 Concordia project: http://5g4iot.vlab.cs.hioa.no/.

promote the use of **GBA** in their Machine-To-Machine (**M2M**) functional architecture but they focus only on using the strong authentication of the **SIM** card. Indeed, they do not provide a comprehensive and flexible cellular **IoT** identity and access management, which enables both easy inclusion of **IoT** devices and strong authentication and confidentiality.

3 Identity Federation

Identity federation provides the means to share or to link users' identities between partners. Such identities are federated between partners when there is an agreement between the providers on a set of identifiers or identity attributes [4]. To share information about a user, partners must be able to identify the user, even though they may use different identifiers for the same user.

When two domains are federated, the user can authenticate to one domain and then access resources in the other domain without having to perform a separate login process.

Single Sign-On (**SSO**) is an important component of identity federation, but it is not the same as identity federation.

Identity federation involves a large set of user-to-user, user-to-application and application-to-application use cases at the browser tier, as well as the service-oriented architecture tier.

3.1 Most Popular Identity Federation Solutions

Although there are multiple Identity Federation solutions only a few popular ones are briefly introduced in this section.

A. Liberty Alliance

To alleviate the identity burden of both the users and the Service Providers, the Liberty Alliance Project, established in 2001 and overtaken by the Kantara Initiative in 2009 introduced the notion of Federated Network Identity [9]. A new actor called Identity Provider is responsible of authenticating the users and federating user accounts at Service Providers that join the Identity Provider's Circle of Trust.

B. OAuth2.0

OAuth 2.0 [10] is a scalable delegation protocol that allows a certain application to do certain tasks on behalf of a user. It establishes the concept of an authorization token, providing information on which services on a server the application is authorized to access to, not overriding the access control decision that the server may take.

C. OpenID Connect

OpenID Connect [11] is a standard built upon OAuth that goes one step further to offer single sign-on and identity provision on the internet. It enables client applications to verify the identity of the user based on the authentication performed by an OpenID Provider, as well as to obtain basic profile information about the user in an interoperable and REST-like manner. OpenID Connect specifies a RESTful HTTP API, using JSON as a data format. Client apps receive the user's identity encoded in a secure JSON Web Token (**JWT**) called ID token.

Currently OpenID Connect is definitely the most popular Identity Management which is used by several identity provider such as Facebook, Google, Twitter and even mobile operators in their Mobile Connect, a secure log-in solution using mobile phones. It is worth noting that OpenID Connect does not provide identity federation, i.e. federation of the **SP**'s identity with the **IDP**'s identity but promotes the usage of the **IDP**'s identity at the service provider. Further, it is not used in identity management for cellular **IoT**.

4   Identity Federation for Cellular IoT

4.1   Current IoT Systems

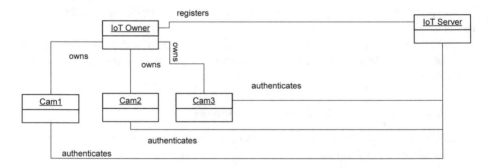

Fig. 1. A typical IoT system object diagram

Let us consider a current typical **IoT** system, for example a home surveillance camera system with 3 **IoT** devices and one **IoT** Server as shown in Fig. 1. The **IoT** Server can be in-house or in the cloud operated by the **IoT** manufacturer.

At the first-time installation all the devices are blank, and the owner will have to carry out a configuration which, although simple for technology professionals might be challenging for common people. First, the owner will have to get his/her devices connected to the wireless local area network and thereafter to Internet. If Wireless LAN [12] or Zigbee [13] is the wireless technology used in the local area network, the owner will have to supply a password or a link key which can be used by devices in the authentication towards the wireless home gateway to get granted connection.

After that the network connection is established, the **IoT** owner will have to personalise the **IoT** Server by registering his/her name, user name and a password which is used for authentication and access control for later sign-on sessions. Next, the owner shall register all devices and define passwords both the ones to be used upon access to the devices and the ones to be used by the devices upon authentication to the **IoT**

Server. Indeed, in order to get granted access the **IoT** devices will have to be authenticated towards the **IoT** Server.

4.2 Current Cellular IoT Systems

Let us now consider a cellular **IoT** system. The cellular mobile network is in this case used to provide Internet connectivity **IoT** devices. The **IoT** owner is also a mobile subscriber and acquires from mobile operator a number of subscriptions corresponding to the number of devices which have to be inserted in the devices as shown in Fig. 2. Upon power on, the **IoT** devices carry out authentication towards the mobile Home Subscriber Server (**HSS**) using their **SIM** cards. The **SIM** authentication is a strong authentication in the mobile network, which ensures that the **IoT** device uses a legal International Mobile Subscription Identity (**IMSI**) belonging to the mobile subscriber i.e. the **IoT** owner. Mobile operators in many countries also query and register the International Mobile Equipment Identity or **IMEI**, which uniquely identifies the device, in order to ban stolen devices.

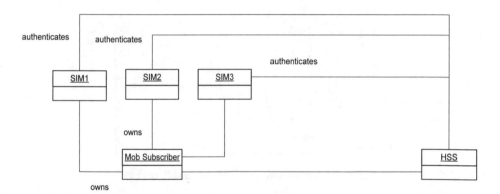

Fig. 2. Mobile connection object diagram

After successful authentication the **IoT** devices are allowed to get connected to the Internet. However, to communicate with the **IoT** server, the **IoT** devices will have to be personalised such that proper authentication with the **IoT** server can be performed as previously described.

4.3 Federation of Mobile and IoT Identities

The main finding from the previous sections is that prior to any communication between **IoT** devices and **IoT** server two authentication procedures must be done as follows:

- *Authentication of the mobile devices via the SIM* by the mobile network to ensure that both the mobile devices belong to the mobile subscriber to get granted access to the mobile network.

- *Authentication of **IoT** devices by the **IoT** server* to ensure that the **IoT** devices do belong to the **IoT** Owner to get granted access to the **IoT** server.

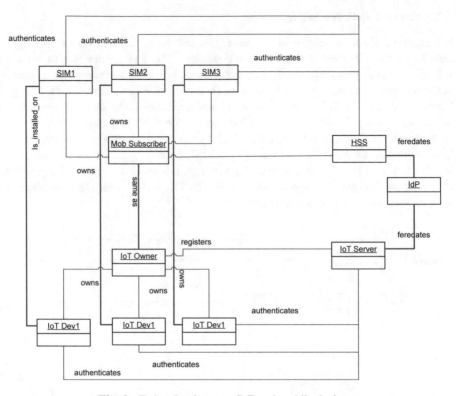

Fig. 3. Federation between IoT and mobile devices

The two authentications are necessary because the **IoT** system and mobile network are completely separate systems that do not know anything about each other.

To remove the later authentication on the **IoT** layer, the concept of federation of identities can come to help. Indeed, as shown in Fig. 3, federation provides the means to specify that the **Mobile_Subscriber** is **same as** the **IoT_Owner** who knows that the SIM_x is installed on IoT_Dev_x. Indeed, when the SIM_x is successfully authenticated by the **HSS**, if the IoT_Dev_x can prove that it carries this SIM_x then it is the IoT_Dev_x belonging to the **IoT_Owner** and can get granted access to the **IoT_Server**. This proof can be in a form of a unique one-time token generated by the HSS and passed through the mobile network to the IoT_Dev_x which presents to the **IoT_Server**.

In order to let the **HSS** and the **IoT_Server** know that the SIM_x containing an $IMSI_x$ is installed on the IoT_Dev_x, an Identity Federation table similar to the one shown in Table 1 has to be installed at both the **IoT_Server** and **HSS**.

Table 1. Identity federation table

Device_ID	IMSI
Camera_Front_Door	$IMSI_1$
Camera_Backyard	$IMSI_2$
Camera_Garage	$IMSI_3$
Smoke_Detector	$IMSI_4$
Contact_Front_Door	$IMSI_5$
Contact_Balcony_Door	$IMSI_6$
Contact_Back_Door	$IMSI_7$

Further, it is necessary to define and implement an authentication protocol between the **IoT_Server** and **HSS**, which enables the **IoT_Server** to ask the **HSS** for authorization upon login request from an **IoT** device. As to make it both simpler and more standardized, we propose to use OpenID Connect and hence to introduce an additional functional entity called **Identity Provider** or **OpenID Provider**, which acts as a middleman managing the identity federation table on behalf of both the **IoT_Server** and **HSS**. In addition, it will on behalf of the **IoT_Server** perform authorization of the **IoT** devices.

Before continuing with the explanation of the authentication procedure let us now present the overall architecture of the solution.

4.4   The Proposed Cellular Identity Federation Solution

As shown in Fig. 4, in the proposed Cellular Identity Federation solution, the current 4G mobile network consisting of standardized network elements such as **eNodeB**s (base stations), **MME** (Mobility Management Entity), **HSS** (Home Subscriber Server) on the control plane and **S-GW** (Serving Gateway), **P-GW** (Packet Data Network Gateway) on the user plane is now interfaced with the **IoT** Server such that information about successfully authenticated **IMSI**s can be shared with the **IoT** Server. Three new entities are introduced in the solution as follows:

- **The Identity Provider (IDP/IDMS):** makes use of the authentication information from the HSS to carry out authentication of the IoT devices on behalf of the **IoT_server**. In our solution, OpenID Connect is selected and used because it is by far the most popular and simple standard.
- **The Secure Database (DB):** To keep the security protection of the **HSS** at the same level as before, instead of introducing an IP interface on the **HSS** and allowing direct interactions with it from the IDP, we introduce a partially mirrored database, which gets transferred from the HSS only relevant parameters about successfully authenticated devices such as **IMSI, IMEI** (International Mobile Equipment Identity), **MSISDN** (Mobile Subscriber ISDN Number), **PDP** (Packet Data Protocol) type, **PDP** address (IP address), **APN** (Access Point Name), etc. These parameters will again be sent to the **IDP** for storage and use in the authentication of the IoT devices.

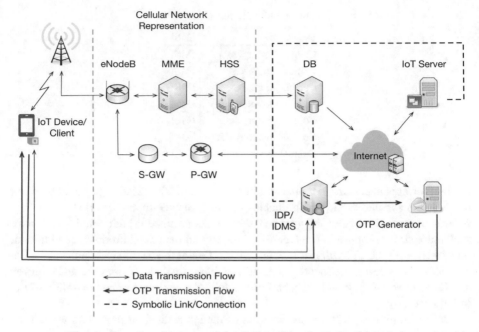

Fig. 4. Overall architecture of the Cellular Identity Federation solution

- **The OTP Generator:** To ensure that an **IoT_Dev$_x$** is really the one belonging to the **IoT_Owner,** it has to prove that it actually is the device carrying a **SIM$_x$** that has been successfully authenticated by the **HSS**. For that, it is not sufficient that it presents its IMSI or IMEI to the **IoT_Server** upon login because both **IMSI** and **IMEI** can be easily sniffed for replay. A more secure token is required. Upon receipt of authentication parameters of an **IoT_Dev$_x$** from the **DB,** the **IDP** will request the **OTP Generator** to produce a one-time password and send it to the **IoT_Dev$_x$** using the **PDP** address such that it can use it to login onto the **IoT_Server**.

4.5   The Authentication Process

To clarify how an IoT device is authenticated let us now consider an **IoT_Dev$_1$** hosting a **SIM$_1$**. The authentication process is as follows:

1. At power on, the **SIM$_1$** on **IoT_Dev$_1$** participates to the authentication of User Equipment
2. Upon successful authentication, the **HSS** stores the state of **IoT_Dev$_1$** as registered and notifies the Replica Database that initiates the duplication and transfer of the parameters of the **IoT_Dev$_1$** to the **IDP**.
3. The **IDP** stores the data and send a request for the generation of a one-time password to be sent to the **PDP** address of the **IoT_Dev$_1$**.
4. The **OTP Generator** generates an **OTP** and sends it to both the **IDP** and **IoT_Dev$_1$**.

5. The **IoT_Client** on the **IoT_Dev₁** fetches the **OTP** and presents it to the **IoT_Server** upon login.
6. The **IoT_Server** redirects the **IoT_Client** to the **IDP**.
7. The **IDP** compares the presented **OTP** and if it matches with the stored one the **IoT_Client** is considered authenticated and directed back to the **IoT_Server**, which grants access to **IoT_Client.**

The authentication process is hence completed without active participation of the **IoT_Server** which does not have to administrate passwords of its **IoT** clients while strong security is still ensured.

5   The Cellular Identity Federation Proof-of-Concept

To validate the proposed Cellular Identity Federation solution a proof-of-concept is built at the Secure 5G4IoT Lab at Oslo Metropolitan University consisting of a 4G/5G mobile network extended with 3 Identity Management entities and 1 IoT entity as shown in Fig. 5.

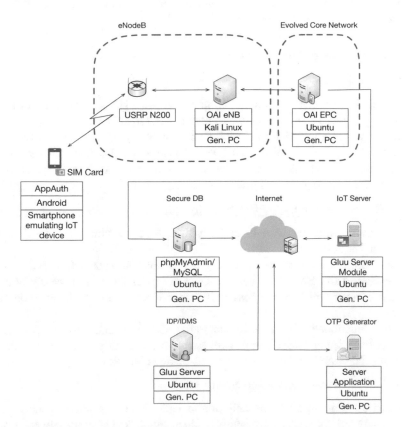

Fig. 5. The Cellular Identity Federation PoC at the Secure 5G4IoT Lab

5.1 4G/5G Mobile Network

To realize an early implementation of a **5G** mobile network OpenAirInterface [14], an open source communication software elaborated by EURECOM are first installed in generic computers and then later virtualized on the OsloMet to achieve a softwarized **5G** mobile network.

The 4G LTE base station **eNodeB** is realized by:

- A generic PC running Kali Linux and OpenAirInterface **eNB** connected to A USRP (Universal Software Radio Peripheral) N200, which is a software-defined radio designed and commercialized by Ettus Research [15].

The whole **Evolved Packet Core** including **HSS** is realized by:

- A generic PC running Ubuntu and OpenAirInterface, which includes an HSS.

5.2 IoT Server

The **IoT Server** is realized by:

- A generic PC running Ubuntu and Gluu Server 3.1.5 and also a lightweight M2M server open source using Eclipse Leshan [16].

5.3 Identity Management Entities

The **IDP** is realized by:

- A generic PC running Ubuntu and Gluu Server 3.1.5 [17], which is an open source identity provider server software compliant with OpenID Connect.
- Gluu Server is adopted because of its fast deployment and flexibility and also because it can act not only as an IDP but also as an IDMS which permits us to review and control all the identities to be issued and to handle basic profiling options, such as user grouping and role assignment.

The **Secure Database** is implemented by:

- A generic PC running Ubuntu and MySQL, an open source relational database management system (RDBMS).
- An extension is implemented allowing the database to establish a read-only SQL connection to the *HSS* database to extract the parameters of successfully authenticated devices which are necessary for identification and authentication of IoT clients.
- Another functionality is added to enable the Secure Database to set up a RESTful request by creating a JSON object for each device identity and then to send it the *IDP* for the establishment or updating of device identities.

The **OTP Generator** is implemented by:

- A generic PC running Ubuntu and an application which is able to generate a one-time password for each successfully authenticated device upon request from the *IDP*. This one-time password is then sent both the *IDP* and the **IoT** Device using its

PDP address. This one-time password will be presented by the **IoT** Device to the **IoT** Server at login.

5.4   IoT Devices

The **IoT Devices** are emulated by:

- Smartphone devices with Android OS equipped with the *AppAuth* [18] Software Development Kit (**SDK**) in order to interact with the *IDP* and the network. The proof-of-concept has been tested with the focus on flexibility and usability for **IoT** applications. The ability of registering and removing new **IoT** owners and their devices at the *IDP* has been verified.

6   Conclusions

In this paper we have introduced a Cellular Identity Federation solution which aims at simplifying the authentication of **IoT** devices in **IoT** applications which could be both technically and economically challenging for the users. By removing the need for **IoT** devices authentication, the proposed solution will contribute to the success of the coming **5G** mobile network.

Although the feasibility of the solution has been verified, the performed tests are still limited, and more diversified tests are needed in order to cover all the relevant scenarios. Most straightforward, the tests with real **IoT** devices such as security sensors, smart locks, e-health, etc. will have to replace the ones with emulated smartphones. This is a more demanding work because a client will have to be implemented for these **IoT** devices. A reasonable approach is to design and implement an open and generic client which can be customised and installed in multiple heterogeneous devices.

In the current solution the **IoT_Client** uses a one-time password to identify and authenticate itself towards the **IoT_Server**. Although it is functional, the solution is not optimal since it requires an **OTP** generator. A better solution would be to establish communications between the **IoT_Client** and its hosted **SIM** such that the **IoT_Client** can query the **TMSI** (Temporary Mobile Subscriber Identity) and use it instead of the one-time password. **TMSI** is temporarily assigned to the device during location registration and can be reallocated at certain intervals determined by the mobile operator. The usage of **TMSI** will prevent sniffing and replay.

Last but not least is to carry out a trial in real **5G** mobile network environment such as the **5G** VINNI Norway **5G** Facility Site at Kongsberg in Norway which will be available by late 2019. Such a trial with real users will make it possible to collect feedbacks that again can be used to improve the proposed solution.

Acknowledgement. This paper is a result of the H2020 Concordia project (https://www. concordia-h2020.eu) which has received funding from the EU H2020 programme under grant agreement No. 830927. The CONCORDIA consortium includes 23 partners from industry and other organizations such as Telenor, Telefonica, Telecom Italia, Ericsson, Siemens, BMW, Airbus, etc. and 23 partners from academia such as CODE, university of Twente, OsloMet, etc.

References

1. GSMA: 3GPP Low Power Wide Area Technologies White Paper, 1 September 2016, Svetlana Grant
2. Dzogovic, B., Do, V.T., Feng, B., Do, T.V.: Building virtualized 5G networks using open source software. In: Proceedings of 2018 IEEE Symposium on Computer Applications & Industrial Electronics (ISCAIE 2018). https://doi.org/10.1109/iscaie.2018.8405499
3. Booch, G., Rumbaugh, J., Jacobson, I.: Unified Modeling Language User Guide, 2nd edn. Addison-Wesley Professional, Boston (2005). ISBN-13: 978-0-321-26797-9
4. Santos,B., Do, V.T., Feng, B., Do, T.V.: Identity federation for cellular internet of things. In: Proceedings of 2018 7th International Conference on Software and Computer Applications (ICSCA 2018). ACM (2018). ISBN 978-1-4503-5414-1
5. 3rd Generation Partnership Project: 3GPP TS 33.220 V8.2.0 (2007-12) Technical Specification Group Services and System Aspects; Generic Authentication Architecture (GAA) Generic bootstrapping architecture (Release 8)
6. Olkkonen, T.: Generic Authentication Architecture, Helsinki University of Technology. http://www.tml.tkk.fi/Publications/C/22/papers/Olkkonen_final.pdf
7. Do, V.T., Jønvik, T., Do, V.T., Jørstad, I.: Enhancing Internet service security using GSM SIM authentication. In: Proceedings of the IEEE Globecom 2006 Conference, San Francisco, USA, 27 November–1 December 2006. ISBN 1-4244-0357-X
8. Do, V.T., Jønvik, T., Feng, B., Jørstad, I.: Simple strong authentication for internet applications using mobile phones. In: Proceedings of IEEE Global Communications Conference (IEEE GLOBECOM 2008), New Orleans, LA, USA, 30 November–4 December 2008. ISBN 978-1-4244-2324-8
9. Liberty Alliance: ID-FF Architecture Overview – vers. 1.2-errata-v1.0
10. IETF Request for Comments: 6749: The OAuth 2.0 Authorization Framework, October 2012
11. OpenID Connect. http://openid.net/connect/. Accessed October 2018
12. IEEE: 802.11 Wireless Local Area Networks. http://www.ieee802.org/11/. Accessed October 2018
13. Zigbee Alliance. https://www.zigbee.org. Accessed October 2018
14. The OpenAirInterfaceTM Software Alliance (OSA). http://www.openairinterface.org/. Accessed October 2018
15. Ettus Research, Inc., USRP N200. https://www.ettus.com/product/details/UN200-KIT. Accessed November 2017
16. Leshan. https://eclipse.org/leshan/. Accessed October 2018
17. Gluu. https://www.gluu.org/. Accessed April 2019
18. AppAuth. https://appauth.io/. Accessed April 2019

Reversible Data Hiding Algorithm in Homomorphic Encrypted Domain Based on EC-EG

Neng Zhou[1], Han Wang[2], Mengmeng Liu[2], Yan Ke[2(✉)], and Minqing Zhang[2]

[1] Key Laboratory of Network and Information Security under Chinese People Armed Police Force (PAP), Engineering University of PAP, Xi'an 710086, China
zn_2019@163.com
[2] Key Laboratory of Network and Information Security under PAP, Engineering University of PAP, Xi'an 710086, China
15114873390@163.com

Abstract. This paper proposes a reversible data hiding algorithm in homomorphic encrypted domain based on EC-EG. First, original image is segmented. The square grid pixel group randomly selected by image owner has one reference pixel and eight target pixels. The n least significant bits (LSBs) of reference pixel and all bits of target pixel are self-embedded into other parts of the image by predictive error expansion (PEE). The n LSBs of reference pixel are reset to zero before encryption for the purpose of avoiding overflow when embedding data. Then, the pixel values of image are encrypted after encoded onto the points of elliptic curve. The encrypted reference pixel replaces the encrypted target pixels surrounding it. Thence, mirror central ciphertext (MCC) is constructed. In a set of MCC, data hider embeds the encrypted extra data into the n LSBs of target pixels by homomorphic addition in ciphertexts, while the reference pixel remains unchanged. Receiver can directly extract extra data by homomorphic subtraction in ciphertexts between target pixels and corresponding reference pixel, and can extract the extra data by subtraction in plaintexts with the directly decrypted image, and can restore the original image without loss. The experimental results show that the proposed algorithm has higher security than the similar algorithms, and the average embedding rate of algorithm is 0.25 bpp.

1   Introduction

Reversible data hiding (RDH) takes data hiding and distortion-free recovery of the original carrier into consideration. At present, RDH mainly includes: lossless compression (LC), histogram shifting (HS) and difference expansion (DE). With the popularity of cloud services, users usually encrypt the uploaded data so that the original data becomes unreadable ciphertext data for the purpose of privacy protection. Reversible data hiding in encrypted domain (RDH-ED) comes into being and has become the latest research hotspot. The RDH-ED requires that

© Springer Nature Switzerland AG 2020
L. Barolli et al. (Eds.): INCoS 2019, AISC 1035, pp. 269–277, 2020.
https://doi.org/10.1007/978-3-030-29035-1_26

the carrier used for embedding be encrypted, and the carrier still be decrypted without error after extracting embedded data.

Zhang [1] firstly proposes RDH-ED, which encrypts images with stream ciphers, and then divides encrypted images into blocks that do not overlap each other. Each block has two groups. By flipping each pixel's 3 least significant bits (LSBs) in the corresponding group, 1 bit information can be embedded. Receiver extracts extra data through the wave function. However, when the block is small, the error rate becomes high. In [2], Zhang proposes a separable RDH-ED, which can extract extra data in both the encrypted domain and the plaintext domain. The above RDH-ED need to vacate room after encryption (VRAE) for data hiding, resulting in lower embedding capacity and higher error rate in the process of data extraction. Ma et al. [3] proposes a RDH-ED by vacating room before encryption (VRBE). Zhang et al. [4] improved reversibility, embedding capacity and image quality by using prediction errors in data embedding. RDH-ED we mentioned above are to encrypt images with symmetric cryptography because symmetric cryptography has lower computational complexity and faster encryption and decryption speed. However, due to using symmetric cryptography, each pair of senders and receivers must have different keys, which is required in the context of multi-party cloud computing services. The amount of keys is huge, which makes it difficult to manage keys. The key amount of public key cryptography is greatly reduced, and the key does not need to be transferred in advance. More importantly, we can perform homomorphic addition or multiplication in the encrypted domain through public key cryptography, which makes the embedding process of extra data more secure.

Chen et al. [5] firstly proposes encrypted image-based reversible data hiding with public key cryptography (EIRDH-P). Chen's method uses the characteristics of public key cryptography to overcome the shortcomings, which symmetric encryption requires the safe channel to pass the key in advance. The disadvantage of the Chen's method is an inherent overflow problem. Subsequently, Shiu et al. [6] and Wu et al. [7] improve Chen's method by solving the overflow problem. The above inseparable algorithms limit the application scenario and scope of the algorithms. The separable algorithms have more application scenarios. Receiver's privilege determines different operations it can perform. For example, receiver can only extract extra data, receiver can only decrypt the image and receiver can both decrypt the image and extract extra data.

Zhang et al. [8] firstly propose a separable EIRDH-P, which combines wet paper code (WPC) with Paillier homomorphic encryption. Zhang's method vacates room by histogram shrinking and uses WPC to embed extra data. Xiang et al. [9] proposed a RDH-ED based on mirroring ciphertext groups (MCGs) by using the homomorphic and probabilistic characteristics of the Paillier cryptosystem. In summary, using the homomorphic characteristics of public key cryptography for RDH-ED is the latest research hotspot. It can be called reversible data hiding in homomorphic encrypted domain (RDH-HED). This paper proposes a RDH-HED algorithm based on EC-EG. The algorithm adopts VABE by predictive error expansion (PEE) and constructs mirror central ciphertext

(MCC). Homomorphic addition in ciphertexts is for data embedding. Receiver can achieve separation of extra data extraction from the original image carrier recovery, extract extra data without any errors and achieve reversibility completely.

2 Related Knowledges

2.1 Construction of Elliptic Curve ElGamal (EC-EG)

When constructing a cryptosystem, the following elliptic curve is mainly used, and the equation is

$$y^2 = x^3 + ax + b(x, y, a, b \in F_P) \tag{1}$$

The set $E(F_P)$ of points on the elliptic curve forms an Abel group.

Definition 1. Suppose that E is an elliptic curve and G is a point on E. If $\exists n \in N^*$ exists, $nG = O$ is established, then n is said to be the order of point G, where O is the infinity point.

Key generation: Select the base field F_P, the elliptic curve E_P. Encode the plaintext information m to the point $M(m)$ on the curve. Select the generator G (base point) of E_P as the public parameter. User selects the secret key $sk \in F_P(sk < n)$ and the public key is $pk = skG$.

Encryption: Bob selects a random integer $r(r < n)$ and calculates the ciphertext as

$$C(m) = (C_1, C_2) = (rG, M(m) + rpk) \tag{2}$$

Decryption: Alice calculates the following equation to complete the decryption.

$$M(m) = C_2 - skC_1 \tag{3}$$

Homomorphic Addition: After encoding the plaintext (m_1, m_2) to the two points $(M(m_1), M(m_2))$ on E_P, the ciphertext is calculated using random integer (r_1, r_2) as.

$$\begin{aligned} C(m_1) &= (C_{11}, C_{21}) = (r_1G, M(m_1) + r_1pk) \\ C(m_2) &= (C_{12}, C_{22}) = (r_2G, M(m_2) + r_2pk) \end{aligned} \tag{4}$$

Then the homomorphic addition can be expressed as

$$\begin{aligned} C(m_1) + C(m_2) &= C_{21} + C_{22} - skC_{11} - skC_{12} \\ &= M(m_1) + r_1pk + M(m_2) + r_2pk - r_1skG - r_2skG \\ &= M(m_1) + M(m_2) \end{aligned} \tag{5}$$

That is to say, the sum of the two ciphertexts is equal to the sum of the plaintexts after decryption.

2.2 Rhombus Pattern Based PEE

The rhombus prediction error extension [10] is to divide the image pixels into overlapping 3×3 pixel groups. Pixels of a group are further divided into "$\times$" and "$\bullet$" sets. Pixels of "$\times$" set are used for data embedding and pixels of "$\bullet$" set are used for prediction. A pixel $p_{i,j}$ in the "$\times$" set is predicted by its four neighboring pixels $(p_{i,j-1}, p_{i,j+1}, p_{i-1,j}, p_{i+1,j})$ in the "$\bullet$" set. The embedding, extraction and recovery procedures are given as follows.

The Embedding Procedure: The predicted value $p'_{i,j}$ is the average of four adjacent pixel values of $p_{i,j}$ and is calculated as

$$p'_{i,j} = \left\lfloor \frac{p_{i,j-1} + p_{i,j+1} + p_{i-1,j} + p_{i+1,j}}{4} \right\rfloor \tag{6}$$

The prediction error $e_{i,j}$ is the difference between the original pixel $p_{i,j}$ and the predicted value $p'_{i,j}$ and is calculated as $e_{i,j} = p_{i,j} - p'_{i,j}$. The difference expansion method [11] is employed to expand the prediction error $e_{i,j}$ for data embedding. $e'_{i,j} = 2 \times e_{i,j} + b$ is used for PEE, where $e'_{i,j}$ is the modified prediction error after data embedding and b is one bit of the data to be embedded. So each group can embed one bit of information. The modified pixel value $P_{i,j}$ after data embedding is calculated as $P_{i,j} = e'_{i,j} + p'_{i,j}$.

Overflow or Underflow Processing: The PEE data embedding may cause overflow or underflow. L_map can be used to avoid embedding into pixels that cause overflow or underflow. Equation 7 can be derived from the above Eq.

$$P_{i,j} = e_{i,j} + p_{i,j} + b \tag{7}$$

According to Eq. 7 it is clear that pixels that do not cause overflow or underflow satisfy the condition in Eq. 8.

$$0 \leq e_{i,j} + p_{i,j} \leq 254 \tag{8}$$

L_map to record the positions of invalid groups can be created by pre-processing the original image. Each group of the original image is checked for Eq. 8 and records the row and column binary data of invalid groups. L_map is compressed and then self-embedded into to the image.

Extraction and Recovery Procedures: The extraction procedure is the inverse of the embedding procedure. Since the "$\bullet$" pixels do not change, the receiver has the same predicted pixel value $p'_{i,j}$ of "$\times$" pixels as the image owner. Given the modified pixel value $P_{i,j}$ and predicted value $p'_{i,j}$, the modified prediction error $e'_{i,j}$ is calculated as $e'_{i,j} = P_{i,j} - p'_{i,j}$. Then the embedded information is calculated as $b = e'_{i,j} \bmod 2$. The original prediction error $e_{i,j}$ is calculated as $e_{i,j} = \left\lfloor \frac{e'_{i,j}}{2} \right\rfloor$. Finally, the original pixel value $p_{i,j}$ is recovered as $p_{i,j} = p'_{i,j} + e_{i,j}$

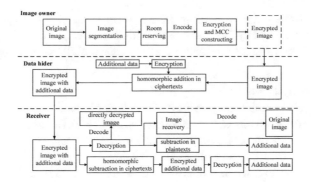

Fig. 1. Flow chart of the proposed algorithm framework

3   Proposed Algorithm

3.1   Room Reserving

3.1.1   Image Partition

The overall framework of RDH-HED based on EC-EG is shown in Fig. 1. Before the original image is encrypted, the pixel values of image must be encoded onto the points of the elliptic curve. Then, it can be encrypted using EC-EG. As is shown in Fig. 2, image owner divides the original image into three non-overlapping blocks A, B, C.

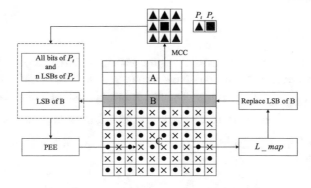

Fig. 2. Image partition and self-reversible embedding

3.1.2   Self-reversible Embedding

Self-reversible embedding is to embed all bits of target pixels and n LSBs of reference pixels in A, the LSB in B into C through PEE. A_{r_map} is the set of locations of reference pixels in A. Image owner sends n and A_{r_map} as shared parameters to data hider and receiver. More importantly, *L_map* is scrambled by the self-embedding key k_1, and is embedded into B by replacing the LSB for data extraction and image recovery to achieve reversibility completely.

3.1.3   Vacating Room

After self-reversible embedding, all bits of target pixels P_t and n LSBs of reference pixels P_r are saved in C. Image owner sets n LSBs of reference pixels P_r to zero, then the extra data is embedded into n LSBs of the target pixels P_t that have been set zero by homomorphic addition in ciphertexts. This operation can also avoid overflow when embedding data into the encrypted domain.

3.2   Image Encryption and Construction of Mirroring Central Ciphertext

After vacating room, image owner randomly selects an integer $r_{i,j}, i = 1, 2, \cdots, M, j = 1, 2, \cdots, N$ for each pixel $P'_{i,j}$, and encrypts image with the public key pk to obtain the ciphertexts.

$$M(C_{P'_{i,j}}) = (C_{1_{i,j}}, C_{2_{i,j}}) = (r_{i,j}G, M(P'_{i,j}) + r_{i,j}pk) \tag{9}$$

After the image is encrypted, image owner constructs MCC by replacing the ciphertexts of target pixels with the ciphertext of reference pixel, which is $c'_t = c_r$. In a square grid, all target pixels become mirror of reference pixel. All target pixels have the same ciphertext and the n LSBs that have been set to zero as reference pixel.

3.3   Data Hiding

Data hider firstly scrambles the extra data with the hidden key k_2. Then groups the scrambled extra data w and each group has n bits, which is recorded as $w_0, w_1, \cdots, w_{n-1}$. From this, the decimal value of each set of data can be calculated as $S_w = \sum_{i=0}^{n-1} w_i \times 2^i, (S_w = 0, 1, \cdots, 2^n - 1)$. To enhance confidentiality, data hider can choose a random integer $r_{S_w}(r_{S_w} < n)$ for each S_w and encrypt it with EC-EG to get ciphertext C_{S_w}.

$$M(c_{S_w}) = (r_{S_w}G, M(S_w) + r_{S_w}pk) \tag{10}$$

According to the method of room reserving, the encrypted image is divided into A_E, B_E, C_E. Data hider embeds extra data into the encrypted domain, which the equation is:

$$\begin{aligned}
M(c''_t) &= M(c'_t) + M(c_{S_w}) \\
&= (r_tG, M(t') + r_tpk) + (r_{S_w}G, M(S_w) + r_{S_w}pk) \\
&= M(t') + r_tpk + M(S_w) + r_{S_w}pk - skr_tG - skr_{S_w}G \\
&= M(t') + M(S_w)
\end{aligned} \tag{11}$$

Among them, C_t'' is the encrypted target pixel with hidden data. Suppose that a group contains 3 bits of data, which is $n = 3, (w_0 w_1 w_2) = (100)$, then we can calculate $S_w = 1, c_{S_w} = E(1, r_{S_w})$. Data hider performs homomorphic addition in ciphertexts to embed 3 bits of extra data into the corresponding MCC $[c_t' = E(160, r_r), c_r = E(160, r_r)]$ and then obtain the corresponding $c_t'' = E(161, r_{S_w} + r_r)$.

3.4 Data Extraction

3.4.1 Extract the Hidden Data from Encrypted Image

When having the shared parameter n, A_{r_map} and the hidden key k_2 receiver extracts extra data from the received encrypted image with hidden data by using homomorphic subtraction in ciphertexts. The steps are as follows:

Step 1: Received image is divided into blocks in the same way as the process of room reserving. Legal receiver extracts the square grid with hidden data.

Step 2: In MCC, the encrypted target pixels with hidden data is $M(c_t'') = M(c_r) + M(c_{S_w})$. $M(c_{S_w})$ is extracted by the homomorphic subtraction, which is $M(c_{S_w}) = M(c_t'') - M(c_r)$.

Step 3: Use $M(S_w) = M(c_{S_w}) + r_{S_w} pk - sk r_{S_w} G$ to decrypt $M(c_{S_w})$ to $M(S_w)$ and decode to get S_w.

Step 4: Receiver can use $w_i = \lfloor \frac{S_w}{2^i} \rfloor, i = 0, 1, \cdots, n-1$ to get $w_0, w_1, \cdots, w_{n-1}$, then use the hidden key k_2 to perform scrambling recovery. At last, we can extract extra data.

3.4.2 Extract the Hidden Data from Decrypted Image

When receiver only has the secret key sk of image owner, a directly decrypted image with hidden data is obtained from $M(P_{i,j}'') = M(C_{i,j}'') + r_{i,j} pk - sk r_{i,j} G$, where $C_{i,j}''$ is the ciphertexts with hidden data and $P_{i,j}''$ is the corresponding plaintexts with hidden data. According to the method of room reserving, the directly decrypted image is divided into three parts of A_D, B_D, C_D, and $M(S_w) = M(P_t'') - M(P_r'')$ can be obtained by plaintext subtraction in the target pixel pair of the square grid, where P_t'' and P_r'' are the target pixels and reference pixel in the directly decrypted image, respectively. Finally, we can obtain extra data with the method of step 3 and step 4 in Sect. 3.4.1.

3.5 Image Restoration

When receiver has the secret key sk, the self-embedded key k_1 and the hidden key k_2, extra data can be extracted after decryption. The original image can be restored. The directly decrypted image is obtained by decrypting with the secret key sk, The directly decrypted image is divided into three parts of A_D, B_D, C_D according to the method of room reserving. Since the extra data is hidden in P_t of A_D, the hidden key k_2 can be used for scrambling recovery after the extra data is extracted. Receiver extracts self-embedded data from B_D and C_D to restore the original image.

L_map is extracted from the LSB of B_D. The self-embedded key k_1 is used to scrambling recovery. Finally, according to the extraction step of PEE, all the self-embedded data can be extracted from C_D, and the data extracted from C_D is used to complete the recovery of the A_D. The original image of reversible recovery has been obtained so far.

4   Simulation Experiment and Theoretical Analysis

The 8-bit gray-scale Lena of 512×512 size is selected in the experiment. Peak signal-to-noise ratio (PSNR) is the most commonly used criterion for evaluating image quality after embedding data. Figure 3 shows PSNR performance comparison between the proposed algorithm and the literature [8, 9, 12] (Table 1).

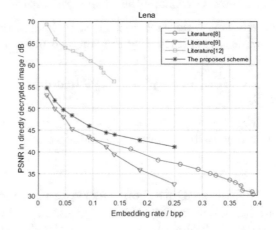

Fig. 3. PSNR performance comparison of Lena

Table 1. Embedding capacity and PSNR values under different n

Lena	Embedding capacity (bits)	PSNR (dB)			
		n = 1	n = 2	n = 3	n = 4
	4095	51.681	53.448	54.719	45.353
	8190	48.765	51.13	51.786	43.472
	12285	46.993	49.022	49.618	41.04
	16380	45.658	48.012	48.32	40.132
	24570	43.034	45.684	45.885	37.97
	32760	40.82	44.301	44.456	36.384
	36855	39.983	43.634	43.92	35.609
	49140	38.219	42.09	42.667	34.353
	65520	35.829	40.03	41.146	32.611

5   Conclusion

The algorithm in this paper uses the method of constructing MCC after EC-EG encryption to perform RDH-ED. The algorithm realizes the separation of extra data extraction and original image restoration, and improves the embedding capacity under the premise of completely recovering the original image. The experimental results show that PSNR of the algorithm is improved in the image due to the self-embedded data by PEE. However, the method of VRBE increases the risk of plaintext leakage. Therefore, it should be improved in the future as RDH-HED that does not require preprocessing of plaintext.

Acknowledgements. This work was supported by National Natural Science Foundation of China under Grant No. 61872384. The authors also gratefully acknowledge the helpful comments and suggestions of the reviewers.

References

1. Zhang, X.P.: Reversible data hiding in encrypted image. IEEE Sig. Process. Lett. **18**(4), 255–258 (2011)
2. Zhang, X.P.: Separable reversible data hiding in encrypted image. IEEE Trans. Inf. Forensics Secur. **7**(2), 826–832 (2012)
3. Ma, K.D., Zhang, W.M., Zhao, X.F., et al.: Reversible data hiding in encrypted images by reserving room before encryption. IEEE Trans. Inf. Forensics Secur. **8**(3), 553–562 (2013)
4. Zhang, W.M., Ma, K.D., Yu, N.H.: Reversibility improved data hiding in encrypted images. Sig. Process. **94**, 118–127 (2014)
5. Chen, Y.C., Shui, C.W., Horng, G.: Encrypted signal-based reversible data hiding with public key cryptosystem. J. Vis. Commun. Image Representation **25**(5), 1164–1170 (2014)
6. Shiu, C.W., Chen, Y.C., Hong, W.: Encrypted image-based reversible data hiding with public key cryptography from difference expansion. Sig. Process.: Image Commun. **39**, 226–233 (2015)
7. Wu, X.T., Chen, B., Weng, J.: Reversible data hiding for encrypted signals by homomorphic encryption and signal energy transfer. J. Vis. Commun. Image Representation **41**, 58–64 (2016)
8. Zhang, X.P., Long, J., Wang, Z.C., et al.: Lossless and reversible data hiding in encrypted images with public-key cryptography. IEEE Trans. Circ. Syst. Video Technol. **26**(9), 1622–1631 (2016)
9. Xiang, S.J., Luo, X.R.: Reversible data hiding in homomorphic encrypted domain by mirroring ciphertext group. IEEE Trans. Circ. Syst. Video Technol. **28**(11), 3099–3110 (2018)
10. Sachnev, V., Kim, H.J., Nam, J., et al.: Reversible watermarking algorithm using sorting and prediction. IEEE Trans. Circ. Syst. Video Technol. **19**(7), 989–999 (2009)
11. Tian, J.: Reversible data embedding using a difference expansion. IEEE Trans. Circ. Syst. Video Technol. **13**(8), 890–896 (2003)
12. Li, Z.X., Dong, D.P., Xia, Z.H., et al.: High-capacity reversible data hiding for encrypted multimedia data with somewhat homomorphic encryption. IEEE Access **6**, 60635–60644 (2018)

Impact of Automation to Event Management Efficiency

Peter Balco[✉] and Andrea Studeničova

Faculty of Management, Department of Information Systems,
Comenius University in Bratislava, Odbojárov 10, P.O. BOX 95,
820 05 Bratislava, Slovakia
{peter.balco,andrea.studenicova}@fm.uniba.sk

Abstract. The intelligent, SMART environments represent a complex, complicated but vibrant technological social ecosystem generating a wide range of events of different types during their lifecycle, i.e., service delivery. These are the results of the multilateral interactions of the individual configuration items that make up the system. Naturally, the greater the number of elements that environment contains, the greater the number of events "can" arise. For each generated event, it is necessary to respond appropriately, as it is the carrier of the information value that "can" affect the correct functionality. In our contribution, we want to demonstrate how well-designed and optimized business processes allow automation and artificial intelligence to be used to manage events to increase productivity and reduce costs.

Keywords: SMART · Cloud · Cloud computing ·
Business process management · BPM · Process automation ·
Artificial intelligence

1 Introduction

Providing the SMART services in an intelligent environment is the goal of many projects. Imagine a school, a manufacturing enterprise, a city service or some other cluster aiming to provide just such services. When we detail this environment, we realize that we are creating a complex, complicated but vibrant technological social ecosystem. This includes a myriad of configuration items that bidirectionally communicate and create an information value that can be used to make decisions and support, or suppress generated events, in certain logical contexts.

Figure 1 presents a simple model of transforming input events into output that does not affect input. It is a transformation where each element presents a configuration item and these properties affect the output with its properties and relationships.

Figure 2 interprets more complicated model of input event transformation. Individual configuration items within the system affect each other, while the output is also influenced by backward and interactions from the environment. Of course, the amount and variety of relationships presented by each business process puts pressure on the management and efficient delivery of services, and with their disproportionate numbers

L. Barolli et al. (Eds.): INCoS 2019, AISC 1035, pp. 278–288, 2020.
https://doi.org/10.1007/978-3-030-29035-1_27

can create risks that are a threat to particular group of participants but appear to be an opportunity for others.

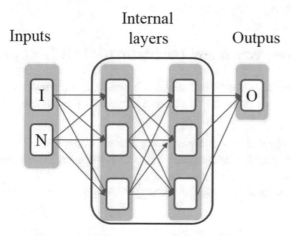

Fig. 1. Transformation of internal event to external outputs without feedback

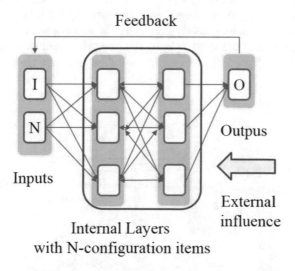

Fig. 2. Transformation of internal event to external outputs with feedback and influence of environment

Therefore, to manage such a complicated, sometimes inhomogeneous system and to meet its objectives, it is necessary to set clear rules and relationships between the configuration items that are linked to the individual business processes of the business processes. From the point of view of practice, we know that many of the events generated by these environments are repetitive, thus creating room for the introduction of intelligent solutions, automation and artificial intelligence into management that have a positive economic impact. "Productivity" is a key concept to which we will pay attention in our contribution. Setting up business processes properly and using intelligent solutions can make it easier to serve customers and eliminate the need for human intervention [4, 10].

2   Event Management and Intelligent (SMART) Environment

Events in an internal, near, or external environment are a result of a clash, a process connection to another process that consists of mutually communicating configuration items. Individual events may take the form of:

- An incident where a downtime or degradation occurs;
- Problem, recurring incident or incident with a wide impact on customers;
- Requirements;
- Scheduled or unplanned activities, changes, configurations;
- Other.

Events can cover a wide range of areas, i.e.: IT or customer services, security, human resources management, logistics, financial services, PR, manufacturing, and others. In SMART environments, service users generate a wide range of events in addition to this, depending on the degree of integration, including several detectors that monitor and support its management. Therefore, for an environment of this type to be effectively managed, it is essential to have an appropriate tool available to ensure the collection, storage and processing of events and data associated with them.

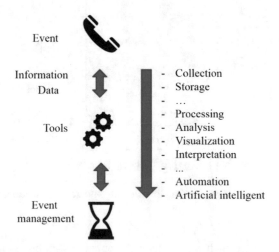

Fig. 3. Even management by using tool for management decision

Subsequent analysis, which must be performed in real time, provides the basis for visualization and decision-making at each management level. In this context, it should be noted that the increase in the number of configuration items presented by users, HW and SW components increases the number of interactions with possible subsequent impacts. If their number exceeds the accepted number, the management and proper functioning of the entire eco-system may fail. To avoid such events, it is necessary to implement an appropriate level of automation as well as management through artificial intelligence using a modern technology portfolio.

Figure 3 presents a communication model where the event is driven by a tool where the output can be a management decision or a basis for automated decision, machine learning, and so on. A suitable event management tool, a technology platform we should include a comprehensive set of tools for managing and managing enterprise-wide processes. It has been designed for all departments that are providers of internal or external services.

3 Process of Automation and Artificial Intelligence

By practice is proven, the introduction of automation in the areas of production or service can be the way to increase the efficiency of management and productivity of individual processes. In this case we are replacing the service provider, responding to the suggestion by the slot machine. The idea of introducing a predefined response to a clear stimulus is not new. This approach has been long in the history, where such approaches have already been implemented in mechanical devices.

The automation presents a degree of development characterized by the implementation of manufacturing, management and other processes with minimal impact of human intervention. These technological changes are associated with the existence of automatic production lines and the introduction of automated plants and plants that utilize elements of modern computing and control technology. Automation does not completely exclude human participation. Its role is important in controlling, managing the work of individual technology units. His participation is "compulsory" in order to set up machines, enter the program, supply materials, maintain, although with the development of machine automation increasingly take over these functions.

In modern IT environments, this machine is presented with an intelligent tool that allows you to automate service management and replace manual events used to manage events. In addition, automation provides two-way outputs in this case. Process automation keeps repetitive and manual activities that are a source of error, limiting employee development and leading to frustration. In such a situation, we need to think about how to deal with different types of events, requests, complaints or changes in the long term, and how to replace manual work with reasonable automation, which allows operational teams to respond promptly to events.

Removing man's influence in decision-making on automation and increasing the autonomy of technological units is realized through artificial intelligence, which is the key to Industry 4.0. Artificial intelligence covers a number of theories aimed at imitating human schemes or, in general, the biological behavior of evaluating and analyzing environmental stimuli or creativity. Artificial intelligence sets goals, that is, the level of intelligence of the autonomous system, the expectations and seeks solutions available in various disciplines [7].

According to the German association Bitkom, this technology combines enormous growth potential; only this year is projected to increase by 92%, whose volume will double by 2020 [6]. In addition to the already discussed productivity in the production or provision of services, in the case of artificial intelligence it is the implementation of specific technological procedures involving the processing of large amounts of data, structured and unstructured, and their complicated links to individual resources. Intelligent manufactured environments provide the space for implementing robotic devices, thus achieving a completely new level of production.

Cognitive computing and machine learning are two other central aspects of artificial intelligence applications that provide the foundation for innovative progress in the context of Industry 4.0. Smart programs make it possible to achieve new solutions and results. Alongside the success of the system. Human-machine co-operation is at the heart of the artificial intelligence and industry 4.0 interconnection and contributes substantially to their success [5].

Artificial intelligence can be used for the following reasons:
Broadband network access;

- Artificial intelligence can be offered as a cloud service, making it available continuously;
- Decreasing costs;
- The degree of application and continuous advancement of intelligent algorithms that have ability to learn.

To correctly assess the possibilities of applying artificial intelligence and the potential for the economy, it is important to understand its feasibility.

The actual degree of maturity of artificial intelligence can be interpreted in autonomous environments, such as cars or other enterprise stock vehicles. Autonomous transport systems present complicated solutions integrating a wide range of configuration items, communicating bidirectionally integrated into many business processes. These processes include the perception and interpretation of the environment, as well as the transformation of these stimuli into concrete actions. Artificial intelligence technologies are progressing steadily despite the existing setbacks, but it can be said that this is a major challenge that cannot be ignored.

4   Process of Automation and Artificial Intelligence

As mentioned in the previous chapter, solutions based on automation and artificial intelligence use platforms built on cloud infrastructure and are available to users as a service. Cloud is a modern and efficient way of delivering information technology (IT) services. The view of IT from the classic model is changing, where investments in infrastructure and human personnel have been dealt with. Where the priority is supplier-client relationships in the form of SLA, i.e. clearly defined contractual relationships and responsibilities [2, 3, 6].

Therefore, to reflect market requirements, i.e. the diverse needs of customers, the following basic requirements for cloud infrastructure have been defined:

- Broadband network access;
- Elasticity;
- Service measurability;
- Self-service character according to service requirements;
- Pooling resources.

With multi-instance and distributed access, IT architecture creates unique solutions for everyone. This approach provides a high degree of flexibility, freedom and security in addressing specific individual needs. Smart Platform provides potential customers with ready to use pre-defined solutions that can be deployed and used in standard mode within a few weeks without designing workflows and programming [8].

From the point of view of the level of management of individual parts, the chain of activities in the processes of automation and artificial intelligence, the most frequently used cloud services are PaaS, SaaS and BPaaS, see Fig. 4.

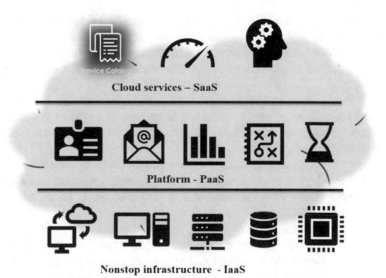

Fig. 4. Structure of cloud services and their hierarchy

A. *Platform as a Service – PaaS*

Solutions are provided to the user by using programming languages, libraries, services, and tools supported by the provider.

B. *Software as a Service – SaaS*

User-provided application-based solutions are deployed on the cloud infrastructure. These applications are available from a variety of client devices, either through a thin client interface, such as a web browser, or through a program interface.

C. *Business Process as a Service – BpaaS*

Today's cloud services are increasingly offering users the whole process that is presented as a combination of multiple services. The service will provide a series of services according to client requirements. The product of this process is data obtained by the service operator, which in turn allows targeting to the client's needs more precisely. The service provider connects the demand and the offer in real time and charges a fee for this service. It is very convenient for clients because the process offers a wide range of services.

In today's conditions it is also called XaaS, which stands for Anything-as-a-Service. It's anything that can be delivered remotely using cloud technology. Typically, this can be Disaster-Recovery-as-a-Service (DRaaS). Naas (Network-as-a-Service), etc. Since the potential of the creators of various services is virtually nothing in the way today, in the future we will see an increase in the variety of still unavailable today cloud-based services [1].

Typical examples of cloud infrastructure services are:

- IT Service Management (ITSM) provides a modern IT management of IT services in the cloud. "A single activity system allows you to consolidate tools and processes, transform service delivery and improve customer service". Processes, workflows can be automated, real-time information gathered, and IT productivity improved.
- Security & Risk Service Management (SRSM) in turn provides security management and automation of security events. It links existing security tools with security operations to respond immediately to events and locations vulnerable to their potential impact and priorities.
- HR Service Management (HRSM) provides employers with the services they need to support employees to meet organizations' strategic goals. Removes frustration and improves employee satisfaction with a single access point for efficient, personalized HR services.
- Customer Service Management (CSM), in turn, increases the value of customer service to distinguish itself from the standard. It creates a space for delivering intelligent customer services where other teams are involved in addition to customer service to quickly and proactively resolve interactions and events.

5   Case Study - Automation Routines and Economic Impacts

Practice shows that even a small increase in efficiency can drastically reduce labor, costs and release budgets for other strategic opportunities. We carry out many activities within individual processes manually without automation, resulting in their inefficiency and low productivity.

In this part of an article, we will present the impacts of implementing automation on efficient process management in self-government environments.

Cities are as a configuration item a complicated economic-social ecosystem where the large number of events take place in parallel and in two-way mode. It is a provider of a wide range of services that are tracked and used by the population. In the event of a service outage, there is a natural dissatisfaction and demand for information. In classic mode, we use Infoline, email communication, web communication. You realize how much time and resources we are losing ineffectively if we realize that on the service removal is already working (Fig. 5).

Fig. 5. Model push/pull services delivery

Did you realize that the error is in the process? Solution? Proactive Control! Imagine an ideal situation, that when an outage is detected, the event is recorded in the system. Proactive management, through a tool where customers are registered, sends them automatic information and, after solving the problem, notifications. In addition, the system can evaluate in which parts of the city the most common events occur and, for example, plan preventive maintenance accordingly. In this way, most of the agenda that involves repetitive activities can be addressed - from requests, through confirmations to various other services. Practice shows that even a small increase in efficiency through process change can significantly reduce labor and costs.

In the Table 1 we summarized the effects of the model - Pull service delivery to customers. As we can see, in the case of a reactive delivery service, there is a natural phenomenon that service consumers are aware of the status of services through existing communication channels where unnecessary efforts are seen on both sides plus we see an unnecessarily burdened communication environment.

Table 1. The pull service delivery model

Our pilot city represents population of 10,000 residents, where 27% from than are using local services, it means: 2700 residents. The type of service represents "reactive" approach. Presented outputs represent the average values from sample of residents

Type of services	Type of communication channel	Time consumption per service and request by customer in "t –Time"	Time consumption per service and request by service provider in "t –Time"
Request for information	Call center request	<3t	<3t
	E mail request	<7t	<4t
	SMS request	<2t	<2t
	WEB browsing	<5t	NA
	Paper sheet	<40t	<20t
	Ticket tool	<2t	<3t
Confirmation request	Call center request	<3t	NA
	E mail request	<7t	<4t
	Paper sheet	<30t	<20t
	Ticket tool	<2t	<2t
Complaint	Call center request	<5t	<5t
	E mail request	<7t	<4t
	Paper sheet	<30t	<20t
	Ticket tool	<2t	<2t
Request for change	Call center request	<10t	NA
	E mail request	<7t	<30t
	Paper sheet	<30t	<60t
	Ticket tool	<2t	<30t

Customers invest time and resources to get information based on their incentives, while suppliers and service operators are burdened with the incentives to respond. The result is non-efficiency of communication, dissatisfaction, unnecessary costs. In addition, recurring activities are a source of error, limiting staff development and leading to frustration. In this situation, we need to think about how to deal with different types of events, requests, complaints or changes in the long term and how to replace manual work with reasonable automation.

Table 2. The push service delivery model

Our pilot city represents population of 10,000 residents, where 27% from than are using local services, it means: 2700 residents. The type of service represents **"proactive"** approach. Presented outputs represent the average values from sample of residents

Type of services	Type of communication channel	Time consumption per service and request by customer in "t –Time"	Time consumption per service and request by service provider in "t –Time"
Request for information	Call center request	<1t	<1t
	E mail request	<7t	<1t
	SMS request	<2t	<1t
	WEB browsing	<5t	NA
	Paper sheet	<40t	<20t
	Ticket tool	<2t	<1t
Confirmation request	Call center request	<1t	NA
	E mail request	<7t	<1t
	Paper sheet	<30t	<20t
	Ticket tool	<2t	<1t
Complaint	Call center request	<2t	<5t
	E mail request	<7t	<4t
	Paper sheet	<30t	<20t
	Ticket tool	<2t	<2t
Request for change	Call center request	<3t	NA
	E mail request	<7t	<13t
	Paper sheet	<30t	<60t
	Ticket tool	<2t	<10t

In the Table 2 we observe the impact of deployment of the model - Push service delivery. As we can see, in the case of a proactive way of delivering services, customers are informed and downsized or degraded, as well as when the incidents are removed. The result is low operating costs and customer consistency. However, it should be borne in mind that there is an automation cost in this latter case and therefore its scope needs to be considered.

By comparing the Tables 1 and 2, reactive and proactive of service delivery processes, including the automation for proactive service management we observe very interesting benefits on the both parts, the customer and the service provider too.

Savings are presented in our case by time, which can be transformed into financial savings. The time saving is presented by the "t" parameter multiplied by the corresponding measured coefficient. In the case of deployment of automation on the customer side, we observe time savings from 20% to 60%. In the case of service suppliers, the time savings were higher, it means from 30% to 90%. The amount of savings depends on the type of service as well as the ability of customers to communicate with smart solutions. Other significant savings are observed on hardware and software infrastructure as well as reduced transmission capacity. We have shown how we can make savings and improve the quality of our services by changing business processes, educating our customers, and using intelligent solutions.

6 Conclusion

The implications of the implementation of automation as well as the elements of artificial intelligence in service delivery are very significant. Our tests, which we conducted as part of a pilot project on several services, show that properly set up automation can save financial resources, but this is tied to suitably set up business processes, the level of automation and the number of service users. It cannot be confirmed that each automation generates savings, so it is necessary to apply an individual approach. We have found that many event responses can be realized immediately without waiting, as they have a high degree of uniformity and large number of users, including a rapid return on investment (ROI) [9, 11]. From the platform's point of view, we recommend going through the cloud services path, as these provide potential customers with pre-defined solutions that can be deployed and used in a few weeks without designing workflows and programming. The team's intention is to continue the project on a wider range of services and processes as well as to measure and optimize their connection to the local SMART CITY infrastructure.

References

1. Zelenay, J., Balco, P., Greguš, M.: Cloud technologies - solution for secure communication and collaboration. In: The 10th International Conference on Ambient Systems, Networks and Technologies (ANT), 29 April–2 May 2019, Leuven, Belgium (2019)
2. Buyya, R., Broberg, J., Goscinski, A.: Cloud Computing: Principles and Paradigms. Wiley, Hoboken (2011). ISBN 978-0-470-88799-8
3. Garg, S.K., Versteeg, S., Buyya, R.: SMICloud: a framework for computing and ranking cloud services. In: Proceeding of the Fourth IEEE International Conference on Utility and Cloud Computing, pp. 210–218 (2011)
4. Garg, S.K., Vecchiola, C., Buyya, R.: Mandi: a market exchange for trading utility and cloud computing services. J. Supercomputing (JOC) (2011)
5. Greguš, M., Kryvinska, N.: Service orientation of enterprises - aspects, dimensions, technologies, Bratislava, Univerzita Komenského, 1. Vyd. (2015). ISBN 978-80-223-3978-0
6. ATP Journal. Umelá inteligencia a Priemysel 4.0. (2017).https://www.atpjournal.sk/novetrendy/umela-inteligencia-apriemysel-4.0.html?page_id=25248
7. Kvasnička, V., Pospíchal, J., Návrat, P., Lacko, P., Trebatický, P.: Umelá inteligencia a kognitívna veda II. Slovenská technická univerzita v Bratislave (2010)
8. Balco, P., Gregus, M.: Process as a service (PraaS) for planning of resources in organizations. In: IEEE 2015 International Conference on Intelligent Networking and Collaborative Systems (2015)
9. Balco, P., Drahosova, M.: The economic aspects of the electronization in Education process. In: IEEE 15th International Conference on Emerging eLearning Technologies and Applications (ICETA) (2017)
10. Zelenay, J., Balco, P., Gregus, M,. Luha, J.: International Conference on Intelligent Networking and Collaborative Systems, pp. 24–33. Springer, Heidelberg (2018)
11. Balco, P., Gregus, M.: The implementation of innovative services in education by using cloud infrastructure and their economic aspects. J. Glob. J. Flexible Syst. Manag. **15**, 69–76 (2014). Springer. Appendix: Checklist of Items to be Sent to Conference Proceedings Editors (see instructions at conference webpage)

A New Fair Electronic Contract Signing Protocol

Xiao Haiyan[1,2(✉)], Wang Lifang[2], and Wei Yuechuan[1]

[1] Engineering University of PAP, Xi'an, China
spritexiao@163.com
[2] Northwestern Polytechnical University, Xi'an, China
466806112@qq.com

Abstract. Electronic contract is an important part of electronic commerce, and the key issue is how to ensure the fairness of contract signing. In view of some electronic contract signing protocols in the network environment can not achieve fairness completely or need the third party involved. This paper proposed a new concurrent signature in virtue of ECDSA scheme and applied it to a fair contract signing protocol. This new protocol based on ECDSA ensure the fairness of contract signing in virtue of ambiguous signatures become valid concurrently, and ensure the confidentiality of the contents of the contract without the third party involved. This protocol is applied to most electronic contract signing in network environment.

1 Introduction

In recent years, the rapid development of e-commerce has attracted the attention of the international community. Electronic contract is an important part of electronic commerce. It poses legal, technical and regulatory challenges to traditional paper transactions with its unique ordering method. With the promulgation of China's electronic signature law and the development of digital signature technology, electronic contracts have attracted more and more attention of scholars.

The key issue of electronic contract signature is how to exchange the signatures of both parties fairly. The fairness means that if both parties are honest, at the end of the transaction, the parties involved in the transaction will get what they want to achieve the purpose of the transaction; if one party is dishonest, the other party will not be deceived and will not cause losses [1]. In recent years, many fair electronic contract signing protocols have been proposed, which are mainly divided into two categories: one is the gradual exchange protocol. Both sides of the transaction need to carry out a lot of complicated interaction steps to achieve the purpose of digital signature exchange, and it is difficult to maintain the fairness of the protocol. The other is to introduce a trusted third party to exchange signatures fairly. However, it is difficult to find a fully trusted third party to conduct transactions on such an open and unreliable network as the Internet, and bottlenecks may arise at the same time. Therefore, more and more people are devoted to the study of optimistic fair exchange protocols with semi-trusted third parties. However, many protocols are lengthy and complex because they contain sub-protocols to resolve disputes, which may cause vulnerability.

© Springer Nature Switzerland AG 2020
L. Barolli et al. (Eds.): INCoS 2019, AISC 1035, pp. 289–295, 2020.
https://doi.org/10.1007/978-3-030-29035-1_28

The new fair electronic contract signature protocol proposed in this paper uses ECDSA to generate verifiable signatures for fair exchange. The initiator first chooses a secret information and hides the secret in its ambiguous signature. After the two parties exchange each other's ambiguous signature and verify it, the initiator disclosures the secret information which can make the signature of both parties valid at the same time. Otherwise, neither party can prove the validity of the signature of the other party to the notary party. (This idea has been reflected in document [2]) In this protocol, no third party is introduced, which greatly improves the efficiency and is suitable for signing most online electronic contracts.

2   ECDSA

The ECDSA (Elliptic Curve Digital Signature Algorithm) is the simulation of DSA algorithm on elliptic curve, it was accepted as IEEE, NIST, and ISO standards, and widely used and plays an important role in cryptography.

And the algorithm is as follows [3]:

- Setup:
  U is the signer and V is the verifier.
  - U establishes elliptic curve parameter $T = (p, a, b, G, n, h)$ and chooses appropriate safety strength.
  - U generates a pair of keys (d_U, Q_U), d_U is his private key, Q_U is his public key, $Q_U = d_u G$.
  - U selects a Hash functions H.
  - U sends the Hash functions H and elliptic curve parameter T which he chosen to V through a secure channel.
- Sign:
  - U chooses a random number $k \in_R [1, n-1]$, computes $(x_1, y_1) = kG$, $r = x_1 \bmod n$, if $r = 0$, U chooses a new random number k.
  - U computes the hash value of message M: $e = H(M)$.
  - U computes $s = k^{-1}(e + d_U \cdot r) \bmod n$.
  - U generates his signature (m, r, s), then sends it to V.
- Verify: Accepts the input (m, r, s), the verity algorithm performs the following.
  - If $r, s \notin [1, n-1]$, verification failed ψ.
  - Else V computes $e = H(M), u_1 = es^{-1} \bmod n, u_2 = rs^{-1} \bmod n$.
  - V computes $(x_1, y_1) = u_1 G + u_2 Q_U$.
  - If $(x_1 \bmod n) = r$, outputs accept. Otherwise, it returns reject.

3   A New Fair Electronic Contract Signing Protocol

Here we use ECDSA construct a concurrent signature scheme and then present a new fair electronic contract signing protocol based on this concurrent signature scheme. In this protocol, when two parties want to sign an electronic contract, user A chooses a secret information which we call it keystone, then uses this keystone to create his ambiguous signature of electronic contract, then sends his ambiguous signature and electronic contract to user B. B verifies this ambiguous signature and uses the matching keystone fix to create his ambiguous signature, then sends it to A. Here both of them can verify the correctness and validity of another's signature, but both of the signature are not valid yet. If both of them are honest and behave correctly, at the last, when A disclosures the keystone, both of the signatures become binding to their respective signers concurrently.

The notations below are used in the description of our protocol:

Alice, Bob: Both parties which signing the contract;
M: Content of the contract;
H: One-way hash function;
T: Timestamp;
ID: Identity of contract signer;
$E = (p, a, b, G, n, h)$: Elliptic curve parameter, which is exchanged between two signers before protocol though a secret channel;
U_A, R_A: The public key and private key of *Alice*, and $U_A = S_A G$;
U_B, R_B: The public key and private key of *Bob*, and $U_B = S_B G$;
$\{\}_k$: Encryption of message with key k;
$A \rightarrow B : M$: principal A dispatches message M addressed to principal B.

The whole protocol is divided into two sub-protocols: Sign-Protocol and Arbitrate-Protocol. If Both parties which signing the contract correctly, they will receive the expected signature of the other's without any involvement of other parties at the final of Sign-Protocol. If there is a dispute between the two parties or one party considers it unfair in implement of protocol, then the protocol goes to Arbitrate-Protocol to arbitrate in the disagreement between them.

Sign-Protocol

Step1. The initial signer *Alice* performs the following:
Establishes elliptic curve parameter $E = (p, a, b, G, n, h)$ and chooses appropriate safety strength, selects a hash functions H, then sends the hash functions H, elliptic curve parameter E which he chosen, and other information to *Bob* through a secure channel.

$$Alice \rightarrow Bob : \{ID_A, ID_B, H, E, T_1\}_{S_A}$$

Step2. *Alice* chooses a random number $k_A \in _R [1, n-1]$, computes $(x_A, y_A) = k_A G$, $r_A = x_A \bmod n$, if $r_A = 0$, chooses a new random number k_A. Picks a random keystone $key \in_R Z_n$, Then computes $S = H(Key)$, $m = (M||S)$, $e = H(m)$, computes $s_A = k_A^{-1}(e + S_A \cdot r_A) \bmod n$. Now *Alice* generates his ambiguous signature (m, r_A, s_A), then sends his signature and other information to *Bob* through a public channel.

$$Alice \rightarrow Bob : \{ID_A, ID_B, M, m, r_A, s_A, T_2\}_{U_B}$$

Step3. When *Bob* receives the information above, performs the following:

 – If $r_A, s_A \notin [1, n-1]$, rejects the signature ψ.
 – Else   computes $e = H(m), u_1 = es_A^{-1} \bmod n, u_2 = r_A s_A^{-1} \bmod n$, $(x_A, y_A) = u_1 G + u_2 U_A$.
 – If $(x_A \bmod n) = r_A$, accepts this contract. Otherwise, returns reject.
 When accept the contract and the ambiguous signature, *Bob* chooses a random number $k_B \in {}_R[1, n-1]$, computes $(x_B, y_B) = k_B G, r_B = x_B \bmod n$, if $r_B = 0$, chooses a new random number k_B, computes $s_B = k_B^{-1}(e + S_B \cdot r_B) \bmod n$. Now *Bob* generates his ambiguous signature (m, r_B, s_B), then sends his signature and other information to *Alice* through a public channel.

$$Bob \rightarrow Alice : \{ID_A, ID_B, M, m, r_B, s_B, T_3\}_{U_A}$$

Step4. When *Alice* receives the information above, performs the following:

 – If $r_B, s_B \notin [1, n-1]$, rejects the signature ψ.
 – Else computes $u_1 = es_B^{-1} \bmod n, u_2 = r_B s_B^{-1} \bmod n, \; (x_B, y_B) = u_1 G + u_2 U_B$.
 – If $(x_B \bmod n) = r_B$, accepts the signature. Otherwise, returns reject.

If *Alice* accepts *Bob's* ambiguous signature, disclosures the keystone *key*, it makes both of the ambiguous signatures become binding to their respective signers and valid concurrently.

If there is a dispute between *Alice* and *Bob*, or one party considers it is unfair in implement of protocol, then the protocol goes to Arbitrate-Protocol to arbitrate in the disagreement between them.

Arbitrate-Protocol: When there is a dispute between *Alice* and *Bob*, if the sign-protocol is not finished, anyone can quit the protocol at any time without any loss, because no one sent the valid signature of his own nor get the valid signature from the other. If the sign-protocol is finished, but someone considers it is unfair, he can have recourse to arbitrator. This Arbitrate-Protocol needs both of them provide the message $ID_A, ID_B, M, m, r_A, s_A, T_2$ which is sent by *Alice* and message $ID_A, ID_B, M, m, r_B, s_B, T_3$ which is sent by *Bob*.

The arbitrator accepts $ID_i, ID_j, M, m, r_{i/j}, s_{i/j}, T_p$, and checks the correctness of these message to make sure that the sign-protocol is finished yet. He computes $e = H(m), u_1 = es_{i/j}^{-1} \bmod n, u_2 = r_{i/j} s_{i/j}^{-1} \bmod n, \quad (x_{i/j}, y_{i/j}) = u_1 G + u_2 U_{i/j}$ checks whether $(x_{i/j} \bmod n) = r_{i/j}$, If all equations are held, the signing process is finished, that means both of them get the correct ambiguous signature of the other's.

If *Alice* think the protocol if unfair, he can keep the keystone *key* secret, so both of the ambiguous signatures can't binding to their respective signers, it means *Alice* quit this contract signing. If *Bob* think the protocol if unfair, this situation means that he doesn't get the right keystone *key* in order to get the valid signature of *Alice's*. In this condition, if *Alice* wants to execute the contract, he must prove the effectiveness of this contract by providing the keystone *key*, so the arbitrator can get this keystone *key* and sends it to *Bob* and resolves this dispute.

4 Security Analysis

The requirements for a fair electronic contract signing protocol were described as follows [4]:

- Effectiveness. If both of the parties are honest and behave correctly, they will receive the expected valid signatures of the electronic contract without the involvement of third party.
- Fairness. If anyone commits fraud, the other party will not suffer loss.
- Timeliness. Either party can choose to quit the protocol in the process of the implement the protocol and without any loss.
- Verifiability. When the keystone is published, anyone can use this keystone to check the validity of the contract by verify both of the signatures.
- Non-repudiation. After signing the electronic contract, no one can deny his signature.

Now we will analyze this new protocol from these five requirements.

1. This New Protocol Satisfies the Effectiveness Requirement
   Under the correct implementation of the protocol, when there is no fraud between the two parties, both parties of the transaction can smoothly get the valid signature of the other party without any involvement of other parties.
   At the last of the Sign-Protocol, when *Alice* disclosures the keystone *key*, it makes both of the ambiguous signatures become binding to their respective signers and valid concurrently. (m, r_A, s_A) is the valid signature for *Alice*, and (m, r_B, s_B) is the valid signature for *Bob*.
   The correctness and unforgeability of ECDSA scheme have proved in [3]. And it ensures the correctness of this new protocol, that is when the keystone *key* is published, both of the ambiguous signatures binding to the contract contents and become valid.

2. This New Protocol Satisfies the Fairness Requirement
   In this protocol, if both two parties are honest, they will send correct information according to the protocol description, and at last, they will receive the valid signature of the other's and get the effective contract without involvement of arbitrator. But if someone is dishonest in the protocol, neither party gets the valid signature of the other', that means this protocol is fair for each party.
   We will consider the possible unfair situations that the protocol may face. If one party commits fraud, the other party will not suffer losses.
   Proof 1: In *Step 3,* when *Bob* receives $\{ID_A, ID_B, M, m, r_A, s_A, T_2\}_{U_B}$, he verifies the correctness of this message, if *Alice's* ambiguous signature (m, r_A, s_A) is invalid, after verification *Bob* will not sends his own signature to *Alice*.
   Proof 2: In *Step 4,* when *Alice* receives $\{ID_A, ID_B, M, m, r_B, s_B, T_3\}_{U_A}$, he verifies the correctness of this message, if *Bob's* ambiguous signature (m, r_B, s_B) is invalid, the secret keystone *key* will not be disclosed by *Alice*, and the ambiguity of the signatures of both parties will not be eliminated, so that the validity of the signature can't be proved to others.

Proof 3: In *Step 4*, If *Alice* accepts *Bob's* ambiguous signature, but he doesn't disclose the keystone *key*, it leads to a situation that *Alice* obtains *Bob's* valid signature on the contract but *Bob* doesn't. In this condition, *Bob* can resolve to arbitrator by execute Arbitrate-Protocol and solve this problem. In Arbitrate-Protocol, after the valid signature is obtained, then *Alice* can't prove the validity of the signature to arbitrator without disclose the keystone *key*.

3. This New Protocol Satisfies thc Timeliness Requirement
   In the process of implementing the Sign-Protocol, either party may unilaterally terminate the protocol without losing fairness.
   Proof 1: In *Step 3*, after received the sent information $\{ID_A, ID_B, M, m, r_A, s_A, T_2\}_{U_B}$ from *Alice*, if *Bob* wants to end the protocol, he will not send his own signature to *Alice*. At this time, the signature of *Alice's* is not valid because the secret keystone *key* is not disclosed.
   Proof 2: In *Step 4*, after received the sent information $\{ID_A, ID_B, M, m, r_B, s_B, T_3\}_{U_A}$ from *Bob*, if *Alice* wants to terminate the protocol, he will not disclose the secret keystone *key*, so that the ambiguous signatures of both parties are not valid.

4. This New Protocol Satisfies the Verifiability Requirement
   After both parties have sent their signatures, the signatures of the contract can be verified by any other party by publishing the secret keystone *key*.
   Bob provides the message $ID_A, ID_B, M, m, r_A, s_A, T_2$ which is sent by *Alice* and *Alice* provides message $ID_A, ID_B, M, m, r_B, s_B, T_3$ which is sent by *Bob*.
   Verifier first uses the keystone *key* check $S \overset{?}{=} H(Key)$.
   If the equation holds, then computes $e = H(m)$,

$$u_1 = es_{i/j}^{-1} \bmod n, u_2 = r_{i/j}s_{i/j}^{-1} \bmod n,$$

$$(x_{i/j}, y_{i/j}) = u_1G + u_2U_{i/j}$$

   and checks whether $(x_{i/j} \bmod n) = r_{i/j}$.
   If all the equations hold, both signatures for the contract are valid.

5. This New Protocol Satisfies the Non-repudiation Requirement
   The idea of concurrent signature in this protocol guarantees the non-repudiation of the signing.
   Proof 1: In *Step 3*, before *Bob* receives the message $\{ID_A, ID_B, M, m, r_A, s_A, T_2\}_{U_B}$ from *Alice*, he will not sent his message $\{ID_A, ID_B, M, m, r_B, s_B, T_3\}_{U_A}$ to *Alice*. The non-repudiation of the signature received is guaranteed.
   Proof 2: In *Step 4*, before *Alice* the receives the message $\{ID_A, ID_B, M, m, r_B, s_B, T_3\}_{U_A}$ and checks it's correctness, he will not disclose the secret keystone *key* to make both of the signatures valid. The non-repudiation of the signature received is guaranteed.
   Proof 3: And this protocol satisfies non-repudiation requirement, neither of them can deny their own signature for the contract because the verifier checks the correctness and the validity by using their own private keys.

5 Conclusion

Due to the rapid development of electronic commerce, fair contract signing turns out to be an important topic nowadays. The key issue is how to exchange both of their valid signatures in a fair way. In this paper, inspired by the idea of concurrent signature, we use ECDSA to construct a new concurrent signature scheme and applies it to a fair electronic contract signing protocol. The nature of simultaneous effective ensures that both parties can fairly exchange their undeniable signatures of the contract content when they sign the contract. This protocol is applicable to electronic contract signing in most networks.

Acknowledgement. The work in this paper is supported by the Natural Science Foundation of Shaanxi Province (No: 2016JQ6030).

References

1. Zhou, J., Deng, R., Bao, F.: Some remarks on a fair exchange protocol. In: PKC 2000. LNCS, vol. 1751, pp. 46–57 (2000)
2. Chen, L., Kudla, C., Paterson, K.G.: Concurrent signatures. In: Advances in Cryptology - EUROCRYPT 2004, LNCS, vol. 3027, pp. 287–305. Springer, Heidelberg (2004)
3. Johnson, D., Menezes, A., Vanstone, S.: The elliptic curve digital signature algorithm (ECDSA). Int. J. Inf. Secur. **1**(1), 36–63 (2001)
4. Asokan, N., Shoup, V., Waidner, M.: Asynchronous protocols for optimistic fair exchange. In: Proceedings of 1998 IEEE Symposium on Security and Privacy, Oakland, California, pp. 86–99, May 1998. pp. 46–52
5. Huang, Z., Lin, X., Huang, R.: Certificateless concurrent signature scheme. In: The 9th International Conference for Young Computer Scientists, ICYCS 2010, pp. 2102–2107 (2010)
6. Hernandez-Ardieta, J.L., Gonzalez-Tablas, A.I., Alvarez, B.R.: An optimistic fair exchange protocol based on signature policies. Comput. Secur. **27**, 309–322 (2010)
7. Wątróbski, J., Karczmarczyk, A.: Application of the fair secret exchange protocols in the distribution of electronic invoices. Procedia Comput. Sci. 31819–1828 (2018)
8. Thammarat, C., Kurutach, W., Aguiar, J.: A secure fair exchange for SMS-based mobile payment protocols based on symmetric encryption algorithms with formal verification. In: Wireless Communications and Mobile Computing (2018)

CAPTCHA Recognition Based
on Kohonen Maps

Yujia Sun[1,2(✉)] and Jan Platoš[1]

[1] Technical University of Ostrava, 17. listopadu 2172/15, 70800 Ostrava,
Poruba, Czech Republic
yujia.sun.st@vsb.cz
[2] Hebei GEO University, No. 136 East Huai'an Road,
Shijiazhuang, Hebei, China

Abstract. CAPTCHA is a security technology commonly used to differentiate between computers and humans. Text-based CAPTCHA is the most widely used method. This paper presents an approach based on the Kohonen maps neural network for recognizing CAPTCHA. This method first preprocesses the given CAPTCHA, segments its characters, extracts character features, and recognizes characters based on the Kohonen maps cluster analysis results. These experimental results show that the proposed method performs well in recognizing CAPTCHA.

1 Introduction

CAPTCHA stands for "Completely Automated Public Turing Test to Tell Computers and Humans Apart" [1], and is a program that distinguishes between humans and computer programs automatically. CAPTCHA has been widely used for internet security. This technology is now almost a standard security mechanism for addressing undesirable or malicious Internet bot programs and has found widespread application on numerous commercial web sites including Google, Yahoo, and Microsoft's MSN [2]. It is generally used to mitigate the impact of Distributed Denial of Service (DDoS) attacks, slow down automatic registration of free email addresses or spam posts to forums, and also defend against automatic web content scraping [1, 3]. In order to verify security and reliability of CAPTCHA, the breaking technology came into being. It involves image processing, pattern recognition, image understanding, artificial intelligence, computer vision, and many other disciplines. The research on CAPTCHA breaking has great value in research and application [4]. Research on CAPTCHA recognition can identify and improve vulnerabilities, improve CAPTCHA security and reliability, and enhance network security. For the principle of parallel distributed operation with a large number of neurons, the efficient learning algorithms, and the ability to imitate human cognitive systems, the neural network algorithm is especially suited to solve problems such as text recognition. The purpose of this paper is to explore the application of Kohonen maps neural networks in CAPTCHA recognition.

L. Barolli et al. (Eds.): INCoS 2019, AISC 1035, pp. 296–305, 2020.
https://doi.org/10.1007/978-3-030-29035-1_29

2 Kohonen Maps

Kohonen maps were developed in 1982 by Teuvo Kohonen, a professor emeritus at the Academy of Finland. Kohonen maps, also called self-organizing maps [5, 6], are special types of neural networks algorithms that learn on their own through unsupervised competitive learning. The work "maps" is used because they attempt to map their weights to conform to the given input data [7].

Kohonen maps consist of two layers of neurons: an input layer and a so-called competitive layer. The connection weight from the input neurons to a single neuron in the competition layer are interpreted as a reference vector in the input space. Thus, Kohonen maps represent a set of vectors in the input space: one vector for each neuron in the competition layer. The structure of Kohonen maps as shown in Fig. 1.

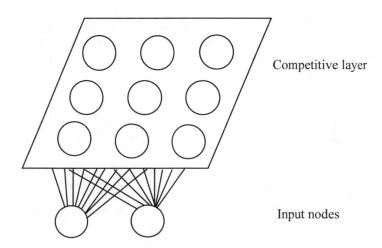

Competitive layer

Input nodes

Fig. 1. The structure of the Kohonen maps

Neurons in the competition layer have excitatory connections to immediate neighbours and inhibitory connections to more distant neurons. All neurons in the competitive layer receive a mixture of excitatory and inhibitory signals from input layer neurons and from other competitive layer neurons.

There are two general types of Kohonen maps. For the Winner Takes All (WTA) algorithm, only the winning neuron adapts the weight. For the Winner Takes Most (WTM) algorithm, neurons that belong to the winner's neighbourhood are allowed to adapt the weights, although with different learning rates. Kohonen maps can be utilized to develop new pattern classification and target recognition systems, performing categorization of the input signal states [8].

3   CAPTCHA Recognition

3.1   CAPTCHA Classification

CAPTCHA can be categorized into three main types [2, 9]:

1. Text-based Schemes: text-based CAPTCHA have been the most widely deployed schemes. These typically rely on a sophisticated distortion of text images that renders them unrecognizable to state-of-the-art pattern recognition programs but recognizable to human eyes.
2. Image-based Schemes: these typically require users to perform an image recognition task.
3. Sound-based Schemes: these typically require users to solve a speech recognition task.

   Text-based CAPTCHA has been the most widely deployed scheme.
   This paper uses a bank of Chinese online bank CAPTCHA as the original sample. This CAPTCHA is a text-based CAPTCHA with a string of randomly generated merged characters, like Fig. 2.

Fig. 2. Sample CAPTCHA

3.2   Process Design

The specific process of CAPTCHA recognition can be shown as Fig. 3. CAPTCHA recognition is completed with the following steps:

1. Analysis of the CAPTCHA image and preprocessing of a series of images.
2. Segmentation of the CAPTCHA image in which the image region of each character in the CAPTCHA image is segmented.
3. Selection of a certain number of single character images as samples and extraction of the features of these images.
4. Features are used as the input of Kohonen maps to train the neural network.
5. The trained neural network is saved and relevant information is recorded.
6. Trained Kohonen maps are used to recognize the CAPTCHA images.

3.3   Image Preprocessing

The CAPTCHA used in this study has the following features: number and characters, some merged. The CAPTCHA must be preprocessed before segmentation in order to highlight the information related to characters in a given image and to weaken or eliminate interfering information. Generally, preprocessing of CAPTCHA includes binarization, denoising, etc. The preprocessed CAPTCHA images are shown in Fig. 4.

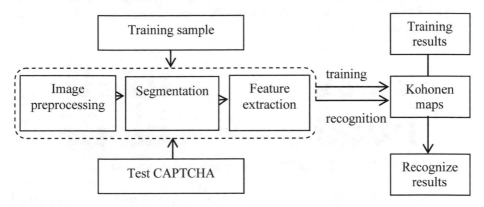

Fig. 3. The process of CAPTCHA recognition

Fig. 4. The preprocessed CAPTCHA image

3.4   Segmentation

The CAPTCHA image after binarization is made of only black and white pixels. It is necessary to cut a single character from the CAPTCHA image to form a binary image of a single character for further processing. The segmentation procedure requires identification of the correct positions for each character. In this paper, character positioning is done to project the binary image in a vertical direction. The projection technique is based upon the idea of projection the image data onto the X-axis. In practice, this is implemented by summing the number of black pixels each column of the image parallel to the Y-axis [10]. By locating a single character, the program can quickly find the best split point between characters. The statistical formula for this vertical projection method is as follows:

$$H(i) = \sum_{j=1}^{n} x(i,j) \quad i = 1 \ldots n .$$ (1)

$$x(i,j) = \begin{cases} 1 & (i,j) \quad \text{black} \\ 0 & (i,j) \quad \text{white} \end{cases}. \tag{2}$$

$H\,(i)$ in formula 1 represents the statistical vertical projection, and (i, j) represents the coordinates of pixel points in the image. The CAPTCHA image after the vertical projection is shown in Fig. 5.

Fig. 5. The CAPTCHA image after vertical projection

As shown in the figure, if the projected area is larger than the width of a single character, it is considered a merged character, and this area can be segmented again. The cropped area is shown in Fig. 6.

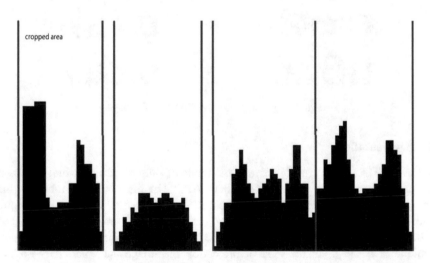

Fig. 6. Cropped area

The single character after segmentation is shown in Fig. 7.

Fig. 7. The single character image after segmentation

3.5   Feature Extraction

Feature extraction is an important component in this approach, and a nice feature has a good effect on clustering. For this study, the character image has been normalized into a 40 * 40 pixels image during character segmentation.

The feature extraction method used in this paper is based on features of grid pixel statistics. As shown in Fig. 8, the image is first divided into equal sized regions. For example, the horizontal and vertical lines are used to divide the image into 16 grids. In a gridding character binary image, the duty cycle refers to the sub-grid region Bj ($j = 1, \ldots, 16$), the ratio of the target pixel number in the region to the total number of pixels in the region. The area of the character sub-grid region Bj is:

$$S = |y_2 \quad y_1| \times |x_2 \quad x_1| \tag{3}$$

The pixel duty ratio of characters in this area is:

$$P_{B_j} = \frac{\sum\limits_{y=y_1}^{y_2} \sum\limits_{x=x_1}^{x_2} g(x,y)}{S}. \tag{4}$$

Where S is the area of the sub-grid region; $g(x, y)$ is the gray value of the pixel point (x, y). The target pixel is set as a black pixel, with a value of 1, the background pixel is set as a white pixel with a value of 0, and $y1$, $y2$, $x1$, $x2$ are set as the upper and lower boundary values of region Bj respectively. The percentage of black pixels in each grid region is calculated, and the statistical results are combined to form a feature vector with 16 dimensions.

3.6   Training

Single character images obtained via segmentation are used inputs for the sample used to train the Kohonen map. A total of 3868 images were processed and 15472 samples were obtained. Each character selects 20 samples for training. The feature vector of each character type in this study has 16 dimensions, so the number of neurons in the

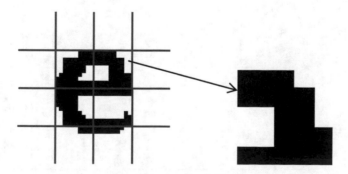

Fig. 8. Feature extraction

input layer is 16. This is a neural network that recognizes 0–9, a–z, and is able to map the input so it is recognized in the output, indicating the type that it belongs to. There are 36 Kohonen map neural network output layers.

The phases of the Kohonen map algorithm is provided below:

Input: character features.

Output: character recognition in an image.

1. Initialization. Initial weight vectors receive random values.
2. Sampling. A vector is chosen at random from the set of training data.
3. Matching. Every node is examined to calculate which weight is most similar to the input vector. The winning node is commonly known as the Best Matching Unit.
4. Updating. Weight is updated using the equation.
5. Continuation. Steps 2–4 are repeated until the feature map halts varying.

Then training neural networks, the learning rate controls how much the weights are adjusted for each update. The lower the value, the slower training will progress as small updates are made to the weights in neural networks. However, if the learning rate is high, the training may not converge and may even diverge. Another parameter is sigma, the radius of the different neighbours in the Kohonen map. In this experiment, the learning rate = 0.2, sigma = 8, and the training results with 5000 training times are shown in Fig. 9.

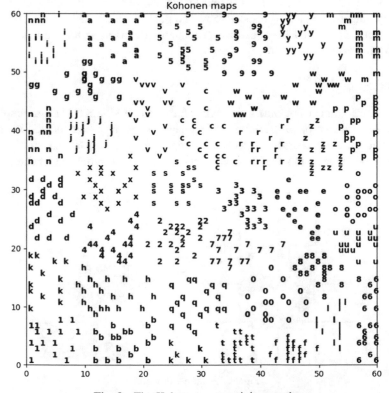

Fig. 9. The Kohonen map training results

3.7   Recognition

The test data set contains 127 images for testing, analysis, and recognition results, of which 100 were normal images and the other 27 were severely adhered and incomplete images. The trained results were saved as templates and the test data set was input into the template for testing.

The CAPTCHA recognition algorithm proposed by this paper was run in the Python environment. Experiment with the CAPTCHA images data set, the experimental results are shown in Table 1. After training, the test image to be recognized was binary and segmented, and then input into the Kohonen map for recognition, with the output being the recognition result. Recognition accuracy was 100%, even for severe adhesion, and 89% for incomplete segmentation of the character image like Fig. 10. The statistics results are shown in Table 2.

Table 1. CAPTCHA experimental treatment effect

	CAPTCHA image	Image preprocessing	Segmentation	Recognition result
1	6xa7	6xa7	6 x a 7	6xa7
2	g4sh	g4sh	g 4 s h	g4sh
3	pxw6	pxw6	p x w 6	pxw6
4	n6py	n6py	m 6 p y	m6py
5	3lmy	3l my	3 l m y	3lmp

Fig. 10. Incomplete segmentation of the character image

Table 2. Experimental testing result statistics

Number of samples	Total characters	Completely recognize the correct CAPTCHA image	Accuracy
720	100	100	100%
720	27	24	89%

The recognition results indicate that characters with recognition errors were generally incompletely segmented due to severe adhesion. Therefore, to improve the recognition accuracy of the CAPTCHA, it is important to focus on the segmentation effect.

4   Conclusion

For samples of CAPTCHA images, the Kohonen map neural network method was adopted to train and recognize single characters obtained by segmentation, achieving high recognition accuracy. The recognition accuracy of proposed method was 100%, even for severe adhesion, and 89% for incomplete segmentation of the character image. The evaluation results prove that the Kohonen map is able to recognition CAPTCHA with a high accuracy. For future works, there is an approach that can be taken to increase recognition accuracy. Segmentation of the CAPTCHA into individual characters is the most difficult part to properly break a CAPTCHA. In order to extend this approach, I am developing redesign the segmentation algorithm based on the Kohonen maps in order to obtain completely character image.

References

1. Ahn, L.V., Blum, M., Langford, J.: Telling humans and computers apart automatically. Commun. ACM **47**, 56–60 (2004)
2. Yan, J., Ahmad, A.E.: Usability of CAPTCHAs or usability issues in CAPTCHA design. In: The 4th Symposium on Usable Privacy and Security, USA, pp. 44–52 (2008)
3. Platos, J., Snasel, V., Kromer, P., Abraham, A.: Detecting insider attacks using non-negative matrix factorization. In: The Fifth International Conference on Information Assurance and Security, pp. 693–696 (2009)
4. Chen, J., Luo, X., Guo, Y., Zhang, Y., Gong, D.: A survey on breaking technique of text-based CAPTCHA. Secur. Commun. Netw. 1–15 (2017)

5. Kohonen, T.: Self-organized formation of topologically correct feature maps. Biol. Cybern. **43**, 59–69 (1982)
6. Kohonen, T.: Self-Organizing Maps, 3rd edn. Springer, New York (2001)
7. Guthikonda, S.M.: Kohonen Self-Organizing Maps. Wittenberg University (2005). http://www.shy.am/wp-content/uploads/2009/01/kohonen-self-organizing-maps-shyam-guthikonda.pdf
8. Kohonen, T., Oja, E., Simula, O., Visa, A., Kangas, J.: Engineering applications of the self-organizing map. Proc. IEEE **84**, 1358–1384 (1996)
9. Osadchy, M., HernandeZ-Castro, J., Gibson, S., Dunkelman, O., Perez-Cabo, D.: No bot expects the DeepCAPTCHA! introducing immutable adversarial examples, with applications to CAPTCHA generation. IEEE Trans. Inf. Forensics Secur. **12**, 2640–2653 (2017)
10. Huang, S.-Y., Lee, Y.-K., Bell, G., Ou, Z.-H.: A projection-based segmentation algorithm for breaking MSN and YAHOO CAPTCHAs. In: International Conference of Signal and Image Engineering(ICSIE 2008), London, UK (2008)
11. Vondrak, I.: ANN. Technical report. http://vondrak.vsb.cz/download/ANN.pdf
12. Snasel, V., Platos, J., Kromer, P., Abraham, A.: Matrix factorization approach for feature deduction and design of intrusion detection systems. In: The Fourth International Conference on Information Assurance and Security, pp. 172–179 (2008)
13. Platos, J., Kromer, P., Snasel, V., Abraham, A.: Searching similar images- Vector Quantization with S-tree. In: IEEE CASoN (2012) 304-388
14. Wang, X.A., Weng, J., Ma, J.F., Yang, X.Y.: Cryptanalysis of public authentication protocol for outsourced databases with multi-user modification. Inf. Sci. **488**, 13–18 (2019)

Effects of Truss Structure of Social Network on Information Diffusion Among Twitter Users

Nako Tsuda$^{(\boxtimes)}$ and Sho Tsugawa$^{(\boxtimes)}$

University of Tsukuba, 1-1-1 Tennodai, Tsukuba, Japan
n.tsuda@mibel.cs.tsukuba.ac.jp, s-tugawa@cs.tsukuba.ac.jp

Abstract. Analyzing the factors that affect information diffusion through social media is an important research topic for several applications such as realizing effective viral marketing campaigns and preventing the spread of fake news. In this paper, we focus on the community structure of a social network of social media users as a factor affecting information diffusion among those users. We extract two types of community structures, a *flow truss* and a *cycle truss*, from the social network of Twitter users and analyze how these structures affect diffusion via cascades of retweets on Twitter. Our results show that tweets disseminated via inter-community retweets have future popularity about 1.2-fold that of tweets disseminated via intra-community retweets. Our results also show that tweets disseminated within a strongly clustered community tend to have less diffusion than tweets disseminated within a weakly clustered community. These results are found both when extracting via cycle truss and flow truss communities, which suggests that our findings are robust against the definitions of community.

1 Introduction

Some information posted on social media is disseminated to many users through a shared social network [1]. Social media users can disseminate other users' posts via functionalities such as *retweeting* on Twitter, and *sharing* on Facebook. These functionalities drive information diffusion on social media.

Many studies have analyzed the factors that affect information diffusion on social media, and various features have been shown to affect information diffusion [2–5]. Among the factors known to strongly affect diffusion are the number of followers of users posting information, whether URLs and hashtags are included in posted information, and the structure of the social network [2,3,6].

This paper focuses on one factor that affects information diffusion: the community structures of the social network among users. Many social networks have a community structure composed of communities of nodes closely connected to each other and links connecting communities (see Fig. 1) [7,8]. It has been suggested that when information is disseminated across multiple communities, the information will be spread more widely [9,10].

© Springer Nature Switzerland AG 2020
L. Barolli et al. (Eds.): INCoS 2019, AISC 1035, pp. 306–315, 2020.
https://doi.org/10.1007/978-3-030-29035-1_30

However, empirical studies analyzing the effects of community structure on information diffusion on social media are still limited. Most prior studies have investigated the effects of community structure on information diffusion by using stochastic information diffusion models [6,9]. In contrast, how the community structure affects actual information diffusion has not been fully explored. In our previous work [10], we have shown that tweets that spread across different communities tend to have higher future popularity than tweets spread within a single community.

Complicating the analysis, there are various definitions of community in social networks and various community detection algorithms, including some that were not been in our previous study [11,12]. Community detection algorithms can be classified into two categories: those that divide an entire social network into multiple communities; and those that extract a single community or multiple communities from (see Fig. 2) [11,13]. In this paper, we call the first type community division algorithms and call the second type community extraction algorithms. To the best of our knowledge, community division algorithms have been used to investigate the relation between community structure and information diffusion but community extraction algorithms have not yet been used [10].

In this paper, we examine the effects of the community structure of a social network on information diffusion by using community extraction algorithms to extract cycle truss and flow truss communities [14]. By using a community extraction algorithm different from the one in our previous study, we strengthen the generalizability of prior findings that have relied on different definitions of communities. Furthermore, in this paper, we analyze the influence of the strength of the community structure on information diffusion, limiting this to tweet diffusion here. The algorithm for extracting the cycle truss and flow truss can control the strength of community structure by parameter tuning. We extract community structures with different strengths and analyze how the strength of community structure affects the extent of tweet diffusion. Our main contributions are summarized as follows.

- We validate our earlier finding that *tweets disseminated via inter-community retweets tend to have higher future popularity than tweets disseminated via only intra-community retweets* by using community extraction algorithms different from those used in our previous work.
- We examine the effects of the strength of community structure on tweet diffusion, finding that tweets disseminated within a strongly clustered community tend to have less diffusion than tweets disseminated within a weakly clustered community.

The rest of the paper is organized as follows. Section 2 explains the research methodology. Section 3 presents the results of the analysis and discusses these results. Finally, Sect. 4 concludes this paper and discusses future work.

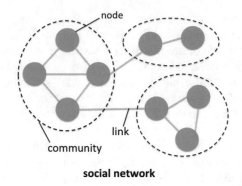

Fig. 1. Example of community structure in a social network

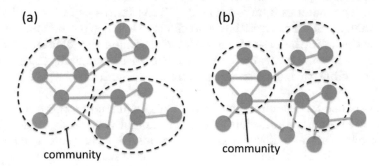

Fig. 2. Illustrative explanations of (a) community division and (b) community extraction in a social network

2   Methodology

For the following analyses, we used the same user set used in our previous work [10]. The users were selected by the following procedures. Twitter API (Application Programming Interface) was used to collect Japanese retweets for one week from the 11th to 17th December, 2013. We used a query "q = RT, lang = ja" in the search Twitter REST API v1.1. As a result, 14,220,864 tweets (original tweets) and 52,129,804 retweets were obtained. Tweets whose number of retweets was 10 to 100 in the above period were extracted and users who retweeted 10 or more of these tweets were extracted as the target users to be analyzed. The number of the target users were 356,453.

We obtained a social network of the 356,453 users as of early January 2016. We collected followers and followees of the target users using the Twitter API. We then constructed a social network representing follower and followee relationships among the target users. In the constructed social network, each target user is represented as a node, and a directed link (u, v) represents that user u follows user v.

Next, we collected the retweets posted by the target users. In this paper, we collected the retweets posted by the target users during January 1–31, 2016 using the Twitter API and we obtained 1,626,183 tweets (original tweets) posted by the target user and 5,496,832 retweets. The retweet and social network data are used in the following analyses.

Next, we applied community extraction algorithm proposed in [14] to the constructed social network for extracting cycle and flow k-truss communities. The cycle k-truss community and the flow k-truss community are defined as generalization of k-truss [15] in undirected networks to directed networks. The three-node relationships shown in Fig. 3a and b are defined as cycle triangle, and flow triangle, respectively. Cycle k-truss communities are extracted as follows. We first obtain all cycle triangles in the given network G. Then, for each link (i, j), we count the number of cycle triangles associated with link (i, j), denoted as $c(i, j)$. We finally remove all links with $c(i, j) < k$ from the network G. Each connected component with at least two nodes of the remaining network is a cycle k-truss community. Note that singleton nodes do not belong to any communities. By replacing "cycle" in the above description with "flow", the flow truss community can be similarly extracted. The truss number k is a nonnegative integer value of $0 \leq k \leq d_{max} - 1$, and d_{max} is the maximum node degree in the network. As k increases, the extracted k-truss communities only contain nodes with a larger number of shared triangles. Therefore, when k is large, strongly clustered and small k-truss communities are obtained whereas when k is small, weakly clustered and large communities are obtained. In this paper, we regard communities extracted with larger value of k as stronger communities than communities extracted with smaller value of k.

The relationship between the parameter k in extracting the cycle and flow truss communities and their characteristics are shown in Fig. 4. Figure 4a shows the fraction of nodes included in the any extracted k-truss communities. Figure 4b shows the fraction of nodes in the largest community among the nodes included in the extracted k-truss communities. Figure 4b shows that most of the nodes belonging to any communities belong to the largest community regardless of the value of k. Also, Fig. 4a shows that different sizes of communities are extracted by changing k. Figure 4a also shows that the size of the cycle k-truss communities are different from the flow k-truss communities even if the value of k is the same.

Finally, for each extracted k-truss community, we calculate the average number of future retweets of the tweets disseminated by the users in the same community and those disseminated by the users outside of the community. Let $r(T, N)$ be the N-th retweet of original tweet T, $u(T)$ be the user who posts original tweet T, and $u(r(T, N))$ be the user who posts retweet $r(T, N)$. If user $u(T)$ and $u(r(T, N))$ belong to the same community, then the retweet $r(T, N)$ is called *intra-community diffusion*, otherwise the retweet $r(T, N)$ is called *inter-community diffusion*. We compare the average number of future retweets after $r(T, N)$ is posted when $r(T, N)$ is intra-community diffusion and when $r(T, N)$ is inter-community diffusion.

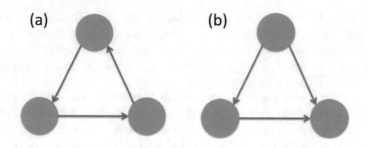

Fig. 3. (a) Cycle triangle (b) Flow triangle

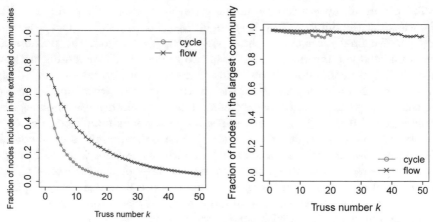

(a) Fraction of nodes included in extracted k-truss community

(b) Fraction of nodes included in the largest community among the nodes included in the extracted k-truss community

Fig. 4. Relation between the parameter k in extracting the cycle and flow truss communities and their characteristics

3   Results and Discussion

3.1   Comparison of Intra-community and Inter-community Diffusion

We compare the future popularity of tweets spread within a community and tweets spread outside the community. For each Nth retweet of tweet T, we investigate the number of future retweets after the Nth retweet is posted. We calculated the average number of future retweets when the Nth retweet was intra-community diffusion and when the Nth retweet was inter-community diffusion. Figures 5a and b show the results for cycle truss communities with $k = 5$ and $k = 10$, respectively. Figures 5c and d show the results for flow truss communities with $k = 10$ and $k = 30$, respectively. Although we obtained the results for other k values, those results are not shown due to space limitations. Note that the sizes of the cycle truss community when $k = 5$ and that of the flow

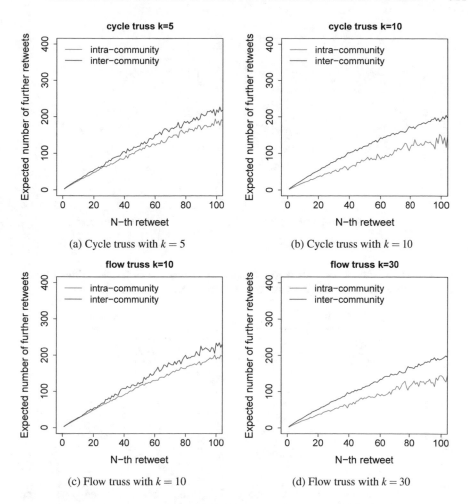

Fig. 5. Comparison of the future popularity of tweets disseminated via intra-community diffusion versus via inter-community diffusion. The number of retweets after the Nth retweet is compared.

truss community when $k = 10$ are similar. The size of the cycle truss community when $k = 10$ and that of the flow truss community when $k = 30$ are also similar.

From Fig. 5, we can see that the future popularity of tweets disseminated via inter-community diffusion is higher than that of tweets disseminated via intra-community diffusion. For instance, looking at $N = 100$ in Fig. 5a, the future number of retweets disseminated via inter-community diffusion is approximately 1.2-fold that of tweets disseminated via intra-community diffusion. This result supports our previous finding that the inter-community diffusion of tweets is correlated with an increase of the future popularity of tweets. This study and our previous study [10] use different types of community detection algorithms,

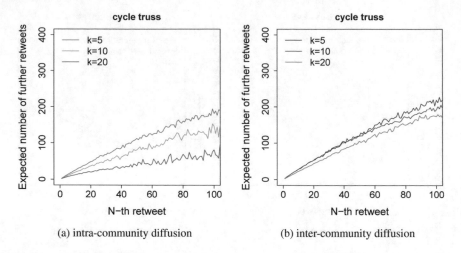

Fig. 6. Comparison of the future popularity of tweets for different values of k. The number of retweets after the Nth retweet is compared (cycle truss).

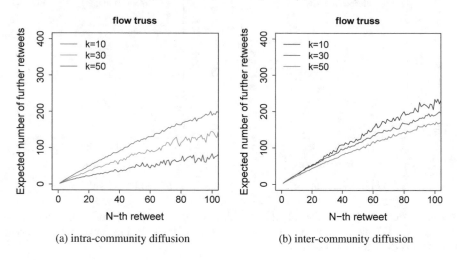

Fig. 7. Comparison of the future popularity of tweets for different values of k. Comparison is the number of retweets after the Nth retweet is posted (flow truss).

and these two studies obtain similar results. This suggests that the finding is robust against the definitions of communities.

3.2 Effect of Strength of Community Structure

To clarify the effects of the strength of community structure on the future popularity of tweets, the results with different values of k are investigated. Figure 6 compares the future number of retweets after the Nth retweet is posted for different values of k with cycle truss communities. Figure 6a shows the results for

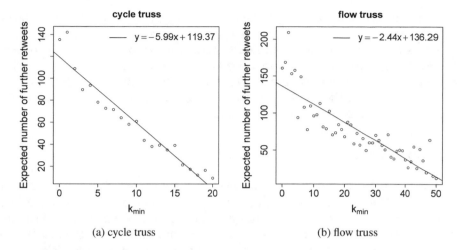

Fig. 8. Average future popularity of tweets disseminated between two users u and v against $k_{min}(u, v)$, the strength of the ties between users u and v.

the cases where the Nth retweet is intra-community diffusion, and Fig. 6b shows the results for inter-community diffusion. Figure 7 shows the analogous results for flow truss communities. As we explained in Sect. 3, more strongly clustered communities are obtained as the value of k increases.

Figures 6 and 7 show that the number of future retweets decreases as k increases. Particularly for intra-community diffusion, the effect of k on the future popularity of tweets is large. These results suggest that tweets spread within a weakly clustered community tend to spread more widely than tweets spread within a strongly clustered community. We speculate that information spread within a strongly clustered community might be interesting or useful only for the community members, which may result in lower future popularity. However, the effects of the contents of the tweets are beyond the scope of this paper, and further investigation is necessary to clarify the cause of this phenomenon.

Finally, further analyses are conducted to more clearly characterize the effects of the strength of the community structure on tweet diffusion. As the value of k increases, communities containing only those users with stronger relationships among them are extracted. Therefore, if user u and user v belong to the same k-truss community for large values of k, these two users u and v are considered to have a strong relationship. In contrast, if user u and user v belong to the same k-truss community only for small values of k, then the relationship between u and v is considered to be weak. Therefore, the strength of the relationship between users u and v can be regarded as $k_{min}(u, v)$, the minimum value of k at which the users u and v do not belong to the same k-truss community. For each pair of users u and v, we investigate the future popularity of tweets posted by user u after user v retweets those tweets. Then, we investigate the relation of this future popularity with $k_{min}(u, v)$. From previous analyses, it is expected that the number of future retweets will be higher as $k_{min}(u, v)$ decreases. Figure 8 shows the relation between $k_{min}(u, v)$ and the future popularity of tweets posted by user u after user v retweets them. From Fig. 8, we can see that as $k_{mim}(u, v)$ increases, the future popularity of tweets posted by u and retweeted by v decreases. This suggests that a tweet disseminated by users who have weak relationships with the user who posts the original tweet tends to spread widely. Namely, tweets spread through weak ties tend to be popular.

4   Conclusion and Future Work

We have investigated the relation between the amount of information diffusion on social media and the community structure of social network among social media users. We focused on cycle and flow truss types, in particular, which are communities that consider link directions. Our results have shown that the future popularity of tweets disseminated via inter-community diffusion is higher than that of tweets disseminated via intra-community diffusion for both cycle and flow truss communities. This supports our previous finding that tweets disseminated via inter-community retweets tend to have higher future popularity than tweets disseminated via intra-community retweets, independent of the definition of community. Moreover, in this paper, we investigated the effects of the strength of the community structure on the future popularity of tweets. We found that tweets spread within a weakly clustered community tend to spread more widely than those spread within a strongly clustered community. In addition, it is suggested that tweets spread through weak ties tend to be more popular than tweets spread through strong ties.

In future work, we intend to analyze how the characteristics of each community affect information diffusion. It is also important to validate the results of this paper using data from other periods and other social media types. Moreover, applying the results of this study to real problems, such as predicting the future popularity of tweets, is also important future work.

Acknowledgments. This work was partly supported by JSPS KAKENHI Grant Number 17H01733.

References

1. Bakshy, E., Hofman, J.M., Mason, W.A., Watts, D.J.: Everyone's an influencer: quantifying influence on Twitter. In: Proceedings of the Fourth ACM International Conference on Web Search and Data Mining (WSDM 2011), pp. 65–74. ACM (2011)
2. Naveed, N., Gottron, T., Kunegis, J., Alhadi, A.C.: Bad news travel fast: a content-based analysis of interestingness on Twitter. In: Proceedings of the 3rd International Web Science Conference (WebSci 2011), pp. 1–7. ACM (2011)
3. Hong, L., Dan, O., Davison, B.D.: Predicting popular messages in Twitter. In: Proceedings of the 20th International Conference Companion on World Wide Web (WWW 2011), pp. 57–58. ACM (2011)

4. Suh, B., Hong, L., Pirolli, P., Chi, E.H.: Want to be retweeted? Large scale analytics on factors impacting retweet in Twitter network. In: Proceedings of the 2nd IEEE International Conference on Social Computing (SocialCom 2010), pp. 177–184. IEEE (2010)
5. Tsugawa, S., Ohsaki, H.: On the relation between message sentiment and its virality on social media. Soc. Netw. Anal. Mining **7**(1), 19:1–19:14 (2017)
6. Nematzadeh, A., Ferrara, E., Flammini, A., Ahn, Y.-Y.: Optimal network modularity for information diffusion. Phys. Rev. Lett. **113**(8), 088701 (2014)
7. Newman, M.E.J., Girvan, M.: Finding and evaluating community structure in networks. Phys. Rev. E **69**(2), 026113 (2004)
8. Ferrara, E.: A large-scale community structure analysis in Facebook. EPJ Data Sci. **1**(1), 9 (2012)
9. De Meo, P., Ferrara, E., Fiumara, G., Provetti, A.: On Facebook, most ties are weak. Commun. ACM **57**(11), 78–84 (2014)
10. Tsugawa, S.: Empirical analysis of the relation between community structure and cascading retweet diffusion. In: Proceedings of the 13th International AAAI Conference on Web and Social Media (ICWSM 2019), vol. 13, no. 1, pp. 493–504 (2019)
11. Fortunato, S.: Community detection in graphs. Phys. Rep. **486**(3), 75–174 (2010)
12. Tsugawa, S.: A survey of social network analysis techniques and their applications to socially aware networking. IEICE Trans. Commun. **102**(1), 17–39 (2019)
13. Miyauchi, A., Kawase, Y.: What is a network community?: A novel quality function and detection algorithms. In: Proceedings of the 24th ACM International on Conference on Information and Knowledge Management (CIKM 2015), pp. 1471–1480. ACM (2015)
14. Takaguchi, T., Yoshida, Y.: Cycle and flow trusses in directed networks. Roy. Soc. Open Sci. **3**(11), 160270 (2016)
15. Wang, J., Cheng, J.: Truss decomposition in massive networks. Proc. VLDB Endow. **5**(9), 812–823 (2012)

The 11th International Workshop on Information Network Design (WIND-2019)

On-Demand Transmission Interval Control Method for Spatio-Temporal Data Retention

Shumpei Yamasaki[1]([envelope]), Daiki Nobayashi[1], Kazuya Tsukamoto[1],
Takeshi Ikenaga[1], and Myung Lee[2]

[1] Kyushu Institute of Technology, Fukuoka, Japan
p232087s@mail.kyutech.jp, {nova,ike}@ecs.kyutech.ac.jp,
tsukamoto@cse.kyutech.ac.jp
[2] CUNY, City College, New York, USA
mlee@ccny.cuny.edu

Abstract. With the development and the spread of Internet of Things (IoT) technologies, various types of data are generated for IoT applications anywhere and anytime. We defined such data that depends heavily on generation time and location as Spatio-Temporal Data (STD). In the previous works, we have proposed the data retention system using vehicular networks to achieve the paradigm of "local production and consumption of STD." The system can provide STDs quickly for users within a specific location by retaining the STD within the area. However, the system does not consider that each STD has different requirements for the data retention. In particular, the lifetime of the STD and the diffusion time to the whole area directly influence to the performance of data retention. Therefore, we propose a dynamic control of data transmission interval for the data retention system by considering the requirements. Through the simulation evaluation, we found that our proposed method can satisfy the requirements of STD and maintain a high coverage rate in the area.

1 Introduction

With the development of IoT (Internet of Things) technology, numerous IoT devices and new applications are spreading. The majority of the existing works employ centralized architecture [1], and computers with high performance CPU and large storage capacity are required to process the enormous application data generated from IoT devices. Therefore, the current network infrastructure is difficult to cope with exponentially increasing data traffic.

Some data generated by IoT devices depend on location and time, such as traffic, weather, disaster, and time-limited store advertisement, and such data are referred to as Spatio-Temporal Data (STD). STDs may not be stored to a remote server connected to the Internet, but they may be sufficient to utilize in the location. Beuchert et al. [2] proposed the platform which makes easy

© Springer Nature Switzerland AG 2020
L. Barolli et al. (Eds.): INCoS 2019, AISC 1035, pp. 319–330, 2020.
https://doi.org/10.1007/978-3-030-29035-1_31

to manage and analyze STDs for location-based applications. The paradigm of "local production and consumption of STD" can solve the problem of current network infrastructure.

Teshiba et al. [3] proposed the data retention system using vehicular network to achieve the paradigm of "local production and consumption of STD" without using existing network infrastructure. This system focuses on the vehicle mobility and the possibility that vehicles are commonly equipped with a high performance CPU and radio communication I/F. Moreover, the system retains STDs within specific area and time by utilizing vehicles as relay nodes (InfoHubs) of data diffusion. It realizes the following three things.

- Collecting and spreading real-time information
- Improving fault tolerance by distributed information management
- Offloading the existing network infrastructure

In addition, the previous works [3,4] proposed the control method of data transmission probability based on node density to avoid collision of radio communication between vehicles. This method improved the performance of the retention system. However, although each STD has different requirements in terms of its diffusion completion time and its lifetime, the previous methods did not consider these requirements at all. Thus, it could be ineffective data retention.

In this paper, we propose a new method that dynamically controls the data transmission intervals by considering the demands for diffusion completion time. Furthermore, the proposed scheme deletes the STD immediately after its lifetime expires. The effectiveness of the proposed method is evaluated by simulation experiments.

In Sect. 2, we review related works. In Sect. 3, we describe the STD retention system proposed in the previous work. Our proposed method is explained in Sect. 4 and Sect. 5 provides the effectiveness of the proposed method through simulation experiments. Finally, Sect. 6 is our conclusion and future problems.

2   Related Work

Li et al. [5] discussed the problem of data diffusion and sharing of vehicular network, and proposed a protocol that uses Geocast Routing based on location within a specific area and sends information from a source to all nodes (Vehicles).

Literature [4,6], Floating Content [7], Locus [8], etc. have been proposed. In literature [6], a node heading to a retention area is specified by switching navigation information of each node, and data is effectively transmitted. In the system of Floating Content and Locus, each node has a list of maintaining data, and exchanges the data list with neighboring nodes. When a node does not have the data, it sends a transmission request to the neighboring node to get the data. Since whether a node performs transmission or not is determined by the transmission probability according to the distance from the center, data acquisition decreases when the node is far from the center, and data collision may occur frequently when the node is biased to near the center.

In order to solve the problem of the related work, in the literature [3], a STD retention system was constructed with the aim of periodically transmitting the data to all receivers existing in the retention area at the set transmission interval using the broadcast based on the positional information. Then, in the next section, the technique devised in the construction of the STD retention system is explained.

3   Spatio-Temporal Data Retention System

In this section, we first describe the assumptions, goals and requirements of the system. Then, we explain the control method of data transmission probability proposed in the previous work [4].

3.1   Assumptions

The STD retention system assumes that each node obtains their position using GPS. Each node broadcasts a beacon containing its own ID at regular intervals. The nodes also broadcasts the data. We assumed that the data includes the information of the retention area (center coordinate, retention area radius R) and the data transmission interval d. Each node determines whether it is located outside or inside the retention area based on its own position information and the retention area information included in the data.

3.2   System Objectives and Requirements

The goal of this system is to periodically spread and retain data to the whole retention area. Therefore, data can be automatically received when the system user enters the retention area. In addition, the system can reduce server load and improve the fault tolerance because of distributed management of STDs without using existing network infrastructure. As a requirement of the system, the *Coverage rate* is defined. It shows the probability which system users can automatically receive STDs when entering the retention area.

$$Coverage\ Rate = \frac{S_{DT}}{S_{TA}}. \tag{1}$$

S_{DT} is the area where can receive data transmitted by nodes within a certain period. Also, S_{TA} is the whole retention area.

The previous work proposed a method of controlling transmission probability based on node density in order to suppress useless data transmission. Section 3.3 shows this technique.

3.3 A Data Transmission Control Based on Node Density

In this system, the radio coverage, the distance from the center coordinates of the retention area to nodes are defined as r and $Distance$, respectively. The transmission target area is defined as follows.

$$\begin{cases} 0 < Distance \leq R + r. & : \text{transmission target area} \\ otherwise. & : \text{unsent area} \end{cases} \qquad (2)$$

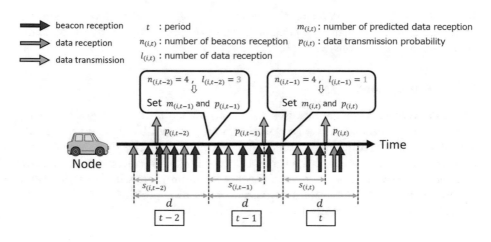

Fig. 1. Setting data transmission probability

Based on the time when data is received for the first time, data is transmitted periodically with the data transmission interval d as one cycle. At this time, data transmission is performed for some nodes which exist outside the retention area in order to improve the coverage rate. The actual transmission timing is transmitted according to the time $s_{(i,t)}$ randomly determined by each node v_i (i: a unique ID given to the node) within d, thereby suppressing the collision of radio channels between nodes. Next, all nodes transmit beacons containing their node IDs at b [s] intervals. By doing so, it is possible to know the number of neighboring nodes around itself. Based on the number of neighboring nodes $n_{(i,t)}$ and the number of beacons received in the $t-1$ cycle, the transmission probability is set in every t cycle (Fig. 1).

When the number of neighboring nodes is 3 or less, the node sets transmission probability $p_{(i,t)} = 1$ in order to transmit data.

When the number of neighboring nodes is 4 or more, $p_{(i,t)}$ is calculated according to the number of neighboring nodes and the surrounding data transmission status. This is to prevent excessive data transmission and data collisions. Although literature [3] shows concrete equations to adjust the data transmission probability, we do not explain the detailed algorithm in this paper due to the lack of space.

In our work, we use this data transmission probability control based on node density to maintain the STD retention after the STD diffusion process (our focus) is completed, as in the literature [3].

3.4   Problems of the Previous Work

The previous work set the allowable waiting time so that nodes deliver data to users. In other words, users can receive the data if they wait as long as the allowable waiting time in the retention area of the data. In the previous work, the allowable waiting time is set to data transmission interval d of nodes, and the operation was evaluated. However, the previous work does not assume the concrete constraint of the allowable waiting time, and does not mention how to set the data transmission interval.

Some STDs with strict constraints, such as a short lifetime and diffusion completion time, will be retained as well as other STDs with loose constraints (e.g. weather or advertising information).

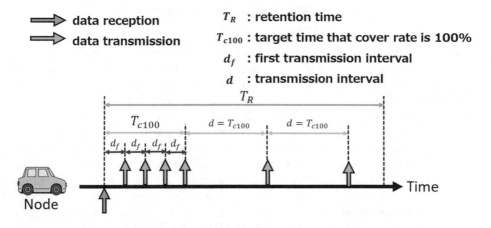

Fig. 2. Data transmission interval during the retention time

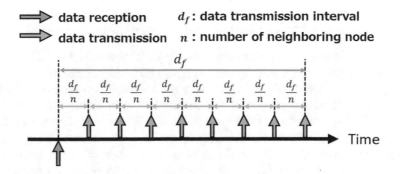

Fig. 3. Random time selection ($n = 8$)

For example, they are radio resource information including available frequency bands, channel interference information, and so on. If the transmission interval of STD is longer than its lifetime, the STD cannot be completely diffused within the area until the expiration of the lifetime (i.e., ineffective retention), and a user cannot get STDs. Therefore, these STDs need to set the transmission interval d as $0 < d < (lifetime\ of\ STDs)$.

When handling the STD retention with severe time-constraint, we need to consider the diffusion completion time-aware data transmission. So, we will propose a method of controlling the data transmission interval in order to satisfy the requirement of diffusion completion time, in the next section.

4   Proposed Method

We propose an on-demand transmission interval control method based on short lifetime and diffusion completion time of STD.

4.1   Decision Policy of Data Transmission Interval

First, we define the data lifetime as *retention time* T_R. We also define the target allowable waiting time for the diffusion completion time as T_{c100}. Since the requirements for T_{c100} differ depending on the content, the possible range of T_{c100} could be $0 < T_{c100} \leq T_R$. We assume when T_R of a STD is short, some users cannot obtain the STD due to the expiration of T_R if T_{c100} is set to a large value but less than T_R. For example, when the available frequency band changes every moment (T_R is short), T_{c100} is need to set to a small value. To solve this problem, we also need to consider the number of neighboring nodes and the size of the retention area for the decision of initial data transmission for the first T_{c100} seconds. If the number of neighboring nodes are small and the retention area is large, the time required to complete the data diffusion tends to be large.

In this paper, we define the initial data transmission interval as $d_f(< d = T_{c100})$ and propose a decision method (Fig. 2). After T_{c100} seconds is over, each node sets the transmission interval to $d = T_{c100}$ in order to reduce the unnecessary data transmissions.

4.2   Dynamic Decision of Initial Data Transmission Interval

4.2.1   Consideration for Random Waiting Time and Hop Count

Assuming that the number of neighboring nodes is n, the number of received data for d is n. Then, each of nodes first checks the transmission interval d included in the received data, and randomly decides the transmission timing $s(0 < s \leq d)$ [s]. Since the average of the random time s is $0.5d$, we assume that each of n nodes re-broadcasts STDs at every intervals of $s = \frac{d_f}{n}$ after receiving the STD (Fig. 3). Furthermore, we define the minimum number of nodes transmitting STD in order to diffuse STD in all direction as γ. In other words, nodes in all direction can receive STD after $\frac{d_f}{n} * \gamma$ seconds. Therefore, the number of possible hops until the expiration of the target time T_{c100} [s] can be expressed by the following equation.

$$\frac{T_{c100}}{\frac{d_f}{n} * \gamma} = \frac{nT_{c100}}{\gamma d_f}.$$ (3)

If x [m] is the extended distance of the radius of the data retention area by one data transmission, we need to set the initial transmission interval d_f by following Eq. (4), while satisfying STD diffusion within the data retention area (Eq. (3)).

$$x \times \frac{nT_{c100}}{\gamma d_f} = R$$

$$d_f = \frac{n}{\gamma} T_{c100} \times \frac{x}{R}.$$ (4)

4.2.2   Consideration for Expansion of Retention Area

In this work, we assume that nodes are uniformly located in the retention area. One hop distance x changes according to location of nodes, so we consider the following three cases.

1. O_case: In the case of Optimistic
2. E_case: In the case of Expected value
3. P_case: In the case of Pessimistic

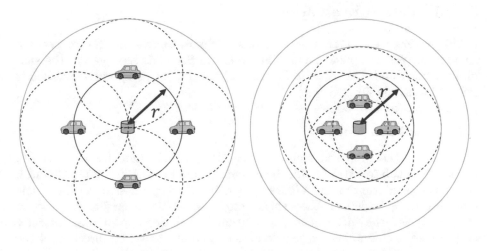

Fig. 4. Node placement for O_case   **Fig. 5.** Node placement for P_case

First, (1) O_case is a state in which a node exists on the edge of the radio coverage r of some node (Fig. 4). In the case of this node arrangement, since the data is always spread with the radius of r by one data transmission. As a result, the required number of hops that can cover the retention area becomes the minimum value. Next, (2) E_case shows that nodes exist uniformly in the radio coverage r. In this case, the expected distance x becomes $\frac{1}{\sqrt{2}}r$. Finally, (3) P_case is defined as the distance $\frac{r}{2}$, which is smaller than the expected value (Fig. 5). In this case, the required number of hops becomes twice compared with that of the optimistic case.

The following initial transmission intervals d_fs (d_{f_o} (O_case), d_{f_e} (E_case), and d_{f_p} (P_case)) are determined by following Eq. (5) to satisfy 100% coverage within T_{c100} seconds.

$$\begin{cases} d_{f\_o} = \frac{n}{\gamma}T_{c100} \times \frac{r}{R}. & (x = r) \\ d_{f\_e} = \frac{1}{\sqrt{2}}\frac{n}{\gamma}T_{c100} \times \frac{r}{R}. & (x = \frac{1}{\sqrt{2}}r) \\ d_{f\_p} = \frac{1}{2}\frac{n}{\gamma}T_{c100} \times \frac{r}{R}. & (x = \frac{1}{2}r) \end{cases} \tag{5}$$

4.2.3   STD Elimination Procedure

If the STD retention is maintained after the expiration of its lifetime T_R, the STD wastes the wireless and storage resources. To solve this problem, in our proposed method, each node checks the start of data transmission (T_s) and the lifetime (T_R) included in the received STD. Then, if $T_n - T_s > T_R$ where T_n shows a present time, the data is discarded, thereby preventing unnecessary data distribution.

5   Simulation Evaluation

In this section, we evaluate the effectiveness of our proposed method by network simulation. We first show simulation models in Sect. 5.1, and present the simulation result and discussion in Sect. 5.2. Note that the comparative method is referred to as the pre_method.

5.1   Simulation Models

We evaluate the proposed method using the network simulator OMNeT++, the traffic simulator SUMO, and Veins implemented in IEEE 802.11p. Figure 6 shows the simulation model. Nodes exist uniformly regardless of location. The lane interval i_l is 200 m, the vehicle interval i_v is 100 m or 200 m ($n = 16$ or $n = 8$) and the speed of nodes is 40 km/h. Vehicles appear intermittently and are set to maintain any vehicle density, and the traveling direction of the nodes alternates between right and left. Moreover, we set the radio range r to 300 m. A beacon transmission interval was set to 1 s, a coefficient for moving average $\alpha = 0.5$, the target number of data transmissions in the t-th cycle γ is 4. Furthermore, retention radius R of all data, a retention time T_R, and T_{c100} is set to 600 m, 4 s, 1 s, respectively. In this simulation, the nodes transmit 10 data at 0.5 s intervals and they retain the data within the retention area. 10 data are sequentially transmitted from the transmission node at 5.5 s intervals, and are accumulated. The packet size of data is 300 bytes.

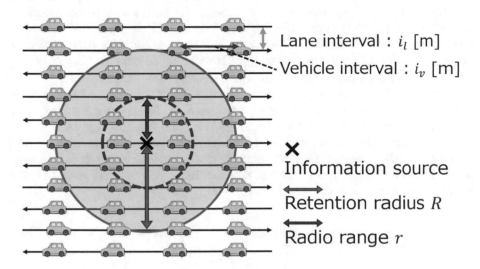

Fig. 6. Simulation model

Table 1. Initial transmission interval by number of neighboring node d_f[s]

	O_case	E_case	P_case	pre_method
$n = 8$	1.000	0.7071	0.5000	5.000
$n = 16$	2.000	1.4142	1.000	5.000

In this case, the number of neighboring nodes is set to $n \geq 8$, and we set the initial transmission intervals of d_{f_o}, d_{f_e}, and d_{f_p} according to Table 1. As in literature [3], we set the transmission interval d of pre_method to 5 s. We evaluate our proposed method from the viewpoint of coverage rate (Eq. (1)) and the average number of transmissions from 0 to T_{c100}.

5.2   Simulation Results and Discussions

Figures 7 and 8 show that the coverage rate is varied with the time series when the number of neighboring nodes is set to 8 and 16, respectively. From these results, the pre_method tends to increase the coverage rate as the number of neighboring nodes increases, but the coverage rate of pre_method at $n = 8$ and $n = 16$ becomes 46.87% and 61.73% at the T_{c100}, respectively. This is because the transmission interval d is set to 5 s ($d = 5$ is longer than T_{c100}). Thus, the initial transmission interval d_f should be set at $0 < d_f < T_{c100}$. The coverage rate of pre_method approaches up to 100% at the time of 5 s. However, since the retention time of the STD (T_R) is 4 s, it can be said that data transmission after T_R is out of demand.

On the other hand, the expected values of d_{f_o}, d_{f_e} and d_{f_p} show high coverage rate at T_{c100}. However, for O_case, the coverage rate decreases to 94.34% when n=16 although the coverage rate is 99.78% when n=8. Because the expected value of distance from the center of retention area to first gamma nodes transmitting data approaches $\frac{1}{\sqrt{2}}r$ (E_case) as the number of neighboring nodes increases. In other words, the data transmission interval to satisfy the requirement of T_{c100} becomes shorter than O_case (d_{f_o}) assumed that first γ nodes exist on r [m] from the center of retention area. Next, the coverage rate at d_{f_e} was maximum 100% and minimum 99.75%, and the coverage rate at d_{f_p} was maximum 100% and minimum 99.99%. This indicates that the coverage rate achieved over 99% regardless of the number of neighboring nodes. It can be seen that the coverage increases as the initial transmission interval decreases. In other words, it indicates that the trend is $d_{f\_o} > d_{f\_e} > d_{f\_p}$.

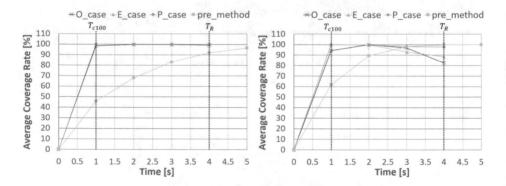

Fig. 7. Coverage rate $(n = 8)$ **Fig. 8.** Coverage rate $(n = 16)$

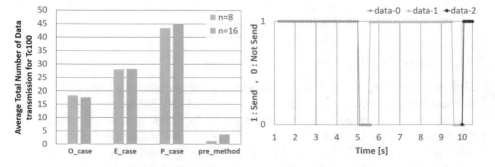

Fig. 9. Average number of data transmissions by method   **Fig. 10.** Time variation of data transmission

Figure 9 shows the average number of data transmissions within T_{c100}. In the proposed method, the average number of data transmissions is that O_case is minimum and P_case is maximum. Since a large number of data transmissions may cause data collisions, it is better that the number of data transmissions is small. Although O_case is the smallest number of data transmission (Fig. 9), the coverage rate is not always more than 99% at T_{c100}. O_case may not be able to satisfy the request because the initial transmission interval is long, especially when the number of neighboring nodes is large. E_case and P_case can always satisfy the coverage of 99% or more at T_{c100}. In terms of the number of data transmission, E_case is 35.41% less than P_case. Thus, E_case can satisfy the requirement while suppressing the number of data transmission.

Finally, Fig. 10 shows the retention time for each data. In this figure, the transmission of each data is indicated by "1", and the non-transmission is indicated by "0." STD1 is retained from 1 s to 5 s, and then the retention of STD2 maintains at 5.5 after 0.5 s interruption until 9.5 s. In this way, we can demonstrate that the proposed method can preserve reliable data retention within T_R.

6   Conclusion

In this paper, we focused on the STD retention system as a way of achieving the paradigm of "local production and consumption of STD." Although the previous method determines the data transmission probability, how to diffuse the STD in the initial phase has not been considered at all. Therefore, in this paper, we propose an on-demand transmission interval control method to complete STD diffusion until the requirement. Through simulations, we showed that the proposed scheme can reliably achieves almost 100% coverage rate, while significantly limiting the increase in the number of STD transmissions. Moreover, we can demonstrate that the STD retention can be controlled in response to the STD lifetime. However, this work does not assume that vehicles move randomly. If the vehicles move randomly, the density of the vehicles in the retention area changes according to the location. Therefore, the STD may not be evenly diffused in the retention area, and we will need to consider that each node changes the data transmission interval by itself.

Acknowledgements. This work was partially supported by JSPS KAKENHI Grant Number 18H03234, NICT Grant Number 19304, and USA Grant number NSF 17-586.

References

1. Wang, S., Zhang, X., Zhang, Y., Wang, L., Yang, J., Wang, W.: A survey on mobile edge networks: convergence of computing, caching and communications. IEEE Access **5**, 1–3 (2017)
2. Beuchert, M., Jensen, S.H., Sheikh-Omar, O.A., Svendsen, M.B., Yang, B.: aSTEP: Aau's Spatio-TEmporal data analytics platform. In: 2018 19th IEEE International Conference on Mobile Data Management (MDM), Aalborg, pp. 278–279 (2018)

3. Teshiba, H., Nobayashi, D., Tsukamoto, K., Ikenaga, T.: Adaptive data transmission control for reliable and efficient spatio-temporal data retention by vehicles. In: The Sixteenth International Conference on Networks, pp. 46–52 (2017)
4. Higuchi, T., Onishi, R., Altintas, O., Nobayashi, D., Ikenaga, T., Tsukamoto, K.: Regional infohubs by vehicles: balancing spatio-temporal coverage and network load. In: Proceedings of IoV-VoI 2016, pp. 25–30 (2016)
5. Li, F., Wang, Y.: Routing in vehicular ad hoc net-works: a survey. IEEE Veh. Technol. Mag. **2**(2), 12–22 (2007)
6. Leontiadis, I., Costa, P., Mascolo, C.: Persistent content based information dissemination in hybrid vehicular networks. In: Proceedings of IEEE PerCom, pp. 1–10 (2009)
7. Ott, J., Hyyti, E., Lassila, P., Vaegs, T., Kangasharju, J.: Floating content: information sharing in urban areas. In: Proceedings of IEEE PerCom, pp. 136–146 (2011)
8. Thompson, N., Crepaldi, R., Kravets, R.: Locus: a location-based data overlay for disruption-tolerant networks. In: Proceedings of ACM CHANTS, pp. 47–54 (2010)

SDN-Based Time-Domain Error Correction for In-Network Video QoE Estimation in Wireless Networks

Shumpei Shimokawa[1(✉)], Takuya Kanaoka[1], Yuzo Taenaka[2], Kazuya Tsukamoto[1], and Myung Lee[3]

[1] Kyushu Institute of Technology, Fukuoka, Japan
{shimokawas,kanaoka}@infonet.cse.kyutech.ac.jp,
tsukamoto@cse.kyutech.ac.jp
[2] Nara Institute of Science and Technology, Ikoma, Japan
yuzo@is.naist.jp
[3] CUNY, City College, New York, USA
mlee@ccny.cuny.edu

Abstract. Our previous study proposed a channel utilization method in Software-Defined Networking (SDN) enabled multi-channel wireless mesh network (SD-WMN), which utilizes all of channel resources efficiently. However, when different types of applications are transferred together, their QoE cannot be maintained because of differences in important factors affecting QoE among these applications. Therefore, in order to handle application flows more efficiently based on QoE, this paper focuses on QoE estimation for every ongoing flows through SD-WMN. Since some parameters required for QoE calculation cannot be obtained from OpenFlow, we estimate QoE based on not only the results from SDN-based measurement but also the estimated values of parameters. Finally, we showed that our proposed method is effective for video QoE estimation, especially in a case where there is no packet loss.

1 Introduction

Efficient resource utilization is one of the important problems in wireless networks. We have been tackling it on a SDN-enabled wireless mesh network (SD-WMN) while maximizing the total throughput [1]. However, it does not always result in the improvement of application performance. Because an application performance, i.e., Quality of Experience (QoE), consists of several factors, throughput may not be important for QoE in some applications. Especially, when multiple different applications coexist on network, their QoE cannot be maintained because of differences in important factors affecting QoE among them. Since diverse applications are increasingly appearing in the Internet, we have to provide high QoE to them by managing the resource utilization efficiently.

Toward efficient resource utilization considering QoE of all flows, we propose a QoE estimation method for ongoing video streaming on the SD-WMN.

L. Barolli et al. (Eds.): INCoS 2019, AISC 1035, pp. 331–341, 2020.
https://doi.org/10.1007/978-3-030-29035-1_32

Although QoE can be easily identified by an offline analysis at an end host, we need to calculate QoE by an online and inside network, while transmitting the flow. However, since some parameters required for QoE calculation cannot be obtained only by OpenFlow, we also propose an online in-network QoE estimation for a video streaming flow by exploiting both OpenFlow-based measurement and parameter estimation to achieve QoE-driven resource utilization.

2   Related Work

QoE-driven network management focusing on video streaming is already studied [2–4]. However, the most of them propose a way to control network on the assumption that QoE is given precisely, and thus a way to measure QoE should be addressed. Reference [5] measures QoE by a measurement agent, which is assumed to handle a receiving video flow in the same way with an end host. However, an intermediate node cannot collect all information of video flow such like an end host. Therefore, this paper focuses on a QoE estimation method that only uses information measured or estimated inside a SDN-enabled network.

From the aspect of SDN-based measurement, reference [6] conducts a delay measurement. This could be useful if a delay is an important factor for QoE calculation. On the other hand, there is a case that a target application may not focus on a delay in QoE calculation. Thus, we focus on QoE estimation, particularly video QoE estimation, based on the measurement and estimation of network performance parameters.

3   QoE Calculation Model for Video Streaming Services

We employ ITU-T G.1071 [7] to calculate the QoE of video streaming services. Section 3.1 provides the brief description of G.1071 and Sect. 3.2 conducts theoretical analysis to clarify the important factors on QoE calculation.

3.1   G.1071-Based QoE Calculation

QoE calculation for a video streaming services is standardized in only ITU-T G.1071 [7]. G.1071 requires several parameters including network quality and video parameters to calculate QoE. Although QoE value is calculated based on both video part and audio part, we focus only on the video part in this study because it is a primary factor of video streaming application. Note that calculated QoE value is ranged from 1 to 5.

G.1071 covers two categories in terms of video resolutions: higher resolution and low resolution, as shown in Table 1. We here describe the QoE calculation model for only higher resolution video due to the lack of space. Those formulas are as follows:

Table 1. The target video settings in G.1071

Category	Lower resolution	Higher resolution
Protocol	RTSP over RTP	MPEG2-TS over RTP
Video codec	H.264, MPEG-4	H.264
Resolution	QCIF (176×114)	SD (720×480)
	HVGA (480×320)	HD ($1280 \times 7201920 \times 1080$)
Video bitrate (bps)	QCIF: 32–1000 k	SD: 0.5–9 M
	HVGA: 192–6000 k	HD: 0.5–30 M
Video framerate	5–30 fps	25–60 fps

$$\text{QoE value} = 1.05 + 0.385 \times Q_V + Q_V(Q_V - 60)(100 - Q_V) \times 7.0 \times 10^{-6}, \quad (1)$$

$$Q_V = 100 - Q_{codV} - Q_{traV}, \quad (2)$$

$$Q_{codV} = A \times e^{B \times b_p} + C + (D \times e^{E \times b_p} + F) + G, \quad (3)$$

$$b_p = \frac{b_r \times 10^6}{r \times f_r}, \quad (4)$$

$$Q_{traV} = H \times \log(I \times p_{lc} + 1), \quad (5)$$

$$p_{lc} = J \times \exp[K \times (L - M) \times \frac{p_{lr}}{M \times (N \times p_{lb} + O) + p_{lr}}] - J. \quad (6)$$

The range of Q_V is from 1 to 100, and Q_V is directly converted to the QoE value by Eq. 1. Parameters of A, B, C, D, E, F, G, H, I, J, K, L, M, N, and O are fixed values defined in G.1071 and take positive value except B and E. Besides, parameters of video bitrate b_r [bps], resolution r [pixel], frame rate f_r [fps], and packet loss concealment (PLC) are pre-determined as the application settings, whereas parameters of packet loss rate p_{lr} [%] and average number of consecutive packet losses p_{lb} are needed to be measured in a reactive manner. The values of fixed parameters in Eq. (6) (i.e., J, K, L, M, N, and O) are determined in accordance with PLC. PLC is one of the technologies in application layer, which corrects a damaged video frame due to packet losses. PLC is classified into Freezing method just ignoring packet losses and Slicing trying to correct packet losses. Since slicing divides a video frame into multiple slices, the correction capability highly depends on the divided number of slices.

3.2   Theoretical Analysis on QoE Calculation

In this section, we investigate the impact of every parameters on QoE. Note that we selectively describe the results of important parameters due to the space limitation. For this purpose, we assume a SD video with the video bitrate of 2.5 Mbps and PLC of Slicing with 1 slice/frame. Also, as a basis of network condition, we use the packet loss rate of 0.1 % and the average number of consecutive packet losses of 1 as the default settings.

Table 2. Impact of packet loss rate on QoE

Packet loss rate [%]	QoE
0	4.654314
0.01	3.333041
0.1	1.874826
1	1.390250

Table 3. Impact of average number of consecutive packet losses on QoE

Average number of consecutive packet losses	QoE
1	1.874826
10	3.110412
100	4.281420

Table 4. Impact of PLC method on QoE

PLC	QoE
Freezing	2.454080
Slicing with 1 slice/frame	1.874826
Slicing with > 1 slice/frame	2.531622

Table 2 shows how the QoE values changes with the increase in the packet loss rates. From this table, we can see that QoE drastically drops when the packet rate is more than 0.1 %. Thus, we can find that packet loss rate is a key factor on QoE for video streaming application. Tables 3 and 4 show how the change in the average number of consecutive packet losses and PLC method impacts on the QoE, respectively. Surprisingly, as the average number of consecutive packet losses becomes larger, the QoE is improved. The increase in the average number of consecutive packet losses means the increase in the number of packets dropped at one. In this case, if the packet loss rate is fixed, the frequency of packet loss event becomes low. That is why the calculated QoE is improved. In short, there is trade-off relationship between packet loss rate and the average number of consecutive packet losses. Regarding to Table 4, we can see that the minimum QoE is brought by Slicing with 1 slice per frame due to the feature of video technology.

In summary, we can remark that packet loss rate, the average number of consecutive packet losses, and the PLC method significantly affect QoE value, and thus we need to obtain these parameters for QoE calculation. However, the latter two parameters cannot be measured in the network because tracking every packets to count consecutive packet losses is quite hard in OpenFlow, and PLC that is an application parameter, which cannot be identified in network. Therefore, we directly measure the packet loss rate, whereas estimate other two parameters.

4 OpenFlow-Based In-Network QoE Estimation

In this section, we propose a QoE estimation method based on the information obtained by OpenFlow. We call this method OpenFlow-based Estimation method (OFE method). OFE method consists of two functions: (1) packet loss rate measurement and (2) video settings estimation.

4.1 OpenFlow-Based Packet Loss Rate Measurement

Although exact packet loss rate (PLR) can be measured only at both end hosts, we try to indirectly measure it based on the information collected in SD-WMN. For this measurement, we use statistic information of each flow (FlowStats), which is collected from APs by a controller (i.e., OFC) on the request basis. Note that in this study, we define a flow as a pair of IP address and port number of source node and destination node. Since FlowStats includes the cumulative number of transmitted packets, the difference between two FlowStats collected in a certain interval are used as the number of transmitted packets. Then, we treat the difference of the number of transmitted packets between two APs (the first AP where a flow enters the SD-WMN (called sender-side AP) and the last AP where it exits (called receiver-side AP)) as the number of packet losses for the flow, thereby calculating PLR based on these values.

However, certain amount of measurement errors cannot be avoided in this simple PLR measurement method because OpenFlow cannot completely synchronize the transmission timing of FlowStats (Fig. 1(i)). That is, since OFC receives a statistic information of the point of when a request arrived at an AP, the condition of on-the-fly packets and/or buffered packets between two APs at the point of FlowStats arrival is different, thereby resulting in the measurement errors. Even a few errors on the number of packet losses significantly affect QoE as discussed in Sect. 3.2.

To solve this issue, we have to correct such kind of errors. Specifically, there are two cases leading to measurement error: the number of transmitted packets on a receiver-side AP is larger than that on a sender-side AP, and vice versa. In the first case, measurement error occurs when a FlowStats request arrives at a receiver-side AP relatively earlier than that to a sender-side AP. Although in most of this kind of case, there is no packet loss, this error causes subsequent measurement errors. Therefore, we hold the difference between the number of transmitted packets sender-side and receiver-side APs as the accumulated surplus packets for correcting subsequent errors (Fig. 1(ii)). In the second case, we expect two possibilities: actual packet losses or errors caused by the timing fault (as in the former case). Since a timing fault frequently happens, we try to correct the errors by taking surplus packets on the receiver-side AP in the first case into account at the subsequent measurements. Specifically, the number of accumulated surplus packets is added to the number of the transmitted packets at the next measurement (Fig. 1(iii)).

In this way, packet loss rate is measured by FlowStats that are periodically transmitted to all APs. G.1071 also requires measurement for 8–16 s as a period to calculate QoE. Therefore, we employ that FlowStats collection and QoE measurement is conducted at the shortest intervals (8 s) because we aim to achieve the QoE-driven network control.

4.2   Parameter Estimation

As described in Sect. 3.2, it is difficult for OFC to measure several parameters related to video image such as video bitrate, frame rate and resolution. Therefore, we try to estimate each of them based on the limited information.

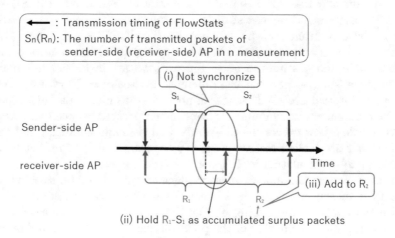

Fig. 1. Image of error correction procedure.

Table 5. Recommended video bitrate for resolution and frame rate.

Resolution	Frame rate [fps]	video bitrate [kbps]
SD (720×480)	24,25,30	2,500
SD (720×480)	48,50,60	4,000
HD (1280×720)	24,25,30	5,000
HD (1280×720)	48,50,60	7,500
HD (1920×1080)	24,25,30	8,000
HD (1920×1080)	48,50,60	12,000

Average Number of Consecutive Packet Losses: Because FlowStats includes statistic information only, OFC cannot understand consecutiveness. As analyzed in Table 3, the value of 1 shows the worst QoE, so, we set it to 1 to avoid the QoE overestimation.

Video Bitrate: Since OFC cannot directly obtain information of video settings, video bitrate is estimated based on the measured throughput. Specifically, we treat the measured throughput as a video bitrate. Note that the throughput is measured based on the number of transmitted bytes of FlowStats at a sender-side AP.

Frame rate/Resolution: Frame rate and resolution are estimated from the estimated video bitrate. Table 5 shows the recommendation of video settings in terms of resolution, frame rate, and video bitrate. Since Table 5 can be expressed as Table 6, we estimate frame rate and resolution based on video bitrate of Table 6. In Table 6, we choose the maximum frame rate from among candidates in Table 5 to avoid QoE overestimation. Also, we choose the range of video bitrate for each entry so that it can be a median of consecutive entries.

PLC: As discussed in Subsect. 3.2, we employ "slicing with 1 slice per frame" as the PLC.

Table 6. Estimation of resolution and frame rate based on video bitrate.

Video bitrate [kbps]	Resolution	Frame rate [fps]
–3,250	SD (720 × 480)	30
3,250–4,500	SD (720 × 480)	60
4,500–6,250	HD (1280 × 720)	30
6,250–7,750	HD (1280 × 720)	60
7,750–10,000	HD (1920 × 1080)	30
10,000–	HD (1920 × 1080)	60

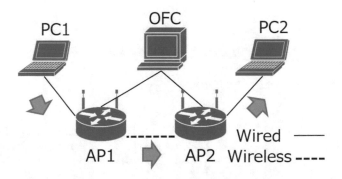

Fig. 2. Experimental environment.

5   Experimental Evaluation

We conduct experiments to evaluate the OFE method in a real wireless environment. The goal of this experiment is to show the effectiveness of the OFE method. We compare the OFE method with a comparative method. Note that the comparative method estimates QoE based on FlowStats like OFE method but does not conduct the error correction for the timing-fault case.

5.1   Experimental Settings

The experimental topology is shown in Fig. 2. We use IEEE 802.11a with fixed 54 Mbps on 120 channel in the wireless settings. Regarding OpenFlow, we use Trema as OFC and install OpenvSwitch on every APs. Networks between the OFC and APs, and PCs and APs, are made by Ethernet cables in order to avoid packet loss and delayin this section. In our experiment, PC1 transmits a video streaming to PC2 for 60 s. The video is made by the H.264 codec with SD (720 × 480), 30 fps and 2.5 Mbps CBR.

5.2   Effectiveness in No Packet Loss Environment

Figure 3 shows the number of transmitted packets and packet loss rate calculated by each method, and Fig. 4 shows the calculated QoE. Measurement errors caused by the timing fault often happen even in no packet loss environment. In the comparative method, QoE drop by more than 2 due to those measurement errors arising from the timing fault. On the other hand, measuring errors between 15 and 23, 31 and 39, and 39 and 47 s are successfully corrected by taking into account the accumulated surplus packets around 7 and 15, and 23–31 s in OFE method. As a result, the occurrence of unnecessary packet losses can be avoided, thereby providing almost same value with the true QoE value, which is measured at end hosts. Therefore, OFE method is effective for estimating packet loss rate and QoE value.

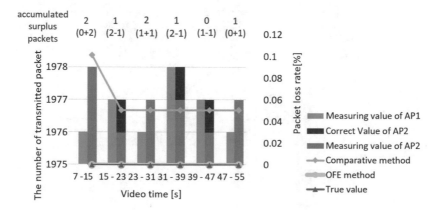

Fig. 3. The number of transmitted packet and packet loss rate in no packet loss environment.

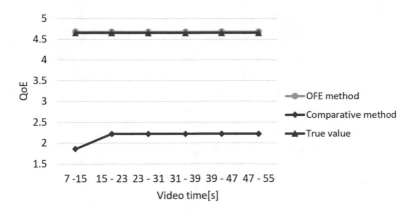

Fig. 4. Measured QoE value in no packet loss environment.

5.3   Effectiveness in Packet Loss Environment

This section conducts an evaluation in case where packet losses inevitably occurs due to the deterioration of the wireless link quality. As for the environment, we intentionally cause packet losses by increasing the distance between AP1 and AP2 up to 20.5 m.

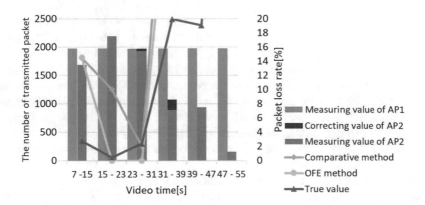

Fig. 5. The number of transmitted packet and packet loss rate in packet loss environment

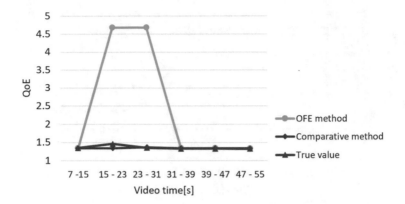

Fig. 6. Measured QoE value in packet loss environment.

Figures 5 and 6 show the results of the estimated packet loss rate and QoE, respectively. In OFE method, packet loss rate between 7 and 15, 31 and 39, 39 and 47, and 47 and 55 s becomes clearly larger than the true value because the OFE method treats the increase of the retransmission delay (i.e., buffered packets) as packet losses and thus the number of transmitted packets at the receiver-side AP decreases. However, the estimated QoE is almost same with the true value because overestimated packet loss rate has little effect on QoE.

On the other hand, packet loss rate between 15 and 23, and 23 and 31 s are lower than true value in OFE method. This is because packets left in the sender AP's queue in previous periods is transmitted late. In this case, actual packet losses are regarded as no packet loss because the number of transmitted packets of the receiver-side AP become larger than those of the sender-side AP or packet losses are corrected by mistake. Therefore, OFE method still has problems in terms of estimation accuracy in case of packet loss environment. This is because we are exploiting only the information of network layer in this study. If we can obtain the information from wireless layer, the accuracy of QoE estimation may be improved. Therefore, we are going to solve this problem by cooperation with wireless layer.

Through experimental results, we showed OFE method was effective under no packet loss environment. On the other hand, the estimation accuracy of the OFE method was clearly degraded under packet loss environment.

6   Conclusion

In this paper, we presented a QoE estimation with time-domain error correction for video streaming aiming to efficiently conduct QoE-driven resource management. At first, we showed that packet loss rate has a significant effect on the QoE value for video streaming through the theoretical analysis. Then, we proposed OFE method that estimates QoE by exploiting FlowStats information, while correcting the measurement errors of packet loss rate. In our experiments, we remarked that OFE method can estimate QoE precisely in the environment where there is no packet loss. On the other hand, we showed that the estimation accuracy by the OFE method dropped in the environment with packet losses. In response to this result, we are going to cooperate with wireless layer to estimate QoE more accurately.

Acknowledgements. This paper was partly supported by JSPS KAKENHI Grant Number 17H03270, NICT Grant Number 19304, and USA Grant number NSF 17-586.

References

1. Tagawa, M., Taenaka, Y., Tsukamoto, K., Yamaguchi, S.: A channel utilization method for flow admission control with maximum network capacity toward loss-free software defined WMNs. In: ICN 2015, February 2016
2. Nam, H., Kim, K.H., Kim, J.Y., Schulzrinne, H.: Towards QoE-aware video streaming using SDN. In: GLOBECOM, December 2014
3. Seppänen, J., Varela, M.: QoE-driven network management for real-time over-the-top multimedia services. In: WCNC, April 2013
4. Erfanian, A., Tashtarian, F., Yaghmaee, M.H.: On maximizing QoE in AVC-based HTTP adaptive streaming: an SDN approach. In: IWQoS, June 2018
5. Farshad, A., Georgopoulos, P., Broadbent, M., Mu, M., Race, N.: Leveraging SDN to provide an in-network QoE measurement framework. In: INFOCOM Workshop CNTCV, May 2015
6. Bouzidi, E.H., Luong, D.H., Outtagarts, A., Hebbar, A., Langar, R.: Online-based learning for predictive network latency in software-defined networks. In: GLOBECOM, December 2018
7. Recommendation ITU-T G.1071: Opinion model for network planning of video and audio streaming applications, November 2016

On Retrieval Order of Statistics Information from OpenFlow Switches to Locate Lossy Links by Network Tomographic Refinement

Takemi Nakamura(✉), Masahiro Shibata, and Masato Tsuru

Computer Science and System Engineering,
Kyushu Institute of Technology, Iizuka, Japan
nakamurat@infonet.cse.kyutech.ac.jp, {shibata,tsuru}@cse.kyutech.ac.jp

Abstract. To maintain service quality and availability in managed networks, detecting and locating high loss-rate links (i.e., lossy links that are likely congested or physically unstable) in a fast and light-weight manner is required. In our previous study, we proposed a framework of network-assisted location of lossy links on OpenFlow networks. In the framework, a measurement host launches a series of multicast probe packets traversing all full-duplex links; and then the controller retrieves statistics on the arrival of those probe packets at different input ports on different switches and compares them to locate high loss-rate links. The number of accesses to switches required to locate all lossy links strongly depends on the retrieval order in collecting the statistics and should be small as much as possible. Therefore, in this paper, to minimize the necessary number of accesses, we develop a new location scheme with an appropriate retrieval order using a Bayesian-based network tomography to refine candidates for lossy links. The results of numerical simulation on a real-world topology demonstrate the effectiveness of the new location scheme.

1 Introduction

The recent proliferation of cloud and edge-could computing technologies requires flexible, dynamic, and reliable networking among geographically distributed but centrally managed servers and sites. Therefore, SDN (Software defined network) in general and OpenFlow technology in particular have been applied not only to data centers but also to enterprise networks and wide area [1].

To maintain service quality and availability in such networks, detecting and locating high loss-rate links that are likely congested or physically unstable in a fast and light-weight manner is essential. In general, network operators constantly monitor the communication performance and the internal status of links by either passive or active measurement. Passive measurement in Open-Flow networks is quite useful. In addition to traditional SNMP monitoring, per-flow statistics can be monitored and collected by FlowStats function. However,

L. Barolli et al. (Eds.): INCoS 2019, AISC 1035, pp. 342–351, 2020.
https://doi.org/10.1007/978-3-030-29035-1_33

although passive measurement itself does not incur additional load on the data plane, frequent accesses to switches for accurate and timely monitoring may incur additional load on the control plane.

Active measurement in OpenFlow networks is also attractive due to the capability of per-flow flexible routing. Probe packets can flow on any designed routes to measure the packet loss, delay, the round-trip-time (RTT), and so on. Furthermore, in networks connecting geographically-wider locations, a "link" between two nodes is not always physical but virtual (e.g., tunneling). So passive measurement only on managed nodes is not enough and active measurement is essential to monitor the entire network. However, probing at a high sending rate for precise and reliable monitoring incurs unnecessary load on switches and the data plane. An infrastructure was proposed to monitor RTT with suppressing the number of flow entries and probe packets [2]. A delay monitoring that covers all links in both directions with minimizing flow entries on switches was also studied [3].

Network-tomographic approaches have been studied to infer network link states without directly monitoring those links [4]. Original network tomography monitors packet-level correlations among measurement paths, which was thought as too costly in practice. Then the Boolean network tomography was proposed that only monitors performance-level correlations among measurement paths to infer the location of bad links [5] and followed by a number of studies because of its practicality (e.g.,[6]). The impact of the capability of routing of probe packets has also been studied in localizing failed nodes based on Boolean network tomography [7].

In our previous study [8], we proposed a framework of detecting and locating lossy links on OpenFlow networks with a light load on both the data and control planes by a collaboration of switches and controller with measurement host. In the framework, the retrieval order in collecting the statistics from switches strongly impacts the number of accesses to switches required to locate all lossy links, i.e., the load incurred on the control plane. Therefore we adopted a simple Boolean network-tomographic inference of highly lossy links to design an appropriate retrieval order. However, this approach is efficient only when the number of lossy links is small or lossy links are commonly shared by many measurement paths.

In this paper, therefore, to minimize the necessary number of accesses, we develop a new location scheme on retrieval order using a Bayesian-based network tomographic refinement of the candidates for lossy links. It uses correlation of loss events of each probe packet that can be monitored by an extension of per-flow statistics of OpenFlow. This extension is feasible and light-weight by using ID field of IP packet. It also requires the prior loss probability of link calculated by using past measurement results. Note that it can be tolerant in inference-error because the aim is not to precisely infer the link loss rates but to promptly locate lossy links. To reduce the computational cost of posterior probabilities, a series-reduced tree is used instead of the original multicast measurement tree.

Section 2 explains the framework and the basic location scheme we previously proposed. Section 3 proposes a new location scheme. Simulation evaluation of both basic and proposed schemes on a real-world network topology is performed in Sect. 4, followed by the concluding remarks in the last section.

2   Overview and the Basic Location Scheme

The framework assumes a network comprising OpenFlow controller (OFC) and OpenFlow switches (OFS). The process starts when the measurement host (MH) sends a request to the OFC as illustrated in Fig. 1. Next, the OFC gets the network topology, calculates probe packet routes, and installs them to OFSs.

Then, the MH launches a series of multicast probe packets traversing all links once and only once (separately in each direction of the full duplex link) to minimize the load on the data plane incurred by probe packets. The probe packets are discarded at "leaf port" on the last OFS of the measurement path. Then, the OFC retrieves statistics on the arrival of those probe packets (i.e., the number of probe packets arriving) at different input ports on different OFSs and compares them to locate high loss-rate links. The number of lost packets on a link (or series of links) between the two switch ports can be calculated by taking the difference in the number of probe packets arriving at those ports.

An example of the route configuration is shown in Fig. 2. The root port is a switch port connected to the MH, and the leaf port is a switch port for discarding probe packets. A path of the probe packet (i.e., measurement flow) from the root port to a leaf port is called "terminal path".

In our framework, measurement routes through which probe packets flow is designed by three steps. Please see more details in [8].

- Generate the shortest path tree in the downward direction from the root (blue dashed lines in Fig. 3).
- Complement unused links not on the shortest path tree (green dotted lines).
- Add return links in the upward direction bound for the root (red lines).

After a series of probe packets flows on all links, the MH informs the OFC of the probing completion. Then, in our previous study, the following algorithm (i.e., the basic location scheme) is used to find lossy links, which determines an appropriate access order to selected OFSs to collect flow statistics. A link is considered as lossy if and only if its loss rate exceeds a threshold value h that is a design parameter representing the target link quality to maintain and depends on the target applications. The packet loss rate of a segment (a series of links) from ports i to j, PLR, can be computed by r_i and r_j that are the numbers of probe packets arriving at switch ports i and j, respectively.

$$PLR = 1 - \frac{r_j}{r_i} \tag{1}$$

First, the OFC accesses to the root port (port 0) and each (j) of all leaf ports to retrieve r_0 and r_j, respectively, in order to calculate the PLR of each

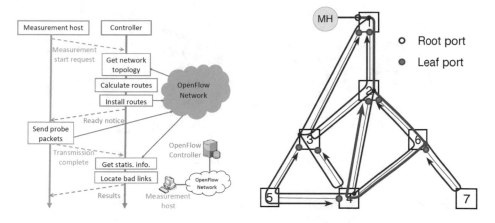

Fig. 1. Measurement process [8]   **Fig. 2.** Route scheme example [8]

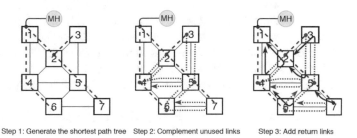

Step 1: Generate the shortest path tree   Step 2: Complement unused links   Step 3: Add return links

Fig. 3. Route scheme design [8]

terminal path, using equation (1). If the PLR of a terminal path is less than h, it does not include any lossy link. If the PLR of a terminal path exceeds h, this terminal path is likely to include one or more lossy links.

Then, we narrow the search range, i.e., the expected locations of lossy links, and determine the retrieval order. If a terminal path is lossy (i.e., its PLR value exceeds threshold h) and there are no other lossy terminal paths, the lossy links are located within a segment between the leaf port and the nearest parent port on the considered lossy terminal path, as illustrated by the dashed line in Case 1 of Fig. 4. The ports along this segment are queried in a binary-search manner until lossy links in this segment are located. If there are multiple lossy terminal paths, the port most commonly shared by those paths is queried first to collect the number r_j of arrival probe packets, as illustrated by the dashed line in Case 2 of Fig. 4. The access and packet loss rate are checked from the most upstream port in the common ports. This procedure generates separated sub-trees, and the same procedure is performed on each sub-tree recursively until all lossy segments are identified in the trees.

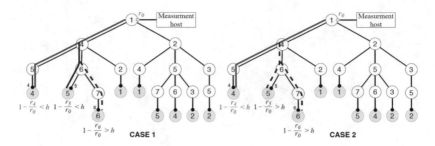

Fig. 4. Basic location scheme

3   Proposed Location Scheme

We propose a new location scheme with an appropriate retrieval order using a Bayesian-based network tomography to probabilistically refine candidates for lossy links. The packet-level loss event correlation among terminal paths is monitored. A unique ID is assigned to each probe packet at the MH, which is essential for the OFC to monitor the packet-level loss event correlation. Then, to compute the posterior loss probability (i.e., expected packet loss rate) on each link based on the monitored events on terminal paths, we need the prior loss probability of link.

Here, a link that often exhibits a high packet loss rate, i.e., at high failure level, lately is assumed to cause a packet loss with a high probability in near future. Therefore, the prior loss probability of link is calculated based on the measured loss rates in the past measurements.

Let l be the current measurement cycle, m be the number of links of the network, and b_{kl} be the prior loss probability on each link k $(1 \leq k \leq m)$ in cycle l. The value b_{kl} reflects the latest measured loss rate p_k of link k by using the Exponential Moving Average (EMA) with smoothing factor α. $b_{k(l+1)}$ is updated as follows.

$$b_{k(l+1)} = (1 - \alpha)b_{kl} + \alpha p_k \tag{2}$$

In case that p_k is not measured in cycle l, $b_{k(l+1)} = b_{kl}$. In the initial cycle 1, b_{k1} is a given initial value identical for any link k.

The posterior loss probability q_k is the conditional probability of packet loss on link k given correlated loss events occurred over measurement paths. Here, L_k is the number of packet losses on link k for n probe packets, Let Y_i be 1 if a probe packet is lost on terminal path i $(1 \leq i \leq z)$ and be 0 otherwise. And $M_j = (Y_1, \ldots, Y_z)$ is the correlated loss events over z paths for the j-th probe packet: $1 \leq j \leq n$.

Based on measured $M_1, \ldots, M_n$ and the prior loss probabilities of all links, q_k is calculated by the following equation:

$$q_k = \frac{E[L_k | M_1, \ldots, M_n]}{n} \tag{3}$$

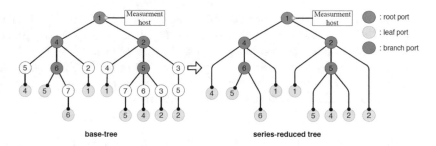

Fig. 5. Making series-reduced tree

By letting X_k be 1 if a probe packet is lost on link k; be 0 if the packet is not lost; and be "*" if unknown (the packet does not arrive at k), the following equation holds.

$$\frac{E[L_k|M_1,\ldots,M_n]}{n} = \frac{\Sigma_{j=1}^n P(X_k = 1|M_j)}{n} = \frac{\Sigma_{j=1}^n \frac{P(X_k=1 \cap M_j)}{P(M_j)}}{n} \quad (4)$$

$P(M_j)$ is calculated based on possible combinations of packet loss occurrences $\mathbf{X} = (X_1,\ldots,X_m)$ consistent with M_j, and $P(X_k = 1 \cap M_j)$ is calculated based on possible combinations of packet loss occurrences consistent with $X_k = 1$ and M_j.

Here, the probability of $X_k = 1$ is set to b_{kl}, and the probability of $X_k = 0$ is set to $(1 - b_{kl})$ to obtain q_k. Link k with a value of q_k exceeding h is regarded as suspected link that is likely a lossy link; it is used for the algorithm.

However, as the number of links inevitably increases in a large-scale topology, the number of combinations of $\mathbf{X}$ to be estimated becomes enormous, and the calculation time of the posterior probability increases, resulting in a large delay to locate lossy links. Therefore, we introduce series-reduced tree to solve this problem.

A path tree for transmitting probe packets is called "base-tree". By directly connecting the root port, leaf ports, and branch ports of a base-tree, a path tree reduced in scale is created and called a series-reduced tree like Fig. 5. Here, the link in the series-reduced tree is called "reduction link" k'. Note the series-reduced tree is used only to calculate the posterior probability and to narrow down the lossy links at first.

After calculating and updating b_{kl}, the prior loss probability $g_{k'}$ on reduction link k' is calculated. If reduction link k' comprises t links $(1 \leq k \leq t)$ of the base-tree, $g_{k'}$ in a given cycle l is calculated from b_{kl} as follows.

$$g_{k'} = 1 - (1 - b_{1l})(1 - b_{2l})\ldots(1 - b_{tl}) \quad (5)$$

Then the posterior loss probability $q_{k'}$ of reduction link k' is calculated from $g_{k'}$, which is used to narrow a possibly-lossy segment as "range".

A reduction link with a value of the posterior loss probability $q_{k'}$ exceeding h is likely to include lossy links, and can be an initial range. We start from the reduction link with the higher posterior loss probability. The actual packet loss rate of the range is checked using equation (1). If it exceeds h, the process proceeds to identify one or more lossy links within the range.

Here, the following two cases of the prior loss probabilities in a range are considered. If all b_{kl} values are the same, binary search is performed. Otherwise, the prior loss probability-based narrowing is performed.

A. *Binary search*
   The statistics is obtained from the center port of the range to be narrowed down, and the range divided based on the port, then the packet loss rate is checked for each range.
B. *The prior loss probability-based refinement*
   The link with the highest prior loss probability within the range is regarded as a suspected link. The statistics is obtained from the port on the downstream side of the suspected link, and the range is divided at the port, then the packet loss rate of each range is checked. When the packet loss rate in the upstream range exceeds h, the statistics is obtained from the upstream port of the suspected link, and the packet loss rate of the link is checked.

After the narrowing within selected reduction link is completed, the process moves to another reduction link which has the next higher posterior loss probability.

4 Simulation Evaluation

To evaluate the proposed location scheme compared with the basic location scheme, numerical simulation are performed on a real-world network topology, from a topology database [9], shown in Fig. 6 (MH is connected).

There are 19 OFSs and 24 links (48 links in both directions); the number n of probe packets is 1000, threshold h of lossy link is 0.01, smoothing factor α is 0.3, initial value b_{k1} of prior loss probabilities is 0.001, and one simulation runs for 10 measurement cycles. The packet loss rate of each link is randomly set to 0.05 0.1 (high failure level), 0.001 0.01 (low failure level), or 0.001(normal link). The path tree is shown in Fig. 7. The number in each circle is an ID of input (downstream) port of a link illustrated by an arrow-line, and the number also identifies the link.

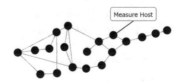

Fig. 6. Simulated network topology

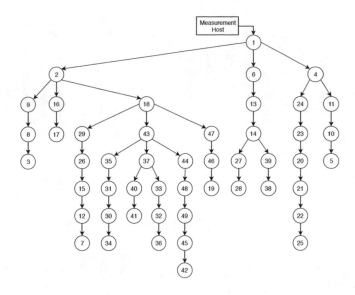

Fig. 7. Generated path tree

Here, low loss-rate links are positioned on (6, 45, 7, 32, 21) in both two patterns. Figure 8 (resp., Fig. 9) shows the number of accesses to switches required to retrieve statistics of all terminal paths (yellow-coloerd); and at additional three (resp., four) stages before all lossy links are detected in each measurement cycle (differently colored). Note that each high loss-rate link always becomes lossy (i.e., its actual loss rate exceeds h) in every measurement cycle and each low loss-rate link unlikely but sometimes becomes lossy.

Pattern 1: high loss-rate links (2, 4, 13) (Fig. 8)
In Pattern 1, a few number of high loss-rate links are set and positioned in the most common parts of all terminal paths. In such case, those terminal paths are observed as lossy and correlated at the packet-level loss. Then the basic scheme always simply checks the common part first, which can result in the most efficient retrieval order of port accesses in this case. Therefore no advantage by the proposed scheme is expected. Furthermore, if a low loss-rate link accidentally becomes lossy in some cycle, it may result in an inefficient retrieval order in the proposed scheme as shown in the 3, 4, and 7th cycles in Fig. 8. This is because the prior loss probability of such a low loss-rate link is generally low and it decreases the search priority of that low loss-rate link (but it is actually lossy in this cycle).

Pattern 2: high loss-rate links (10, 20, 39, 27, 18, 16, 8) (Fig. 9)
In Pattern 2, a more number of high loss-rate links are set and positioned in less-common and downstream parts of terminal paths. In such case, those terminal paths are observed as lossy but almost independent (uncorrelated) at the packet-level loss. Then, again, the basic scheme always simply checks the common part

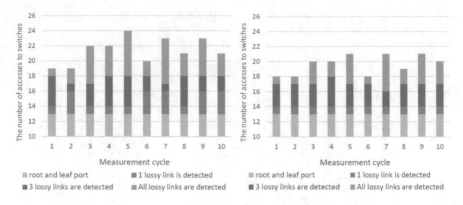

Fig. 8. Pattern 1 (left: proposed, right: basic)

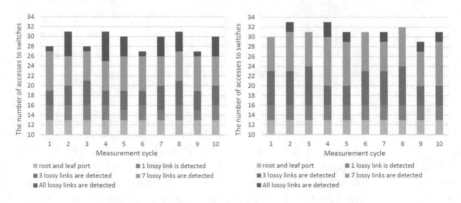

Fig. 9. Pattern 2 (left: proposed, right: basic)

first but it is not the actual lossy part in this case. On the other hand, even in the first measurement cycle, the proposed scheme utilizes the packet-level loss event correlation among terminal paths to infer which reduction link is more likely to be lossy. It computes a high posterior loss probabilities of the reduction links on which the high loss-rate links are included, and likely results in a better retrieval order. In the second and later cycles, since the positions of high loss-rate links are unchanged, both the prior and posterior loss probabilities of high loss-rate links are kept high, and thus a better retrieval order is kept as well.

However, similarly to pattern 1, if a low loss-rate link accidentally becomes lossy, it may result in a disadvantage of the proposed scheme as shown in the 2, 4, 5, 7, and 10th cycles in Fig. 9.

Table 1 shows the average number of retrievals to find all lossy links in pattern 2 over 10 trials of simulation; the proposed scheme outperforms the basic at any cycle.

Table 1. Average number of retrievals to find all lossy links in pattern 2

Measurement cycle	1	2	3	4	5	6	7	8	9	10
Propose	30.2	29.2	30.0	28.9	30.1	29.4	29.6	28.6	28.3	29.2
Basic	31.3	31.6	32.2	31.1	32.0	32.0	31.2	31.3	30.6	31.2

5   Concluding Remakes

Based on our previously proposed framework to monitor and locate lossy links on OpenFlow networks, we have developed a new location scheme to reduce the necessary number of accesses in collecting the statistics from switches. A Bayesian-based network tomography on a series-reduced tree is adopted to refine the search range with an acceptable computational cost. The simulation results show the new scheme can efficiently locate all lossy links with a fewer number (less than the 60% of the total number of links) of accesses to switches even in case that the basic scheme is not efficient.

Acknowledgements. The research results have been achieved by the "Resilient Edge Cloud Designed Network (19304)," NICT, and by JSPS KAKENHI JP16K00130 and JP17K00135, Japan. We thank Mr. Suguru Goto and Mr. Yuki Fujimura for assistance.

References

1. Jain, S., Kumar, A., et al.: B4: experience with a globally-deployed software defined WAN. In: Proceedings of ACM SIGCOMM 2013, pp. 3–14 (2013)
2. Atary, A., Bremler-Barr, A.: Efficient round-trip time monitoring in OpenFlow networks. In: Proceedings of IEEE INFOCOM, pp. 1–9 (2016)
3. Shibuya, M., Tachibana, A., Hasegawa, T.: Efficient active measurement for monitoring link-by-link performance in OpenFlow networks. IEICE Trans. Commun. **E99B**(5), 1032–1040 (2016)
4. Castro, R., Coates, M., Liang, G., Nowak, R., Yu, B.: Network tomography: recent developments. Stat. Sci. **19**(3), 499–517 (2004)
5. Duffield, N.: Network tomography of binary network performance characteristics. IEEE Trans. Inf. Theory **52**(12), 5373–5388 (2006)
6. Tachibana, A., Ano, S., Hasegawa, T., Tsuru, M., Oie, Y.: Locating congested segments over the internet based on multiple end-to-end path measurements. IEICE Trans. Commun. **E89–B**(4), 1099–1109 (2006)
7. Ma, L., He, T., Swami, A., Towsley, D., Leung, K.K.: Network capability in localizing node failures via end-to-end path measurements. IEEE/ACM Trans. Netw. **25**(1), 434–450 (2017)
8. Tri, N.M., Tsuru, M.: Locating deteriorated links by network-assisted multicast proving on OpenFlow networks. In: Proceedings of the 24th IEEE Symposium on Computers and Communications (ISCC 2019), 6 p, to appear in July 2019
9. The Internet Topology Zoo: http://www.topology-zoo.org/. Accessed 15 Feb 2019

Problem of Determining Weights of Edges for Reducing Diameter

Kaito Miyanagi and Hiroyoshi Miwa$^{(\boxtimes)}$

Graduate School of Science and Technology, Kwansei Gakuin University,
2-1 Gakuen, Sanda-shi, Hyogo, Japan
{miyanagi,miwa}@kwansei.ac.jp

Abstract. It is necessary to reduce delay in an information network. The total delay time between two nodes is the sum of the delay times in all links and nodes contained in the path between the nodes; therefore, we can reduce the total delay time by reducing the delay times in some links and nodes in the path. Since the additional cost such as the increase of the link speed is necessary, the links whose speed is increased must be determined by considering the trade-off between performance and cost. In this paper, we formulate this network design problem as an optimization problem of determining the weight of each edge so that the diameter of a network is less than or equal to a given threshold. The objective of this optimization problem is to minimize the sum of the costs under the condition that the number of edges whose weight is changed is restricted. We prove that this problem is NP-complete, and we propose a polynomial-time algorithm to the problem that the number of edges whose weights are changed is restricted to one.

1 Introduction

In an information network delay time is an important performance measure. Since the total delay time between two nodes is the sum of the delay times in all links and nodes contained in the path between the nodes, we can reduce the total delay time by reducing the delay times in some links and nodes in the path. However, in order to reduce the delay time of links and nodes, it is necessary to increase the link speed and to add buffers, and so on and it increases the additional cost; therefore, the links whose speed is increased must be determined by considering the trade-off between performance and cost.

In this paper, we address this network design problem of determining the weight of each edge so that the diameter of a network is less than or equal to a given threshold. The diameter of the network is the maximum of the distance between two vertices among all pairs of two vertices, where the distance between the vertices is defined by the shortest path length. We prove that this problem is NP-complete, and we propose a polynomial-time algorithm to the problem that the number of edges whose weights are changed is restricted to one.

© Springer Nature Switzerland AG 2020
L. Barolli et al. (Eds.): INCoS 2019, AISC 1035, pp. 352–359, 2020.
https://doi.org/10.1007/978-3-030-29035-1_34

2   Related Research

Since diameter is related to delay in an information network, many network design problems considering diameter have been extensively studied so far.

A Moore graph is defined as a graph with the largest number of vertices under the constraint of a diameter and a degree. The diameter of a Moore graph is "small" in comparison with the number of vertices. Generally, it is not known how to design a Moore graph with any diameter and any degree; however, it is known how to design a graph with small diameter and bounded degree such as de Bruijn graphs [1] and Kautz graphs [2].

A de Bruijn graph with d-ary k-digits is constructed as follows. As an example, let us explain the de Bruijn graph of $d = 2$ and $k = 3$. The number of binary 3 digits is eight, $000, 001, 010, 011, 100, 101, 110, 111$ and 00010111 is the shortest sequence including all of them. We assume that the beginning and the end of this sequence are connected. It means that 00101110, which shifts this number sequence by 1 bit, can be regarded as the same number sequence. By extracting one 3-digit number from the beginning of this sequence, all of the 8 binary 3-digit numbers appear. Let these numbers be vertices, respectively, and let a vertex be connected to the vertex whose number is shifted by 1 bit. The diameter of a de Bruijn graph is small in comparison with the number of vertices. A generalized de Bruijn graph is defined whose number of vertices is arbitrarily positive integer [3], although the number of vertices of a de Bruijn graph with d-ary k-digit is d^k. There are also studies using de Bruijn graph for data compression [4].

A Kautz graph $K(d, k)$ is similarly defined as the subgraph of de Bruijn graph. The number of the vertices of a Kautz graph $K(d, k)$ with the maximum degree is d and the diameter is k is $d^k + d^{k-1}$. A generalized Kautz graph is similarly defined [5,6]. The consecutive-d graph is also known as a class of graphs combining generalized de Bruijn graphs and generalized Kautz graphs [7].

The problem of finding a spanning subgraph with the smallest cost whose diameter is restricted is known [8]. It is still NP-hard, even if all edges have the same length and all the edge costs are the same; however, when the diameter is less than or equal to three and a spanning subgraph is restricted to a spanning tree, it can be solved in polynomial time [9]. When the cost of all edges is the same and a spanning subgraph is restricted to a spanning tree, it can be solved also in polynomial time [9].

There is the problem of determining the lengths of the edges in a graph whose diameter is less than or equal to a given positive value [10]. When an undirected graph and a set of positive integers C whose number of integers is equal the number of edges are given, the problem asks whether each integer can be assigned to the length of an edge so that the diameter of the graph is less than or equal to a given positive value. This problem is generally NP-hard.

3   Problem of Determining Edge Weights to Reduce Diameter

Let $G = (V, E)$ where V is a set of vertices and E is a set of edges be a connected undirected graph representing an information network. Each vertex corresponds to a node such as a router, and each edge corresponds to a link between nodes. Let the edge weight function be $w : E \to \mathbb{R}_+$. The weight of an edge corresponds to the delay time of the link corresponding to the edge. Let the edge cost function be $c : E \to \mathbb{R}_+$. The cost of an edge corresponds to the cost required to reduce the weight by one unit, which corresponds to the cost required to reduce the link delay by a unit time. Let a network be denoted by $N = (G, w, c)$. Let a path set be $Q = \{q_1 = \{s_1, t_1; p_1\}, q_2 = \{s_2, t_2; p_2\}, \ldots, q_r = \{s_r, t_r; p_r\}\}$, where $s_i, t_i \in V (i = 1, 2, \ldots, r)$ and p_i is a path between vertex s_i and vertex t_i. For the sake of simplicity, we assume that there is at most one path between a pair of vertices. The weight of the path p_i $(i = 1, 2, \ldots, r)$ is defined as the sum of the weights of the edges contained in path p_i. Let the weight of the path p_i be denoted by $|p_i|$. When the weights of all paths in Q are less than or equal to a given threshold, we say that (N, Q) satisfies the diameter constraint. When Q is the set of the shortest paths between two vertices of each pair for all pairs of vertices, if the weights of all paths in Q are less than or equal to a given threshold, it means that the diameter is less than or equal to the threshold. Therefore, we use the term of diameter constraint also for a general case of Q.

Note that any path in Q does not change regardless of weights and costs of edges. Only the weight of a path can change depending on the change of weights and costs of edges.

Problem of determining edge weights to reduce diameter (DWP)
INSTANCE: *Network $N = (G = (V, E), w, c)$, path set Q, positive integer p, positive real number B, D.*
QUESTION: *Are there a set of real numbers $g(e)(\geq 0)$ $(e \in E)$ that satisfies the following constraints:*

- $\sum_{e \in E} g(e)c(e) \leq B$.
- *The weight of each path p_i $(i = 1, 2, \ldots, r)$ in Q is less than or equal to D in the resulting network after the weight of edge e is updated to $w(e) - g(e)$ for all edges $e \in E$.*
- *The number of edges e such that $g(e) > 0$ is less than or equal to p.*

□

This problem corresponds to a network design problem that determines the links to reduce delay and the quantity of the reduction so that the maximum delay is less than or equal to a threshold. The set of links to reduce delay corresponds to the set of edges e such that $g(e) > 0$; the quantity of the reduction corresponds to $g(e)$. The number of links to reduce delay must be less than or equal to p in order to reduce work load. The total cost to reduce delay, $\sum_{e \in E} g(e)c(e)$, must be less than or equal to B in order to reduce the cost.

Theorem 1. *DWP is generally NP-complete.*

Proof of Theorem 1. We prove the NP-completeness by reducing the Hitting Set Problem (HSP), which is a known NP-complete problem, to DWP in polynomial time. HSP is defined as follows: For set $U = \{u_1, u_2, \ldots, u_n\}$ and the family of sets $C = \{C_1, C_2, \ldots, C_k\}(C_i \subseteq U(i = 1, 2, \ldots, k))$ on U, a positive integer $z(\leq n)$, HSP asks whether there is a subset of U, U' ($|U'| \leq z$), where U' contains at least one element from each set of C.

We transform an instance of HSP $(U = \{u_1, u_2, \ldots, u_n\}, C = \{C_1, C_2, \ldots, C_k\}, z)$ to an instance of DWP $(N = (G = (V, E), w, c), Q, p, B, D)$ as follows (Fig. 1).

For each element u in U, we make an edge u and connect them in a row. That is, we make a linear graph whose vertex set is $\{a_0, a_1, \ldots, a_n\}$ and whose edge set is $u_1 = (a_0, a_1), u_2 = (a_1, a_2), \ldots, u_n = (a_{n-1}, a_n)$. For each set $C_i = \{u_{i_1}, u_{i_2}, \ldots, u_{i_{q(i)}}\}$ $(i = 1, 2, \ldots, k)$, we make two vertices s_i and a vertex t_i, then we add edges $(s_i, a_{i_1-1}), (a_{i_1}, a_{i_2-1}), \ldots, (a_{i_{q(i)}}, t_i)$. Thus, we construct graph $G = (V, E)$. Let $w(s_i, a_{i_1-1})$ be defined as $w(s_i, a_{i_1-1}) = \max_j(2|C_j| + 1) - 2|C_i| + 1$ $(i = 1, 2, \ldots, k)$ and let the weight of the other edges be one. Let $c_{u_1} = c_{u_2} = \ldots = c_{u_n} = 1$ and let the cost of the other edges be $k(n + 2)$. Let $p_i = \{(s_i, a_{i_1-1}), u_{i_1}, (a_{i_1}, a_{i_2-1}), u_{i_2}, \ldots, u_{i_{q(i)}}, (a_{i_{q(i)}}, t_i)\}$ $(i = 1, 2, \ldots, k)$ and $Q = \{(s_i, t_i; p_i) \mid i = 1, 2, \ldots, k\}$. Let $p = z, B = z, D = \max_j(2|C_j| + 1)$. The transformation from an instance of HSP to an instance of DWP can be executed in polynomial time.

The weight of each p_i is $\max_j(2|C_j| + 1) + 1$, which is one more than D. Since the costs of edges except $u_1, u_2, \ldots, u_n$ is large, we cannot reduce the weight of one unit within $B = z(\leq n)$. Therefore, it is necessary to reduce the weights of $u_1, u_2, \ldots, u_n$.

First, we show that, if an instance of HSP, $(U = \{u_1, u_2, \ldots, u_n\}, C = \{C_1, C_2, \ldots, C_k\}, z)$, has a solution, the instance of DWP, $(N = (G = (V, E), w, c), Q, p, B, D)$, has also a solution. Let the weights of the edges in G corresponding to the vertices of the solution of HSP be zero. Since the number of the edges is less than or equal to z, the cost is z or less. The weight of p_i is D or less, because the weight of at least one edge in p_i is zero. Therefore, we have the solution of DWP.

Conversely, we show that, if an instance of DWP, $(N = (G = (V, E), w, c), Q, p, B, D)$, has a solution, the instance of HSP, $(U = \{u_1, u_2, \ldots, u_n\}, C = \{C_1, C_2, \ldots, C_k\}, z)$, has a solution. As mentioned above, we have no choice but to reduce the weights of edges $u_1, u_2, \ldots, u_n$. The weight of at least edge in p_i must be zero, and it must be chosen among $u_1, u_2, \ldots, u_n$. The set of the elements corresponding to the edges whose weight is updated to zero is a solution of HSP.

Thus, HSP can be reduced to DWP in polynomial time. Moreover, DWP clearly belongs to class NP. Consequently, DWP is NP-complete. □

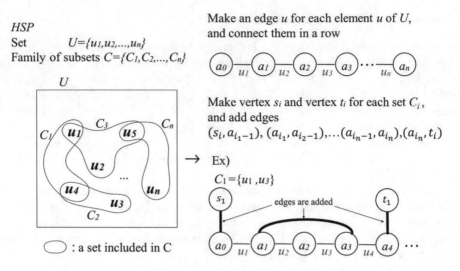

Fig. 1. Transformation from an instance of HSP to an instance of DWP

4   Polynomial Time Algorithm for Restricted DWP

In the rest of this paper, we consider the optimization version of DWP to minimize the sum of the weights of edges. For the sake of simplicity, we assume that the edge cost of all edges is one.

As shown in the previous section, since DWP is generally NP-complete, we cannot expect to solve DWP in polynomial time. However, when $p = 1$, that is, the number of edges whose weights are reduced is restricted to one, DWP can be solved in polynomial time.

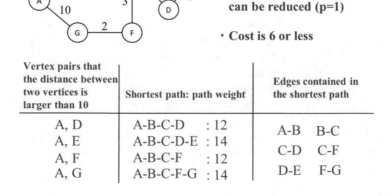

- **Diameter is 10 or less**

- **Weight of only an edge can be reduced (p=1)**

- **Cost is 6 or less**

Vertex pairs that the distance between two vertices is larger than 10	Shortest path: path weight	Edges contained in the shortest path
A, D	A-B-C-D : 12	A-B B-C
A, E	A-B-C-D-E : 14	C-D C-F
A, F	A-B-C-F : 12	D-E F-G
A, G	A-B-C-F-G : 14	

Fig. 2. Example of DWP (1).

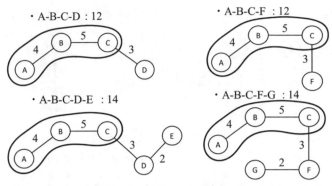

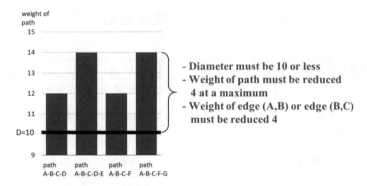

Edge (A,B) or edge (B,C) is included in all paths

Weight of either edge (A,B) or edge (B,C) must be reduced

Fig. 3. Example of DWP (2).

- Diameter must be 10 or less
- Weight of path must be reduced 4 at a maximum
- Weight of edge (A,B) or edge (B,C) must be reduced 4

Fig. 4. Example of DWP (3).

We show a polynomial-time algorithm for DWP where the weight of only a specified edge can be reduced. The algorithm for DWP where $p = 1$ executes this algorithm for every edge and finds edge e that the value of $g(e)$ is the minimum.

The basic idea of the algorithm is described as follows. Let $e \in E$ be the edge whose weight can be reduced. If there is a path in Q that does not include e and the weight of the path is larger than D, it is infeasible, because, even if the weight of e is changed to any value, the weight of the path cannot be reduced to D or less. Therefore, we assume that there are not such an edge. Let $\{p'_1, p'_2, \ldots, p'_z\}$ be the set of the paths including e whose weights are more than D. Let the weight of path p'_i ($i = 1, 2, \ldots, z$) be denoted by $|p'_i|$. When $\max_{i=1,2,\ldots,z}(|p'_i| - D)$ is less than or equal to $w(e)$, the weight of e is updated to $w(e) - \max_{i=1,2,\ldots,z}(|p'_i| - D)$ by reducing $\max_{i=1,2,\ldots,z}(|p'_i| - D)$ from $w(e)$. Thus, the weights of all paths in $\{p'_1, p'_2, \ldots, p'_z\}$ become D or less. The quantity of reduction, $\max_{i=1,2,\ldots,z}(|p'_i| - D)$, is the minimum, because there remains at least one path whose weight is more than D, if the quantity less than $\max_{i=1,2,\ldots,z}(|p'_i| - D)$ is reduced.

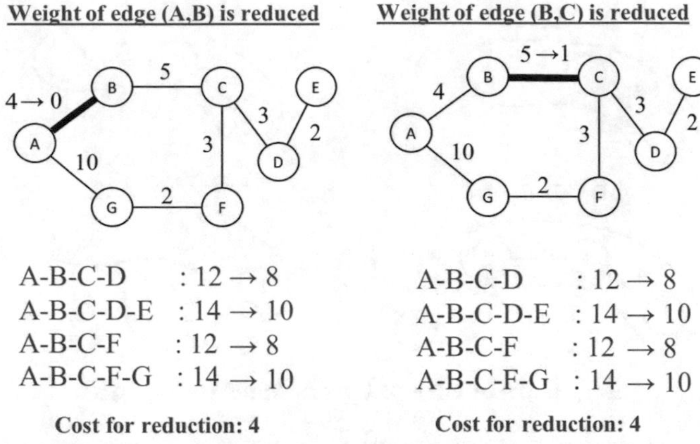

Weight of edge (A,B) is reduced   **Weight of edge (B,C) is reduced**

A-B-C-D   : $12 \rightarrow 8$
A-B-C-D-E : $14 \rightarrow 10$
A-B-C-F   : $12 \rightarrow 8$
A-B-C-F-G : $14 \rightarrow 10$

A-B-C-D   : $12 \rightarrow 8$
A-B-C-D-E : $14 \rightarrow 10$
A-B-C-F   : $12 \rightarrow 8$
A-B-C-F-G : $14 \rightarrow 10$

Cost for reduction: 4   **Cost for reduction: 4**

Fig. 5. Example of DWP (4).

We describe the algorithm in detail in Algorithm 1.

Algorithm 1. DWPp1

Input: Network $N = (G = (V, E), w, c)$, set of paths Q, a positive real number D, $p = 1$.

Output: Edge e^* whose weight is reduced, weight $w(e^*)$ of edge e^*, reduced quantity r^*.

1   $e^* \leftarrow \emptyset$, $r^* \leftarrow \infty$
2   **for** *all* $e \in E$ **do**
3   **if** *there is a path in Q that does not include e and the weight of the path is larger than D* **then**
4   Let e be an infeasible edge
5   **else**
6   Let $\{p'_1, p'_2, \ldots, p'_z\}$ be the set of the paths including e whose weight is more than D
7   **if** $\max_{i=1,2,\ldots,z}(|p'_i| - D) \leq w(e)$ **then**
8   **if** $\max_{i=1,2,\ldots,z}(|p'_i| - D) \leq r^*$ **then**
9               $r^* \leftarrow \max_{i=1,2,\ldots,z}(|p'_i| - D)$
10              $e^* \leftarrow e$
11              $w(e^*) \leftarrow w(e) - \max_{i=1,2,\ldots,z}(|p'_i| - D)$
12   **else**
13   Let e be an infeasible edge
14 **if** *all edges are infeasible edges* **then**
15   **return** This instance is infeasible
16 **else**
17   **return** e^*, $w(e^*)$, r^*

We show an example in Figs. 2, 3, 4 and 5.

From the above examination, we can prove the following theorem.

Theorem 2. *Algorithm 1 outputs the solution of DWP where $p = 1$ in polynomial time.*

5 Conclusions

In this paper, we addressed the network design problem of determining the weight of each edge so that the diameter of a network is less than or equal to a given threshold. We proved that this problem is NP-complete, and we proposed a polynomial-time algorithm to the problem when the number of edges whose weights are reduced is restricted to one.

For the future work, it remains that the problem when the number of edges whose weights are reduced is restricted to a constant can be solved in polynomial time.

Acknowledgements. This work was partially supported by the Japan Society for the Promotion of Science through Grants-in-Aid for Scientific Research (B) (17H01742) and JST CREST JPMJCR1402.

References

1. de Bruijn, N.G.: A combinatorial problem. In: Proceedings of Koninklijke Nederlandse Academie van Wetenschappen, vol. 49, pp. 758–764 (1946)
2. Kautz, W.: Bounds on directed (d,k)-graphs. In: Theory of Cellular Logic Networks and Machines (1968)
3. Du, D.Z., Gao, F., Hsu, D.F.: De Bruiin digraphs, Kautz digraphs, and their generalizations. In: Combinatorial Network Theory, vol. 1, pp. 65–105 (1996)
4. Boucher, C., Bowe, A., Gagie, T., Puglisi, S.J., Sadakane, K.: Variable-order de Bruijn graphs. In: Proceedings of IEEE Data Compression Conference, 7–9 April 2015
5. Bermond, J.-C., Peyrat, C.: De Bruijn and Kautz networks: a competitor for the hypercube? In: Hypercube and Distributed Computers, pp. 279–293 (1989)
6. Imase, M., Itoh, M.: A design for directed graphs with minimum diameter. IEEE Trans. Comput. **C–32**(8), 782–784 (1983)
7. Du, D.Z., Hsu, D.F., Peck, G.W.: Connectivity of consecutive-d digraph. Discret. Appl. Math. **37/38**, 169–177 (1992)
8. Plesnik, J.: The complexity of designing a network with minimum diameter. Networks **11**(1), 77–85 (1981)
9. Miwa, H., Ito, H.: Sparse spanning subgraph preserving connectivity and distance between vertices and vertex subsets. IEICE Trans. Fundam. **E81–A**(5), 832–841 (1998)
10. Garey, M.R., Johnson, D.S.: Computers and Intractability, A Guide to the Theory of NP-Completeness (1978)
11. CAIDA. http://www.caida.org/

Network Design Method Resistant to Cascade Failure Considering Betweenness Centrality

Yuma Morino and Hiroyoshi Miwa[✉]

Graduate School of Science and Technology, Kwansei Gakuin University,
2–1 Gakuen, Sanda-shi, Hyogo 669–1337, Japan
{morino-m,miwa}@kwansei.ac.jp

Abstract. There are many cases in which a single failure in a network, such as congestion, large-scale power failure, and traffic congestion in an information network, causes a chain of failures and load increases by concentrating the load on a part of nodes or links, and the damage expands. In order to avoid cascade failures, it is necessary to prevent the load from concentrating on some nodes and links. Cascade failures are likely to occur from nodes or links with high betweenness centrality. Even if the original network was designed so that there were no nodes or links with high betweenness centrality, the betweenness centrality would change due to a failure. On the other hand, if links and nodes are properly protected, nodes and links with high betweenness centrality can be prevented from appearing even after a failure; consequently, further cascade failures can be prevented. In this paper, we formulate the optimization problem to determine such nodes and links and prove the NP-hardness. In addition, we propose a polynomial-time approximation algorithm and evaluate the performance.

1 Introduction

There are many cases in which a single failure in a network, such as congestion, large-scale power failure, and traffic congestion in an information network, causes a chain of failures and load increases by concentrating the load on a part of nodes or links, and the damage expands. In order to avoid cascade failures, it is necessary to prevent the load from concentrating on some nodes and links.

Cascade failures are likely to occur from nodes or links with high betweenness centrality. Generally, if the betweenness centrality of a node (resp. the link) is high, the number of shortest paths passing through the node (resp. the link) is large. Therefore, a failure of nodes or links with high betweenness centrality influences extensively. Even if an original network is designed so that there are no nodes or links with high betweenness centrality, it is not easy to design considering the influence, because betweenness centrality changes due to a failure.

There are many methods to design a robust network. For example, there is a method based on protection. Even if a failure occurs, it is desirable to

L. Barolli et al. (Eds.): INCoS 2019, AISC 1035, pp. 360–369, 2020.
https://doi.org/10.1007/978-3-030-29035-1_35

protect all links so that the failure probability of a link is sufficiently small by sufficient backup resource and fast rapid recovery system. We can make the failure probability of a link in IP layer sufficiently small by fast switch function and backup resource reserved in advance in a lower layer. If such a recovery system is provided, we can expect that a link between two IP routers does not fail, because a failure of the link is rapidly recovered and the failure cannot be detected in the IP layer. Such a link is called a protected link. Link protection needs much cost, because it needs much network resource. Since it is not practical to protect all links, only some links whose failures significantly degrade the performance of a network must be protected. Consequently, it is necessary to find the smallest number of links to be protected so that a network resulting from failures of any non-protected links provides users connectivity. Node protection is similarly defined.

When links and nodes are properly protected, nodes and links with high betweenness centrality can be prevented from appearing even after a failure; consequently, we can expect that further cascade failures are prevented.

In this paper, we deal with a network design by protection so that cascade failures are avoided. First, we formulate the problem to determine protected nodes and links. Next, we prove the NP-hardness and we propose a polynomial-time approximation algorithm. Furthermore, we evaluate the performance and show the effectiveness of the proposed algorithm.

2   Related Research

The reference [1] proposes a mathematical model for cascade failure and shows that even a single node failure can degrade the performance of the entire network significantly. The reference [2] analyzes the dynamics by cascade failure.

Some methods to avoid cascade failures have been proposed so far. For example, some methods preventing cascade failures by removing nodes and links after the first failure or attack in order to avoid the propagation of the influence are proposed [3,4].

As a result from the viewpoint of the protection, the reference [5] deals with the problem of determining protection links such that the diameter of the resulting network after some links are broken must be a given threshold or less and an approximation algorithm was proposed.

3   Problem of Determining Protected Set Resistant to Cascade Failure Considering Betweenness Centrality

Let $G = (V, E)$ where V is a set of vertices and E is a set of edges be a connected undirected graph representing an information network. Each vertex corresponds to a node such as a router, and each edge corresponds to a link between nodes. We associate the weight and capacity with each edge by weight function $w : E \to \mathbb{R}_+$ and capacity function $c : V \cup E \to \mathbb{R}_+$. Let $\{q_1 = \{s_1, t_1\}, q_2 = \{s_2, t_2\}, \ldots, q_r =$

$\{s_r, t_r\}\}$ be a demand set. For each pair $q_i = \{s_i, t_i\}$ of vertices, let the demand level between s_i and t_i be $d_i \in \mathbb{R}_+$ $(i = 1, 2, \ldots, r)$. The length of a path is defined as the sum of the weight of all the edges contained in the path. The shortest path between two vertices is defined as the path with the smallest path length among all paths between the vertices. When all vertex pairs are connected, we say that the connection constraint is satisfied. Let $m(e)$ be the sum of the demand levels of all the shortest paths passing edge e. Let $m(v)$ be the sum of the demand levels of all the shortest paths passing vertex v. For network $N = (G = (V, E), c, w)$ and demand set $Q = \{(q_1, d_1), (q_2, d_2), \ldots, (q_r, d_r)\}$, when $m(e) \leq c(e)$ for all edges $e \in E$ and $m(v) \leq c(v)$ for all vertices $v \in V$, we say that the capacity constraint is satisfied.

In this paper, when the demand level between two vertices is d and there are some shortest paths with the same path length between them, we assume that the same demand level d is loaded on all the edges contained in these paths.

For the network $N = (G = (V, E), c, w)$ and $F \subseteq V \cup E$, let N_F be the resulting network that all edges in F are removed and that all vertices included in F and all edges incident to the vertices are removed.

Let F_i $(i = 1, 2, \ldots, t)$ be a set of vertices and edges whose size is k or less. Each set of F_i $(i = 1, 2, \ldots, t)$ is called k-removal set. Let $K = \{F_1, F_2, \ldots, F_t\}$ be a set of k-removal sets. We call K a family of k-removal sets. A k-removal set corresponds to a set of nodes and links broken simultaneously in a failure, and a family of k-removal sets corresponds to a set of possible simultaneous failures.

For, K, a family of k-removal sets, we call $P(\subseteq V \cup E)$ a protection set against $K = \{F_1, F_2, \ldots, F_t\}$, if there is set P that all networks $N_{F_i \setminus P}$ $(i = 1, 2, \ldots, t)$ satisfy the capacity and connection constraints. For any i $(i = 1, 2, \ldots, t)$, even if the elements of F_i not included in the protection set P are removed, the connection constraint and the capacity constraint are always satisfied. This means that, even if any simultaneous failure in K occurs, all vertex pairs in Q are always connected and the capacity constraint is always satisfied. Thus, when the nodes and the links corresponding to vertices and edges in P are protected, even if any simultaneous failure occurs, the concentration of load on some nodes and links are always avoided; consequently, cascade failures are avoided.

Problem of Determining Protected Set Resistant to Cascade Failure Considering Betweenness Centrality (BCPP)

INSTANCE: *Network* $N = (G = (V, E), c, w)$, *demand set* $Q = \{\{q_1, d_1\}, \{q_2, d_2\}, \ldots, \{q_r, d_r\}\}$, *a family of k-removal sets,* $K = \{F_1, F_2, \ldots, F_t\}$, *a positive integer p.*
QUESTION: *Is there a protection set against* $K = \{F_1, F_2, \ldots, F_t\}$ *whose size is less than or equal to p?*   □

An example of this problem is shown in Figs. 1, 2 and 3. In Fig. 1, the capacity constraint cannot be satisfied by removing the edge between (v_3, v_6). However, by protecting this edge, the capacity constraint is always satisfied even if any other edge is removed as shown in Figs. 2 and 3.

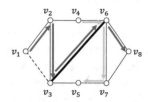
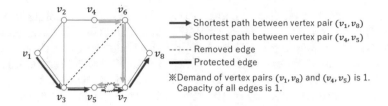

Fig. 1. Example of concentration of demand levels.

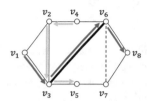

Fig. 2. Edge (v_3, v_6) is protected and (v_1, v_3) is removed.

Fig. 3. Edge (v_3, v_6) is protected and (v_6, v_7) is removed.

Theorem 1. *BCPP is NP-hard, even if $k = 2$ and the weight of all edges is restricted to one.*

(Proof of Theorem 1). We show that the vertex cover problem known as NP-hard, is reduced to BCPP in polynomial time.

The vertex cover problem (VC) asks whether there is a vertex subset $V' \subseteq V$ ($|V'| \leq K$) such that each edge of graph $G = (V, E)$ is incident to at least one vertex of V'.

We show the transformation from an instance (G, K) of VC to an instance $(N = (G_P, c, w), Q, p, k)$ of BCPP as follows. First, we make graph H_{uv} for all edges (u, v) in G (Fig. 4). Vertices u and v in G correspond to edges u and v in H_{uv}, respectively. When edges (a, b) and (b, c) have the common vertex b in G, H_{ab} and H_{bc} have the common edge b (Fig. 5). Let the capacities of edges u and v in H_{uv} be the degrees of vertices u and v in G, and let the capacities of the other edges in H_{uv} be 1. We make the path $\{e_1, e_2, e_3, e_4\}$ with length 4 between vertices S and T. Let the capacity of edges e_1, e_2, e_3, e_4 be 0. We add the edges between S and s_{uv} for all edges (u, v) in G and the edges between T and t_{uv} for all edges (u, v) in G. Let the capacities of these edges be one. Let the weights of all edges be one. Let the demand set Q be $Q = \{(\{s_{uv}, t_{uv}\}, 1) | (u, v) \in E\}$, and let $p = K$ and $k = 2$. Thus, an instance (G, K) of VC is transformed to an instance $(N = (G_P, c, w), Q, p, k)$ of BCPP. This transformation is executed in polynomial time. We show an example in Fig. 6.

We prove that, if and only if an instance (G, K) of VC has a solution, the instance $(N = (G_P, c, w), Q, K, 2)$ of BCPP transformed as above mentioned has a solution.

First, we show that, if an instance (G, K) of VC has a solution V', the instance $(N = (G_P, c, w), Q, K, 2)$ of BCPP has a solution. Let the set of the edges of G_P corresponding to V' be the protected edge set. For all edges (u, v) of G, as at least one of vertices u or v is included in V', the corresponding edges u or v of H_{uv} in G_P is protected. Even if any two edges except the protected edge from H_{uv} are removed, since the resulting graph is connected, s_{uv} and t_{uv} is connected. Therefore, the graph resulting from the removal of any two edges from G_P except the protected edges satisfies the connection constraint. For all pairs of s_{uv} and t_{uv} in G_P, even if any two edges except the protected edge are removed from F_{uv}, the shortest path between s_{uv} and t_{uv} traverses edge u or v in H_{uv}. The number of shortest paths through edge u or v does not exceed the capacity of the edge, because the maximum number of the paths is equal to the number of the edges incident to vertex u (resp. v) in G and the capacity of the edge is equal to the degree of vertex u (resp. v) in G. Therefore, the capacity constraint is also satisfied. Thus, the protected edge set is the solution of the instance $(N = (G_P, c, w), Q, K, 2)$ of BCPP.

Conversely, we show that, if the instance $(N = (G_P, c, w), Q, K, 2)$ of BCPP has a solution, the instance (G, K) of VC has a solution. If neither edges u nor v of H_{uv} are the protected edges, when these edges are removed, the shortest path between s_{uv} and t_{uv} is changed to $s_{uv} \rightarrow S \rightarrow e_1 \rightarrow e_2 \rightarrow e_3 \rightarrow e_4 \rightarrow T \rightarrow t_{uv}$. Since the capacities of the edges e_1, e_2, e_3, e_4 is 0, the capacity constraint is not satisfied. Therefore, if the instance $(N = (G_P, c, w), Q, p, k)$ of BCPP has a solution, at least one of the edges u or v for all H_{uv} must be protected. Let V' be the set of the corresponding vertices of G to the protected edges of G_P. $|V'| \leq K$, since the number of the protected edges is K or less. In addition, for all edges (u, v) in G, at least one of vertices u or v is included in V', because at least one of edges u or v of H_{uv} in G_P must be protected. It follows that V' is the solution of the instance (G, K) of VC.

Consequently, VC is reduced to BCPP in polynomial time; therefore, BCPP is NP-hard.   □

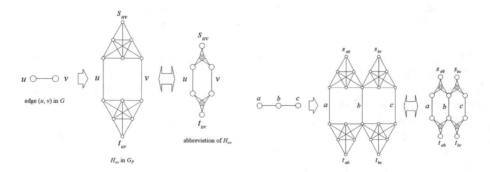

Fig. 4. Structure of H_{uv}.   **Fig. 5.** Structure of G_P.

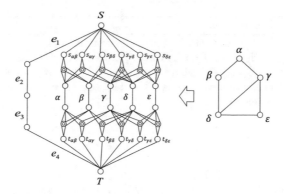

Fig. 6. Example of G_P and G.

4 Polynomial Time Approximation Algorithm

In this section, we deal with the optimization version of BCPP whose objective is to minimize the number of protected edges. Since BCPP is NP-hard, we cannot expect a polynomial-time algorithm for BCPP.

For the sake of simplicity, we assume that the elements of a removal set and a protection set are restricted to edges.

We propose a polynomial-time approximation algorithm for BCPP in this section.

First, we describe the algorithm when $k = 1$. When $N' = (G' = (V, E - \{e\}), c, w)$ and demand set Q do not satisfy the connection constraint or the capacity constraint, edge e must be protected. We can check in polynomial time whether these constraints are satisfied. We execute this check for all edges $e \in E$ and add the edge e to the protected edge set P if at least one of these constraints are not satisfied. Even if an edge not included in P is removed, since the resulting network satisfies both of these constraints, the edge is not necessary to be protected. Consequently, P is the protected edge set with the smallest size.

We describe the basic idea of the algorithm when $k(\geq 2)$ is a constant integer. The basic idea is the polynomial-time reduction from BCPP to the vertex cover problem (VC) of a hyper graph. First, the algorithm **Protect-k-apx** converts an instance of BCPP to an instance of VC. Then, the algorithm **Approx-VC(H)** outputs an approximation solution of the instance of VC. The algorithm **Approx-VC(H)** is an approximation algorithm of VC based on a maximal matching algorithm for a hyper graph. The maximal matching algorithm guarantees the approximation ratio of k, the size of a hyper edge. Finally, the algorithm **Protect-k-apx** converts the solution to the solution of the instance of BCPP. Since the approximation ratio does not change by this conversion, we get the approximation solution of BCPP with the approximation ratio of k.

When $k = 1$, the algorithm **Protect-k-apx** output the optimum solution, because all vertices are isolated in the hyper graph and the algorithm **Approx-VC(H)** outputs the optimum solution of VC.

Algorithm 1. Protect-k-apx

Input: Network $N = (G = (V, E), c, w)$, the demand set
  $Q = \{\{q_1, d_1\}, \{q_2, d_2\}, \ldots, \{q_r, d_r\}\}$, a family of k-removal sets,
  $K = \{F_1, F_2, \ldots, F_t\}$ where $F_i \subseteq E$ $(i = 1, 2, \ldots, t)$, a positive integer p.
Output: P, a protection set against $K = \{F_1, F_2, \ldots, F_t\}$ whose size is less
  than or equal to p.

1 $P \leftarrow \emptyset$
2 $E_H \leftarrow \emptyset$
3 $i \leftarrow 1$
4 **while** $i \leq t$ **do**
5   **for** *All subsets $E'(\in F_i)$* **do**
6   **if** *Network $N' = (G' = (V, E \setminus E'), c, w)$ and Q do not satisfy the connection constraint or the capacity constraint* **then**
7             $E_H \leftarrow E_H \cup E'$

8     $i \leftarrow i + 1$
9 $P \leftarrow Approx - VC(H_i = (E, E_H))$
10 **return** P

Theorem 2. *Algorithm* **Protect-k-apx** *outputs an approximation solution with the approximation ratio of k in polynomial time, when k is a positive constant integer.*

5 Performance Evaluation

We evaluate the performance of the algorithm proposed in Sect. 4 using the actual network topology which is presented by CAIDA(Cooperative Association for Internet Data Analysis) [6].

We assume that the weight of all edges is one. In addition, we assume that there is the same demand level of one unit between all vertex pairs and that the capacity of an edge is g times of the sum of the demand levels passing the edge.

We define a family of removal sets, $K = \{F_1, F_2, \ldots, F_t\}$ as follows. For vertex $v_i(\in V)$, let F_i be the union of the set of the edges incident to v_i and the set of the edges between vertices adjacent to v_i. We show an example of a removal set in Figs. 7 and 8. A family of removal sets K is determined by a set of removal sets made from randomly chosen vertices.

We evaluated the average of the number of the protected edges and the ratio of the number of the protected edges to the number of edges in a network. When the number of the removal sets in a family of removal sets is one (resp. two, three, four, five, six), we show the average of five (resp. eleven, fourteen, thirteen, nine, five) families of k-removal sets.

We show the result of Network 1 in Tables 1, 2, 3 and 4. In each Table, "number" is the number of removal sets. The values of $p1.5$, $p2$, $p3$, and $p4$ are the average number of the protected edges when $g = 1.5, 2, 3, 4$. The values of $p1.5opt$, $p2opt$, $p3opt$, and $p4opt$ are the average of the optimum solutions

---

Algorithm 2. Approx-VC(H)

Input: Hypergraph $H = (V, E)$.
Output: Vertex subset V'.
1 $V' \leftarrow \emptyset$
2 Label *"unscanned"* for all vertices $v \in V$
3 Let $\{e_1, e_2, \ldots, e_h\}$ be the set of the hyper edges e_i $(i = 1, 2, \ldots, h)$ sorted in descending order of the number of other hyper edges sharing at least one vertex with the hyper edge e_i.
4 **for** $i = 1$ **to** h **do**
5 | **if** *All vertices in hyper edge e_i are "unscanned"* **then**
6 | | Label all vertices in the edge e_i as *"scanned"*
7 | | $V' \leftarrow V' \cup \{\text{All vertices in } e_i\}$

8 **return** V'

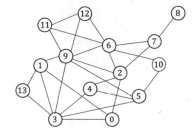

Fig. 7. Network1.

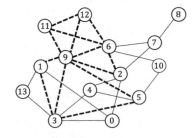

Fig. 8. Removal set F_9.

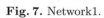

Table 1. Edge capacity:$g = 1.5$.

Number	p1.5	p1.5opt	Average of approximation ratio	p1.5ratio(%)	p1.5ratio_opt(%)
1	3.4	3.4	1	13.08	13.08
2	7.27	7.27	1	27.97	27.97
3	10.57	10.57	1	40.66	40.66
4	14.46	14.46	1	55.62	55.62
5	16.22	16.22	1	62.39	62.39
6	19.4	19.4	1	74.62	74.62

when $g = 1.5, 2, 3, 4$. In addition, *p1.5ratio, p2ratio, p3ratio,* and *p4ratio* are the average ratio of the number of the protected edges to the number of edges in Network 1. *p1.5ratio_opt, p2ratio_opt, p3ratio_opt* and *p4ratio_opt* are the average of the minimum values of *p1.5ratio, p2ratio, p3ratio,* and *p4ratio*.

In Table 1, the average of the approximation ratio is one for any number of removal sets when $g = 1.5$. In this case, since the capacity is too small in comparison to the demand levels when some edges are removed, the capacity constraint cannot be satisfied; therefore, almost all edges must be protected.

Table 2. Edge capacity:$g = 2$.

Number	p2	p2opt	Average of approximation ratio	p2ratio(%)	p2ratio_opt(%)
1	2.4	2.4	1	9.23	9.23
2	5.55	5.55	1	21.33	21.33
3	8.5	8.36	1.01	32.69	32.14
4	11.62	11.08	1.04	44.67	42.6
5	13.11	12.56	1.04	50.43	48.29
6	15.8	15.4	1.03	60.77	59.23

Table 3. Edge capacity:$g = 3$

Number	p3	p3opt	Average of approximation ratio	p3ratio(%)	p3ratio_opt(%)
1	2	2	1	7.69	7.69
2	4.36	4.27	1.02	16.78	16.43
3	7.07	6.57	1.07	27.2	25.27
4	10.38	8.85	1.17	39.94	34.02
5	11.44	9.89	1.14	44.02	38.03
6	14	11.8	1.19	53.85	45.38

Consequently, the difference between the approximation solution and the optimal solution is sufficiently small. The average of the approximation ratio is 1, 1, 1.01, 1.04, 1.04, 1.03 when $g = 2$; 1, 1.02, 1.07, 1.17, 1.14, 1.19 when $g = 3$; 1, 1.02, 1.08, 1.24, 1.25, 1.23 when $g = 4$. In these cases, the approximation solution is close to the optimum solution. The number of the protected edges tends to decrease according to the increase of the value of g, because, since the capacity is larger in comparison to the demand levels when some edges are removed, the capacity constraint tends to be satisfied; therefore, smaller number of edges are protected.

Table 4. Edge capacity:$g = 4$.

Number	p4	p4opt	Average of approximation ratio	p4ratio(%)	p4ratio_opt(%)
1	1.4	1.4	1	5.38	5.38
2	3.09	3	1.02	11.89	11.54
3	5.21	4.79	1.08	20.05	18.41
4	8.46	6.69	1.24	32.54	25.74
5	9.22	7.22	1.25	35.47	27.78
6	11	9	1.23	42.31	34.62

6 Conclusion

In this paper, we addressed a network design problem determining protection links that is resistant to cascade failures by preventing the occurrence of nodes and links with high betweenness centrality even after a failure.

There is a method based on protection to design a robust network. When links and nodes are properly protected, nodes and links with high betweenness centrality can be prevented from appearing even after a failure; consequently, we can expect that further cascade failures are prevented.

First, we formulated the problem to determine protected nodes and links. Next, we proved the NP-hardness by the polynomial-time reduction from the Vertex Cover problem to the problem. Furthermore, we designed a polynomial-time approximation algorithm when the number of the simultaneous broken elements are restricted to a constant positive integer. We evaluated the performance of the proposed algorithm and the results indicated that the approximation ratio of the proposed algorithm is sufficiently small in the topology of an actual network.

Acknowledgements. This work was partially supported by the Japan Society for the Promotion of Science through Grants-in-Aid for Scientific Research (B) (17H01742) and JST CREST JPMJCR1402.

References

1. Crucitti, P., Latora, V., Marchiori, M.: Model for cascading failures in complex networks. Phys. Rev. E **69**–4, 045104 (2004)
2. Dobson, I., Carreras, B.A., Lynch, V.E.: Complex systems analysis of series of blackouts. Chaos **17**, 026103 (2007)
3. Motter, A.E., Lai, Y.C.: Cascade-based attacks on complex networks. Phys. Rev. E **66**–6, 065102 (2002)
4. Motter, A.E.: Cascade control and defense in complex networks. Phys. Rev. Lett. **93**–9, 098701 (2004)
5. Imagawa, K., Fujimura, T., Miwa, H.: Approximation algorithms for finding protected links to keep small diameter against link failures. In: Proceedings of International Conference on Intelligent Networking and Collaborative System (INCoS2011), Fukuoka, Japan, 30 Nov–1–2 Dec 2011 (2011)
6. CAIDA. http://www.caida.org/

Real-Time Multi-resource Allocation via a Structured Policy Table

Arslan Qadeer[1], Myung J. Lee[1(✉)], and Kazuya Tsukamoto[2]

[1] Department of EE, City College of CUNY, New York, USA
aqadeer000@citymail.cuny.edu, mlee@ccny.cuny.edu
[2] Department of ECE, Kyutech, Japan
tsukamoto@cse.kyutech.ac.jp

Abstract. Mobile Edge Cloud endures limited computational resources as compared to back-end cloud. Semi-Markov Decision Process (SMDP) based Multi-Resource Allocation (MRA) work [6] introduces optimal resource allocation for mobile requests in the resource constrained edge cloud environments. In this study, we scale existing SMDP MRA work for real-world scenarios. First, we structure the policy tables in a two dimensional matrix such that columns represent states of the system and rows for the actions. Second, we propose an index based search technique over structured policy tables. Simulation results demonstrate that our approach outperforms the legacy method and retrieves an optimal action from the policy tables in the order of microseconds, which meets the delay criteria of real-time applications in edge cloud based systems.

1 Introduction

With the ever-increasing need of IT infrastructure for computing and storage in tremendously growing mobile communications technologies, the emerging Internet of Things (IoT) and Machine Type Communication (MTC) are expected to instigate a huge number of device connections, which undergo low storage capacity, high energy consumption, low bandwidth and high latency [1]. Mobile Edge Cloud (MEC), leveraging the traits of both cloud computing and mobile computing, has provided considerable capabilities to mobile devices to alleviate the above inherent network limitations [2,3].

As compared to back-end clouds, where unlimited pools of computational and storage resources are provided, edge clouds are supposed to posses limited amount of resources [4]. Therefore, efficient resource allocation techniques are required to improve the overall system capacity. Several solutions including the partitioning approach [5–7] are proposed to determine which module of a mobile application should be offloaded and how efficiently allocated resources in the edge cloud. A Semi-Markov Decision Process (SMDP) based Multi-Resource Allocation scheme is devised [6], which works on average reward criterion and proposes a strategy to determine whether to offload a service request on edge cloud or back-end cloud. However, aforementioned works are purely theoretical

© Springer Nature Switzerland AG 2020
L. Barolli et al. (Eds.): INCoS 2019, AISC 1035, pp. 370–379, 2020.
https://doi.org/10.1007/978-3-030-29035-1_36

and lack any practical implementations in the real-world scenarios. We scale MRA (Multi-Resource Allocation) [6] work in order to implement it in practical use cases.

A Resource Manager (RM) is introduced at the Edge Cloud (EC) to help manage all computing and bandwidth resources, which will be responsible for meeting the QoS requirements inherent to IoT applications and services, and acts as a coordinator for the end-devices, EC, and back-end Cloud (BC). This work seeks to aid the RM to find an optimal action for resource allocation either in the EC or BC in real-time to minimize the latency for delay sensitive IoT and real-time mobile applications. The policy tables calculated and stored at EC [5,6] grow exponentially which contain optimal resource allocation actions responding to a range of service requests. As such, it is impractical to search large size policy tables in linear fashion due to the time limitation of delay sensitive applications. Therefore, how to search the policy tables while complying the end-to-end delay constraint should be carefully considered by the RM.

The rest of the paper is organized as follows: Sect. 2 describes our system model for policy tables, while index based policy table search techniques are described in Sect. 3. The performance evaluation is discussed in Sect. 4, and conclusions are presented in Sect. 5.

2   System Model

For our study, we consider the COSMOS test-bed environment which consists of edge and core cloud computing infrastructure in which multiple mobile or IoT devices can connect to the EC through wireless access point [8,9]. These devices can run applications locally, or offload some modules of the application to EC or to the BC cloud for faster execution and better energy conservation.

Upon arrival of a new service request, it can be decided whether to run it on the native device or to be offloaded to the edge cloud based on the network performance and application characteristics [7]. For the offloaded module, the policy table is searched for optimal action, which has already been calculated by MRA algorithm [6]. MRA strategy not only adaptively determines the location (EC or BC) for the execution of service request but also determines the optimal amount of wireless bandwidth i and computing resource j to allocate to the accepted service request. This MRA strategy achieves the maximum system benefits (Eq. (11) in [6]) in terms of throughput and blocking probability while maintaining required latency requirement. This MRA problem is formulated as SMDP and solved as a linear programming problem to calculate an optimal policy which is composed of all the probabilities of randomly selecting the actions (Eq. (13) in [6]). This approach has a predictive ability of future state that lies in the transition probabilities (p_i^j) from the current state x_i^j to all potential next states if action a_i^j is chosen upon receiving a new service request. These policy tables are stored at the EC to be consumed by the Resource Manager (RM). The provisioning of the resources can be sped-up if tailored approaches are applied while arranging and searching the policy tables.

In case of real-world 5G environments, the linear search for large size policy tables is an inappropriate approach as it violates the latency requirement. Therefore, fast and efficient mechanisms have to be devised to meet the latency requirement for real-time applications. Another aspect could be to support massive machine type communication (mMTC), where hundreds of thousands of devices are expected to be connected with the EC [10]. Resource provisioning for such large number of devices should be fast enough to minimize the blocking probability of service requests.

The formulation of policy tables is classified into two different ways. The first one is comparatively large table and the second one is significantly smaller and contains adequate information to satisfy an incoming service request. The size of the policy tables is proportional to the available resources in terms of the total number of VM units (M) at edge cloud and total number of bandwidth units (B) on the wireless channel[1]. This also corresponds to the number of total possible states (S_T) of the system. The notations used in this paper are given in Table 1.

The flow of policy table search during resource provisioning is illustrated in Fig. 1. Upon arrival of a new service request the RM searches the policy table and finds an optimal action. Here, 0 means the chosen action was a reject. On the other hand, if chosen actions were $\{a_i^j, a_i\}$, then the RM allocates resources on the EC or BC accordingly. The objective is to provide a mechanism to construct and search huge size policy table for the RM such that, an optimal action is retrieved in minimal possible time.

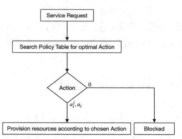

Fig. 1. Policy Table Search-Flow for resource provisioning by Resource Manager in Edge based Cloud Computing System

3   Index Based Policy Table Search

Aforementioned Policy Tables are stored as a database which contains, to name a few, all the possible states of a system, occurrences (values) of those states and their corresponding actions. The searching time of a value is proportional to the growth of such databases. Efficient and effective ways of amalgamating sporadic

---

[1] The wireless bandwidth and VM units are defined in [6]. Where *"bandwidth refers to the wireless connections between the end devices and the EC, and one wireless bandwidth unit refers to the minimum bandwidth required to support mobile computing offloading, for example, 50, 100 Kbps, etc. Similarly, VM refers to the minimum computational resource required to execute a service request, e.g. 1 core of CPU with 1 GB memory. Then, the total bandwidth/VM available can be expressed as the integer multiple of the bandwidth and VM unit. For the simplicity of computation, we assume a single service request requires at least one basic unit of wireless bandwidth and VM units, and only the integral numbers of basic bandwidth units and VM units are allocated"*.

Table 1. Notations

Name	Description
B	The total number of wireless bandwidth units available
M	The total number of VM units available on the edge cloud
W	The maximal number of wireless bandwidth units that the system provides to one service request
T	The maximal number of VM units that the system provides to one service request
α_W	Limiting factor for W
α_T	Limiting factor for T
x_i^j	The number of ongoing services that are allocated i units of wireless bandwidth and j units of VM on edge cloud
y_i	The number of ongoing services that are allocated i units of wireless bandwidth and T units of VM on back-end cloud
a_i^j	The action to accept the request by allocating i units of wireless bandwidth and j units of VM on edge cloud
a_i	The action to accept the request by allocating i units of wireless bandwidth and T units of VM on back-end cloud
S_T	Total number of states
S_{Ec}	Edge Cloud based states
S_{Bc}	Back-end Cloud based states
A_s	Set of allowable actions at state s
M_x	Maximum occurrences (value) of state x_i^j
M_y	Maximum occurrences (value) of state y_i
X_i^j	Size of Policy Table of state x_i^j
Y_i	Size of Policy Table of state y_i

data and retrieving useful information are indispensable in any databases. This assertion is even more critical for real-time applications. The standard linear search is not adequate as the time complexity of such search is $O(m \times n)$, where m and n are the number of rows and the number of columns respectively of the policy table matrix [11]. In our case, we see that the total number of elements approaches to 70 millions which takes search time on the order of milliseconds (Sect. 4). Therefore, index based search like [12,13] is devised which drastically diminishes the search time.

3.1 Apply Limit to W and T

A limit can be set with an integer variable (α) to define the maximum number of bandwidth units W and maximum number of VM units T which can be assigned to one service request. In order to accommodate multiple instances of a state which require maximum allowed bandwidth and VM units W and T respectively, we limit W and T by an integer variable α such that:

$$W \leq \frac{B}{\alpha_W} \tag{1}$$

$$T \leq \frac{M}{\alpha_T} \tag{2}$$

where B is total available wireless bandwidth units and M is total available VM units. This also allows us to generate dynamic state models according to a desired environment instead of having a fixed upper limit for both bandwidth and VM units as suggested in previous study [6].

3.2   Total Possible Number of States (S_T)

The states for an EC is given by

$$\{x_1^1, x_1^2, x_1^3, ..., x_2^1, x_2^2, x_2^3, ..., x_W^T\}$$

and BC is given by

$$\{y_1, y_2, y_3, ..., y_W\}.$$

The sizes of both EC and BC states are different and calculated separately:

$$S_{Ec} = W \times T \tag{3}$$

$$S_{Bc} = W \tag{4}$$

where W and T represent the maximum bandwidth units and VM units allowed for a service request, respectively. Thereafter, the total number of states are:

$$S_T = S_{Ec} + S_{Bc} \tag{5}$$

Thus, the set of allowable actions A_s at EC for any given state x_i^j becomes

$$\{a_1^1, a_1^2, a_1^3, ..., a_2^1, a_2^2, a_2^3, ..., a_W^T\}$$

and at BC for any given state y_i becomes

$$\{a_1, a_2, a_3, ..., a_W\}.$$

Note that, for certain occurrences of a given state there may be some actions which are prohibited. For example, if $M = 10, B = 10$, the current state of the system is x_2^2, and occurrence of such state is 5 then further resource allocations cannot be done. Hence, such actions are replaced with 0, meaning the action of reject. Accordingly, it makes the total number of possible actions for any given state as:

$$A = S_T + 1 \tag{6}$$

where one extra action represents the reject action i.e. 0 in Fig. 1.

3.3 Policy Tables

Policy tables (PT) are composed in the form of two dimensional matrix of variable size, which contains the states arranged by occurrences as columns and corresponding actions as rows. Upon arrival of a new service request, RM consults the policy table for an optimal action and provisions resources at EC or BC. The size of PT depends upon Eqs. (5-6) and the maximum values (occurrences) of all the states which are calculated later. We propose two kinds of policy tables which are concatenated differently from small policy tables calculated for all the individual states at each possible occurrence [6].

3.3.1 Policy Table-I

This is a two dimensional matrix where rows demonstrate the actions A_s, and columns show the occurrences of a state x_i^j and their corresponding probabilities of actions. For any given state x_i^j the occurrences range from 0 to M_x which represent the maximum possible occurrence of the state. Here, it is important to calculate M_x, which is an integer value, to make sure that the resource demand by a state x_i^j does not exceed the available edge cloud VM units M and bandwidth units B. Therefore, M_x is given by:

$$M_x \leq \begin{cases} \frac{M}{j}, & \forall M < B \\ \frac{B}{i}, & \forall B < M \\ \frac{M+B}{i+j}, & \forall M = B \end{cases} \tag{7}$$

where $i \leq W$ and $j \leq T$. Similarly, for any given state y_i the values range from 0 to M_y. Here, we assume that the available number of VM units on back-end cloud are unlimited and occurrence of state y_i is only constrained by available bandwidth units B. Therefore, M_y is given by:

$$M_y \leq \frac{B}{i} \tag{8}$$

where $i \leq W$. Therefore, the size (X_i^j) of a matrix for any given state x_i^j becomes $R \times M_x$ and size (Y_i) of a matrix for any given state y_i becomes $R \times M_y$, where R represents the number of rows of matrix of policy table and corresponds to the actions A Eq. (6). We construct Policy Table-I by concatenating these tiny matrices of all states. Thus, the size of PT-I becomes $(R \times N)$, where $R = A$ and N is given as:

$$N = \sum_{j=1}^{T} \sum_{i=1}^{W} X_i^j + \sum_{i=1}^{W} Y_i \tag{9}$$

Note that rows R of PT-I remain the same, which makes it a perfect rectangular matrix. From Fig. 2 we can see that if the current state and its value in the system are x_1^2 and 2 respectively, the RM retrieves the whole column (circled in red) and decides an optimal action among all the probabilistic actions.

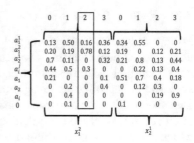

Fig. 2. Policy Table I for service requests. Elements depict the probability of actions with respect to current state and value of the state.

3.3.2 Policy Table-II

Just like PT-I, PT-II is also a two dimensional matrix. This time states are placed in columns and rows represent the occurrence (value) of each state. As compared to PT-I in which a range of possible actions are extracted and then RM decides which action to choose, PT-II is given only with single action. We can say that PT-II is structured in more of a normalized [14] form and debars the redundant data of PT-I. The size of PT-II is $R \times N$ where R ranges from 0 to $Max(M_x, M_y)$. The reason to choose the maximum among the maximum occurrences of all the states is to create a matrix of a regular shape. In this case, the maximum occurrence constraints (M_x, M_y) are not applied. Therefore, actions for such prohibited occurrences are replaced by 0 as explained in Sect. 3.2. Here, $N = S_T$, this makes the size of PT-II remarkably small as compared to PT-I.

Figure 3 illustrates an example of PT-II in which states are arranged column wise and rows represent occurrences of states. For example, If the current state is x_2^2 and its occurrence in the system is 2, then a_1^1 (circled in red) is taken with corresponding probability, and other actions are also possible with their corresponding probabilities.

$$
\begin{array}{c}
\scriptstyle Max(M_x, M_y) \\ \scriptstyle : \\ \scriptstyle 95 \\ \scriptstyle 94 \\ \scriptstyle : \\ \scriptstyle 3 \\ \scriptstyle 2 \\ \scriptstyle 1 \\ \scriptstyle 0
\end{array}
\begin{bmatrix}
0 & a_2 & a_i & 0 & 0 & a_1^1 & 0 & 0 \\
a_2^1 & 0 & a_2 & a_2 & a_1 & 0 & a_1^1 & a_i \\
a_1^1 & a_1^1 & a_1 & a_1^1 & 0 & a_1 & a_1^2 & a_1 \\
a_1^1 & a_1^2 & a_2^1 & a_1^1 & a_2^1 & a_2 & a_2^1 & a_1^2 \\
a_1^1 & a_2^1 & a_1^2 & a_1^1 & a_1^2 & a_2^1 & a_2^2 & a_2^1 \\
a_2^1 & a_2^1 & a_2^1 & a_2^1 & a_1^1 & a_1^1 & a_1^1 & a_2^1 \\
a_1^2 & a_2^1 & \boxed{a_1^1} & a_2^1 & a_2^1 & a_2^1 & a_1^1 & a_2 \\
a_i^j & a_i^j & a_2^2 & a_i^j & a_2^1 & a_1^2 & a_1^2 & a_i^j
\end{bmatrix}
$$

$$x_1^1 \quad x_1^2 \quad x_2^2 \ldots x_i^j \quad y_1 \quad y_2 \quad \ldots \quad y_i$$

Fig. 3. Policy Table II for service requests. Elements depict optimal actions with respect to current state and value of the state.

3.4 Index Based Search on Policy Tables

In order to search the policy table and find an optimal action in real-time, we propose an index based search on the policy tables. We assume that system have information about the current state and its occurrence. For index based search we store the indices of states, actions and max occurrences of states as $1 \times N$ vectors. The number of operations that are executed to find an optimal action differ for both policy tables. We also apply linear search for both policy tables and compare the results in Sect. 4.

3.4.1 Searching Through PT-I

The following operations involved while searching PT-I:

- Get the index of currently known state: To find that we search through the vector of states.
- Get the maximum possible occurrence of that state: use above index to find the maximum possible occurrence of given state from the vector of the maximum values.
- Consume the currently known occurrence of the state to calculate index of exact column in PT-I: add all occurrences of previous states and the current occurrence of the state.
- Retrieve the column using column index found in above step.
- RM decides actions depending on the probabilities of such actions of service requests.
- Find the index of chosen probability.
- Retrieve action using above index: search through vector of actions.

For a particular state, only relevant column is retrieved and rest of the columns are eliminated during search.

3.4.2 Searching Through PT-II

As mentioned before that PT-II includes only normalized data and discards the redundant information. Thus, the reduced number of steps required to fetch the optimal action are listed below:

- Get the index of currently known state: To find that we search through the vector of states.
- Consume the currently known occurrence of the state and above found index to find action (e.g. *Action* = PT-II$(2,1)$)

PT-I is large, which provides flexibility over choosing an action but takes more search time. Conversely, PT-II takes less search time with no flexibility.

4 Evaluation

In this section, we evaluate the performance of proposed techniques. The performance is calculated under various values of available wireless bandwidth units (B) and VM units (M). For the manageability of the model computation, we use $B = M$. However, the model works for other cases $(B < M, M < B)$ as well (Eq. (7)). The advantages of proposed techniques are clearly revealed by comparing with legacy linear search method.

The simulations are written in MATLAB and ran on Windows PC (Dell, Intel Core i7 (7th Gen) 7700T/2.9 GHz Quad-Core, DDR3 16 GB SDRAM). The system values (B, M) are scaled up with the purpose of replicating real-world scenarios depending on the size of edge cloud [2]. Other system parameters (α_W, α_T) are set to default values (i.e. 4). The limiting factors α_W and α_T can be modified to replicate different traffic models according to the need. For example, α_W and α_T can be set closer to W and T to support massive number of users requiring less resources (a few Kbps and few MB memory), while they can be set closer to 1 for broadband users [10].

No. of States: As we can see in Fig. 4, number of states grow linearly upon increasing the edge cloud resources. If we double the number of bandwidth and VM units from 200 to 400, the number of states increase approximately four times, which causes growth of policy tables likewise.

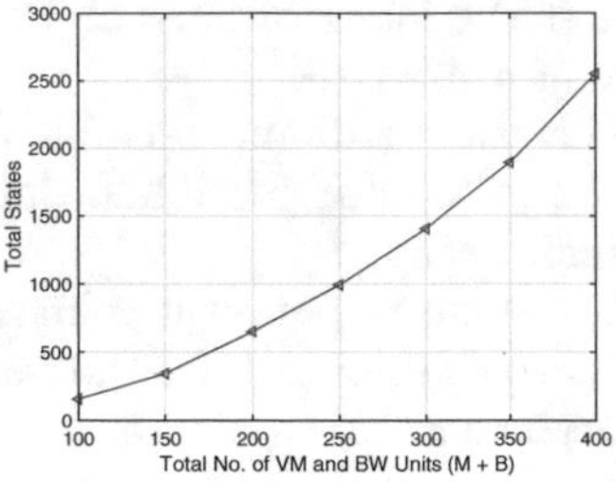

Policy Tables: Figure 5 shows that PT-I grows exponentially and reaches up to 72 million (2551×28318) elements in the matrix. Whereas PT-II hardly crosses half million (201×2550) elements. Both have their own

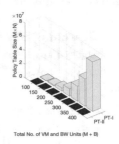

Fig. 4. No. of States for $\alpha_W = 4$, $\alpha_T = 4$ and $M = B$

advantages over the other which are discussed at the end of Sect. 3.4.2.

Search Time: The simulations are run 1000 times and average time for both linear and indexed search of PT-I and PT-II is shown in Fig. 6. We observe that linear search time for PT-I approaches to $90 \, ms(milliseconds)$ which is beyond the end-to-end delay criteria (in the order of $1 \, ms$) of ultra-reliable low-latency Machine Type Communication (uMTC) [10]. Linear search time for PT-II remains under $10 \, ms$, which is due to the remarkably reduced size of the table. However, this also does not meet the minimum delay requirement of uMTC as network and application processing delays are yet to be added which further increase the end-to-end latency. Whereas, index based

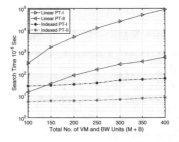

Fig. 5. Size of Policy Tables for $\alpha_W = 4$, $\alpha_T = 4$ and $M = B$

search time for PT-I and PT-II remains under $70 \, \mu s$ and $10 \, \mu s$ respectively. This can be clearly seen that index based search time for both policy tables qualify the end-to-end delay criteria of uMTC and real-time applications.

5   Conclusion

In this paper, we scale SMDP Multi-Resource Allocation work and present an index based search approach on large size policy tables to speed up admission control of service requests in the edge cloud based systems. In deriving the optimal action from a policy table, we consider end-to-end delay constraints of uMTC and real-time applications. Our proposed technique for structuring the policy tables and searching through them, not only outperforms

Fig. 6. Linear and indexed Search Time comparison between PT-I and PT-II

the legacy linear search method, but also meets the delay requirement of real-time applications in the growing edge cloud systems.

The index based search technique can help Resource Manager (RM) in EC based systems to search an optimal action for a service request from large size policy tables in order of microseconds (μs). We intend to continue to study further and implement our proposed work for real traffic in the real-world experiments like COSMOS test-bed.

Acknowledgements. This work is partially supported by USA Grant number NSF Award 181884.

References

1. Taleb, T., Samdanis, K., Mada, B., Flinck, H., Dutta, S., Sabella, D.: On multi-access edge computing: a survey of the emerging 5G network edge cloud architecture and orchestration. IEEE Commun. Surv. Tutor. **19**(3), 1657–1681 (2017)
2. Abbas, N., Zhang, Y., Taherkordi, A., Skeie, T.: Mobile edge computing: a survey. IEEE Internet Things J. **5**(1), 450–465 (2018)
3. Borgia, E., Bruno, R., Conti, M., Mascitti, D., Passarella, A.: Mobile edge clouds for information-centric IoT services. In: Proceedings of IEEE Symposium Computers and Communications (ISCC), Messina, Italy, June 2016, pp. 422–428 (2016)
4. Jararweh, Y., et al.: The future of mobile cloud computing: integrating cloudlets and mobile edge computing. In: Proceedings of 23rd International Conference on Telecommunications (ICT), Thessaloniki, Greece, May 2016, pp. 1–5 (2016)
5. Yang, L., Liu, B., Cao, J., Sahni, Y., Wang, Z.: Joint computation partitioning and resource allocation for latency sensitive applications in mobile edge clouds. In: 2017 IEEE 10th International Conference on Cloud Computing (CLOUD), Honolulu, CA, 2017, pp. 246–253 (2017)
6. Liu, Y., Lee, M.J., Zheng, Y.: Adaptive multi-resource allocation for cloudlet-based mobile cloud computing system. IEEE Trans. Mobile Comput. **15**(10), 2398–2410 (2016)
7. Liu, Y., Lee, M.J.: An adaptive resource allocation algorithm for partitioned services in mobile cloud computing. In: IEEE Symposium on Service-Oriented System Engineering, San Francisco Bay, CA, pp. 209–215 (2015)
8. Piri, E., et al.: 5GTN: a test network for 5G application development and testing. In: 2016 European Conference on Networks and Communications (EuCNC), Athens, 2016, pp. 313–318 (2016)
9. NSF, PAWR: COSMOS: Cloud Enhanced Open Software Defined Mobile Wireless Testbed for City-Scale Deployment. https://cosmos-lab.org
10. Osseiran, A., Monserrat, J.F.: 5G Mobile and Wireless Communications Technology, pp. 240–270. Cambridge University Press, Cambridge (2016)
11. Shah, S., Shaikh, A.: Hash based optimization for faster access to inverted index. In: 2016 International Conference on Inventive Computation Technologies (ICICT), Coimbatore, 2016, pp. 1–5 (2016)
12. Lin, C., Hsu, C., Hsieh, S.: A multi-index hybrid trie for lookup and updates. IEEE Trans. Parallel Distrib. Syst. **25**(10), 2486–2498 (2014)
13. Bulysheva, L., Bulyshev, A., Kataev, M.: Visual database design: indexing methods. In: 2018 Sixth International Conference on Enterprise Systems (ES), Limassol, 2018, pp. 25–29 (2018)
14. Database Normalization. https://support.microsoft.com/en-us/help/283878/description-of-the-database-normalization-basics

Efficient Migration of Large-Memory VMs Using Private Virtual Memory

Yuji Muraoka and Kenichi Kourai[✉]

Kyushu Institute of Technology, 680-4 Kawazu, Iizuka, Fukuoka, Japan
{murayu,kourai}@ksl.ci.kyutech.ac.jp

Abstract. Recently, Infrastructure-as-a-Service clouds provide virtual machines (VMs) with a large amount of memory. Such large-memory VMs can be migrated to other hosts on host maintenance, but it is costly to always preserve hosts with sufficient free memory as the destination of VM migration. Using virtual memory in destination hosts is a possible solution, but the performance of VM migration largely degrades because traditional general-purpose virtual memory causes frequent paging during the migration. This paper proposes *VMemDirect*, which achieves efficient migration of large-memory VMs using *private virtual memory*. VMemDirect creates private swap space for each VM on fast NVMe SSDs. Then it transfers likely accessed memory data to physical memory and the other data to the private swap space *directly*. This *direct memory transfer* can completely avoid paging during VM migration. We have implemented VMemDirect in KVM and showed that the performance of VM migration and the migrated VM was improved dramatically.

1 Introduction

As Infrastructure-as-a-Service (IaaS) clouds are widely used, they also provide VMs with a large amount of memory. For example, Amazon EC2 provides VMs with 12 TB of memory and plans those with 24 TB of memory in 2019. Such large-memory VMs are used for big data processing. When a host running a VM is maintained, the execution of the VM can be continued by migrating the VM to another host in advance. VM migration transfers the state of a VM, e.g., virtual CPUs and memory via the network and restarts the VM at the destination host. If the memory of the VM is updated during the transfer, VM migration retransfers the updated memory data. As such, VM migration requires free memory that can accommodate the entire memory of a VM in a destination host. However, it is costly to always preserve such hosts as the destination of VM migration, particularly, for large-memory VMs. If VM migration is not possible, VMs have to be stopped during host maintenance.

One possible solution is to use virtual memory in a destination host. Virtual memory enables part of the memory of a VM to be stored in secondary storage and performs paging. When a VM requires memory data in the storage, the data is paged in from the storage to physical memory. In exchange for

© Springer Nature Switzerland AG 2020
L. Barolli et al. (Eds.): INCoS 2019, AISC 1035, pp. 380–389, 2020.
https://doi.org/10.1007/978-3-030-29035-1_37

this, unnecessary memory data in physical memory is paged out to the storage. However, traditional general-purpose virtual memory is incompatible with VM migration. Since frequent paging occurs during VM migration, the performance of VM migration degrades largely. After the migration, the execution performance of the VM is also largely affected by paging because necessary memory data is often paged out.

This paper proposes *VMemDirect* for efficient migration of large-memory VMs using *private virtual memory*. VMemDirect creates *private swap space* for each VM on fast NVMe SSDs, which are becoming rapidly inexpensive, and integrates private virtual memory with VM migration. It directly transfers memory data of a VM to either physical memory or private swap space in a destination host. Since this *direct memory transfer* does not cause any paging during VM migration, the performance degradation can be avoided. VMemDirect also predicts future memory access of a VM and locates likely accessed memory data in physical memory as much as possible. Therefore, the execution performance of the VM can be preserved after the migration.

We have implemented VMemDirect in KVM to achieve efficient VM migration using private virtual memory. Private swap space is created using a special file called a *sparse file* to enable direct memory transfer to the swap space. Upon VM migration, VMemDirect determines the destination of memory data according to the memory access history of a VM and directly transfers the data to the same location even on retransfers. After VM migration, it performs paging using the userfaultfd mechanism in Linux. Our experimental results show that VMemDirect could reduce the migration time and the downtime and improve the performance of the migrated VM dramatically.

The organization of this paper is as follows. Section 2 describes issues on VM migration using traditional virtual memory. Section 3 proposes VMemDirect for efficient VM migration using private virtual memory and Sect. 4 explains its implementation. Section 5 shows experimental results. Section 6 discusses related work and Sect. 7 concludes this paper.

2 Migration of Large-Memory VMs

Even if a destination host does not have sufficient free memory, VM migration is possible by using virtual memory, as illustrated in Fig. 1. Using virtual memory, VM migration can transparently store part of the memory of a VM in swap space on secondary storage and use a necessary amount of memory. When the VM requires memory data in the swap space, a page-in is performed and the data is moved to physical memory. In exchange for this, a page-out is performed and memory data unlikely accessed in physical memory is moved to the swap space. Since the performance of storage is much lower than that of memory, the execution performance of the VM degrades. Fortunately, using recent NVMe SSDs can reduce this overhead.

However, traditional general-purpose virtual memory is incompatible with VM migration. VM migration first transfers the entire memory data of a VM

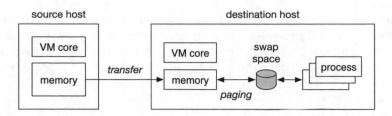

Fig. 1. VM migration using virtual memory.

to physical memory in the destination host. After the physical memory becomes full, the following transfers always cause page-outs from physical memory to swap space because transferred memory data has to be first stored in physical memory. This large number of page-outs lead to the increase in migration time. Also, retransfers of updated memory data cause paging. If the memory data to be updated exists in swap space, VM migration has to first page in that memory data to physical memory and then update it. At the same time, page-outs are necessary because physical memory is already full. For large-memory VMs, the first memory transfers take a long time and therefore the amount of memory data updated during the transfer increases.

In the final phase of VM migration, the downtime of the VM increases if paging occurs frequently. This is because the final phase stops the VM and transfers the rest of the state consistently. In particular, if the memory of virtualization software is managed by the same virtual memory as the memory of VMs, as in KVM, it is often paged out during VM migration. At the destination host, virtual devices provided by virtualization software are not used until the final phase of VM migration. Therefore, the memory for them is unlikely accessed data, which is a target of page-outs. When the state of virtual devices is restored, page-ins occur frequently. After VM migration, frequently updated memory data is often stored in physical memory. However, memory data that are just read frequently can be stored in swap space. Since such data is transferred only once, it is often paged out by following memory transfers. This can lead to frequent paging after the VM is resumed at the destination host.

To counteract these problems, split migration [8,9] has been proposed using multiple destination hosts. Split migration divides the memory of a VM and transfers the memory fragments to multiple smaller hosts. It transfers the state of virtual CPUs and devices and likely accessed memory data to a main host and the other memory data to sub-hosts. If the VM requires memory data in sub-hosts after VM migration, the data is transferred to the main host by remote paging. Since no paging occurs during VM migration, split migration can improve the migration performance. Also, it can increase the execution performance of the migrated VM because necessary memory data is transferred to the main host in advance. However, split migration is more costly than VM migration using virtual memory. It requires small but multiple sub-hosts and, for efficient remote

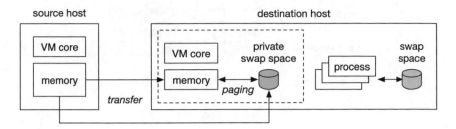

Fig. 2. VM migration using private virtual memory in VMemDirect.

paging, high-speed network such as InfiniBand [5,6]. In addition, migrated VMs are subject to host and network failures.

3   VMemDirect

For efficient migration of large-memory VMs, this paper proposes *VMemDirect* using *private virtual memory* and the technology of split migration. Figure 2 illustrates VM migration in VMemDirect. VMemDirect creates *private swap space* for each VM on fast NVMe SSDs in a destination host and integrates private virtual memory with VM migration. Since recent NVMe SSDs are becoming more inexpensive, the cost can be lower than using multiple sub-hosts and expensive high-speed network as in split migration. In addition, the execution of VMs are not affected by host or network failures after VM migration.

Since private virtual memory targets only the memory of a VM, VMemDirect can prevent performance degradation due to paging of the memory of virtualization software itself. During memory transfers in VM migration, the memory of virtualization software is not paged out at the destination host. When a VM is created, VMemDirect creates private swap space on an NVMe SSD with the same size as the memory of the VM. This swap space is optimized for the memory of the VM. Memory areas in the VM correspond to blocks in the swap space. Memory data of the VM is stored in blocks in the swap space only when it does not exist in physical memory. The other blocks do not have actual data in the swap space.

Upon VM migration, VMemDirect directly transfers memory data to either physical memory or private swap space, instead of relying on paging of traditional virtual memory. At the source host, VMemDirect determines locations where each memory data is stored at the destination host when it starts VM migration. The locations are not changed during VM migration and VMemDirect retransfers memory data to the same location. This *direct memory transfer* can prevent paging from occurring during VM migration. No data in physical memory is paged out, while no memory data in private swap space is paged in. When VMemDirect retransfers memory data, it directly updates data in physical memory or private swap space. Since the structure of private swap space is designed for this direct access, it is easy to find the corresponding blocks.

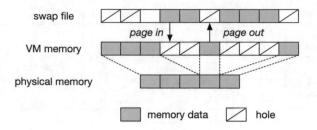

Fig. 3. The memory management of a VM using private swap space.

The locations of memory data in the destination host are determined on the basis of the memory access history of a VM. VMemDirect transfers likely accessed memory data to physical memory and the other data to private swap space. This increases the probability that retransferred memory data is stored in physical memory. As a result, the overhead due to overwriting data in private swap space can be reduced. Unlike VM migration using traditional virtual memory, not only frequently updated memory data but also frequently read data is stored in physical memory. Therefore, the occurrence of paging can be suppressed after the VM is resumed.

4   Implementation

We have implemented VMemDirect in KVM using QEMU-KVM 2.4.1 and Linux 4.11. QEMU-KVM is virtualization software that runs on top of Linux.

4.1   Swap File for Private Swap Space

For private swap space, VMemDirect creates a swap file with the same size as the memory of a VM using a *sparse file*. A sparse file can contain blocks that have no actual data, which are called *holes*. Using this swap file, it can make one-to-one relation between memory areas of the VM and blocks of the swap file. Memory data of the VM is stored in either physical memory or the swap file (Fig. 3). When it is paged in to physical memory, VMemDirect makes the corresponding block of the swap file a hole to reduce storage usage. To create a hole, it writes a metadata standing for an empty block to storage.

VMemDirect accesses the swap file using direct I/O so that the page cache is not allocated for the file blocks. Direct I/O enables data to be directly read from and written to storage without storing it in the page cache managed by the operating system. In VMemDirect, most of the physical memory is used to store memory data of a VM. If the page cache were created for paged-out data, traditional virtual memory would have to page out data in physical memory. To use the bandwidth of NVMe SSDs as much as possible, VMemDirect accesses the swap file by the chunk larger than the 4-KB page. In the current implementation, VMemDirect uses 256-page chunks.

4.2 Direct Memory Transfer

To manage the locations where memory data is transferred on VM migration, VMemDirect creates a *location bitmap*. This bitmap allocates one bit for each 4-KB page and the value is determined according to the memory access history of a VM. The memory access history is obtained as described in the previous work [8, 9]. When memory data is transferred to the swap file, VMemDirect sets 1 to the corresponding bit in the bitmap. Otherwise, it sets 0 to the bitmap. VMemDirect transfers memory data of each page with the value of the corresponding bit. At the destination host, VMemDirect stores received memory data in either physical memory or the swap file according to the specified location.

To write memory data to the swap file by the chunk larger than the page, VMemDirect temporarily stores received memory data in a buffer on physical memory. When the buffer becomes full, VMemDirect writes data to storage at once. Since memory data is transferred sequentially in the first phase of VM migration, VMemDirect can write buffered data to contiguous blocks in the swap file. In contrast, received memory data is not contiguous when memory data is retransferred. For such data, VMemDirect writes data to the swap file by the page.

4.3 Paging for Private Virtual Memory

VMemDirect achieves paging for private virtual memory using the userfaultfd mechanism in Linux. After VM migration, it registers the entire memory of the VM to userfaultfd. When the VM accesses a non-existent memory page and a page fault occurs, that event is notified to QEMU-KVM by userfaultfd. Next, VMemDirect reads memory data from the block corresponding to the faulting address in the swap file. Then, it writes the data to the corresponding memory page of the VM. In addition, it performs the same operations for the other pages in the same chunk. At the same time, it removes these blocks in the swap file and makes the blocks holes to complete page-ins.

At the same time, VMemDirect selects a chunk including the most unlikely accessed memory pages to perform page-outs. It writes the data of the selected pages to the corresponding blocks in the swap file and removes the mapping of the pages. To obtain the memory data and remove the mapping atomically, we used the extension of userfaultfd, which is developed for remote paging [8]. To manage the locations of memory data for these page-outs, VMemDirect creates a location bitmap in the destination host as well as the source host of VM migration. The bits are set when VMemDirect receives memory data from the source host. When a page-in occurs, the corresponding bit is set to 0. For a page-out, that is set to 1.

5 Experiments

We conducted several experiments to examine the performance of VM migration and a migrated VM in VMemDirect. For comparison, we examined the performance for *ideal* VM migration with sufficient memory and *naive* VM migration

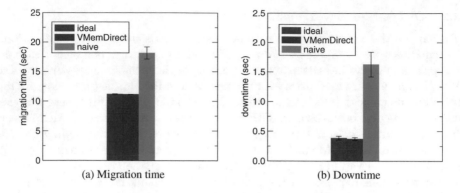

Fig. 4. The comparisons of migration performance.

with traditional virtual memory. We used 1 TB of Samsung NVMe SSD 970 PRO as swap space. For the source and destination hosts, we used two PCs with an Intel Xeon E3-1226 v3 processor and 16 GB of memory. These PCs were connected using 10 Gigabit Ethernet. We ran Linux 4.11 as the host operating system and QEMU-KVM 2.4.1 as virtualization software. We used a VM with one virtual CPU and 12 GB of memory. When using traditional virtual memory, we adjusted the size of free memory to 6 GB, which was the half of the memory size of the VM.

5.1 Migration Performance

To examine migration performance, we first measured the time needed for VM migration. Figure 4(a) shows the migration time. Compared with the ideal migration, the migration time increased by 2.1 times in the naive migration. For VMemDirect, in contrast, the migration time was almost the same as the ideal migration.

Next, we measured the downtime during VM migration. As shown in Fig. 4(b), the naive migration suffered from 2.2 s longer downtime than the ideal migration. The root cause was that many page-ins occurred when virtual devices were resumed in the final phase. In contrast, VMemDirect reduced the downtime by 15 ms. This is because QEMU-KVM could not accurately estimate the downtime in VMemDirect. QEMU-KVM enters the final phase when it is estimated that the rest of the memory data will be transferred in 300 ms. In VMemDirect, the final memory transfers were completed earlier than the estimation because only a smaller amount of memory data was transferred to the slow swap space than during the estimation.

5.2 Performance of Private Virtual Memory

To examine the performance of private virtual memory, we ran in-memory database called memcached [4] in a VM and measured the performance using

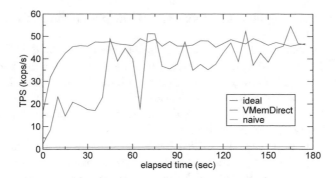

Fig. 5. The memcached performance in a migrated VM.

the memaslap benchmark. We allocated 5 GB of memory to memcached and ran
the benchmark before and after VM migration. Figure 5 shows the comparison
of the transactions per second after VM migration. VMemDirect achieved 39
times higher performance on average than after the naive migration. Compared
with after the ideal migration, the performance in VMemDirect was 11% lower
on average in a stable state and largely fluctuated. This is because paging sometimes occurred and the performance degraded by that paging. The reason why
the performance was much lower during the first 45 s is that paging occurs frequently. This is probably due to the implementation issue of the memory access
history.

6   Related Work

vMotion provides two different migration methods in terms of swap space [2].
Unshared-swap vMotion uses different swap spaces between the source and destination hosts and transfers memory data stored in swap space to the destination
host. In contrast, shared-swap vMotion stores a swap file in shared storage and
transfers no memory data in swap space. To support paging during VM migration, the destination host uses temporary swap space and integrates that swap
space into shared one after the migration. This migration method can be more
efficient than VMemDirect, but network paging is always necessary.

Like shared-swap vMotion, Agile live migration [3] locates swap space for each
VM in the network. It pages out memory data except for the current working
set aggressively. Upon VM migration, it transfers no memory data in the swap
space but only data in the working-set memory. This method can further improve
migration performance, but the paging overhead is much larger because most of
the data is paged out.

FlashVM [7] is virtual memory using paging based on SSDs. It pages out
more memory pages at once than when using HDDs. Since random reads of
SSDs are fast, FlashVM prefetches more useful pages to reduce page faults. In
addition, it adjusts the rate of writeback to SSDs to reduce the latency of page

faults. This prefetching technique can improve the performance of private virtual memory in VMemDirect.

Swap space using SSDs and ExpEther has been proposed [10]. ExpEther extends PCI Express using Ethernet and enables SSDs to be used without the limitation of physical locations. High-speed communication using DMA is performed between local memory and remote SSDs. Paging with this swap space can extend computer memory. Using this system, VMemDirect can use remote SSDs as private swap space flexibly.

VSwapper [1] improves the performance of VMs using virtual memory. It monitors storage I/O and prevents unmodified pages from being written to swap space on page-outs. Also, it stores data written to paged-out pages in a temporal buffer and prevents data from being read from swap space if the entire page is written. These optimizations achieve 10 times performance improvement. They can be also applied to private virtual memory in VMemDirect.

7   Conclusion

This paper proposed VMemDirect for achieving efficient VM migration by cooperating with private virtual memory. VMemDirect creates private swap space for each VM on fast NVMe SSDs. It directly transfers likely accessed memory data to physical memory and the other data to the private swap space. After VM migration, VMemDirect performs paging between physical memory and the private swap space for the memory of each VM. Our experimental results show that VMemDirect could improve the performance of VM migration and a migrated VM dramatically, compared with VM migration using traditional virtual memory.

One of our future work is to evaluate the performance of VMemDirect using VMs with various numbers of virtual CPUs, various amounts of memory, and various applications. Another direction is to clarify trade-offs between using high-speed network and NVMe SSDs for VM migration. We will compare various performance of VMemDirect with that of S-memV with split migration and remote paging [8,9] when we run various applications in VMs. In addition, we are planning to support N-to-one migration [9] in VMemDirect. The original N-to-one migration integrates memory data across multiple hosts into one large host. Similarly, VMemDirect needs to efficiently transfer memory data in both physical memory and private swap space to one host.

Acknowledgements. The research results have been achieved by the "Resilient Edge Cloud Designed Network (19304)," the Commissioned Research of National Institute of Information and Communications Technology (NICT), Japan.

References

1. Amit, N., Tsafrir, D., Schuster, A.: VSwapper: a memory swapper for virtualized environments. In: Proceedings of ACM International Conference on Architectural Support for Programming Languages and Operating Systems, pp. 349–366 (2014)

2. Banerjee, I., Moltmann, P., Tati, K., Venkatasubramanian, R.: VMware ESX memory resource management: swap. VMware Tech. J. **3**(1), 48–56 (2014)
3. Deshpande, U., Chan, D., Guh, T., Edouard, J., Gopalan, K., Bila, N.: Agile live migration of virtual machines. In: Proceedings of IEEE International Parallel and Distributed Processing Symposium (2016)
4. Fitzpatrick, B.: memcached – a distributed memory object caching system. http://memcached.org/
5. Gu, J., Lee, Y., Zhang, Y., Chowdhury, M., Shin, K.: Efficient memory disaggregation with Infiniswap. In: Proceedings of USENIX Symposium Networked Systems Design and Implementation (2017)
6. Liang, S., Noronha, R., Panda, D.: Swapping to remote memory over InfiniBand: an approach using a high performance network block device. In: Proceedings of IEEE Cluster Computing (2005)
7. Saxena, M., Swift, M.: FlashVM: virtual memory management on flash. In: Proceedings of USENIX Annual Technical Conference (2010)
8. Suetake, M., Kashiwagi, T., Kizu, H., Kourai, K.: S-memV: split migration of large-memory virtual machines in IaaS clouds. In: Proceedings of IEEE International Conference Cloud Computing, pp. 285–293 (2018)
9. Suetake, M., Kizu, H., Kourai, K.: Split migration of large memory virtual machines. In: Proceedings of ACM SIGOPS Asia-Pacific Workshop of Systems (2016)
10. Suzuki, J., Baba, T., Hidaka, Y., Higuchi, J., Kami, N., Uchida, S., Takahashi, M., Sugawara, T., Yoshikawa, T.: Adaptive memory system over Ethernet. In: Proceedings of USENIX Workshop on Hot Topics in Storage and File Systems (2010)

Transmission Control Method to Realize Efficient Data Retention in Low Vehicle Density Environments

Ichiro Goto[1(✉)], Daiki Nobayashi[1], Kazuya Tsukamoto[3], Takeshi Ikenaga[1], and Myung Lee[2]

[1] Kyushu Institute of Technology, Kitakyushu, Japan
`goto.ichiro959@mail.kyutech.jp`, `{nova,ike}@ecs.kyutech.ac.jp`
[2] City College of New York, New York, USA
`mlee@ccny.cuny.edu`
[3] Kyushu Institute of Technology, Iizuka, Japan
`tsukamoto@cse.kyutech.ac.jp`

Abstract. With the development and spread of Internet of Things (IoT) technology, various kinds of data are now being generated from IoT devices, and the number of such data is expected to increase significantly in the future. Data that depends on geographical location and time is commonly referred to as spatio-temporal data (STD). Since the "locally produced and consumed" paradigm of STD use is effective for location-dependent applications, the authors have previously proposed using a STD retention system for high mobility vehicles equipped with high-capacity storage modules, high-performance computing resources, and short-range wireless communication equipment. In this system, each vehicle controls its data transmission probability based on the neighboring vehicle density in order to achieve not only high coverage but also reduction of the number of data transmissions. In this paper, we propose a data transmission control method for STD retention in low vehicle density environments. The results of simulations conducted in this study show that our proposed scheme can improve data retention performance while limiting the number of data transmissions to the lowest level possible.

1 Introduction

With the development and spread of Machine-to-Machine (M2M) and Internet of Things (IoT) technologies, various kinds of data are now being generated from IoT devices. In the current era, the data generated from IoT devices are mostly collected and analyzed by Internet cloud servers that then provide useful data to applications. According to Cisco Systems, Inc., the number of M2M connections can be expected to grow from 6.1 billion in 2017 to 14.6 billion by 2022 [1]. Therefore, a considerable traffic volume increase will result because of the increasingly enormous amounts of data flowing into the Internet. In order to cope with this, it will be necessary to increase link bandwidth and improve

L. Barolli et al. (Eds.): INCoS 2019, AISC 1035, pp. 390–401, 2020.
https://doi.org/10.1007/978-3-030-29035-1_38

router performance levels to prevent the burden on network infrastructure from skyrocketing due to unrestrained growth.

From the viewpoint of data content, some data generated from IoT devices, such as traffic, weather, and disaster-related information, are highly dependent on location and time. We define such information as "spatio-temporal data (STD)." The most effective way to use STD is to provide it directly to the users who are in the vicinity of the STD generation location. However, since servers connected to the Internet collect IoT device data, users often receive data from servers located far away. Therefore, to facilitate the "local production and consumption of data", a network infrastructure that can retain data within a specific area is crucial.

Previously, we proposed a novel Geo-Centric Network (GCN) as an infrastructure that can facilitate collection, analysis, and provision of STD based on geographical proximity. We have also proposed a vehicle-based STD retention system equipped with large capacity storage modules, high-performance computing resources, and short-range wireless communication equipment, as a means of distributing STD based on geographical proximity [2]. In this system, vehicles capable of wireless communication are defined as regional information hub (InfoHub), and the purpose of this system is to retain STD within a specific area. STD management by InfoHub vehicles has three particular advantages. First, the user can acquire the most up-to-date date in real-time since the data is retained in the vehicle network around the information source. The second is its ability to improve fault tolerance by distributing the data to each vehicle because all other vehicles will retain the data received even if one vehicle fails. Third, it allows data to be collected, analyzed, and distributed by each vehicle without using the server, which leads to improved server load distribution.

However, since all vehicles use the same communication channel, channel competition occurs when the number of vehicles increases, which in turn leads to deterioration in communication quality. In order to solve this problem, we have proposed a method for controlling vehicle data transmission probabilities based on the density of neighboring vehicles [2]. However, since vehicle density levels are always changing due to the mobility of the vehicles participating in the network, areas where the vehicle density levels become low may occur. In such case, this method may not be able to retain data effectively.

In this paper, we proposed a transmission control method for low vehicle density environments. In such environments, since the area that can be covered only by the own vehicle increase, the importance per transmission increases. Therefore, it is necessary to increase the transmission frequency. Accordingly, in this system, each vehicle adjusts its transmission period based on the degree of contribution to the coverage rate improvement. The effectiveness of this method is shown in our simulation results.

The rest of this paper is organized as follows. In Sect. 2, we describe studies related to data retention, while we outline our previous STD retention system and discuss the problems to be addressed in Sect. 3. In Sect. 4, we describe our proposed method to achieve efficient data retention in low vehicle density environments, while simulation models and evaluation results are provided in Sect. 5. Finally, we give our conclusions in Sect. 6.

2   Related Works

For data utilization using vehicular networks, Maihofer et al. proposed the method in which all vehicles in the retention area hold the data and position information of all other vehicles [3]. This method, which have been the subject of many previous studies, is suitable for vehicles that have a wide range of practical uses because no outside infrastructure is required. In a separate study, Maio et al. proposed a method using a software-defined network (SDN) as a technique for setting an optimum retention area range [4]. In that method, the server also operates an SDN controller that collects and analyzes mobility information such as speed and position from the vehicles, and then calculates the maximum radius of the retention area. Additionally, Rizzo et al. proposed an application for exchanging information between vehicles based on the assumption that infrastructure would become unusable in a disaster [5], while Leontiadis et al. proposed a method in which navigation information is exchanged to facilitate prediction of a vehicle heading into a retention area and in which data is delivered efficiently [6]. In the Floating Content [7] and Locus [8] system, each vehicle has a data list and exchanges it with passing vehicles. Next, each vehicle determines their data transmission probability based on their distance to the center where the data was generated. As the distance from the center increases, data acquisition probabilities decline. On the other hand, when there are numerous vehicles in the vicinity of the center, channel contention occurs, and the communication quality deteriorates because all vehicles transmit data with high transmission probabilities. Furthermore, even if data can be stored in the vicinity of the place where it was generated, there is an overhead in the data acquisition process because the users acquire data by query/response type information distribution such as query transmission, data discovery, and transfer to user. With these points in mind, we propose a novel network base capable of passively acquiring data as part of efforts to improve overhead and promote local production and consumption of data.

3   STD Retention System

In this section, we describe the assumptions, requirements, and outline of the retention system [2], and then discuss the related problems.

3.1   Assumption

In this system, STD includes not only data for an application but also area information such as the center coordinates, radius R of the retention area, length r of an auxiliary area, and frame period d. Each vehicle can obtain location information using a Global Positioning System (GPS) receiver and broadcasts a beacon that includes a unique identifier (ID). Furthermore, all vehicles are equipped with the same antenna and transmit at equal power levels.

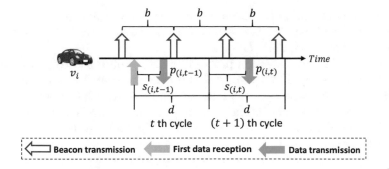

Fig. 1. Data transmission procedure

3.2   System Requirements

In this paper, we defined coverage rate as an indicator of the data retention state. The coverage rate formula is as follows:

$$Coverage\ Rate = \frac{S_{DT}}{S_{TA}} \tag{1}$$

where S_{TA} is the size of retention area and S_{DT} is the size of the total area where a user can obtain the data transmitted from InfoHub vehicles within the one cycle. A high coverage rate means that users can automatically receive STD from anywhere within the retention area, so it is important to maintain a high coverage rate within the retention system.

3.3   Previous Work

In this section, we will provide an outline of our previous work [2] in which we proposed a method to control data transmission probability. This method uses simple information, such as the number of neighboring vehicles and the data reception number in order to improve the overhead of processing location information [9]. We begin by defining the data transmission area. When each vehicle v_i transmitted data, it confirms the retention area center coordinates from the received data and then calculates its distance y from the center. Data transmission areas are classified into the following two categories by y

$$\begin{cases} 0 < y \le R + r : & data\ transmission\ area \\ otherwise : & out\ of\ area \end{cases} \tag{2}$$

where R is the retention area radius, and r is the auxiliary area length. The vehicles within the auxiliary area also transmit data to improve the coverage rate. Next, we will describe the data transmission timing. Figure 1 shows the data transmission procedure. When each vehicle v_i receives data from other vehicles for the first time, it first checks the frame period d included in the received data.

Then, the vehicle randomly determines the next transmission time $s_{(i,t)}$ within d in order to avoid data transmission collisions. Furthermore, $s_{(i,t)}$ is calculated at the beginning of the cycle t.

The actual data transmission probability is controlled according to the neighboring vehicle density. Each vehicle detects the number of neighboring vehicles $n_{(i,t)}$ from the number of beacons transmitted by those vehicles. When the number of neighboring vehicles is more than four, the vehicle's transmission range has the potential to completely cover all the neighboring vehicles. For example, if the neighboring four vehicles are located to a vehicle's north, south, west, and east, they can cover the vehicle's transmission range. Therefore, the data transmission probability $p_{(i,t)}$ is determined based on the number of neighboring vehicles $n_{(i,t)}$ as described below.

case1 $n_{(i,t-1)} \leq 3$:
Since the vehicle's transmission area cannot be completely covered by that of the neighboring vehicles, the vehicle has to set its transmission probability individually $p_{(i,t)}$ to 1.

case2 $n_{(i,t-1)} \geq 4$:
In high vehicle density environment, since the potential for data transmission collisions increases with increases in the number of data transmissions, it is necessary to use the minimum number of vehicles to maintain high coverage rate. On the other hand, if the neighboring vehicles are clustered in some directions, coverage within the vehicle's transmission range may be spotty even if there are a large number of neighboring vehicles. To solve these problems, we defined $m_{(i,t)}$ as the estimated value of the number of received data during t-th cycle and adjusted the transmission probability $p_{(i,t)}$ based on the $m_{(i,t)}$. The predicted value $m_{(i,t)}$ is given as following equation:

$$m_{(i,t)} = \alpha * l_{(i,t-1)} + (1 - \alpha) * m_{(i,t-1)} \tag{3}$$

where $m_{(i,t-1)}$ is the predicted value of the previous cycle, $l_{(i,t-1)}$ is the number of received data during the previous cycle, and α is the moving average coefficient. The vehicle adjusts the data transmission probability by the following equation so that the number of received data becomes β in the t-th cycle.

$$p_{(i,t)} = \begin{cases} p_{(i,t-1)} + \frac{\beta - l_{(i,t-1)}}{n_{(i,t-1)}+1} & (0 < m_{(i,t)} < \beta) \\ p_{(i,t-1)} & (m_{(i,t)} = \beta) \\ p_{(i,t-1)} - \frac{l_{(i,t-1)} - \beta}{n_{(i,t-1)}+1} & (m_{(i,t)} > \beta) \end{cases} \tag{4}$$

Here, the transmission probability in the first cycle (initial value of transmission probability) is set to $\frac{\beta}{n_{(i,t-1)}+1}$. This means that the average value is set in order to control the number of received data from all vehicles (not only the number of neighboring vehicles, but also its own) to β. If $m_{(i,t)}$ is less than β, the vehicle increases the data transmission probability by $\frac{\beta - l_{(i,t-1)}}{n_{(i,t-1)}+1}$ in order to compensate for the shortage $\beta - l_{(i,t-1)}$ created by all $n + 1$ vehicles. On the other hand, if $m_{(i,t)}$ is more than β, the vehicle decreases the data transmission probability by

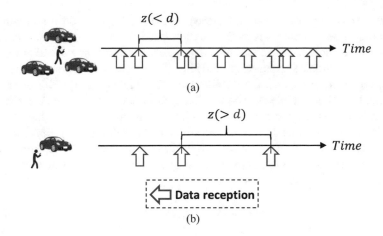

Fig. 2. Problems with the previous method: (a) high vehicle density, (b) low vehicle density

$\frac{l_{(i,t-1)} - \beta}{n_{(i,t-1)} + 1}$ in order to reduce the excess amount $l_{(i,t-1)} - \beta$ produced by all $n + 1$ vehicles. If $m_{(i,t)}$ is equal to β, $p_{(i,t)}$ is set to $p_{(i,t-1)}$ because the current data transmission probability is appropriate. If the value of $\frac{\beta - l_{(i,t-1)}}{n_{(i,t-1)} + 1}$ or $\frac{l_{(i,t-1)} - \beta}{n_{(i,t-1)} + 1}$ is negative, $p_{(i,t)}$ is set to $p_{(i,t-1)}$.

3.4 Problems with the Previous Method

As shown in Fig. 2(a), if the vehicle density is high, the user's data reception interval z rarely exceeds the frame period d because the number of data transmissions is large. However, as shown in Fig. 2(b), if the vehicle density is low, z can exceed d. This is because the transmission interval within two consecutive cycles can exceed d due to the randomness of the transmission timing determination and the influence per vehicle can become especially large in low vehicle density environments. Therefore, since z can exceed d, method that is suitable for such environments is necessary.

4 Proposed Method

In this section, we describe a method for facilitating efficient data retention in low vehicle density environments. We first present a setting the minimum transmission period and then introduce the data transmission period control.

In order to solve the problems related to the previous method, it is necessary to make the maximum transmission interval randomly determined smaller than d. Hence, the frame period must be set to half of d. This frame period d_{min} is then defined as the minimum data transmission period.

By setting the frame period to d_{min}, the data transmission interval can be prevented from exceeding d. However, the number of data transmissions increases drastically. In order to suppress an increase in the number of data transmissions, our proposed method controls each vehicle's data transmission period according to its degree of contribution to the coverage rate improvement. The degree of contribution is calculated based on the size of the area S in which the communications range does not overlap with the nearest vehicle (Fig. 3). This area cannot be covered by other vehicles in the retention area. Therefore, the larger S is, the larger the influence on the coverage rate improvement becomes. Then, based on the results of our previous study, a case where the number of neighboring vehicles n is three or less is defined as a low vehicle density environment, and the data transmission period is calculated according to the size of the non-overlapping area S using the following equation:

$$d_{ctl} = d - d_{min} * \frac{S}{S_{max}} \qquad (5)$$

where d_{ctl} is the frame period set in a low vehicle density environment and S_{max} is the maximum value of the non-overlapping area, which is the value achieved when the distance between vehicles x equals the maximum communication distance. As the distance between vehicles increases from this equation, that is, as the non-overlapping area increases, the data transmission period is controlled to approach the minimum data transmission period. The area S is calculated by the following equation:

$$S = \pi r^2 - 2\left\{ r^2 \cos^{-1}(1 - \frac{h}{r}) - (r - h)\sqrt{h(2r - h)} \right\} \qquad (6)$$

where r is the communication range radius and h is the height of two identical arcs appearing at the overlap area.

For the distance calculation between vehicles, if the transmission power and antennas of all vehicles are the same, the distance between vehicles is calculated by measuring the loss power from the received radio wave strength of the receiving side. Loss L is calculated by the free space propagation loss and is given by the following equation:

$$L = 20 \log(\frac{4\pi f l}{c}) \qquad (7)$$

where f is the frequency, l is the distance, and c is the wave velocity. By solving this equation for l, we can calculate the distance. Since signals can be blocked by objects (such as buildings) in real environments, the calculated distance may be different from the theoretical value. When the received radio wave strength from neighboring vehicles is smaller than the theoretical value, the proposed method calculates the distance between vehicles as larger than the actual value so that the data transmission period becomes shorter and the data transmission frequency becomes higher. However, this increase compensates for areas where radio waves do not reach certain vehicles due to the decrease in signal strength from neighboring vehicles within communication range. Consequently, our proposed method can realize effective data retention by calculating distances between vehicles from their measured power losses even when a real environment is assumed.

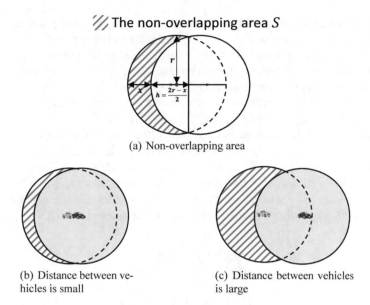

(a) Non-overlapping area

(b) Distance between ve-
hicles is small

(c) Distance between vehicles
is large

Fig. 3. Difference in non-overlapping area with the change in the distance between vehicles

5 Simulation

In this section, we evaluate the performance of our proposed method using a simulation.

5.1 Simulation Model

We evaluated our proposed method using the Veins [12] simulation framework, which simultaneously implements both the IEEE 802.11p specification for wireless communications and the vehicular ad-hoc network (VANET) mobility model. Veins also includes a Two-Ray interference model and a simple obstacle shadowing model that has been calibrated and validated against real world measurements. Veins can combine the Objective Modular Network Testbed in C++ (OMNeT++) [10] network simulator with the Simulation of Urban MObility (SUMO) road traffic simulator [11].

In our simulations, which are based on our previous work [2], the frame period d and the beacon interval b were set to 5 s, the moving average coefficient α was set to 0.5, and the target value of the number of received data β was set to 4.

For comparison purposes, three methods, naive (2.5s), naive (5s), and our previous method [2] were used. The naive method always set the transmission probability $p_{(i,t)}$ of all vehicles to 1. The numbers 2.5 and 5 indicate the transmission interval d. Since the naive (2.5s) method's transmission interval is set to d_{min}, the naive (2.5s) method can achieve the maximum coverage rate discussed in this study.

To show the effectiveness of our proposed method, we used random topology (Fig. 4) in which vehicles with randomly generated starting and end points ran on a road grid at a spacing of 50 m. A traffic signal that switches short time (in 5 s intervals) in order to focus on the change in the coverage rate due to the movement of the vehicle was installed at the intersection. Furthermore, in order to create vehicle density differences, we made routes through all areas (A, B, C, and D), the left area (B and C), bottom area (C and D), and combined them. This made it possible to create environments in which the density of vehicles in area A is low. We then created and evaluated 10 kinds of movement models. The communication range of the vehicle was set to 300 m, and the speed was set to 40 km/h. In addition, the retention area radius R was set to 750 m, and the auxiliary area length r was set to 250 m. In this simulation, the vehicle numbers were set to 20, 40, and 60, and the simulation time was set to 300 s.

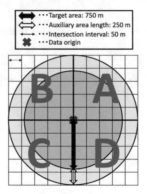

Fig. 4. Simulation Model

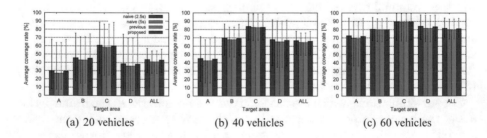

(a) 20 vehicles   (b) 40 vehicles   (c) 60 vehicles

Fig. 5. Coverage rate

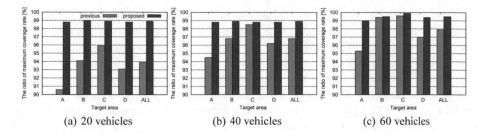

(a) 20 vehicles   (b) 40 vehicles   (c) 60 vehicles

Fig. 6. The ratio of maximum coverage rate

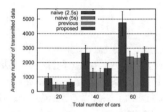

Fig. 7. The number of data transmissions

Fig. 8. Increase rate

5.2 Performance Evaluation

Figure 5 shows the average coverage rate, by area, for each number of vehicles. The horizontal axis represents the area, the vertical axis represents the coverage rate, and the error bar represents the maximum and minimum values of 10 simulation trials. The average coverage rate of the naive (5s) method and the previous method is lower than of the naive (2.5s) method, especially in the environment where the vehicle density is low (area A). This result indicates that the maximum coverage rate cannot be achieved using the current transmission period setting in low vehicle density environment. In contrast, our proposed method can achieve a coverage rate close to the naive (2.5s) method regardless of the vehicle density. To show the coverage rate in detail, Fig. 6 provides the ratio of the coverage rate of our previous and proposed methods to the naive (2.5s) method. Here, the coverage rate of the naive (2.5s) method is presented as 100. In our previous method, the ratio decreases as the vehicle density decrease (area A). However, in our proposed method, the ratio approaches 99% at any vehicle density. Thus, we can conclude that our proposed method can achieve data retention close to that of the naive (2.5s) method, regardless of vehicle density.

Next, we evaluated the number of data transmissions, the average number of which is shown Fig. 7. Here, it can be seen that the naive (2.5s) method has significantly increased the number of data transmissions compared with the previous method. However, we see also that our proposed method can suppress the increase in the number of data transmissions. To evaluate the increase in the number of data transmissions in detail, Fig. 8 shows the increase rate in the number of data transmissions for the naive (2.5s) and proposed methods over the previous method. It shows that the number of data transmissions produced by the naive (2.5s) method has increased approximately twice, while the proposed method can suppress that increase up to approximately 40%. Therefore, we can conclude that the proposed method can suppress the increase in the number of data transmissions up to about 40% while achieving a the coverage rate of approximately 99% against the naive (2.5s) method.

6   Conclusions

In this paper, we proposed a system that facilitates the retention of STD in a specific area by using an ad-hoc network constructed solely by InfoHub vehicles. Additionally, our proposed STD retention system improves coverage rates in low vehicle density environments by controlling data transmission periods based on the size of the area in which the communications range of one vehicle does not overlap with the nearest vehicle (practically, the vehicle with the highest beacon intensity). Through simulations, we clarified that the proposed method could suppress the increase in the number of data transmissions up to approximately 40% and achieve a coverage rate of approximately 99% against the maximum coverage rate. In our future work, as part of efforts for further coverage rate improvements, we will verify a novel STD retention system that cooperates with Mobile Edge Computing (MEC).

Acknowledgements. This work supported in part by JSPS KAKENHI Grant Number 18H03234, NICT Grant Number 19304, and USA Grant number NSF 17-586.

References

1. Cisco: Cisco Visual Networking Index: Forecast and Trends, 2017-2022, Cisco White Paper (2019). https://www.cisco.com/c/en/us/solutins/collateral/service-provider/visual-networking-index-vni/white-paper-c11-741490.pdf
2. Teshiba, H., Nobayashi, D., Tsukamoto, K., Ikenaga, T.: Adaptive data transmission control for reliable and efficient spatio-temporal data retention by vehicles. In: Proceedings of ICN 2017, Italy, pp. 46–52, April 2017
3. Maihofer, C., Leinmuller, T., Schoch, E.: Abiding geocast: time-stable geocast for ad hoc networks. In: Proceedings of ACM VANET, pp. 20–29 (2005)
4. Maio, A., Soua, R., Palattella, M., Engel, T., Rizzo, G.: A centralized approach for setting floating content parameters in VANETs. In: 14th IEEE Annual Consumer Communications & CCNC 2017, pp. 712–715, January 2017
5. Rizzo, G., Neukirchen, H.: Geo-based content sharing for disaster relief applications. In: International Conference on Innovative Mobile and Internet Services in Ubiquitous Computing, Advance in Intelligent System and Computing, vol. 612, pp. 894–903 (2017)

Transmission Control Method to Realize Efficient Data Retention 401

6. Leontiadis, I., Costa, P., Mascolo, C.: Persistent content-based information dissemination in hybrid vehicular networks. In: Proceedings IEEE PerCom, pp. 1–10 (2009)
7. Ott, J., Hyyti, E., Lassila, P., Vaegs, T., Kangasharju, J.: Floating content: information sharing in urban areas. In: Proceedings of IEEE PerCom, pp. 136–146 (2011)
8. Thompson, N., Crepaldi, R., Kravets, R.: Locus: a location-based data overlay for disruption-tolerant networks. In: Proceedings of ACM CHANTS, pp. 47–54 (2010)
9. Higuchi, T., Onishi, R., Altintas, O., Nobayashi, D., Ikenaga, T., Tsukamoto, K.: Regional InforHubs by vehicles: balancing spatio-temporal coverage and network load. In: Proceedings of IoV-VoI16, pp. 25–30 (2016)
10. OMNeT++. https://omnetpp.org/
11. SUMO. http://www.dlr.de/ts/en/desktopdefault.aspx/tabid-9883/16931_read-41000/
12. Veins. http://veins.car2x.org/

The 7th International Workshop on Frontiers in Intelligent Networking and Collaborative Systems (FINCoS-2019)

Cloud CRM System for Mobile Virtual Network Operators

Maria Jedrzejewska[1], Adrian Zjawiński[1], Vincent Karovič[2(✉)],
and Iryna Ivanochko[3]

[1] Lodz University of Technology, Lodz, Poland
[2] Faculty of Management, Comenius University, Bratislava, Slovakia
vincent.karovic2@fm.uniba.sk
[3] School of Business, Economics and Statistics, University of Vienna,
Vienna, Austria
iryna.ivanochko@univie.ac.at

Abstract. This work concerns of some aspects of the customer relationship management system for mobile virtual network operators. The aim of this work is to present a model of a cloud based customer relationship management system dedicated to mobile virtual network operators that will be integrating and using customer information obtained from mobile network operators. The system is developed into cloud environment, which allows for easy flow of data between mobile virtual network operators and a mobile network operators.

1 Introduction

Customer Relationship Management (CRM) is a broad notion with multiple definitions that evolved throughout years. One of the proposed definitions states that CRM is "a technology and system that sustains sales, marketing and customer service activities. It is designed to capture and interpret customer data, both structured and unstructured, and to sustain the management of the business side of customer related operations. CRM technology automates processes and workflows and helps organize and interpret data to support a company in engaging its customers more effectively" [1]. CRM systems are usually software applications providing tools to effectively collect, organize and present information about company's customers [2]. It is important to model CRM system according to the structure of the company and characteristics of its customers in order to maximize benefits of using such solution.

Mobile Virtual Network Operators (MVNO) share resources with corresponding Mobile Network Operators (MNO), which makes them naturally inclined to use cloud-based information systems, such as cloud CRM. Building an effective solution is of utmost importance in telecom industry, because of intense competition and resulting continuous decrease in loyalty of customers [3]. The aim of this report is to present a model of a cloud based CRM system dedicated to MVNO that will be integrating and using customer information obtained from MNO.

© Springer Nature Switzerland AG 2020
L. Barolli et al. (Eds.): INCoS 2019, AISC 1035, pp. 405–414, 2020.
https://doi.org/10.1007/978-3-030-29035-1_39

2   Literature Review

2.1   Mobile Virtual Network Operators

MNO is the owner of the physical network and radio access infrastructure. It usually realizes all of the telecommunication services by itself, however it can share parts of it with MVNO. Depending on the amount of share, the MVNO can take care of these telco services and he may, or may not manage the routing. Such cooperation is mutually beneficial, because MNO can focus on improving the physical network and increasing its availability, while MVNO attracts new clients and manages the current ones, possibly offering broader range of services than it is possible to the MNO alone [4–6]. The main problem is how to organize collaboration, as both the constant flow of data and high security must be maintained. A cloud solution achieving both goals could be proposed, however it must be carefully planned to meet a high availability requirement characteristic to the telecommunication industry [22, 23]. Common data and communication protocols could be stored in the cloud [7, 8, 20]. What is more, MNO could share such cloud to encourage new MVNO into collaboration, potentially increasing profits and network usage.

2.2   Customer Relationship Management in Telecom Industry

The main role of the CRM systems in telecom industry and other industries is to build strong relationship with customers by knowing them better and therefore recognizing their true needs. Telecom companies should develop their truth worthiness by answering to customer expectations and offering tailored services and improving their experiences to prevent customers leaving because of a better offer of a competitor [9, 10]. It is important to mention that gaining customers is much more expensive than retaining existing ones [11]. This is why understanding existing customers and responding to their needs is crucial for companies in such competitive market.

Important CRM Features
According to SelectHub interview [12], in which 529 companies of varied size were interviewed, the most requested features in good CRM systems. It can be observed that one of the most important property of a good system is integration with other tools and environments [18, 19]. Secondly, the marketing operations should be automated as much as it is possible. The third most important feature is the possibility of contact management. It can be noticed that clients are looking for rather complex, integrated and multi-purpose systems that would work well with their current infrastructure and facilitate marketing and general organization within the company.

CRM Features for Telecom Industry
In order to design a solution that will be specifically tailored to telecom companies, it is necessary to consider the character of their operation. Telecom companies gather a lot of information about their customers that could be used for targeted marketing. This information refers to both individual users (personal data, location, etc.) and

interactions between them (visible in their call history). Availability of such data allows to build a CRM solution that concentrates more on relationships between individuals and in particular, show how such relationships can impact our products. Such CRM systems are sometimes referred to as "Social CRM" [13, 14]. It is important to recognize importance of interactions in telecom industry, as people can easily influence other people in both positive and negative way. For example, a family or a company are most likely using services of only one mobile phone operator. Moreover, within such entities as families or companies, it is possible to recognize the most influential users. For example, in a company this would be a company's CEO or someone responsible for directing phone operation services within the firm. Such users can be called alpha users [15]. On a more analytical level, on a graph constructed from users (nodes) and phone calls (edges), alpha users are usually points that are "connection hubs" or concentration points. In order to develop cost-effective marketing campaign, alpha users are usually targeted first, as they can share and influence other users to use a new product or service [16].

One of the possible methods of user segmentation are self-organizing maps (SOM) [15]. Such maps concentrate on finding and isolating people of similar behavior (for example with similar calling and receiving calls ratio) instead of analyzing sociological backgrounds. The results obtained by such maps can appear arbitrary in the beginning, however can be successfully used to detect alpha users and other characteristic groups of users. Mobile operators can largely benefit from such automated methods of group segmentation, as they deal with massive amount of data that would be very difficult to analyze without automatic algorithms.

Competitive nature of telecom market makes it easy to lose customers. Because telecom users influence each other, losing clients can be potentially dangerous. Therefore, another crucial functionality for the telecom CRM is the analysis of customer lifecycle and group lifecycle to be able to successfully retain and improve loyalty. When considering communities of users instead of individuals, it is possible to acquire better understanding of their lifecycle by observing the evolution process of given group [14]. As the network grows, new connections between nodes are created. The opposite of this process can happen - the contraction of social network, where nodes are losing connections and some nodes are left not connected (those customers that no longer use offered services).

A useful CRM system for telecom industry should gather and present user data that can be used to observe differences between customers and groups of customers. This will facilitate understanding users and in the result build strong and loyal customer base [17, 21].

3 The Model of the CRM for MVNO

Proposed CRM system is a hybrid cloud CRM system that can be easily integrated with the data provided by MVNO's corresponding MNO.

3.1   Deployment

The architecture of the proposed CRM system presented in Fig. 1 is based on the one described by Yrjo and Rushil [7] and uses a hybrid cloud solution. There is a private MNO cloud connected with the public telecommunication cloud, which consists of publicly shared CRM controllers hosted by MNO and all MVNOs. The CRM controller is responsible for backend computations and is used by CRM web server to perform all computationally heavy operations. The network usage data registered by MNO are shared with each controller in the form of call data records and IP data records. The public cloud enables the use of a common mediation mechanism to standardize the data exchange process. The primary benefit of such standardization is the improvement in the scalability and manageability.

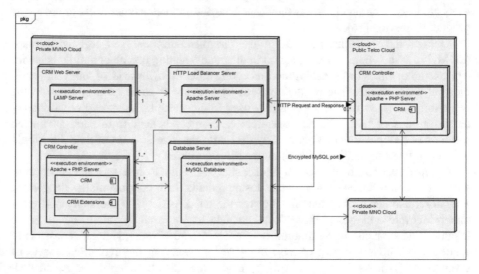

Fig. 1. Deployment diagram. CRM system structure in the cloud [7].

Each MVNO has its own private cloud with a database, a private copy of the CRM controller, and a CRM web server hosting the web application, which provides the UI layer for management of the whole CRM system. What is more, the private controller can be extended by additional modules and functions in the form of extensions. An extension is a custom component, defining a set of its operations and the visual presentation of those and sharing them with the CRM controller. This allows each MVNO to implement its own, private extensions usable inside its CRM system. As the web server is managed by the MVNO, modifications in the user interface to serve such extensions should be trivial. The data is processed by the extension, so that a custom presentation model could be prepared either directly in the CRM controller, or defined along the extension and passed to the controller indirectly. The HTTP load balancer distributes the computation of operations issued from the web application to multiple CRM controllers in order to optimize the process.

The system should be additionally highly secure. The data are stored only in the private cloud of MVNOs and the web application is hosted privately, eliminating many trust concerns. The risk of information theft is limited only to the moment of a data transfer between a web application and a CRM controller, thus the connection itself must be secured. For a more confidential data some subset of operations could be limited to be realized only on the private controller, losing the benefits of computational distribution, however keeping the whole process private.

What is more, both the web application and CRM controller could be stored in the form of an application image (e.g. a Docker image with all the required functionalities and programs already installed and preconfigured) offered by the MNO. This could simplify the initial installation process and thus should increase the interest of potential MVNOs in such solution.

3.2   Components

CRM system modules and integration with custom extensions are presented in the component diagram of the proposed CRM system. Figure 2 shows the system modules and integration with custom extensions. The whole CRM system is enclosed inside a single CRM component. All functionalities of the system are divided into a set of subcomponents to ensure proper modularization and thus to simplify the process of development and testing. The first of the proposed modules is a customer module, responsible for accessing, processing and updating all the personal data about customers. Along with the transaction module it creates the basis of the data layer of the whole system. A statistics module uses those data to perform statistical analysis and prepare metrics describing all customers. Its separation from the data layer helps isolating the data and the business logic. Segmentation module is used primarily to help understand the collected data and allows the customers groups analysis, which is especially important in the telecom industry. Feedback module is used to track customers satisfaction (by creating and performing surveys) and providing real world insights into the data. All of this information can be used inside marketing module to run targeted marketing campaigns and special offers, as well as advertise the new services and use recognized alpha users to maximize the customers reach. Security module authorizes all operations, controlling the access to the CRM system and the database. To simplify the database access and optimize it, a data persistence mechanism is used: the database can be cached locally and data can be mapped to objects, allowing direct access in the code instead of forcing to write SQL queries.

To customize the system for individual needs, each MVNO could use CRM extensions component. It includes the extensions controller, which is responsible for connection and mediation with the main CRM controller. Multiple extensions could be developed, specifying the backend operations. Separation of extensions from the main CRM allows the MNO to share to its MVNOs a single, common CRM application image, simplifying the update process. MVNO can keep its private extensions unchanged, having automatically updated CRM system.

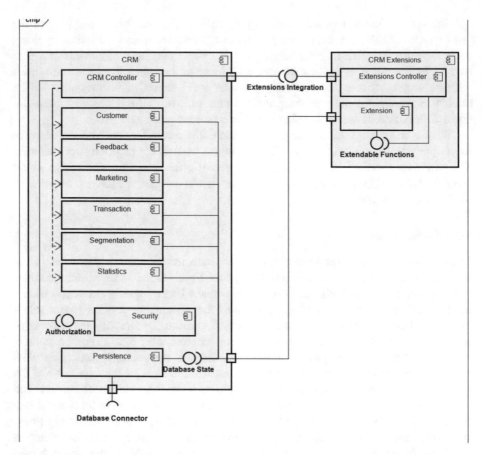

Fig. 2. Component diagram. CRM system modules and integration with custom extensions.

3.3   Functionalities

One of the most important functionalities of the proposed CRM system is user segmentation that includes analysis of social interactions between operator's users and dividing them into groups according to those interactions. Apart from group segmentation by such social networks, traditional segmentation by personal details such as age, location and social status should also be included. Figure 3 presents the process diagram showing how social networks are extracted from call data delivered by MNO.

The criteria for dividing social networks should be developed by marketing specialists. Such criteria are used to implement an appropriate clustering algorithm that can be targeted to the needs of the operator. MVNO can utilize call detail records shared by MVO to form call graphs that present interactions between operator's clients. Then, such graph can be analyzed using previously described criteria, which results in segmentation of customers into groups.

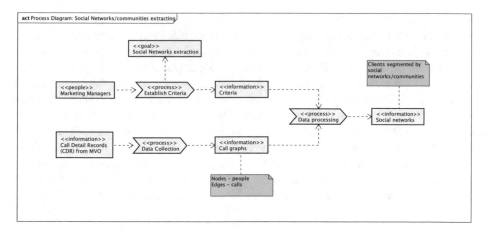

Fig. 3. Process diagram. Social networks extraction.

The exemplary view of the user interface shown in Fig. 4 presents community clusters marked on call graphs. The dots in the diagram represent users. Apart from the view of social networks (communities), employee would have standard graphs presenting segmentation by, for example, gender and age as visible in Fig. 4.

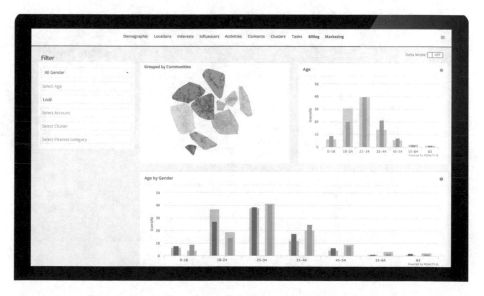

Fig. 4. Proposition of user interface - customer segmentation.

CRM could provide a range of other functionalities other than segmentation such as those presented in the use case diagram in Fig. 5. This report has no intention on presenting whole range of functionalities included in a general CRM systems, as this range is usually very wide. A user of the CRM system is able to examine each individual client. The system will be able to present basic information about him/her, including demographic data, billing information, used services, results of any surveys in which he/she took part. Each client has its own importance factor which indicates how important he/she is for the company. An employee can display segmentation graphs and diagrams in a variety of ways, one of which is segmentation by social network/community and view alpha users (and other types of users if they were defined) in each segment. It is also possible to view evolution graphs of social networks and predictions on how they will evolve in the future. Furthermore, the system allows for automated marketing; for example, by sending and automatically analyzing surveys or some promotional materials and offers that were chosen basing on the information on given customer - his/her history, lifecycle status, used products etc.

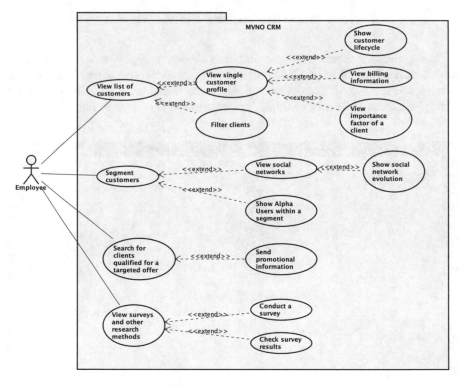

Fig. 5. Use case diagram.

4 Conclusion

This report presented some aspects of the proposed CRM system for MVNO. The system would be integrated into cloud environment, which allows for easy flow of data between MVNO and a MNO. The call detail records that are obtained from MNO can be utilized in CRM system for various research, analysis and marketing purposes, such as segmentation of users. In the case of telecom industry, it is important to look at customers not only as individuals, but also as members of social networks that evolve over time. Analysis of such social networks, not only individual users, can help in better understanding of the dynamics and behavior of customers. In the result, it can allow for more tailored and effective approach when dealing with clients. Additionally, a sample structure of components for the CRM system was proposed, enabling to create its application image and deploy it directly to the MVNOs, lowering an entry threshold. This should significantly simplify the eventual future updates of the system. What is more, a customization of the system is possible in the form of private CRM extensions, preserving the personalization capabilities.

References

1. Greenberg, P.: The clarity of definition: CRM, CE and CX. Should we care? [WWW Document]. ZDNet (2015). https://www.zdnet.com/article/the-clarity-of-definition-crm-ce-and-cx-should-we-care/. Accessed 18 Apr 19
2. Winer, R.S.: A framework for customer relationship management. Calif. Manag. Rev. **43**, 89–105 (2001). https://doi.org/10.2307/41166102
3. Raina, D.I., Pazir, D., Pazir, D.: Customer relationship management practices in telecom sector: comparative study of public and private companies introduction, pp. 2231–2528 (2017)
4. Banerjee, A., Dippon, C.M.: Voluntary relationships among mobile network operators and mobile virtual network operators: an economic explanation. Inf. Econ. Policy **21**, 72–84 (2009). https://doi.org/10.1016/j.infoecopol.2008.10.003
5. Cricelli, L., Grimaldi, M., Ghiron, N.L.: The competition among mobile network operators in the telecommunication supply chain. Int. J. Prod. Econ. **131**, 22–29 (2011). https://doi.org/10.1016/j.ijpe.2010.02.003
6. Kaczor, S., Kryvinska, N.: It is all about services - fundamentals, drivers, and business models. J. Serv. Sci. Res. **5**(2), 125–154 (2013). The Society of Service Science
7. Yrjo, R., Rushil, D.: Cloud computing in mobile networks - Case MVNO. In: 2011 15th International Conference on Intelligence in Next Generation Networks. Presented at the 2011 15th International Conference on Intelligence in Next Generation Networks (ICIN): "From Bits to Data, from Pipes to Clouds", pp. 253–258. IEEE, Berlin (2011). https://doi.org/10.1109/ICIN.2011.6081085
8. Gregus, M., Kryvinska, N.: Service orientation of enterprises - aspects, dimensions, technologies. Comenius University in Bratislava, ISBN 9788022339780 (2015)
9. Rajini, G., Sangamaheswary, D.V.: An emphasize of customer relationship management analytics in telecom industry. Indian J. Sci. Technol. **9**, 1–5 (2016)
10. Kryvinska, N., Gregus, M.: SOA and its business value in requirements, features, practices and methodologies. Comenius University in Bratislava, ISBN 9788022337649 (2014)
11. Prakash, S.: AI 101: understanding customer churn management. Data Sci. (2018). https://towardsdatascience.com/ai-101-understanding-customer-churn-management-514416c17643. Accessed 19 Apr 2019
12. Adair, B.: CRM Features List in 2019—CRM Functionality & Capabilities Checklist (2019). https://selecthub.com/customer-relationship-management/crm-features-functionality-list/. Accessed 17 Apr 2019

13. Greenberg, P.: CRM at the Speed of Light: Social CRM Strategies, Tools, and Techniques for Engaging Your Customers. McGraw-Hill, New York (2010)
14. Wu, B., Ye, Q., Wang, B., Yang, S.: Group CRM: a New Telecom CRM Framework from Social Network Perspective (2009)
15. Ahonen, T.T., Kasper, T., Melkko, S.: 3G Marketing: Communities and Strategic Partnerships. Wiley, Hoboken (2005)
16. Kryvinska, N.: Building consistent formal specification for the service enterprise agility foundation. J. Serv. Sci. Res. **4**(2), 235–269 (2012). The Society of Service Science
17. Molnár, E., Molnár, R., Kryvinska, N., Greguš, M.: Web intelligence in practice. J. Serv. Sci. Res. **6**(1), 149–172 (2014). The Society of Service Science
18. Poniszewska-Marańda, A.: Modeling and design of role engineering in development of access control for dynamic information systems. Bull. Polish Acad. Sci. Tech. Sci. **61**(3), 569–580 (2013)
19. Majchrzycka, A., Poniszewska-Marańda, A.: Secure development model for mobile applications. Bull. Polish Acad. Sci. Tech. Sci. **64**(3), 495–503 (2016)
20. Poniszewska-Marańda, A., Rutkowska, R.: Access control approach in public software as a service cloud. In: Zamojski, W., et al. (eds) Theory and Engineering of Complex Systems and Dependability. Advances in Intelligent and Soft Computing, vol. 365, pp. 381–390. Springer, Heidelberg (2015). ISSN 2194-5357, ISBN 978-3-319-19215-4
21. Tkachenko, R., Izonin, I.: Model and principles for the implementation of neural-like structures based on geometric data transformations. In: Hu, Z., Petoukhov, S., Dychka, I., He, M. (eds) Advances in Computer Science for Engineering and Education. ICCSEEA 2018. Advances in Intelligent Systems and Computing, vol. 754, pp. 578–587. Springer, Cham (2019)
22. Poniszewska-Marańda, A.: Selected aspects of security mechanisms for cloud computing – current solutions and development perspectives. J. Theor. Appl. Comput. Sci. **8**(1), 35–49 (2014)
23. Smoczyńska, A., Pawlak, M., Poniszewska-Maranda, A.: Hybrid agile method for management of software creation. In: Kosiuczenko, P., Zieliński, Z. (eds) Engineering Software Systems: Research and Praxis. Advances in Intelligent Systems and Computing, vol. 830, pp. 101–118. Springer, Heidelberg (2019)

A Basic Framework of Blockchain-Based Decentralized Verifiable Outsourcing

Han Wang[1], Xu An Wang[1(✉)], Wei Wang[2], and Shuai Xiao[1]

[1] Key Laboratory for Network and Information Security of the PAP,
Engineering University of the PAP, Xi'an 710086, Shaanxi, China
aca_wang@163.com, wangxazjd@163.com
[2] Engineering University of the PAP, Xi'an 710086, Shaanxi, China

Abstract. With the development and application of blockchain technology, more and more fields are beginning to use blockchains for industrial innovation, In recent years. At the same time, the data integrity verification techniques based on blockchain have also attracted wide attention of researchers. However, the state-of-the-art schemes are only using the structure of the blockchain on centralized server which cannot leverage the characteristics of the blockchain and resist malicious modification attacks. In addition, the centralized blockchain structure introduces additional storage and computing costs. In this paper, we propose a decentralized verifiable outsourcing framework based on blockchain networks, in which no administrators are required to be in charge of auditing and recording data changes.

1 Introduction

Cloud storage has attracted the attention of many researchers, in recent years. Since Ateniese proposed Provable Data Possession (PDP) [1] in which TPA is allowed to check the integrity of data at untrusted stores, it solves the problem of low user audit efficiency and huge data transmission. So far, more and more organizations and individuals have chosen to store data in the cloud [2]. There are many cloud storage systems such as Google Drive [3], OneDrive [4] and Dropbox [5]. The emergence of these products has significantly affected the way of people work.

Cloud collaboration is an application scenario of cloud storage. Users in the same group can share and modify files with each other. However, because the file exists in the cloud server, the user loses the physical control of the data, the integrity of the outsourced data cannot be guaranteed. Ateniese proposed Provable Data Possession (PDP) that solves the problem. The client divides its data into blocks and generates a label for each block. Once the integrity of the outsourced data is checked, the client sends a challenge message to review some randomly selected blocks. In response to this challenge, the CSP will generate proof of the required block and be verified by the customer. Based on this, many data integrity auditing schemes shared by support group users were proposed.

© Springer Nature Switzerland AG 2020
L. Barolli et al. (Eds.): INCoS 2019, AISC 1035, pp. 415–421, 2020.
https://doi.org/10.1007/978-3-030-29035-1_40

However, these schemes all have the flaw that the group manager has the potential of modifying the user record at will.

With the introduction and application of blockchain technology in recent years [6], more and more researchers are beginning to explore the use of blockchains. Its non-stringable modification and zone-centered features are very suitable for cloud data sharing. And researchers have also made corresponding attempts in the field of cloud computing. Zheng et al. [8] proposed a privacy protection scheme based on paillier encryption, which stores the hash value of the data in the block after the data is uploaded. Although the scheme achieves decentralized storage, encrypted data is not conducive to sharing. Huang et al. [7] proposed a blockchain-based secure data sharing scheme. After the user uploads the file, the cloud server packages and generates the block. Huang et al. [9] proposed a privacy hunt auditing solution without group administrators. When users need to operate on data, the server records the user's operations and stores them in the blockchain. However, there is a problem with the above schemes. The blockchains used by these schemes are completely controlled by group management or cloud servers, and they are often semi-trustworthy. Therefore, these solutions cannot meet the security needs.

In this paper, we propose a blockchain-based decentralized verifiable outsourcing framework. The user's operation record is completely generated by the blockchain network, and non-administrator needs to be responsible for auditing and recording data changes. The performance comparsion is shown in Table 1.

Table 1. Performance comparison

Schemes	Data dynamics	Group manager	Blockchain store
Zheng et al. [8]	No		Decentralization
Huang et al. [7]	Yes	Yes	Centralization
Huang et al. [9]	Yes	No	Centralization
Our proposed scheme	Yes	No	Decentralization

2   Preliminaries and Background

2.1   Bilinear Group

Let $\mathbb{G}_1$ *and* $\mathbb{G}_2$ be two multiplicative cyclic groups of large prime order p, and g be a generator of $\mathbb{G}_1$. A bilinear map is a map $e : \mathbb{G}_1 \times \mathbb{G}_1 \rightarrow \mathbb{G}_2$ with the following properties:

1. *Bilinearity*: for $\forall h_1, \forall h_2 \in \mathbb{G}_1$ and $\forall u, v \in \mathbb{Z}_p$, $e(h_1^u, h_2^v) = e(h_1, h_2)^{u,v}$.
2. *Computability*: there exists an efficiently computable algorithm for computing map e.
3. *Non-degeneracy*: $e(g, g) \neq 1$.

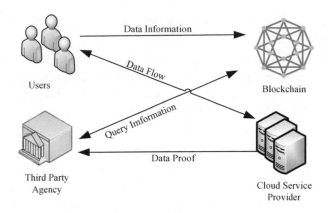

Fig. 1. System model.

2.2 Blockchain

Blockchain is a data structure that links many blocks by a hash pointer [6]. In the blockchain, each block stores a pointer consisting of the hash value of the previous block, so that each block ensures the existence and correctness of the previous block. Bitcoin [12] is the world's first project to apply blockchain to decentralization, and more attention has been paid to this blockchain. With the deepening of research, the application of blockchain technology is no longer limited to cryptocurrencies, and is more applied to data security, distributed storage and other fields, such as InterPlanetary File System (IPFS) [10] and health records sharing [11].

Smart Contract. The smart contract was first proposed by Nick Szabo in 1995. A special protocol used in the development of contracts within the blockchain, which contains a code function that can interact with other contracts, make decisions, store data, and transfer digital cash. The Smart Contracts provide the conditions for verification and execution of the contract. Smart contracts allow trusted transactions without third parties, which are traceable and irreversible. The smart contract of this program consists of three parts: user registration contract, user file relationship contract, document modification contract, document audit contract.

Blockchain Network. A blockchain network is a network of peer-to-peer (P2P) protocols in which each entity in the network has the same right. All nodes in the network participate in checksum broadcast transactions and block information, discover and maintain connections with other nodes.

2.3 System Model

User. User can be defined as $U = \{u_1, u_2, \ldots, u_n\}$. As the owner of the data, user can upload and modify the data. Users within a group can initiate data integrity challenges for the shared files of the group. When a user joins or launches a group, registration and logout operations are required (Fig. 1).

TPA. As Third-party organization, TPA responses the data integrity audit request of the corresponding user. The information of the relevant data block is obtained through the blockchain. TPA is responsible for users and fulfill the task of data check for the files stored in cloud server.

CSP. Manages and coordinates a lot of cloud servers to provide scalable and on-demand data storage services to group users.

Blockchain. Responsible for recording records of user action files and file data tags. In our scenario, each user's operation of the data is treated as a transaction process, which is packaged into a blockchain through a smart contract. The security of user data can be improved by utilizing the non-tamperable and distributed features of the blockchain.

3   Our Proposed Scheme

In this section, we will give the detail of Blockchain-based Decentralized Verifiable Outsourcing Framework. Our proposed scheme is mainly composed of three parts: registration, data outsourcing, and shared data integrity audit. Figure 2 shows the process of our scheme.

3.1   Public Auditing for Shared Data

Assume $\mathbb{Z}_p$ is a finite field in a large prime order p and $\mathbb{G}_1$ and $\mathbb{G}_2$ are two multiplicative cyclic groups. $e : \mathbb{G}_1 \times \mathbb{G}_1 \rightarrow \mathbb{G}_T$ is a bilinear map. The scheme is secure in the plain public key model, and assumes hash functions $H : \{0,1\}^* \rightarrow \mathbb{G}_1$, g is a generators of $\mathbb{G}_1$.

Registration. The user must be registered as a legitimate user, then the user obtains a triple which includes the unique address, private key, and public key. It can be expressed as $address, sk, pk$.

Data Outsourcing. In order to outsource data, a file F is divided into n data blcoks $m_1, m_2, \cdots, m_n$. Then the user calculates a hash value $h(m_i)$ and signature as follows for each file block for each data blocks.

$$\sigma_i = (h(m_i) \cdot g^{m_i})^x \tag{1}$$

The user then submits the changes as a transaction to the blockchain network via the smart contract, waiting for confirmation by the entire network node. After the transaction is determined, the user uploads the signature and data to the CSP.

Shared Data Integrity Auditing. The user can post an audit request through the smart contract. Then the program automatically generates a random audit sequence $\{(l, \lambda_l)\}_{l \in L}$, where L can be divided into d set subset of $L_1, L_2, \cdots, L_d$

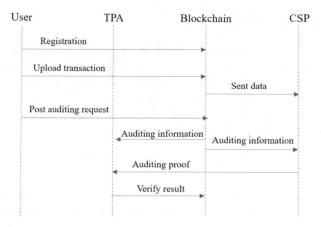

Fig. 2. Process of our proposed scheme.

by different signers and λ_l is a random number. The audit sequence is then sent to the TPA and CSP respectively.

When the audit sequence is received, the CSP calculates the evidence P as following:

1. For each subset L_i calculate $\mu_i = \sum_{l \in L_i} \lambda_i m_i$ and $\beta_i = \prod_{l \in L_i} \sigma_i^{\lambda_l}$.
2. The CSP sents an auditing proof $P = \{\mu_i, \beta_i\}_{l \in L}$

When the proof P is received, the TPA verifies the following equation using the user's public key, data hash value.

$$e(\prod_{i=1}^{d} \beta_i, g) = \prod_{i=1}^{d} e(\prod_{l \in L_i} \sigma_l^{\lambda_l}, g)$$

$$= \prod_{i=1}^{d} e(\prod_{l \in L_i} (h(m_l) \cdot g^{m_l})^{x_l \lambda_l}, g)$$

$$= \prod_{i=1}^{d} e(\prod_{l \in L_i} (h(m_l)^{\lambda_l} \cdot \prod_{l \in L_i} g^{m_l \lambda_l}, g^{x_l})$$

$$= \prod_{i=1}^{d} e(\prod_{l \in L_i} H(m_l)^{\lambda_l} \cdot g^{\mu_i}, pk_i) \tag{2}$$

3.2 Group Management

If a user want to exit from the group, it must log out. A legitimate user can submit a launch request through a smart contract, then the smart contract cancels the user's upload modification permission and revokes the relationship with the file by setting the user status to exit. The user's previous modification record has been solidified into the blockchain whice cannot be tampered with, so that the cloud data need not be audited immediately.

4   Security Analysis

In response to the audit request of the smart contract, the CSP must provide a proof $P = \{\mu_i, \beta_i\}_{l \in L}$. According to the paper [13], BLS-HVAs cannot be forged without knowing the signer's private key. Therefore, we only need to prove that the data of the challenged block is unforgeable.

We designed a game in which CSP provides a forged proof $P\prime = \{\mu_i\prime, \beta_i\}_{l \in L}$ in response to the TPA challenge.

$$\mu \neq \mu\prime \Rightarrow \sum_{l \in L} \lambda_l m_l \neq \sum_{l \in L} \lambda_l m_l\prime \tag{3}$$

If the CSP passes the verification with the forged proof, the CSP wins; otherwise, the CSP fails. Assuming the CSP wins, then we can get equation as following:

$$e(\prod_{i=1}^{d} \beta_i, g) = \prod_{i=1}^{d} e(\prod_{l \in L_i} H(m_l)^{\lambda_l} \cdot g^{\mu_i'}, pk_i) \tag{4}$$

But the correct auditing proof is $P = \{\mu_i, \beta_i\}_{l \in L}$, we have

$$e(\prod_{i=1}^{d} \beta_i, g) = \prod_{i=1}^{d} e(\prod_{l \in L_i} H(m_l)^{\lambda_l} \cdot g^{\mu_i}, pk_i) \tag{5}$$

According to the properties of bilinear mapping, we can get

$$\sum_{l \in L} \lambda_l m_l \neq \sum_{l \in L} \lambda_l m_l\prime \tag{6}$$

Equation 6 contradicts the assumptions, so the CSP cannot pass the data integrity auditing.

5   Conclusion

In this paper, we propose a decentralized public auditing framework based on blockchain. We point out the shortcomings of the traditional data sharing auditing scheme with group management and the existing auditing scheme using the blockchain. We use the smart contract and blockchain automation, the non-tamperable features, to build the whole solution. In the paper, we give the system model and running process of our solution. The overall feasibility of the scheme is proved theoretically. In the future, we will further give the specific implementation process and efficiency analysis of the system.

Acknowledgements. This work was supported by the National Cryptography Development Fund of China (grant no. MMJJ20170112), Natural Science Basic Research Plan in Shaanxi Province of China (grant no. 2018JM6028), National Natural Science Foundation of China (grant no. 61772550, U1636114, and 61572521), and National Key Research and Development Program of China (grant no. 2017YFB0802000). This work is also supported by Engineering University of PAP's Funding for Scientific Research Innovation Team (grant no. KYTD201805).

References

1. Ateniese, G., Burns, R., Curtmola, R., et al.: Provable data possession at untrusted stores. In: Proceedings of the 14th ACM Conference on Computer and Communications Security, CCS 2007, pp. 598–609. ACM, New York (2007). https://doi.org/10.1145/1315245.1315318
2. Cai, H., Xu, B., Jiang, L., Vasilakos, A.V.: IoT-based big data storage systems in cloud computing: perspectives and challenges. IEEE Internet Things J. 4(1), 75–87 (2017). https://doi.org/10.1109/JIOT.2016.2619369
3. Google Drive. http://www.google.cn/intl/en/drive/
4. Onedrive. https://onedrive.live.com
5. Dropbox. https://dropbox.com
6. Narayanan, A., Bonneau, J., Felten, E., et al.: Bitcoin and Cryptocurrency Technologies. Princeton University Press, Princeton (2016)
7. Huang, L., Zhang, G., Yu, S., Fu, A., Yearwood, J.: SeShare: secure cloud data sharing based on blockchain and public auditing. Concurr. Comput.: Pract. Exp. e4359 (2018)
8. Zheng, B., Zhu, L., Shen, M., et al.: Scalable and privacy-preserving data sharing based on blockchain. J. Comput. Sci. Technol. 33(3), 557–567 (2018)
9. Huang, L., Zhang, G., Fu, A., et al.: Privacy-preserving public auditing for non-manager group. In: 2017 IEEE International Conference on Communications (ICC), pp. 1–6. IEEE, Paris (2017). https://doi.org/10.1109/ICC.2017.7997370
10. Li, M., Weng, J., Yang, A., et al.: CrowdBC: a blockchain-based decentralized framework for crowdsourcing. IEEE Trans. Parallel Distrib. Syst. 30, 1251–1266 (2019). https://doi.org/10.1109/TPDS.2018.2881735
11. Liang, X., Zhao, J., Shetty, S., et al.: Integrating blockchain for data sharing and collaboration in mobile healthcare applications. In: 2017 IEEE 28th Annual International Symposium on Personal, Indoor, and Mobile Radio Communications (PIMRC), pp. 1–5 (2017). https://doi.org/10.1109/PIMRC.2017.8292361
12. Nakamoto, S.: Bitcoin: a peer-to-peer electronic cash system (2009). https://bitcoin.org/bitcoin
13. Qian, W., Cong, W., Kui, R., et al.: Enabling public auditability and data dynamics for storage security in cloud computing. IEEE Trans. Parallel Distrib. Syst. 22(5), 847–859 (2011)

An Efficient Mobile Cloud Service Model
or Tactical Edge

Bo Du[✉], Nanliang Shan, and Sha Zhou

Engineering University of PAP, Xian, Shaanxi, China
726682174@qq.com

Abstract. With the explosive growth mobile cloud service, emergency military tasks on the tactical edge gradually show the characteristics of regional dispersion, dynamic change, multi-point concurrency and marginalization. At the same time, military operations show the trend of migrating information-oriented, decision-making and control to the tactical edge. In order to solve the problems of military action under the situation of tactical edge network, limited equipment access, lack of edge of tactical maneuver and edge information fusion processing capacity, an efficient mobile cloud service model for tactical edge is put forward. The model can provide flexible tactical edge information exchange and information processing ability. Also, it is suitable to self-adaptive to the frontline battlefield environment on the aspects of collaborative perception, decision making, time delay and energy consumption demand.

Keywords: Mobile cloud service · Tactical edge · Edge computing

1 Introduction

In recent years, there have been frequent terrorist incidents at home and abroad, and the fight against terrorism is in an urgent and grave situation. Among them, the "March 15 terrorist attack" in New Zealand and the "April 21 bombing" in Sri Lanka in 2019 shocked the whole world. In this context, military operations show a new trend of information oriented, decision-making and control moving to the tactical edge [1], that is, the tactical edge has become an important direction of military operations. Grace Lewis, lead researcher for the institute of software engineering (SEI) at Carnegie Mellon university, points out that tactical edge environments are characterized by battlefield dynamics, limited computing resources, high task pressure and high demand for network connectivity. Moreover, in the tactical marginal environment, front-line commanders, combatants and support personnel widely use handheld devices to enhance the situational awareness of the battlefield against terrorism, and complete the mission planning, combat decision-making and other key links. Tactical edge cloud service architecture represented by Tactical small cloud (Tactical Cloudlet [2]) can only realize the mobility of Tactical edge computing unloading through the existing battlefield network environment, without considering the Tactical edge's characteristics of sudden task area dispersion, dynamic changeful, multi-point concurrency, marginalization, etc. [3]. In addition, with the advent of emerging technological revolutions such as mobile cloud computing and edge computing [4], military tactics not only provide

© Springer Nature Switzerland AG 2020
L. Barolli et al. (Eds.): INCoS 2019, AISC 1035, pp. 422–430, 2020.
https://doi.org/10.1007/978-3-030-29035-1_41

key guidance for military operations, but also put forward higher requirements for the completion of military tasks:

- Military operations are usually in a highly dynamic tactical environment, and changes in tactical tasks, types of collaboration, and operational rhythms will have a great impact on established operational plans and command and control relationships on the tactical edge, and even affect the success or failure of military terrorist missions. Therefore, it is urgent to take into account the needs of the military operation mode on the tactical edge of maneuverability to adapt to the dynamic changes of missions and situations in the course of operations.
- Edge due to military tactics for the growing need for data transmission rate and quality of service, compute-intensive applications (such as virtual reality, augmented reality, face recognition, etc.) are widely used, traditional tactical edge mobile equipment limited resources, storage and computing power is insufficient, therefore, how to build in the challenging environment of the tactical edge mobile cloud service system, to provide military tactical edge organization coordinated, flexible, distribution of resources on demand elasticity of mobile cloud services will be important research object under the new situation of military action.
- As traditional command and control center mode, military action to realize data processing to the tactical edge down, facing the tactical edge resources integration scheduling problems such as inadequate, command communication time delay is too large, need to improve the mobile devices in the condition of limited access to the edge of the tactical information support, implementation tactics of the edge information quick and autonomic computing.

To sum up, based on the existing research results [5] and the specific requirements of the military tactics, this paper edge computing technology and mobile cloud computing architecture, proposes a service model oriented to mobile cloud on the edge of the military tactics and on the edge of the military tactics to solve military operations environment network and equipment limited access [6], the data processing mode single, lack of edge information fusion processing power, command communication time delay is too large, so as to provide flexible for military tactical edge information exchange and processing ability, to adapt to the military battlefield environment on the perception, coordination, decision making, time delay, energy consumption of special needs.

2 Mobile Cloud Service Model Construction Oriented to the Edge of Military Tactics

Counter terrorism Tactical Edge Oriented Mobile Cloud Service Model (MCS) proposed in this paper consists of Mobile Cloud Service Nodes (MC), Counter terrorism Tactical Edge Nodes (TE) and infrastructure. Among them, mobile cloud service nodes are composed of mobile base stations with deployed computing and storage resources. Through the infrastructure, mobile cloud computing capabilities with high availability are provided to a large number of military tactical edge nodes nearby. Meanwhile, military tactical edge nodes extend mobile cloud services to the tactical edge through edge computing capabilities. In general, military tactical edge nodes are composed of the edge computing and storage resources of personal handheld or wearable smart devices, vehicle-mounted and airborne devices. They realize real-time edge computing by using their own computing resources and mobile cloud computing by collaborative scheduling and task unloading. Therefore, the MCS model is an important expansion and application of mobile cloud computing and edge computing on the edge of military tactics.

2.1   Overall Architecture of MCS Model

As shown in Fig. 1, in the overall architecture of the MCS model, the MCS model is deployed on the edge of military tactics, and the key parts include mobile cloud service nodes and tactical edge nodes. According to the distance between tactical edge nodes and mobile cloud service nodes, the military tactical edge can be divided into mobile cloud computing area and edge computing area.

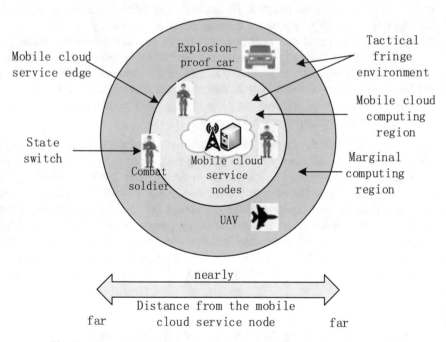

Fig. 1. Overall architecture diagram of the mobile cloud service model

The strategy of mobile cloud service nodes is: (1) resource virtualization. MC nodes efficiently integrate various computing and storage resources, hide the differences between resource attributes and operations, make mobile cloud services independent of specific hardware, and realize the abstract unified access to resources. Meanwhile, it ensures that TE nodes can use mobile cloud services on demand. (2) flexible allocation of resources. MC nodes fully integrate the physical resources of cloud services, adopt various scheduling, allocation and balancing methods, and flexibly configure, increase and release virtual resources according to the usage of MC nodes.

The strategy of tactical edge node is: (1) edge calculation. In the edge computing area, TE nodes are usually far away from MC nodes or have limited access. By virtue of their own computing resources and military battlefield situational awareness, they can achieve rapid acquisition and processing of tactical edge information. (2) task unloading. In the mobile cloud computing area, TE nodes, within the coverage of mobile cloud services, can gather the computing-intensive tasks being performed to MC nodes and realize task unloading by utilizing the mobile cloud computing capacity of MC nodes, so as to make up for the shortage of computing and storage resources of TE nodes. Therefore, the military operation mode combining edge computing and task unloading makes tactical edge nodes flexibly adapt to the dynamic changes of tasks and situations in the process of military operations.

From the perspective of MC nodes, the overall architecture of MCS model can be abstracted into the fully networked node cloud as shown in Fig. 2b, that is, MC nodes are realized based on the virtual network structure, with advantages of multi-node control, robustness and good destructiveness. Specifically, through task unloading, MC node provides corresponding virtual nodes for TE node and represents corresponding virtual resources, thus realizing flexible allocation of virtual resources. At the same time, each virtual node can share information and communicate with each other through mobile cloud service nodes, rapidly transforming the traditional hierarchical accusation mode as shown in Fig. 2a into a flexible and robust network structure. Each TE node continuously gathers information to the MC node and obtains the Shared information of other TE nodes. This continuous interaction mode between TE node and MC node is conducive to real-time acquisition of the overall situation of the battlefield and efficient decision-making.

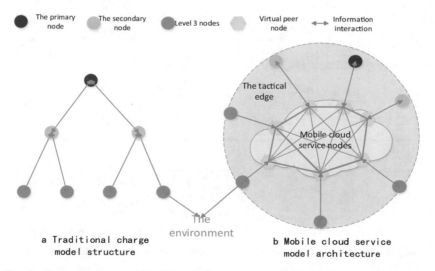

Fig. 2. Structure diagram of traditional charge model and mobile cloud service model

From the perspective of TE nodes, TE nodes have both server-side and client-side attributes. As a server, TE node can quickly acquire and process tactical edge information through its own computing resources and situational awareness, which can effectively overcome the resource access restriction problem in the harsh military tactical environment and realize the edge computing mode with low delay and independent control of information processing services. As the client, TE nodes are mobile cloud service requesters, through task uninstall will gather in MC tactical edge information collected by the node processing, effectively overcome TE node computing and storage capacity is insufficient, the problem of limited energy consumption [7], and receive task processing results, for flexible and dynamic flexibility of mobile cloud computing services.

2.2   Functional Architecture of MCS Model

The functional architecture of MCS model is shown in Fig. 3. Its main components include edge computing functional module and mobile cloud computing functional module.

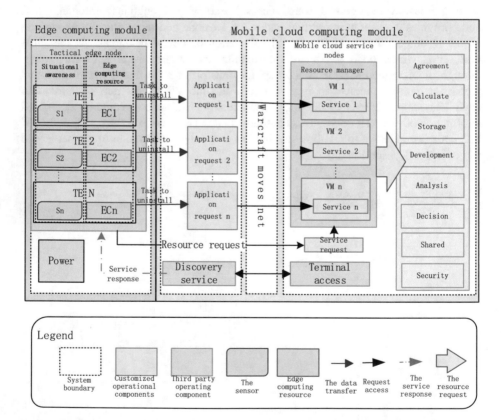

Fig. 3. Functional architecture diagram of the tactical mobile cloud service model

The edge computing function module is located at the tactical edge node, which can achieve the following functions:

- Perception. Tactical edge node sensors $\{S1, S2, \ldots, Sn\}$ can perceive the surrounding environment, quickly acquire battlefield situation information and convert and store it in real time.
- Borderline functionality. The information collected by the sensor can be processed in real time and respond to the user's request timely.

The mobile cloud computing function module involves tactical edge nodes and mobile cloud service nodes, which can achieve the following functions:

- Discovery service. Realized by the service discovery module of tactical edge node, the network address and port of mobile cloud service node can be queried, and the connection mode can be established as the interface for tactical edge node to receive service response.
- Terminal access function. It is realized by the terminal access module of the mobile cloud service node, broadcasting the network address and port of the mobile cloud service node, confirming the establishment of the connection between the mobile cloud service node and the tactical edge node, and returning the service response to the tactical edge node.
- Application functions. Provided by the application module of the tactical edge node, the task is unloaded through the application request function, and the corresponding applications and virtual resources in the mobile cloud service node are activated.
- Mobile communications. It is provided by the battlefield basic network environment to guarantee the stable and high-quality network services of tactical edge nodes and mobile cloud service nodes.
- Resource elastic distribution function. Provided by the resource manager of the mobile cloud service node, it involves the computation-intensive task components and virtual machine resources of the corresponding mobile application program, virtualizes the cloud service resources such as protocol, computing and storage, so as to realize the flexible distribution of cloud service resources on demand.
- Service request function. Provided by the service request module of the mobile cloud service node, it is applicable to the implementation of the function extension of tactical edge node. In particular, when tactical edge node does not have such application, it can also request the basic resources of the mobile cloud service node.
- Move cloud computing functions. Provided by computing and storage resources of mobile cloud service nodes, relying on powerful data processing and data storage capacity, it can quickly complete computationally intensive unloading tasks of tactical edge nodes and provide other functional services, such as programmable development functions.

3   MCS Model Operation Mechanism

3.1   Mobile Cloud Service Discovery

The process of finding and utilizing cloud services in mobile cloud services oriented to the edge of military tactics is crucial. Therefore, discoverability is the key feature that distinguishes MCS model from traditional model [8], and it is also the precondition for mobile cloud service nodes to be unloaded by nearby tactical edge nodes for calculation and temporary data storage. In the MCS model, the process of mobile cloud service discovery can be summarized as the "cloud edge finding – edge finding cloud" interactive discovery service as shown in Fig. 4.

"Cloud find edge" refers to the mobile cloud service node will own hardware and software resources in service, through radio mobile cloud service node of the network address and port find the edge nodes need to access the mobile cloud service tactics, at the same time, the tactical edge node resource request confirm and establish a secure connection, so as to realize the two-way communication of clouds to the edge. Therefore, the essence of "cloud edge finding" is the process of discovering the software and hardware capabilities provided by mobile cloud service nodes.

"Edge finding cloud" refers to the process in which tactical edge nodes dynamically query the network address and port of available mobile cloud service nodes nearby and request to establish a connection. Its essence is that tactical edge nodes actively discover available mobile cloud service resources nearby, and establish connection through multicast discovery and unicast query, two core mechanisms. Among them, multicast discovery USES single-hop link local messaging mechanism for tactical edge nodes to dynamically discover service announcements broadcast by nearby available mobile cloud service nodes. Unicast query messaging is used to connect specific service directories that are known to contain the available mobile cloud service node lookup tables. The mechanism mentioned above plays an important role in the rapid discovery, optimization and secure connection of nearby tactical mobile cloud nodes.

In the MCS model, mobile cloud service discovery is the first step of application and resource interaction. In order to ensure that tactical edge nodes can query relevant service nodes and the availability of mobile cloud service resources, the realization of mobile cloud service discovery needs to meet the consistency of operation steps, data format and semantics.

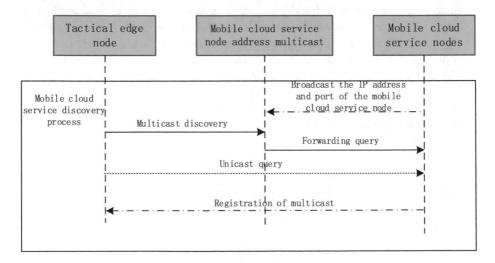

Fig. 4. Mobile cloud service discovery process

3.2   Mobile Cloud Service Supply

Mobile cloud service supply refers to the ability of mobile cloud service nodes to integrate all computing and storage resources and provide mobile cloud service based on infrastructure. After the tactical edge node completes the process of service discovery and establishes service connection with the mobile cloud service node, it flexibly allocates resource allocation and deploys virtual service resources according to the tactical edge node's demand for mobile cloud service resources, and establishes a one-to-one mapping relationship between the virtual service code and the tactical edge node. In addition, when resource requests of tactical edge nodes are completed or resource requests are made by tactical edge nodes with higher priority, mobile cloud service nodes can dynamically increase and release virtual service resources according to pre-set service rules.

In the mode of mobile cloud service supply, mobile cloud service resources are pre-supplied together with virtual service resources. The mobile cloud service resource allocation is completed in advance according to the task needs, and the corresponding virtual service resource is formed in the resource manager in advance. Virtual resources in the resource manager are encapsulated as a service and matched to the capabilities of tactical edge nodes. In addition, the service virtualization of resources can reduce energy consumption and simplify the supply process of mobile cloud services for military tactics when the task unloading of tactical edge nodes is realized.

3.3   Mobile Cloud Service Operation

Mobile cloud service operation mode can be divided into mobile cloud computing of mobile cloud service nodes and autonomous computing of tactical edge nodes. When the tactical edge node completes the process of service discovery and successfully obtains the service supply of the mobile cloud service node, the tactical edge node can realize mobile cloud computing through task unloading. When the tactical edge node service finds that it fails or does not need to receive the service supply from the mobile cloud service node, the tactical edge node USES its own computing resources to realize autonomous edge computing. For example, when soldiers perform military tasks such as explosive detection and terrorist face recognition near mobile cloud service nodes, they can perform mobile cloud computing of mobile cloud service nodes. Tactical unmanned aerial vehicles (tavs) can perform autonomous computing of tactical edge nodes when performing military tasks such as target tracking and image processing.

4   Summarizes

Based on the new trend of tactical fringe military operations, this paper comprehensively analyzes the capability requirements of tactical fringe military missions, and proposes a mobile cloud service model for military fringe. Thus provides a flexible model performs the task of military forces, the tactical edge information processing ability, resources on demand elasticity distribution, on the edge of the military tactics to solve military operations environment data processing mode single, the lack of edge information fusion processing capacity, the tactical edge resources integration scheduling problems such as inadequate, command communication time delay is too high, to adapt to the military battlefield environment for collaborative perception, decision making, time delay, energy consumption of development demand.

References

1. Lewis, G., Echeverria, S., Simanta, S., et al.: Tactical cloudlets: moving cloud computing to the edge. In: Military Communications Conference (2014)
2. Satyanarayanan, M., Bahl, P., Caceres, R., et al.: The case for VM-based cloudlets in mobile computing. IEEE Pervasive Comput. **8**(4), 14–23 (2009)
3. Wang, Y.: Characteristics of terrorist activities and discussion on medical service support mode of anti-terrorism operations. People's Mil. Doct. **46**(8), 455–456 (2003)
4. Zhang, K., Gui, X., Ren, D., Li, J., Wu, J., Ren, D.: Research review on computing migration and content caching in mobile edge networks. J. Softw. (5), 1–26 (2019)
5. Wei, G.: Opportunistic peer-to-peer mobile cloud computing at the tactical edge. In: IEEE Military Communications Conference (2014)
6. Konstantinos, P., George, I., Leandros, T.: SDN-enabled tactical Ad Hoc networks: extend programmable control to the edge. IEEE Commun. Mag. **56**(7), 132–138 (2018)
7. Jin, Y., Liu, Q.: Entropy and self organizing in edge organizations. Center for edge Power Publications (2009)
8. Cheng, S.: Application research of us military tactical cloud computing. Command Control Simul. (06), 139–147 (2017)

Benefits from Engineering Projects Implementation

Oleg Kuzmin[1], Volodymyr Zhezhukha[1(✉)], Nataliia Gorodyska[1],
and Eleonora Benova[2]

[1] Lviv Polytechnic National University, Bandera Street, 12, Lviv, Ukraine
{oleh.y.kuzmin, volodymyr.y.zhezhukha,
nataliia.a.gorodyska}@lpnu.ua
[2] Faculty of Management, Comenius University in Bratislava,
831 04 Bratislava, Slovakia
eleonora.benova@fm.uniba.sk

Abstract. The key methods for establishing the enterprises outcome from engineering projects implementation, which are the most widespread ones within the international engineering activities, have been determined and characterized. The main types of costs that business entity may incur in relation to the engineering project realization have been highlighted. Based on examining the most common concepts of managerial decision-making at risk (as it is under those circumstances that managers of industrial enterprises make decisions on engineering services cost), it was proposed to use the game theory instruments for establishing the structure of engineering payments during the engineering services delivery.

1 Introduction

Industrial enterprises, together with engineering companies, are active participants of the engineering market of any country. Due to this fact, they are an important link in a technological chain of building the competitiveness of products in certain sectors. While participating in the engineering relationships, these business entities act as the direct agents of industrial production modernization and address the shortcomings of innovation cycle. Regrettably, it had to be recognized that engineering as the business direction of industrial enterprises is only at the formation stage in most of countries in transition. Engineering departments of these entities are frequently lack of information regarding the relevant technologies and methods for designing the industrial facilities.

Analysis of each of the methods for measuring the income of industrial enterprises from engineering projects implemented makes it possible to single out a number of classification attributes of engineering payments (Fig. 1).

As we can see from Fig. 1, it is advisable to distinguish the basic and accompanying engineering payments by the nature. The main engineering payments associated with the "basic" subject of the engineering project, which, for example, can include architectural design works, project management, drafting of technical tasks, development of design and estimate documentation, design works and the like. All other, so-called additional work, provided for the engineering project implementation, associated

L. Barolli et al. (Eds.): INCoS 2019, AISC 1035, pp. 431–441, 2020.
https://doi.org/10.1007/978-3-030-29035-1_42

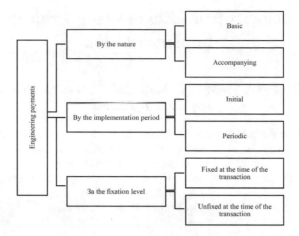

Fig. 1. Typology of engineering payments of industrial enterprises during the engineering projects implementation (Note: developed by the authors)

with the accompanying engineering payments (for example, technical consulting, monitoring the production process, conducting training courses, paying the cost of material assets provided etc.).

The existence of the combined method of determining the income of an industrial enterprise for the engineering services provided in practice of engineering activity allows us to argue that it is advisable to single out the initial and periodic engineering payments classified by the period.

2   Structure of Engineering Payments

2.1   Simulating the Structure of Engineering Payments

The structure of engineering payments for the engineering project implementation depends on a number of different factors, among which are the provisions of the contract for the engineering services delivery, the company's willingness to take risks, its policies and objectives, the financial condition of the engineering services customer, the expected effects of creating facilities on the findings of the engineering project and the like.

The study of literary sources on riskology allows us to argue that the most common concepts of management decision-making under risk are the following [1–7]: the fuzzy sets theory, game theory, the theory of statistical decisions, stochastic programming, etc. Game theory can be considered as one of the most widespread in modern conditions of risk management and managerial decision-making. A careful study of a content of such a theory suggests that its provisions are fully and sufficiently suited to establishing the structure of an industrial enterprise income from an engineering project implementation.

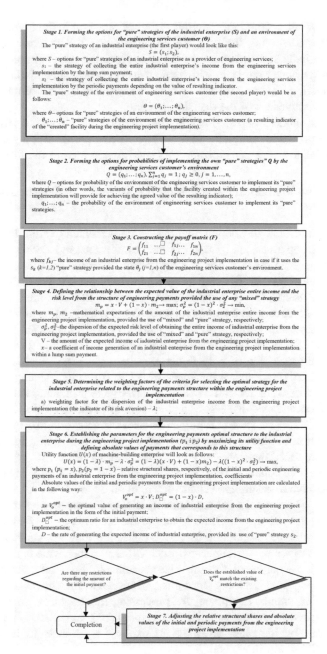

Stage 1. Forming the options for "pure" strategies of the industrial enterprise (S) and an environment of the engineering services customer (Θ)

The "pure" strategy of an industrial enterprise (the first player) would look like this:

$$S = (s_1; s_2),$$

where S – options for "pure" strategies of an industrial enterprise as a provider of engineering services;

s_1 – the strategy of collecting the entire industrial enterprise's income from the engineering services implementation by the lump sum payment;

s_2 – the strategy of collecting the entire industrial enterprise's income from the engineering services implementation by the periodic payments depending on the value of resulting indicator.

The "pure" strategy of the environment of engineering services customer (the second player) would be as follows:

$$\Theta = (\theta_1; \ldots; \theta_n),$$

where Θ – options for "pure" strategies of an environment of the engineering services customer;

$\theta_1; \ldots; \theta_n$ – "pure" strategies of the environment of the engineering services customer (a resulting indicator of the "created" facility during the engineering project implementation).

Stage 2. Forming the options for probabilities of implementing the own "pure" strategies" Q by the engineering services customer's environment

$$Q = (q_1; \ldots; q_n), \ \textstyle\sum_{j=1}^{n} q_j = 1; \ q_j \geq 0, j = 1, \ldots, n,$$

where Q – options for probability of the environment of the engineering services customer to implement its "pure" strategies (in other words, the variants of probability that the facility created within the engineering project implementation will provide for achieving the agreed value of the resulting indicator);

$q_1; \ldots; q_n$ – the probability of the environment of engineering services customer to implement its "pure" strategies.

Stage 3. Constructing the payoff matrix (F)

$$F = \begin{pmatrix} f_{11} & \ldots\square & f_{1j}\ldots & f_{1n} \\ f_{21} & \ldots\square & f_{2j}\ldots & f_{2n} \end{pmatrix},$$

where f_{kj} – the income of an industrial enterprise from the engineering project implementation in case if it uses the s_k $(k=1,2)$ "pure" strategy provided the state θ_j $(j=1,n)$ of the engineering services customer's environment.

Stage 4. Defining the relationship between the expected value of the industrial enterprise entire income and the risk level from the structure of engineering payments provided the use of any "mixed" strategy

$$m_p = x \cdot V + (1-x) \cdot m_2 \to \max; \ \sigma_p^2 = (1-x)^2 \cdot \sigma_2^2 \to \min,$$

where m_p, m_2 – mathematical expectations of the amount of the industrial enterprise income from the engineering project implementation, provided the use of "mixed" and "pure" strategy, respectively;

σ_p^2, σ_2^2 – the dispersion of the expected risk level of obtaining the entire income of industrial enterprise from the engineering project implementation, provided the use of "mixed" and "pure" strategy, respectively;

V – the amount of the expected income of industrial enterprise from the engineering project implementation;

x – a coefficient of income generation of an industrial enterprise from the engineering project implementation within a lump sum payment.

Stage 5. Determining the weighting factors of the criteria for selecting the optimal strategy for the industrial enterprise related to the engineering payments structure within the engineering project implementation

a) weighting factor for the dispersion of the industrial enterprise income from the engineering project implementation (the indicator of its risk aversion) – λ;

Stage 6. Establishing the parameters for the engineering payments optimal structure to the industrial enterprise during the engineering project implementation $(p_1 \ i \ p_2)$ by maximizing its utility function and defining absolute values of payments that corresponds to this structure

Utility function $U(x)$ of machine-building enterprise will look as follows:

$$U(x) = (1-\lambda) \cdot m_p - \lambda \cdot \sigma_p^2 = (1-\lambda)(x \cdot V) + (1-x)m_2 - \lambda((1-x)^2 \cdot \sigma_2^2) \to \max,$$

where p_1 $(p_1 = x)$, $p_2(p_2 = 1-x)$ – relative structural shares, respectively, of the initial and periodic engineering payments of an industrial enterprise from the engineering project implementation, coefficients

Absolute values of the initial and periodic payments from the engineering project implementation are calculated in the following way:

$$V_o^{opt} = x \cdot V; \ D_\square^{opt} = (1-x) \cdot D,$$

де V_o^{opt} – the optimal value of generating an income of industrial enterprise from the engineering project implementation in the form of the initial payment;

$D_\square^{opt}$ – the optimum ratio for an industrial enterprise to obtain the expected income from the engineering project implementation;

D – the rate of generating the expected income of industrial enterprise, provided its use of "pure" strategy s_2.

Are there any restrictions regarding the amount of the initial payment?

Does the established value of V_o^{opt} match the existing restrictions?

Completion

Stage 7. Adjusting the relative structural shares and absolute values of the initial and periodic payments from the engineering project implementation

Fig. 2. Generalized sequence of establishing the structure of engineering payments to an industrial enterprise during the engineering project implementation (Note: developed by the authors)

Thus, taking into account the generally accepted notation in the framework of game theory [8–14], and solving the problem of determining the structure of engineering payments to an industrial enterprise during an engineering project implementation, we can argue that we have a game, in which two players participate: an industrial company and the engineering services customer's environment, that is, a set of uncertain factors affecting the effectiveness of the decision-making.

A principal model for establishing the engineering payments structure to an industrial enterprise when implementing an engineering project using game theory for the purpose of clarity and generalization should be presented in the form of Fig. 2.

2.2 Testing the Method of Establishing the Structure of Engineering Payments

The application of the developed method is implemented within the engineering activity of Termoplastavtomat OJSC (Ukraine) for an engineering product - the establishment of a production line for the production of heat-resistant film LP 63-1500 M for production facilities of Arbix-Ukraine LLC. As it is known from the words of Termoplastavtomat OJSC management, in 2011 the customer representatives addressed the enterprise with a commercial request to establish this production line and purchase the necessary equipment. As a result, Termoplastavtomat OJSC supplied Arbix-Ukraine LLC with all the necessary equipment for the technological line intended for the production of polyethylene, including and heat-resistant, film with a sleeve width of 1500 mm and a thickness of 0.2 to 0.12 mm. Among the equipment supplied were: a machine for cutting through disposable packaging from cardboard, a blown semi-automatic machine MR-5, a pneumoform machine APF-1 and an extrusion blow aggregate K24-114, several rotary shredders, a granulation line for plastic waste, and the installation of film sintering unit UAP-06. It is worth noting that all the above equipment is produced independently by Termoplastavtomat OJSC.

According to the contract, Termoplastavtomat OJSC is not only obliged to supply equipment, but also to ensure the establishment of an appropriate production line, that, in fact, indicates the formation of engineering relations between the parties. It is worth noting that Termoplastavtomat OJSC is one of the leaders of the Ukrainian machine-building industry in the production of various injection molding machines.

The product range of Termoplastavtomat OJSC is rather wide and includes: canning and packing equipment for the pharmaceutical, food, oil refining and other industries, forging and pressing machines, equipment for the chemical industry, household boilers, molds, equipment for processing thermoplastic materials waste of various forms (waste smelting, scrap, defective products), and others. The high level of technology used by the plant allows it to export its products to many European countries and the world in general.

After analyzing the content of the contract between Termoplastavtomat OJSC and Arbix-Ukraine LLC, as well as the results of a survey of plant specialists, we have got an opportunity to form the content of an engineering package regarding the supply and establishment of a technological line for the production of polyethylene film, including the heat-resistant one, with a sleeve width of 1500 mm and a thickness of 0.2 to 0.12 mm. First, it was already mentioned above about the list of all the equipment that was supplied by Termoplastavtomat OJSC to the customer as part of the engineering project. Secondly, it is worth noting that within the transaction condition it was provided some adjusting of the extrusion-blown unit K24-114 in accordance with the features of Arbix-Ukraine LLC technological processes. Thirdly, according to the agreement of the parties, the specialists of Termoplastavtomat OJSC were obliged to carry out the training of workers of the Khmelnitsky plant Arbix-Ukraine LLC related to operating a new technological line and this was provided by the content of the prime contract. Fourthly, the content of the prime contract was also ensuring the Termoplastavtomat OJSC support to functioning of the established processing line. Fifth, the total value of the engineering contract amounted to UAH 3.89 million. Sixth, considering the typology of engineering payments of industrial enterprises within the engineering projects implementation that is shown in Fig. 1, it is worth noting that the engineering contract between the parties was characterized by the basic engineering payments by nature, and regarding the implementation period - by the initial ones. And they were set at the time of the transaction (typology by the level of fixation). Thus, it can be argued that the transaction between Termoplastavtomat OJSC and Arbix-Ukraine LLC is characterized by a method of fixed (undifferentiated) payment as a method for determining an income from engineering services delivered.

Considering the peculiarities of the engineering contract result between Termoplastavtomat OJSC and Arbix-Ukraine LLC (the production line was installed and adjusted), we now examine the possibility of using the combined payment method of determining an industrial enterprise income for the engineering services provided to the customer instead of the fixed (non-differentiated) one. It will be based on a combination of a fixed payment and a percentage of the result. Therefore, a combination of initial and periodic engineering payments for the period of the commission (and not only the use of initial payments) will be characteristic of the transaction.

It is worth paying attention to the fact that the project calculations of Termoplastavtomat OJSC and Arbix-Ukraine LLC provided that the production capacity of a new production line for manufacturing polyethylene film, including the heat-resistant one, with a 1500-mm-wide sleeve film and a thickness from 0.2 to 0.12 mm will be 60 kg/h. But, the specialists of both companies expected (and it turned out to be true) an average line load for the coming years only by 50% (30 kg/h).

Given the above-noted input information about the engineering agreement, and through implementing the stages of game theory application, presented in Fig. 2, let us establish the income structure of Termoplastavtomat OJSC from the implementation of an engineering project for Arbix-Ukraine LLC.

Stage 1. Forming the options for "pure" strategies of Termoplastavtomat OJSC *(S)* and the environment of engineering services customer (Arbix-Ukraine LLC) *(Θ)*. There are two "pure" strategies for Termoplastavtomat OJSC *(S)*:

(1) a strategy of collecting the lump sum payment for the engineering project implementation (s_1) by means of the initial payment (UAH 3.89 million);

(2) a strategy of collecting the entire income of Termoplastavtomat OJSC from the engineering project implementation by means of periodic payments (s_2), depending on the value of resulting indicator. Since 30 kg of polyethylene film with a width of 1500 mm and a thickness of 0.2 to 0.12 mm were expected to be produced per hour on the production line, the annual production capacity will be the multiplication of the film

production per hour by the number of working hours of the production line per day, as well as by the average number of working days per year. Given the information provided by Arbix-Ukraine LLC (it should be noted that the operation mode of the company's plant is double-shifting), we obtain the following results: 30 kg/hour. * 16 h. * 250 days. = 120 thousand kg. Consequently, with an average price of one kg of polyethylene film in 21 UAH, in value terms, the expected annual cost of plastic film (the result of an engineering project) will be UAH 120 thousand. * UAH 21 = UAH 2.52 million. Taking into account the conditional period of agreement between the parties regarding the period of charging periodic payments for 4 years, we can calculate the interest rate of these payments: (UAH 3.89 million./(4 years * UAH 2.52 million)) * 100% = 38, 59%. It is this value that allows Termoplastavtomat OJSC, subject to achieving the expected hopes regarding the implementation of the amount of plastic film produced by the production line of Arbix-Ukraine LLC, to receive the expected income from the engineering project implementation (4 years * UAH 2.52 million. * 0. 3859 = UAH 3,89 million).

The "pure" environment strategies of Arbix-Ukraine LLC " (Θ), in contrast to the "pure" strategies of Termoplastavtomat OJSC, are much larger. Each of them will in fact be in a certain limit: between the least and the most expected volume of sales of the Arbix-Ukraine LLC polyethylene film, produced on the technological line, which is the subject of the engineering contract. Therefore, an important task according to the theory of games is to establish the extreme limits of this implementation: minimal and maximal. Obviously, the maximum value of a possible sales will not exceed the maximum value of the technological line production capacity (60 kg/h). But according to the state of market conditions, the upper maximum limit may be less than the production capacity.

The results of a surveying the specialists of Termoplastavtomat OJSC and Arbix-Ukraine LLC made it possible to establish that the minimum expected limit for the polyethylene film sale by a customer of a production line can be 20% less than expected, that is, 0.8 * (4 years * UAH 2.52 million.) = UAH 8.06 million, instead of the expected UAH 10.08 million. The main reason for this is the European integration processes of Ukraine, because it is known that in the EU there are separate bans on the use of polyethylene, primarily because of its negative impact on the environment. Consequently, the implementation of a pessimistic forecast is possible with a certain level of probability.

Thus, while summarizing, it can be argued that the possible volume of polyethylene film sales of Arbix-Ukraine LLC, manufactured on the production line, will be in the range of UAH 8.06 million. (minimum limit) to UAH 20.16 million. (maximum limit) for 4 years. Therefore, as a result, let us form a set of "pure" strategies of the Arbix-Ukraine LLC (Θ, million UAH) (we will consider different values between the maximum and minimum values with an interval of UAH 2 million):

$$\Theta = (8.06; 10.00; 12.00; 14.00; 16.00; 18.00; 20.16).$$

Stage 2. Forming the options for probabilities of implementing the own "pure" strategies (Q) by the environment of Arbix-Ukraine LLC. At this stage, as already noted above, it is necessary to establish options for the probability that, due to the facility formed during the engineering project implementation, it will be possible to achieve the set value of the resulting indicator, or in other words, due to the established and streamlined production line, it will be possible to product and sale annually within four years of plastic film in the range of maximum and minimum expected limits.

As a result of surveying the specialists of Termoplastavtomat OJSC and Arbix-Ukraine LLC, the options for probability of implementing its "pure" strategies by Arbix-Ukraine LLC have been established, and they are presented below:

$$Q = (0.12; 0.31; 0.26; 0.18; 0.06; 0.04; 0.03).$$

Stage 3. Constructing the payoff matrix (F). To build the payoff matrix, let us calculate the values of its elements for "pure" strategies (s_1) and (s_2):

$$f_{21} = 0.3859 * 8.06 = 3.11 \text{ million UAH};$$
$$f_{22} = 0.3859 * 10.00 = 3.86 \text{ million UAH};$$
$$f_{23} = 0.3859 * 12.00 = 4.63 \text{ million UAH};$$
$$f_{24} = 0.3859 * 14.00 = 5.40 \text{ million UAH};$$
$$f_{25} = 0.3859 * 16.00 = 6.17 \text{ million UAH};$$
$$f_{26} = 0.3859 * 18.00 = 6.95 \text{ million UAH};$$
$$f_{27} = 0.3859 * 20.16 = 7.78 \text{ million UAH};$$
$$f_{11} = f_{12} = f_{13} = f_{14} = f_{15} = f_{16} = f_{17} = 3.89 \text{ million UAH}.$$

Obtained from the results of preliminary calculations, the payoff matrix of Termoplastavtomat OJSC will look as follows (in million UAH):

$$F = \begin{matrix} 3.89 & 3.89 & 3.89 & 3.89 & 3.89 & 3.89 & 3.89 \\ 3.11 & 3.86 & 4.63 & 5.40 & 6.17 & 6.95 & 7.78 \end{matrix}.$$

Stage 4. Defining the relationship between the expected value of Termoplastavtomat OJSC entire income and the risk level from the structure of engineering payments provided the use of any "mixed" strategy. Let us remind that the expected value of Termoplastavtomat OJSC entire income from the engineering project implementation is calculated by the mathematical expectation, and the level of risk – within a dispersion. When using the formulas, we will obtain the following results for the "pure" strategy (s_2):

$$\begin{aligned} m_2 &= 0.12 * 3.11 + 0.31 * 3.86 + 0.26 * 4.63 + 0.18 * 5.40 + 0.06 * 6.17 + 0.04 \\ &\quad * 6.95 + 0.03 * 7.78 \\ &= 4.63 \text{ million UAH}. \end{aligned}$$

$$\begin{aligned} \sigma_2^2 &= 0.12 * (3.11 - 4.63)2 + 0.31 * (3.86 - 4.63)2 + 0.26 * (4.63 - 4.63)2 + 0.18 \\ &\quad * (5.40 - 4.63)2 + 0.06 * (6.17 - 4.63)2 + 0.04 * (6.95 - 4.63)2 + 0.03 \\ &\quad * (7.78 - 4.63)2 \\ &= 1.22 \text{ million UAH} \end{aligned}$$

While using the formulas, we can construct an equation for the expected value of the Termoplastavtomat OJSC entire income and the risk level for any "mixed" strategy:

$$m_p = 3.89x + 4.63(1 - x) = 3.89x + 4.63 - 4.63x = 4.63 - 0.74x$$
$$\sigma_p^2 = 1.22(1 - x)^2 = 1.22x^2 - 2.44x + 1.22.$$

Stage 5. Determining the weighting factors of the criteria for selecting the optimal strategy for Termoplastavtomat OJSC related to the engineering payments structure within the engineering project implementation. At this stage, the one should actually assign the weighting factors to the mathematical expectation and variance for any "mixed" strategy. Attention should be paid to the fact that in this case such factors are set by an industrial enterprise depending on its willingness to take risks in order to generate additional income. For the considered engineering project of Termoplastavtomat OJSC, it is worth noting that there is no information on the probability distribution during this transaction (and cannot be), because, as noted above, there was a lump sum payment in the framework of using the undifferentiated payment method. As an example, we will carry out the appropriate calculations for the set of such weighting factors of the dispersion of an industrial enterprise income from the engineering project implementation (its risk aversion ratios) λ:

$$\lambda = (0; 0.2; 0.4; 0.6; 0.8; 1).$$

Stage 6. Establishing the parameters for the engineering payments optimal structure to Termoplastavtomat OJSC during the engineering project implementation (p_1 i p_2) by maximizing its utility function and defining absolute values of payments that corresponds to this structure.

Provided $\lambda = 0$, Termoplastavtomat OJSC is ready to take a maximum risk, when establishing the income from the engineering project implementation to Arbix-Ukraine LLC, and vice versa, when $\lambda = 1$. Using the constructed equations, we can represent the utility functions of Termoplastavtomat OJSC under various λ (Table 1).

Table 1. Utility functions of Termoplastavtomat OJSC for different values of its propensity and aversity to risk within the engineering payments structure establishment*

Indicators of propensity and aversity of Termoplastavtomat OJSC to risk, λ	Utility functions of Termoplastavtomat OJSC
0	$-0.74 \cdot x + 4.63$
0.2	$-0.244 \cdot x^2 - 0.104 \cdot x + 3.46$
0.4	$-0.488 \cdot x^2 + 0.532 \cdot x + 2.29$
0.6	$-0.732 \cdot x^2 + 1.168 \cdot x + 1.12$
0.8	$-0.976 \cdot x^2 + 1.804 \cdot x - 0.05$
1	$-1.22 \cdot x^2 + 2.44 \cdot x - 1.22$

* an example of constructing the utility function for $\lambda = 0.2$. In accordance with ratios, we will obtain: $U(x) = (1 - 0.2) \cdot (4.63 - 0.74x) - 0.2 \cdot (1.22x^2 - 2.44x + 1.22) = -0.244 \cdot x^2 - 0.104 \cdot x + 3.46$.

It is worth noting that in most cases, the utility function of Termoplastavtomat OJSC has a quadratic expression (Table 1). An exception occurs only for $\lambda = 0$. As is well known, any quadratic function reaches its maximum at the value of x, at which its first derivative is 0. The linear function for $\lambda = 0$ have its maximum at x = 0.

Let us set the parameters of the engineering payments optimal structure of Termoplastavtomat OJSC in the implementation of the engineering project for Arbix-Ukraine LLC, that is, the ratio of lump sum and periodic payments at a certain level of risk. In particular, for $\lambda = 0.4$, when the utility function of Termoplastavtomat OJSC is $-0.488 \cdot x^2 + 0.532 \cdot x + 2.29$, we equate its first derivative to 0:

$$(-0.488 \cdot x^2 + 0.532 \cdot x + 2.29)' = 0$$
$$-0.976x + 0.532 = 0$$
$$x = 0.545 \text{ or } 54.5\%$$

Considering the above-mentioned, the parameters of the optimal structure of Termoplastavtomat OJSC engineering payments during the implementation of the engineering project for Arbix-Ukraine LLC at the level of risk $\lambda = 0.4$ will be:

$$p_1 = 54.5\%, \ p_2 = 45.5\%,$$

The results of such calculations for all the λ are represented at Table 2.

Table 2. Parameters of the optimal structure of Termoplastavtomat OJSC engineering payments during the implementation of the engineering project by means of maximizing its utility function

Indicators of propensity and aversion of Termoplastavtomat OJSC to risk, λ	Parameters for the income optimal structure	
	Relative structural share of the initial payment, p_1	Relative structural share of the periodic payments, p_2
0	0	1
0.2	0	1
0.4	0.545	0.455
0.6	0.798	0.202
0.8	0.924	0.076
1	1	0

Thus, the calculations show that only for $\lambda = 0.4$, $\lambda = 0.6$ and $\lambda = 0.8$, it was expedient for Termoplastavtomat OJSC to foresee in the agreement a clause that would provide for the existence of initial and periodic payments when delivering Arbix-Ukraine LLC engineering project for the formation and establishment of a technological line for the polyethylene film production.

At the same stage, according to the generalized model for establishing the engineering payments structure of Termoplastavtomat OJSC during the implementation of an engineering project using the game theory, represented at Fig. 2, it is necessary to determine the absolute values of payments corresponding to the optimal structure (such as corresponding to the level of λ selected by an industrial enterprise). Since in the given example attention is not focused on a particular λ, we will carry out the corresponding calculations for all λ (Table 3).

Table 3. The established absolute values of engineering payments during the engineering projects implementation that correspond to the level of λ, selected by Termoplastavtomat OJSC

Indicators of propensity and aversity of Termoplastavtomat OJSC to risk, λ	An amount of the initial payment, million UAH	Periodic payments, %
0	0	38.59
0.2	0	38.59
0.4	2.12	17.56
0.6	3.10	7.83
0.8	3.59	2.93
1	3.89	0

* an example for establishing the absolute values of payments for $\lambda = 0.4$. According to ratios, we will obtain as follows:

$V_o^{opt} = x \cdot V = 0.545 \cdot 3.89 = 2.12$ million UAH
$D^{opt} = (1-x) D = 0.455 \cdot 38.59 = 17.56\%$

If the obtained value of the initial payment of Termoplastavtomat OJSC during the engineering project implementation for Arbix-Ukraine LLC is in the interest of the latter, the calculations should be completed according to Fig. 2. Otherwise, proceed to step 7, that is, to carrying out the adjustment of the relative structural shares and the absolute values of the initial and periodic payments.

3 Conclusion

The main methods for determining the industrial enterprises income of from the implementation of engineering projects, which are the most common in engineering practice, are identified and characterized. They are, in particular: time-based payment, payment of actual costs plus a fixed part, fixed (undifferentiated) payment, from the position of percentage of the result and combined. On the basis of this, a typology of engineering payments of industrial enterprises was carried out during the implementation of engineering projects, according to the results of which engineering payments were distinguished by nature, implementation period, and fixation level.

References

1. Kryvinska, N.: Building consistent formal specification for the service enterprise agility foundation. J. Serv. Sci. Res. **4**(2), 235–269 (2012). The Society of Service Science
2. Kaczor, S., Kryvinska, N.: It is all about services - fundamentals, drivers, and business models. J. Serv. Sci. Res. **5**(2), 125–154 (2013)
3. Wong, K.: The role of risk in making decisions under escalation situations. Appl. Psychol. **54**(4), 584–607 (2005)
4. Herrmann, J.: Engineering Decision Making and Risk Management. Wiley, Hoboken (2015)
5. Beroggi, G., Wallace, W.A.: Operational Risk Management: The Integration of Decision, Communications, and Multimedia Technologies. Kluwer Academic, Boston (1998)
6. Chapman, R.: Institute of risk management. In: Simple Tools and Techniques for Enterprise Risk Management, 2nd edn. Wiley, Hoboken (2011)
7. Moran, T., West, G.T.: Multilateral investment guarantee agency. In: MIGA-Georgetown University Symposium on International Political Risk Management, (International political risk management; 3). World Bank, Washington, D.C. (2005)
8. Bauso, D.: Game Theory with Engineering Applications. Society for Industrial and Applied Mathematics, Philadelphia (2016)
9. Binmore, K.G.: Game Theory: A Very Short Introduction. Oxford University Press, New York (2007)
10. DeVos, M.J., Kent, D.A.: Game Theory: A Playful Introduction. American Mathematical Society, Providence (2016)
11. Kryvinska, N., Gregus, M.: SOA and its business value in requirements, features, practices and methodologies. Comenius University in Bratislava (2014)
12. Molnár, E., Molnár, R., Kryvinska, N., Greguš, M.: Web intelligence in practice. J. Serv. Sci. Res. **6**(1), 149–172 (2014)
13. Chander, P.: Game Theory and Climate Change. Columbia University Press, New York (2018)
14. Friedman, D.: On economic applications of evolutionary game theory. J. Evol. Econ. **8**(1), 15–43 (1998)

The 5th International Workshop on Theory, Algorithms and Applications of Big Data Science (BDS-2019)

GRNN Approach Towards Missing Data Recovery Between IoT Systems

Ivan Izonin[1(✉)], Natalia Kryvinska[2,3], Pavlo Vitynskyi[1],
Roman Tkachenko[1], and Khrystyna Zub[1]

[1] Lviv Polytechnic National University, Lviv, Ukraine
ivanizonin@gmail.com, pavlo.vitynsky@gmail.com,
roman.tkachenko@gmail.com, khrystyna.zub@gmail.com
[2] Comenius University in Bratislava, Bratislava, Slovakia
natalia.kryvinska@univie.ac.at
[3] University of Vienna, Vienna, Austria

Abstract. This paper describes the main reasons for the problem of filling missed values in IoT devices. The method of solving the missing data recovery task using the regression approach is proposed. It is based on the use of artificial intelligence tool. Authors describes the main procedures for applying GRNN to solving this task. A number of practical investigations were carried out on the restoration of missed data in the real environment monitoring dataset. For this purpose, the dataset for assessing the air quality in the city, which contains a lot of passes, was selected. The high accuracy of the proposed method was experimentally investigated. A comparison of the work of the neural network with a number of existing machine learning algorithms is carried out. It has been found that GRNN provides at least 5% higher accuracy compared to existing methods. Prospects for further research on GRNN modifications for improving the accuracy of its work are outlined.

1 Introduction

The intellectual analysis of information [1] obtained from a variety of devices based on the Internet of Things (IoT) is becoming more and more actual today's task [2]. The development of such methods is required both for diverse industrial systems [3–5] and for research [6–8]. However, a number of problems determine the searching of effective solutions to the missing data recovery tasks. These include [9–13]:

- connection errors;
- external attacks;
- sensing errors;
- owing to data packet collision;
- signal attenuation;
- wave shadowing;
- hardware failures;
- channel fading;

© Springer Nature Switzerland AG 2020
L. Barolli et al. (Eds.): INCoS 2019, AISC 1035, pp. 445–453, 2020.
https://doi.org/10.1007/978-3-030-29035-1_43

- synchronization issues;
- environmental blockages;

Missing values in the dataset in most cases result in the removal of such vectors from it for further intellectual analysis [6, 14]. On the other hand, incorrectly predicted values of missed data, allowing the use of vectors for modelling, may cause inaccurate results of the analysis of the required information [15]. Both problems predetermine the need to apply existing or develop new artificial intelligence methods that would allow high prediction accuracy of missed values [16, 17].

This paper proposes the use of an artificial neural network to fill the missing values in datasets received by IoT device. The GRNN-based regression approach to solving a stated task should ensure high accuracy of the solution of the task, taking into account all the independent variables of each input vector [7].

2 GRNN Approach for Missing Data Recovery Task

The main task, solved by General Regression Neural Network (GRNN) is regression task.

GRNN has three layers: input, radial and output. Each of the radial elements of the hidden layer contains a Gaussian function with a center in a specific example. The value of the output signals is proportional to the nuclear ratings of the probability of belonging to the corresponding classes.

The main steps of the GRNN application are:

1. Let's suppose, that we have a training sample of m vectors. Each vector contains a set of input (always known components) $X_{i,j}$, where $i = 1,\ldots,m$ is the number of vector, $j = 1,\ldots,n$ is the number of the vector's component and output component Y_i, which is known only for the vectors of the training sample.
2. The aim of the neural network is to predict the output component Y for the given input components of the vector.
3. The search of the Euclidean distances of the input vector to all vectors of the training sample.

$$R_i = \sqrt{\sum_{j=1}^{n} (X_{i,j} - X_j)^2}. \tag{1}$$

4. To move from it to the Gaussian distances

$$D_i = \exp\left(-\frac{(R_i)^2}{\sigma^2}\right). \tag{2}$$

where σ is the magnitude of the function, which is selected experimentally.

5. The evaluation of the desired output component is based on the formula:

$$Y = \frac{\sum\limits_{i=1}^{m} (D_i \times Y_i)}{\sum\limits_{i=1}^{m} D_i}.$$ (3)

The GRNN network trains almost instantly, but in some cases, it may turn out to be large and slow. It depends on a variety of factors, in particular from:

- the training samples, their size and structure;
- the software solution;
- the quality of the algorithm;
- the parallelism;
- the use of the GPU;
- etc.

3 Modeling and Results

3.1 Dataset Descriptions

A dataset containing real data collected by the IoT device was used for modelling [18]. It contained a set of data on the chemical composition of air near an Italian city [19]. Sensors of the device collected information about the hourly chemical composition of atmospheric air based on the set of indicators given in Fig. 2.

Since the dataset contained a large number of spaces, for the simulation all vectors with missing values at least one of the 12 indicators were removed [20]. In this way, the simulation was carried out at a date size of 6950 vectors.

It was divided into two samples in the ratio of 80% to 20%. The first data sample was used for training, the second for testing.

Figure 1 gives a brief description of each variable: the name of the chemical air quality index, its the mean, the minimum and maximum values, which were calculated for the entire sample of data.

Since the most missed values contained a CO variable, the simulations were performed to restore the lost data of this column.

3.2 Quality Indicators

In order to comprehensively analyze the results of the GRNN effectiveness, four different indicators were used for the calculating the error [21–25]:

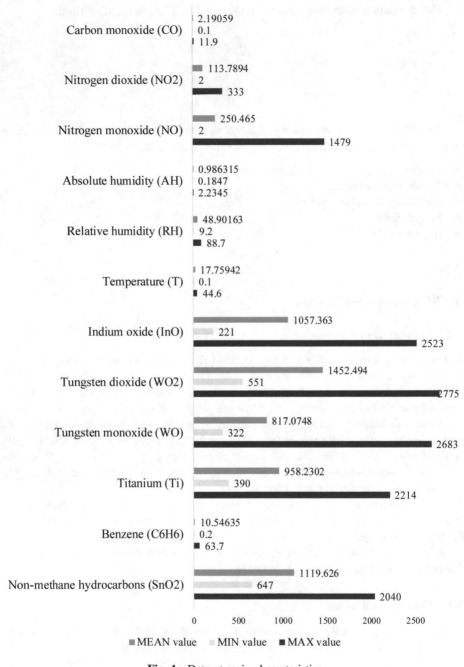

Fig. 1. Dataset main characteristics

- *Mean Absolute Percentage Error* (MAPE):

$$MAPE = \frac{1}{N} \sum_{i=1}^{n} |\frac{y_i^p - y_i}{y_i}| 100. \tag{4}$$

- *Root Mean Squared Error* (RMSE):

$$RMSE = \sqrt{\sum_{i=1}^{N} (y_i^p - y_i)^2}. \tag{5}$$

- *Mean Absolute Error* (MAE):

$$MAE = \frac{1}{N} \sum_{i=1}^{n} |y_i^p - y_i|. \tag{6}$$

- *Symmetric Mean Absolute Percentage Error* (SMAPE):

$$SMAPE = \frac{\sum_{i=1}^{N} |y_i - y_i^p|}{\sum_{i=1}^{N} (y_i^p + y_i)}. \tag{7}$$

where y_i is the true value for the i observation; and y_i^p is the obtained by method value; N is the dimension of the test sample.

3.3   GRNN Method Results

As a result of experimental investigations, the errors of the neural network operation were calculated for the restoration of the missing values of the CO parameter in the data set. Indicators (4)–(7) have been used for this purpose. Their values are shown in Table 1.

Table 1.   GRNN method's results.

Method	Indicators			
	MAPE, %	SMAPE	RMSE	MAE
General Regression Neural Network (GRNN)	23,00494	0,062234	0,482687	0,2965907

The visualization of the proposed approach result's effectiveness in the form of the scatter plot are shown on the Fig. 2. On the *ox* axis are true values, on *oy* axis are values, obtained by GRNN method.

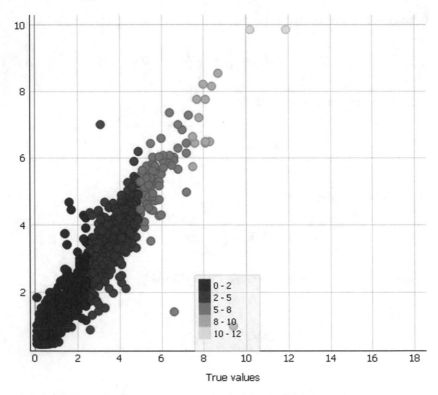

Fig. 2. Visualization of the GRNN method results

3.4 Comparison

Comparison of the GRNN (s = 0.1) effectiveness in the task of missing data recovery occurred using a number of known methods: AdaBoost; SVR; SGD regressor (linear regression using SGD algorithm). The parameters of the existing methods are as in [26].

The results of this comparison based on (4) in the test mode for all methods are shown in Fig. 3.

All other indicators show similar results. We have chosen MAPE since it, from (4)–(7), most clearly shows the difference between the magnitude of the errors of all the investigated methods.

As can be seen from Fig. 3, the use of GRNN for solving a task shows the highest accuracy of work among all methods. It is 77%. Despite the high speed of SGD regressor, this method shows 5% less accuracy than GRNN.

The worst results in terms of the accuracy of filling out missed values are two known methods: AdaBoost and SVR. The accuracy of their work is less than 70%.

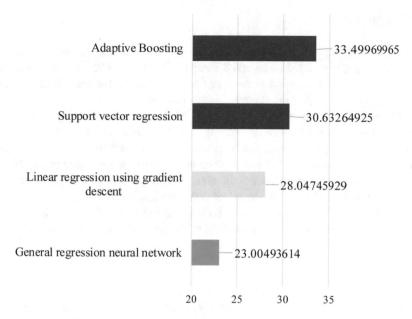

Fig. 3. MAPE, % for different methods (*ox* axis represents the MAPE, %)

In particular, the MAPE error of the AdaBoost method is greater by 13% for the error of the GRNN, and the SVR error is higher by more than 7%. The further intelligent analysis of air pollution based on data obtained by these methods will not provide reliable results.

3.5   Further Work

Further work for increasing GRNN accuracy for the missing data recovery will be carried out in the following directions:

- to design and detailed analysis of the GRNN ensemble for solving the missing data recovery task in cases of large dimensional data;
- to apply schemes of polynomial expansion of GRNN inputs, in particular using the Kolmogorov-Gabor polynomial, and experimental studies on its effectiveness. In particular, from the point of view of precision-training time, because due to high approximation properties this polynomial will increase the accuracy of the filling the missed data. On the other hand, due to a significant increase in the dimension of the input GRNN layer, the training time of the latter will significantly increase;
- to design and investigations of the Multi-GRNN for taking into account the local features of the input data approximation.

4 Conclusion

This paper considers the use of the General Regression Neural Network to solve the task of filling gaps in the data received by the IoT device. The main reasons for the occurrence of gaps in such data are described, the advantages and disadvantages of the existing toolkit for solving the problem are analyzed.

Experimental studies have been carried out to eliminate missing values in the monitoring data on air pollution in the city, which are obtained by a real IoT device. The high accuracy of the proposed neural network for solving the stated task based on different metrics is established. An experimental comparison was made of the efficiency of the proposed network with existing machine learning methods. The highest accuracy of the GRNN is established among all of considers methods.

The directions of further researches concerning the search for optimum the values of the prediction accuracy and speed of machine learning procedures are defined.

References

1. Lenhard, T.H., Gregus, M.: An unusual approach to basic challenges of data mining. In: 2015 International Conference on Intelligent Networking and Collaborative Systems, Taipei, pp. 105–109 (2015). https://doi.org/10.1109/incos.2015.40
2. Han, H., Kang, H., Son, J.: Intelligent operation monitoring IoT tag for factory legacy device. In: 2017 International Conference on Information and Communication Technology Convergence (ICTC), pp. 781–783 (2017)
3. Ivanochko, I., Urikova, O., Gregus, M.: Modern mobile technologies for collaborative e-business. In: 2014 International Conference on Intelligent Networking and Collaborative Systems, Salerno, pp. 515–520 (2014). https://doi.org/10.1109/incos.2014.20
4. Mandorf, S., Gregus, M.: The e-business perspective as a solution for inertia against complexity management in SME. In: 2014 International Conference on Intelligent Networking and Collaborative Systems, Salerno, pp. 237–241 (2014)
5. Babichev, S., Lytvynenko, V., Gozhyj, A. Korobchynskyi, M., Voronenko, M.: A fuzzy model for gene expression profiles reducing based on the complex use of statistical criteria and shannon entropy. In: Advances in Computer Science for Engineering and Education, pp. 545–554 (2018)
6. Kaminskyi, R., Kunanets, N., Rzheuskyi, A.: Mathematical support for statistical research based on informational technologies. In: Ermolayev, V., et al. (eds) Proceedings of the 14th International Conference on ICT in Education, Research and Industrial Applications. Integration, Harmonization and Knowledge Transfer. Main Conference Kyiv, Ukraine, 14–17 May, vol. 1, pp. 449–452. CEUR-WS.org (2018)
7. Ivanochko, I., Gregus, M., Urikova, O., Alieksieiev, I.: Synergy of services within SOA. Procedia Comput. Sci. **98**, 182–186 (2016)
8. Kut, V., Kunanets, N., Pasichnik, V., Tomashevskyi, V.: The procedures for the selection of knowledge representation methods in the "virtual university" distance learning system. In: Advances in Intelligent Systems and Computing, vol. 754, pp. 713–723 (2019)
9. Fekade, B., Maksymyuk, T., Kyryk, K., Jo, M.: Probabilistic recovery of incomplete sensed data in IoT. IEEE Internet Things J. **5**(4), 2282–2292 (2018)
10. Wu, H., Xian, J., Wang, J., Khandge, S., Mohapatra, P.: Missing data recovery using reconstruction in ocean wireless sensor networks. Comput. Commun. **132**, 1–9 (2018)

11. Parker, L.E.: A spatial-temporal imputation technique for classification with missing data in a wireless sensor network. In: 2008 IEEE/RSJ International Conference on Intelligent Robots and Systems, pp. 3272–3279 (2008)
12. PCI-MDR: Missing Data Recovery in Wireless Sensor Networks using Partial Canonical Identity Matrix. https://arxiv.org/abs/1810.03401. Accessed 01 Apr 2019
13. Droniuk, I., Fedevych, O.: Forecasting of the trend of traffic based on Ateb-functions theory. In: 2015 Xth International Scientific and Technical Conference "Computer Sciences and Information Technologies" (CSIT), pp. 139–141 (2015)
14. Kaminskyi, R., Kunanets, N., Pasichnyk, V., Rzheuskyi, A., Khudyi, F.: Recovery gaps in experimental data. In: Lytvyn, V., et al. (eds) Proceedings of the 2nd International Conference on Computational Linguistics and Intelligent Systems. Main Conference, Lviv, Ukraine, 25–27 June, vol. 1, pp. 170–179. CEUR-WS.org (2018)
15. Kovalchuk, A., Lotoshynska, N.: Assessment of damage to buildings in areas emergency situations. In: 2016 IEEE First International Conference on Data Stream Mining & Processing (DSMP), Lviv, pp. 152–156 (2016). https://doi.org/10.1109/dsmp.2016.7583528
16. Lytvyn, V., Vysotska, V., Peleshchak, I., Rishnyak, I., Peleshchak, R.: Time dependence of the output signal morphology for nonlinear oscillator neuron based on Van der Pol model. Int. J. Intell. Syst. Appl. **10**, 1–8 (2018)
17. Lytvyn, V., Peleshchak, I., Vysotska, V., Peleshchak, R.: Satellite spectral information recognition based on the synthesis of modified dynamic neural networks and holographic data processing techniques. In: 2018 IEEE 13th International Scientific and Technical Conference on Computer Sciences and Information Technologies, pp. 330–334 (2018)
18. Vito, S., Massera, E., Piga, M., Martinotto, L., Di Francia, G.: On field calibration of an electronic nose for benzene estimation in an urban pollution monitoring scenario. Sens. Actuators B: Chem. **129**(2), 750–757 (2008)
19. UCI Machine Learning Repository: Air Quality Data Set. http://archive.ics.uci.edu/ml/datasets/air+quality. Accessed 08 Feb 2019
20. Mishchuk, O., Tkachenko, R., Izonin, I.: Missing data imputation through SGTM neural-like structure for environmental monitoring tasks. In: Hu, Z.B., Petoukhov, S., (eds) Advances in Computer Science for Engineering and Education. Advances in Intelligent Systems and Computing, vol. 938, pp. 142–151. Springer, Cham (2019)
21. Ivanochko, I., Urikova, O., Greguš, M.: Mobile technologies enabling collaborative services management. Int. J. Serv. Econ. Manag. **6**(4), 310–326 (2014)
22. Lytvyn, V., Vysotska, V., Dosyn, D., Burov, Y.: Method for ontology content and structure optimization, provided by a weighted conceptual graph. Webology **15**(2), 66–85 (2018)
23. Rashkevych, Y., Peleshko, D., Kovalchuk, A., Kupchak, M., Figura, R.: Time series partitional clustering. In: Perspective Technologies and Methods in MEMS Design, Polyana, pp. 170–171 (2011)
24. Dronyuk, I., Fedevych, O., Lipinski, P.: Ateb-prediction simulation of traffic using OMNeT ++ modeling tools. In: 2016 XIth International Scientific and Technical Conference Computer Sciences and Information Technologies (CSIT), Lviv, pp. 96–98 (2016)
25. Nazarkevych, M., Oliarnyk, R., Troyan, O., Nazarkevych, H.: Data protection based on encryption using Ateb-functions. In: 2016 XIth International Scientific and Technical Conference Computer Sciences and Information Technologies (CSIT), Lviv, pp. 30–32 (2016). https://doi.org/10.1109/stc-csit.2016.7589861
26. 1.11. Ensemble methods—scikit-learn 0.20.3 documentation. https://scikit-learn.org/stable/modules/ensemble.html#adaboost. Accessed 25 Mar 2019

Research and Application of Big Data Correlation Analysis in Education

Du Bo$^{(\boxtimes)}$, Li Ai, and Yuan Chen

Engineering University of PAP, Xian, Shanxi, China
726682174@qq.com

Abstract. How to adjust the training and construction plan for the college faculty and improve their capacity or education and instruction in Chinese universities is an urgent problem to be explored and solved under the background of education big data. In the era of big data, relevant analysis has received extensive attention because of its ability to discover the intrinsic associations between things quickly and efficiently, and has been effectively applied to various fields. In this paper, we are based on the data of 2018 senior professional title defense in several Chinese universities to analyze the correlation between the respondent's grades and ages, education and majors by using big data correlation mining technology. The purpose is to explore the status quo and problems of teachers' talent team construction, and then find a suitable way for the construction and training of college teachers in the era of big data.

Keywords: Correlation analysis · Education big data ·
Teachers team construction · Senior title defense

1 Introduction

In 2012, the United Nations issued a white paper titled *"Big Data for Development: Challenges & Opportunities"*, which pointed out that "the era of big data has arrived, and the emergence of big data will have a profound impact on all sectors of society." [1] Big data is promoting great changes in human work, life and thinking, and its "power" is also strongly impacting the entire education system, becoming a subversive force to promote innovation and reform of the education system. College education is the main channel of cultivating high-quality talents, and building a team of high-quality teachers is the key to building first-class colleges and universities. This study attempts to discuss the current status and characteristics of talent team construction in a certain field in Chinese colleges by analyzing the data of "2018 higher vocational defense of several domestic colleges", so as to provide scientific reference for the establishment of the next talent training program in colleges.

L. Barolli et al. (Eds.): INCoS 2019, AISC 1035, pp. 454–462, 2020.
https://doi.org/10.1007/978-3-030-29035-1_44

2   Research Objects and Characteristics

We selected several large-scale defense data of higher vocational colleges, which is highly representative and typical for the comprehensive coverage. It has the following characteristics:

(1) the data sources cover the major colleges and universities in the selected field, and the respondents are all excellent talents of the unit in recent years, which to some extent can represent the overall quality of the recent senior talents of the colleges and universities in the field.

(2) the total number of people participating in the defense reached over 1,000, among whom 27.4% participated in the defense of senior professional title and 72.6% of deputy senior professional title.

(3) the majors involved include the major majors at present, the respondents' educational background ranges from bachelor to doctor (postdoctor) and the age span ranges from over 30 to over 50, which covered a comprehensive range.

(4) the evaluation team is composed of several famous experts in various fields of academia. The evaluation follows the mode of large group evaluation, and the respondents of different majors are graded and ranked in a uniform standard and fair competition.

(5) besides respondent's thesis, the evaluation by the experts also combines their achievements, comprehensively considering the basic knowledge, cutting-edge theory, practical skills and foreign language level, so that the defense results can basically reflect the true level of the respondent.

3   The Research Methods

3.1   Sample Distribution

This study mainly explores the influence of multiple independent variables on the ability and quality of respondents. The final defense scores of respondents are selected as the dependent variable of this research, and several factors with relatively analytical value are selected as independent variables, including age, education background and major. In order to meet the needs of statistical analysis and make the conclusions more analyzable, three independent variables are grouped and dimensionalized, among which the majors are divided into four categories: economics and management (EM), science and engineering (SE), literature, history, philosophy and law (LHPL), and others. The basic information of data samples is shown in Table 1.

Table 1. The basic information of data samples

Variable categories	Variable name		Percentage (%)		
			Associate professor	Lecturer	Totality
Independent variables	Age	<40	4.11	41.10	45.21
		40–49	16.44	31.50	47.94
		≥ 50	6.85	0	6.85
	Degree	Doctor (post doctor)	5.48	19.18	24.66
		Master	9.59	34.24	43.83
		Bachelor	12.33	19.18	31.51
	Major	EM	5.48	13.70	19.18
		SE	4.11	24.66	28.77
		LHPL	2.74	8.22	10.96
		Others	15.07	16.02	41.09
	Paper relevance	High correlation	20.55	54.79	75.34
		Moderate correlation	4.11	15.07	19.18
		Weak correlation	1.37	2.74	4.11
		Uncorrelation	1.37	0	1.37
Dependent variable	Defense core		88.759 (average)	1.694 (standard deviation)	

3.2 Research Ideas and Methods

Although this defense data has a strong representativeness and typicality, this paper does not simply rely on the results of data analysis to draw a conclusion, but combines quantitative analysis and qualitative analysis to analyze according to the actual situation of talent training and development. After this activity, related management departments organized a seminar on issues about the selection and training of high-level scientific and technological innovation talents. In order to make the analysis results more scientific and convincing, this paper makes full use of the brainstorming of experts in various fields to conduct comparative analysis with the existing data, achieving a better analysis effect.

Firstly, a causal graph is established to analyze the variance of age, education background, major and defense score. Then, according to the significant degree of difference obtained, we judge whether it is necessary to further analyze the influence of the three factors on defense scores. For the factors with significant differences, through calculating mean, variance and other statistical values combined with drawing, one-dimensional plus two-dimensional analysis method is adopted to obtain more intuitive analysis results. Finally, according to the conclusion of data analysis and the actual situation of current talent training, we expect to find out the characteristics with guiding significance and provide scientific reference for the formulation of the next talent training program.

4   Statistical Analysis of Data

4.1   Univariate Analysis

Firstly, we use univariate analysis of variance was used to explore the differences in defense scores among different ages, educational backgrounds and majors (Table 2). The results show that, at the significant level of 5%, changes in age and education background have an extremely significant impact on defense performance. Therefore, we focus on these two factors for a more specific statistical analysis. Considering the situation that senior and deputy titles are in the same large group, in addition to the overall analysis, we also distinguish between associate professors and lecturers for analysis and discussion, to eliminate the differences between different titles.

Table 2. Results of bivariate variance

Independent variables	F	P-value	F crit
Age	6.106	0.004	3.128
Degree	11.763	0.000	3.128
Major	1.025	0.387	2.737

Age - Scores Analysis
The younger age of the backbone force is a development trend of the talent team in recent years. The statistical analysis of the correlation between age and achievement can reflect the recent age distribution of high-quality talents. As can be intuitively seen from Fig. 1, 75% of the respondents, whose age is under 40, scored above 88.5, and with the increase of age the scores tend to decline, which can be seen from that, for the respondents between 40 and 49 years old, over 50% scored below 89, while for the respondents over 50 years old scored below 88.5. These reflect that young talents have more advantages in terms of ability and quality, and also verify that the policy of making the talent team younger has achieved good results in new period. The age-scores distribution of different job titles (Table 3) shows the same trend as the general distribution that young people scored higher on average. In addition, in the same age group, the average score of associate professors is higher than that of lecturers, reflecting that the professional title is in direct proportion to comprehensive quality in the same age group.

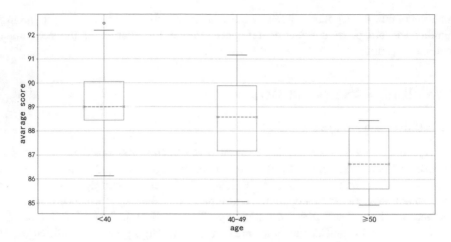

Fig. 1. Age - scores boxplot

Table 3. Age - scores distribution

Age	Associate professor		Lecturer		Totality	
	Average score	Standard deviation	Average score	Standard deviation	Average score	Standard deviation
<40	91.733	1.079	89.045	1.367	89.289	1.543
[40, 49]	88.921	2.101	88.350	1.375	88.545	1.652
≥ 50	86.748	1.524	0	0	86.748	1.524

Degree - Scores Analysis

It can be seen from the boxplot shown in Fig. 2 that 50% of the doctoral (post-doctoral) defense scores are above 90 points, while more than 50% of the faculty with bachelor's degree scores are below 88 points. And the one with the highest score has doctor's degree, while the one with the lowest score is a bachelor. The former reflects that the higher the degree, the higher the score, which means that the comprehensive quality of highly educated talents is stronger. The results of associate professors and lecturers with different degrees are consistent with the general trend, and the average score of associate professors was higher than that of lecturers (Table 4).

Majors - Scores Analysis

In the process of this defense review, the large-organization review mode is adopted, which takes into account the different advantages of various professional teachers in academic research and practical skills. The overall score tends to be balanced by comprehensive consideration of basic knowledge, cutting-edge theory, practical skills and foreign language level. By analyzing the average score and standard deviation of respondents who are engaged in different majors, as shown in Table 5, we find that science and engineering faculty have the highest overall average scores, which reflects

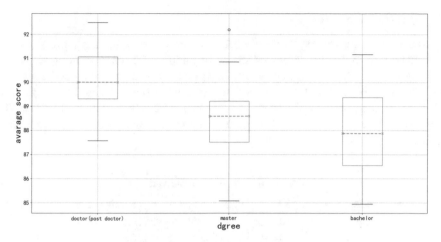

Fig. 2. Degree - scores boxplot

Table 4. Degree - scores distribution

Degree	Associate professor		Lecturer		Totality	
	Average score	Standard deviation	Average score	Standard deviation	Average score	Standard deviation
Doctor (post doctor)	91.235	0.958	89.885	1.186	90.185	1.253
Master	89.093	2.323	88.362	1.126	88.522	1.456
bachelor	87.489	2.010	88.281	1.482	87.971	1.710

that the cultivation and introduction of talents in this direction bring good results to some extent. However, the difference of the overall average score of different majors is small, which indicates the fairness and rationality of the large group system of higher vocational assessment.

Table 5. Majors – scores distribution

Major	Associate professor		Lecturer		Totality	
	Average score	Standard deviation	Average score	Standard deviation	Average score	Standard deviation
EM	90.145	1.433	88.233	1.174	88.779	1.494
SE	89.977	3.264	89.131	1.690	89.252	1.894
LHPL	88.485	2.850	88.858	0.692	88.765	1.238
Others	88.046	2.319	88.608	1.354	88.402	1.752

4.2   Bivariate Analysis

In order to do a further exploration whether defense scores have a more significant distribution rule under the combined action of multiple factors, according to the univariate variance results of age, education background and major mentioned above, age and education background, we select age and education, which have significant influence on academic performance, to conduct bivariate variance (Table 6). The results show that age, education background and the interaction between them have a significant impact on performance.

Table 6. Results of bivariate variance

Difference source	SS	df	MS	F	P-value	F crit
Degree	103.766	2	51.881	53.926	0.000	3.059
Age	49.361	2	24.680	25.653	0.000	3.059
Degree & Age	24.136	4	6.034	6.272	0.0001	2.435
Interior	138.540	144	0.962			
Total	315.798	152				

Figure 3 shows the correlation statistics of age, education background and achievement. The average score of doctoral teachers under the age of 40 is the highest (90.393), followed by that is master teachers under the age of 40 (89.859). It can be concluded that those with higher education background and younger teachers have higher scores. We also find by observing the original data that the top 4 with scores were all under the age of 40, and 3 of them are doctor in science and technology, and the lowest score was teachers with bachelor's degree above 50 years old.

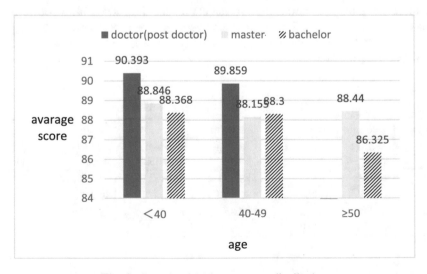

Fig. 3. Age & education – scores distribution

The above analysis of 2018 annual defense data of higher vocational colleges, to some extent, reflects the current situation and characteristics of talent team construction in the selected field, and mainly has the following conclusions.

Firstly, the construction of talents in colleges and universities is on the rise. The analysis results of the relationship between age, education background and defense scores in the previous chapters of this paper, show that young talents with advanced education background have a comparative advantage. Among the teachers with different degrees of education, the average score of doctoral (post-doctoral) teachers is the highest, 75% of the doctoral (post-doctoral) teachers have the average score of above 89.2; the average score of those under 40 years old is the highest; 50% of the defense scores are above 89; and the highest score is those under 40 years old (post-doctoral). All these indicate that the country has made some achievements in training and attaching importance to talents with high academic qualifications in recent years, and the gradient construction of talent teams in colleges and universities has achieved initial results. What's more, higher education and younger age is also a key direction of talent team construction in colleges and universities under the new situation.

Secondly, the training and introduction of professional and technical personnel have achieved remarkable results. In this senior title defense, science and technology teachers scored the highest average score of 89.252, reflecting the higher comprehensive quality of professional and technical teachers. In recent years, efforts have been made to strengthen the training and introduction of high-level talents in this field, and remarkable results have been achieved. In addition to the talents independently cultivated by colleges and universities in this field, a large number of professional and technical talents with high education and young age have been introduced from outside.

Third, the advantages of large group system of higher vocational assessment are obvious. The evaluation team, whose composition is reasonable and proper, is composed of several influential experts in various fields of academia. Fowling the mode of large group assessment, the way of unified organization and fair competition is adopted for respondents of different majors in this defense. With comprehensive consideration of basic knowledge, cutting-edge theory, practical skills and foreign language level, experts are able to balance the different advantages of different professional teachers. According to the research and analysis of this paper, the average scores of economics and management, science and engineering, literature, history, philosophy and law and other subjects are respectively 88.779, 89.252, 88.765 and 88.402. There is no big difference between the defense scores of different majors, and the average scores tend to be the same, which verifies the rationality and scientificity of the large group system of senior professional title evaluation.

Finally, it is necessary to build a platform for high-level talents. Seeing from the proportion of doctors (post doctors), 25% is not enough to meet the current demand for high-level talents in the field. In the analysis of the correlation between the thesis defense and the major, it is also found that most of the doctoral (postdoctoral) teachers' thesis, which is not highly related to the actual work, was completed during the school period, which can be considered to be caused by the lack of follow-up research platform after their graduation. The traction of platform and task is very critical to the role of talent cultivation. It is particularly important for the cultivation of high-level talents and the application of complex research methods to independently train more doctors and build a layered and focused platform for talent promotion and integration.

References

1. Big Data for Development: Challenges & Opportunities [EB/OL]. http://www.unglobalpulse. org/sites/default/files/Big-DataforDevelopment-UNGlobalPulseJune2012.pdf. Accessed 01 May 2012
2. Han, J., Kamber, M.: Data Mining: Concepts and Techniques. Morgan Kaufmann, San Francisco (2001)
3. Baldwin, J.N., Bootman, J.L., Carter, R.A.: Big data and analytics in higher education: opportunities and challenges. Am. J. Pharm. Educ. (2015)
4. Thompson, B.: Factor analysis. In: The Blackwell Encyclopedia of Sociology (2007)
5. Thompson, B.: Canonical correlation analysis. In: Encyclopedia of Statistics in Behavioral Science (2005)
6. Huda, M., Anshari, M., Almunawar, M.N., et al.: Innovative teaching in higher education: the big data approach. TOJET (2016)
7. Cui, H., Zhang, D.: Strategies on teacher professional development in big data era. In: 2018 4th International Conference on Education Technology, Management and Humanities Science (ETMHS 2018). Atlantis Press (2018)
8. Fan, X., Fan, M.: Precise teacher management based on big data. J. Teach. Educ. (1), 10 (2018)

Towards a Computational Model of Artificial Intuition and Decision Making

Olayinka Johnny, Marcello Trovati[✉], and Jeffrey Ray

Department of Computer Science, Edge Hill University, Ormskirk, UK
{24023868,trovatim,rayj}@edgehill.ac.uk

Abstract. The ability to perform a detailed decision-making approach based on large quantities of parameters and data is at the core of the majority of sciences. Traditionally, all possible scenarios should be considered, and their outcomes assessed via a logical and systematic manner to obtain accurate and applicable methods for knowledge discovery. However, such approach is typically associated with high computational complexity. Moreover, it is non-trivial for researchers to develop and train models with deep and complex model structures with potentially large number of parameters. However, there are compelling evidence from psychology and cognitive research that intuition plays an important role in the process of intelligence extraction and the decision-making process. More specifically, by using intuitive models, a system is able to take subsets from networks and pass them through a process to determine relationship that can be used to predict future decision without a deep understanding of a scenario and its corresponding parameters. When an artificial agent manifests human intuition properties, then we can describe this as artificial intuition. In this article, we discuss some requirements of artificial intuition and present a model of artificial intuition that utilises semantic networks to improve a decision system.

1 Introduction

This work focuses on modeling of artificial intuition in decision making. A major challenge in decision making is coping with a large unstructured data-sets based on partial unknown knowledge, and with multiple states which can be modeled by complex networks such as dependency network and Bayesian network. Making decision from such networks would require using the rational decision-making process which involve the need for deeper analysis to allow the investigation of their corresponding scenarios. Most of these models are logic driven and are time dependent [1]. Moreover, there are potentially a large number of parameters, which increase the overall computational complexity in some scenarios [2]. However, during critical and time based decision making scenarios such as medical diagnosis in an emergency, human use intuition to solve problems [3–5]. In our approach we particularly interested in the connection between artificial

© Springer Nature Switzerland AG 2020
L. Barolli et al. (Eds.): INCoS 2019, AISC 1035, pp. 463–472, 2020.
https://doi.org/10.1007/978-3-030-29035-1_45

intuition and decision making and to develop a computational based model of artificial intuition in decision making. The driving question is how to utilise a model based on Artificial Intuition to improve a decision system. The theoretical underpinning of this study is drawn from the study of psychology and cognitive research. Human intuition is an unconscious mental process aimed to solve problems without using a rational decision-making process. Loosely speaking, intuition is the ability to see connections between concepts or properties that can be interpreted as valid with respect to some networks. That is by using intuitive models, a system is able to take subsets from two networks and pass them through a process to determine relationship that can be used to predict future decision. When an artificia l agent manifests human intuition properties, then we can say that there is Artificial Intuition [6]. More specifically, Artificial Intuition is an automatic process, which does not search rational alternatives, jumping to useful responses in a short period of time, and is mainly focus in providing responses without iterative search of solutions for a given problem [6]. The work is organised as follows: In Sect. 2, we present a background and an overview of the current research in the field. In Sect. 3, we present discussion of the approaches. In Sect. 4, we present a description of the model. In Sect. 5, we conclude our work.

2   Background and Related Work

From a psychology and cognitive research perspective, Simon [7] presents a descriptive research on managerial decision making and problem solving where he provides insights into the nature of intuition. He noted that executives who attend to real-time information are actually developing their intuition and aided by intuition, they can react quickly and accurately to changing stimuli in their firm or its environment. Although the data are limited, the executives who relied most heavily on real-time information were also most frequently described as being intuitive. Simon [7] explained that intuition is dependent on past knowledge and experience for better recall of solutions to the given problems or normal logical process. Kahneman [8] have identified two fundamental and distinct modes of decision making in humans. The theory distinguished between intuition and reasoning. Stanovich and West [9] labeled them as System 1 (intuition) and System 2 (reasoning). System 1 is an automatic, fast and often unconscious way of thinking. It is autonomous and efficient, requiring little energy or attention and is dependent on some known information. System 2 is an effortful, slow and deliberately controlled way of thinking. While system 1 is likely to be affected by recognition of patterns, system 2 is based on rational choices where humans use logic in its best sense to perform a cost/benefit analysis that will provide the best possible choice. Studies have shown that human beings do not have the natural ability to do two things at the same time, hence some have argued that it is important to move from system 2 (consciousness/rationality/using logic) to system 1 (subconsciousness, intuition). Moving from system 2 to system 1 requires the agent/human to acquire a lot of skills and experiences over a period.

This is very useful during critical and time based decision making scenarios such as the weather forecast and medical diagnosis in an emergency [3–5]. Studies indicate most clinical decisions are made using the fast, hardwired intuitive System 1 approach that depends heavily on the inductive reasoning experience of the decision maker [3,4]. They further emphasized that the experience and recognition skills of the decision maker will determine how well the presented information is interpreted. Most System 1-based decisions are correct, but also subject to bias-induced error in cases where atypical signs/symptoms present, or when a non-specific pattern is mistakenly associated with the wrong diagnosis. Essentially, they recognized that intuition is accomplished through recognition of significant information and it synthetizes immediate decisions or actions without the need of a rational process [10]. He also acknowledged that this depends on the use of prior knowledge and experiences and the context of the problem. He [11] described experiential learning theory as the process whereby knowledge is created through the transformation of experience. According to Tao and He [12], experiential learning emphasizes the role that appropriate environments and experiences play in the learning process as the learner is directly in touch with the realities being studied rather than merely thinking about the encounter or studying the experience of others with such phenomena. The most basic conclusion from the theory is that people do learn from their past experiences. A fundamental requirement that facilitate learning is an appropriate environment where learners can have experiences of intuition decision. This emphasizes the role that appropriate pattern of intuition and experiences play in the learning process. An important style of knowledge acquisition reflects the model to the characteristics of the heuristics learning behavior [11]. Heuristic refers to any techniques that improves the average-case performance on a learning task but does not necessarily improve the worst-case performance [12]. It uses knowledge of previously tried solutions to guide the search into fruitful areas of the search space. The conclusion from this is that heuristics uses estimations based on domain knowledge which are represented in the form of patterns, networks, trees or graphs. Frantz [13] explains the approach for handling intuition in the form of the novice user and the expert user. He described intuition as subconscious pattern recognition and adds that knowledge and past experience are very important for intuition to be accurate. The drawback of this work is that he did not explain the concept in terms of how it mapped to the problems and the evolving nature of entities as well as that of the environment. Therefore, lacks practical implementation. Dundas and Chik [1] present an implementation of human-like intuition that used series-based model and the principles of connectivity and unknown entities. In this approach, they represented the problems and experience as sets. In particular, the space of intuition contains relational mappings between experience set elements and their associated attributes. A mapping function associating elements from the knowledge set with the experience set is at the core of any problem solving activity based on artificial intuition. In general, the algorithm goes through the following stages:

1. Obtaining initial conditions,
2. Obtaining an equation from recollected pieces of human experiences,
3. Obtaining of weight values of importance,
4. Recognition of waste information,
5. Application of adjustment factors on all considered processes to calculate the final answer,
6. Check if there are any external influences and then present the answer to the user.

Srdanov et al. [14] presents an artificial intuition method that searches for optimal path by combining trial and error with a random choice. More specifically, they combined logic and randomness. The approach first forms a large set with multiple repetition of some elements which contains possible candidates for the solution. It then searches by using logic and applying random choice to select from many of its elements, which enables the solution to be reached. They argued that using random choice reduced the search space and enable the system to converge to the solution in fewer steps than without the random choice. Similarly, Srdanov et al. [15] described the same trial and error with randomness algorithm as in Srdanov et al. (2016) and in addition used symmetry properties in pattern. They applied this to a case where the digits in a quintuplet can be repeated as against Srdanov et al. [14] where digits in the quintuplets cannot be repeated. Diaz-Hernandez and Gonzalez-Villela [6] proposed that human intuition is embodied in three stages: Inputs, Processing, and Outputs and correlated each stage to its equivalent in artificial intuition. They presented an artificial intuition approach that focused on synthetizing algorithms that improve robot performance in pick and place tasks. The approach focused on simplifying the processing stage of decision-making process, by reducing the complexity of the set of instructions needed to solve the task. The algorithm presented by Diaz-Hernandez and Gonzalez-Villela [6] is embodied as mathematical expressions that directly model obstacle avoidance in robots. Tao and He [12], suggested a learning system based on intuition through artificial intuition networks as a mechanism of cooperative learning which will provide to the user a game-like experience, making situations more obvious and easier to learn. They developed a general instrument for measuring trusted intuition in the context of intuitive learning system. The drawback of the system is that they only described principles but did not describe an implementation. A related research area is computational creativity. It involves exploring computational approaches to simulate and evaluate creativity, using AI techniques such as semantic networks [16]. Zhang et al. [17] stated that creativity blends seemingly disparate ideas often embodied in existing knowledge. Boden [16] hinted that the creative mind searches through a search space. According to Gilovich et al. [18], the search space can be traversed during creativity by using different thinking styles strategy defined by rules and constraints that makes analogical pattern matching possible. This include unconscious, associative heuristics such as intuition, rules of thumb, trial and error and common sense. They stated that a conscious thinking style can be constructed from a logical rule such as find as many as possible or find the most unusual, combined with a constraint. An important aspect of creative models is the use of conceptual blending [19] that explores how concepts can be imagined and combined into new meaning. Such combination can be based on similarity measures. De Smedt [20] presents a model that exhibit creativity in an artistic context, that are capable of generating or evaluating an artwork. He implemented a concept search space as a semantic network of related concepts, and search heuristics to traverse the network in order to generate creative concepts. Pease [21] applied a combinational meta search approach that uses search over the space of all possible models for the class of artifact desired to implement computational creativity. They used the technique of concept blending, amalgamation, and compositional adaption to allow the recombination of items from within a knowledge base, while retaining some of the structure from the parent items. Kelly et al. [22] showed an interactive genetic algorithms (IGAs) approach that simulated creativity through the use of divergent and convergent thinking processes.

3   Discussion of the Approaches

The studies reviewed in the precious section recognize that knowledge and past experience are very important for intuition to be accurate. The concept of intuition, as suggested by the early studies, focuses largely on the concept itself, rather than on the representation and use of entities in the process. While the earlier studies took a psychological and cognitive approach, some recent studies discussed above attempt to study artificial intuition from the computational point of view. To the best of our knowledge, there are four main studies [1, 6, 12, 14], which have considered the computational model of artificial intuition. However, an important limitation of these approaches is that they did not present a detailed representation of algorithm and use of intuitive entities in the process. Based on the forgoing discussion, artificial intuition essentially identifies the presence of the following.

1. Knowledge and Experience, which include prior knowledge of events, concepts, patterns as well as variables, which have been acquired over time from experience. This knowledge of patterns has been created and is held in a mental map about the subject of interest. These patterns are relevant information pieces that help to create the intuitive decision, which is used to recognize and act rapidly. Decisions that are made in circumstances similar to previous experience, and whose outcome could be potentially harmful, or potentially advantageous, induce a somatic response used to mark future outcomes that are important to us, and to signal their danger or advantage [23]. In general, somatic marker is concerned about the outcome of the choices that is made about the decision options that are presented. Intuitive feeling is created based on such experiences and this aids the decision process in an automatic manner. Thus, when intuition assesses a decision context and juxtaposed a negative somatic marker to the outcome of the decision option, it sends an automatic response to drop that option.

2. Common-sense understanding and identification of subtle trends and the selective attention to certain aspects or events; the preference for viewing situations from a broader perspective.
3. Similarity recognition, which is a comparison of similar and dissimilar characteristics; the ability to recognize subtle likenesses to cues found in past episodes despite many differences in the current situation.
4. Partial information about the subject of interest or context of the problem.

4   Description of the Model: The Main Architecture

The aim of this section is to provide the main architecture and the essential requirements of the model. In this study, we consider creativity as a curious mind that searches a path to an objective. More specifically, we consider the curiosity as a scenario-based search space that searches the path based on the scenario. It consciously or unconsciously searches through the paths connecting the concepts that are related to each other. The more paths are searched, the more the creative idea grows and short paths to nearby concepts yield commonly available associations [24]. For example, "Car reminds us of a Bicycle" is a closely related concept to Car. In fact, this generates a search process in our mind on how each of these concepts are related. However, the way each concept relates to other concepts depends on education, environment and personality [20]. Concepts can be related to other concepts, and this relatedness can be as deep as the representation requires. This process is often referred to as spreading activation [25]. An important aspect to consider is the similarity measurements of the concepts. More specifically, how do the concepts compare with respect to their semantic properties.

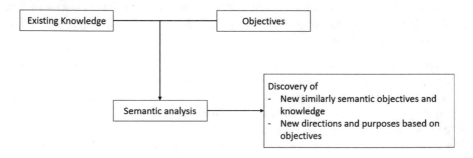

Fig. 1. Modelling curiosity

More specifically, as depicted in Figs. 1 and 2, the overall approach is the decomposition and analysis of the objectives within a specific scenario. In particular,

• An objective is defined as a collection of semantic properties and specific features.

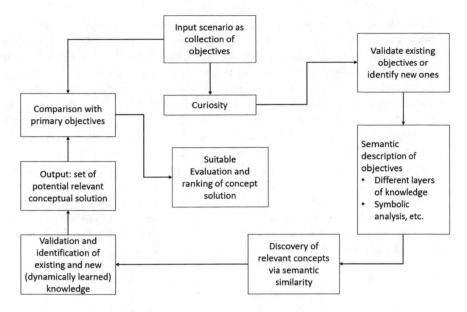

Fig. 2. The main components of the method

- Curiosity is the process to discover new knowledge
- "random curiosity" is searching or observing without clear objectives
- "targeted curiosity" (which is what we consider here, and we will refer to it as simply "curiosity") is based on specific objectives and we are looking for any potential new way to do things
- In any such process, there is a waste of energy or cost involved
- Therefore, there will be a trade-off stage, which will need to be addressed in future (future work).

4.1 Requirements of Artificial Intuition Model

The essential properties of artificial intuition model are to have the capacity for:

1. Pattern recognition: Recognize patterns which are relevant information pieces that enables humans to synthetize decisions. This help to create the intuitive decision.
2. Have multidimensionality capacity. This means that the model has the capability to integrate multiple threads of information simultaneously without diminishing the quality or accuracy of the response. More specifically, the system has the capacity to map or replay the patterns and adapt to uncertain and new environments.
3. Performance in terms of the capacity to obtain answers much faster.
4. An essential aspect of the development of intuitive models is the ability for knowledge network representation, which includes a network of concepts with

their corresponding properties and attributes. These concepts and their corresponding attributes are the central data structure of the model, which can be defined as semantic networks. These are a graphical knowledge representation of concepts and their mutual connections within the context of the domain that is described by the concepts. Fundamentally, the process of building a semantic network requires that documents are all restricted to a domain that can be characterized by well-defined, inter-related concepts. These concepts form the basis for the scientific terminology of the domain [26]. Furthermore, each concept in the network is represented by a node and the hierarchical relationship between them is depicted by connecting appropriate concept nodes via is-a or instance-of links [27]. Nodes at the lowest level in the is-a hierarchy denote tokens, while nodes at higher levels denote classes or categories of types. Properties in the network are also represented by nodes and the fact that a property applies to a concept is represented by connecting the concept and property nodes via an appropriately labelled link.

Essentially, modeling of artificial intuition is accomplished through recognition of significant patterns and properties that are made available by prior knowledge and experience, given the context of the problem. Therefore, given the above, we define artificial intuition as the ability of a system to assess a problem context and use pattern recognition or properties from a dataset to choose a course of action or aid the decision process in an automatic manner.

5   Conclusion

In this article, we have presented an artificial intuition and the related field of computational creativity. We have introduced a modeling approach to artificial intuition. This is part of a wider line of inquiry with the aim to define a computational model of artificial intuition and by providing an analysis of objectives within some specific scenarios. Such model can find some potential real-world application in Simulating weather prediction; integration with Clinical Decision Support (CDS) system to reduce diagnostic errors; simulating the decision of a dealer in the business of buying and selling of stocks; simulating the decision of a player in a game. In the future, we aim to provide a rigorous evaluation of scenarios captured by semantic networks.

References

1. Dundas, J., Chik, D.: Ibsead: - a self-evolving self-obsessed learning algorithm for machine learning. Int. J. Comput. Sci. Emerg. Technol. **1**(4), 74 (2011). (E-ISSN 2044-6004)
2. Trovati, M.: Reduced topologically real-world networks. Int. J. Distrib. Syst. Technol. **6**, 13–27 (2015)
3. Payne, L.K.: Intuitive decision making as the culmination of continuing education: a theoretical framework. J. Contin. Educ. Nurs. **46**, 326–332 (2015)

4. Graber, M.L., Kissam, S., Payne, V.L., Meyer, A.N.D., Sorensen, A., Lenfestey, N., Tant, E., Henriksen, K., LaBresh, K., Singh, H.: Cognitive interventions to reduce diagnostic error: a narrative review. BMJ Qual. Saf. **21**, 535–557 (2012)
5. Hams, S.P.: A gut feeling? Intuition and critical care nursing. Intensive Crit. Care Nurs. **16**, 310–318 (2000)
6. Diaz-Hernandez, O., Gonzalez-Villela, V.J.: Analysis of human intuition towards artificial intuition synthesis for robotics. Mechatron. Appl. Int. J. (MECHATROJ) **1**, 23 (2015)
7. Simon, H.A.: Making management decisions: the role of intuition and emotion. Acad. Manag. Exec. (1987-1989) **1**(1), 57–64 (1987)
8. Kahneman, D., Frederick, S.: Representativeness revisited: attribute substitution in intuitive judgment. In: Gilovich, T., Griffin, D., Kahneman, D. (eds.) Heuristics and Biases: The Psychology of Intuitive Judgment, pp. 103–119. Cambridge University Press, Cambridge (2002)
9. Stanovich, K.E., West, R.F.: Individual differences in reasoning: implications for the rationality debate? Behav. Brain Sci. **23**, 645–665 (2000)
10. Kahneman, D.: Maps of bounded rationality : a perspective on intuitive judgment and choice (2002)
11. He, J., He, P.: Fuzzy relationship mapping and intuition inversion: a computer intuition inference model. In: IEEE 2008 International Conference on MultiMedia and Information Technology, December 2008
12. Tao, W., He, P.: Intuitive learning and artificial intuition networks. In: IEEE 2009 Second International Conference on Education Technology and Training, December 2009
13. Frantz, R.: Herbert simon. Artificial intelligence as a framework for understanding intuition. J. Econ. Psychol. **24**(2), 265–277 (2003)
14. Srdanov, A., Kovacevic, N.R., Vasic, S., Milovanovic, D.: Emulation of artificial intuition using random choice and logic. In: IEEE 2016 13th Symposium on Neural Networks and Applications (NEUREL), November 2016
15. Srdanov, A., Milovanović, D., Vasić, S., Ratković Kovačević, N.: The application of simulated intuition in minimizing the number of moves in guessing the series of imagined objects, February 2017
16. Boden, M.A.: The Creative Mind: Myths and Mechanisms. Routledge, Abingdon (2003)
17. Zhang, Z.-X., Zhong, W.: Barriers to organizational creativity in Chinese companies, pp. 339–367, April 2016
18. Bottom, W., Gilovich, T., Griffin, D., Kahneman, D.: Heuristics and biases: the psychology of intuitive judgment. Acad. Manag. Rev. **29**, 695 (2004)
19. Koestler, A.: The Act of Creation. Arkana, London (1989). OCLC: 22220796
20. Blancke, S., De Smedt, J.: Evolved to be irrational? Evolutionary and cognitive foundations of pseudosciences, pp. 361–379, January 2013
21. Pease, A., Corneli, J.: Chapter 10 Evaluation of Creativity, pp. 277–294. Springer, Cham (2018)
22. Kelly, J., Papalambros, P.Y., Seifert, C.M.: Interactive genetic algorithms for use as creativity enhancement tools. In: AAAI Spring Symposium: Creative Intelligent Systems (2008)
23. Bechara, A.: Emotion, decision making and the orbitofrontal cortex. Cereb. Cortex **10**, 295–307 (2000)
24. Schilling, M.A.: A "small-world" network model of cognitive insight. Creat. Res. J. **17**(2–3), 131–154 (2005)

25. Collins, A.M., Loftus, E.: A spreading activation theory of semantic processing. Psychol. Rev. **82**, 407–428 (1975)
26. Brasethvik, T., Gulla, J.A.: A conceptual modeling approach to semantic document retrieval. In: Pidduck, A.B., Ozsu, M.T., Mylopoulos, J., Woo, C.C. (eds.) Advanced Information Systems Engineering, pp. 167–182, Springer, Heidelberg (2002)
27. Shastri, L.: A connectionist approach to knowledge representation and limited inference. Cognit. Sci. **12**(3), 331–392 (1988)

Twitter Analysis for Business Intelligence

Tariq Soussan and Marcello Trovati[(✉)]

School of Computing, Edge Hill University, Ormskirk, UK
tariqsoussan@gmail.com, marcello.trovati@edgehill.ac.uk

Abstract. The evolvement of social media has made it an important and valuable part of people's daily life all over the world. Many businesses use social media in different ways that benefit their business, such as to advertise their products and services, and as a way of strengthening relationships. Institutes utilize social media to promote programs and courses to current and prospective students, to advertise important events such as career fairs and to interact with various institute members to provide them with up to date news and information regarding their wellbeing, health, security, comfort and satisfaction. Throughout this work, web mining and opinion mining will be applied to a general business Twitter account to explore the developing themes of discussions amongst businesses' online communities. The account will be analysed through text mining, social network analysis and sentiment analysis. It is important to monitor what topics and words are trending in high volume on specific online twitter communities as the objective is to try to conclude the sentiments of the posts posted on the Twitter account.

1 Introduction

Social networks allow people or entities to exchange ideas and data in virtual communities online [13]. An essential advantage of social media is that it holds mobile technologies containing platforms that allow users to create user-generated content. The content can be altered, promoted, and discussed online [14].

As social media began to develop and thus reach different countries around the world, more and more entities became more involved in including social media as part of their online communities, marketing campaigns, and discussion boards. One of the main types of institutes that have a crucial interest in these technologies are the higher education institutes who aspire for their students to become more dynamic learners in the most innovative ways possible [11].

Throughout this work, a Twitter account will be examined to see how its community members interact using it. The aim is to explore the developing themes of discussions amongst online communities using a Twitter data set. It is important to monitor what topics and words are trending in high volume on the online Twitter communities and what requests, comments and suggestions are being retweeted. The tweets on the Twitter account will be grouped and categorized into positive and negative feelings through opinion mining or sentiment analysis. There is a need to attain the opinion or sentiment of the online community speakers as well as a need to track and monitor how they feel about specific topics related to the general businesses and their work.

© Springer Nature Switzerland AG 2020
L. Barolli et al. (Eds.): INCoS 2019, AISC 1035, pp. 473–480, 2020.
https://doi.org/10.1007/978-3-030-29035-1_46

Overall, the knowledge that can be gained will give recommendations that will support general businesses' board officials in making decisions to improve their resources. It is important to turn data into decisions since the knowledge that will be extracted can help the board officials assess the topics and issues that were mostly highlighted from analysing customer's, staff's, and people's perceptions and points of views. In the coming sections, Sect. 2 will give a critical discussion and literature review about social media and data mining social media. Section 3 will discuss the methods used to analyse this data. Section 4 will discuss the results and analysis of the results. Finally, Sect. 5 will give a conclusion of what was found and will make potential recommendations that may be used by the general businesses.

2 Literature Review

The concept of social media is so complex that it was defined in various literatures in many ways. It is considered as the clearest, appealing and collaborative shift in education [16]. It was described as applications built on Web 2.0 foundations, allowing the generation of user content and exchanging it [4]. Therefore, it helps convert internet users from just being readers to being participators, by permitting them to produce content and communicate online to form private and business relationships [9]. It is also regarded as "a give-to-get environment" because online communication is based on content swap [22]. Social media can also be regarded as a Two-Way platform where users can share and discuss information [12].

Essentially, the platforms of social media were created to connect individuals together and not companies, since they were not intended to be marketing platforms [19, 3]. However, in recent years, social media started to be used for social media marketing and this changed. The importance of social media can be highlighted through its effect on companies, institutes, or firms and how their users utilize it. It enables their users to connect and to expand their marketing opportunities through a growth of communication. The democratization of information has permitted companies to communicate with customers and for customers to communicate with each other making this model a Two-Way communication model [17]. In addition, institutes can listen and reply to their customers, motivating them to help each other and cooperate to improve their services and products [15].

One of the most popular and most used social media websites these days is Twitter, which was developed by Obvious Corporation in 2006 [7]. It was defined as an online social network that allows millions of individuals to stay in touch with their family, friends, and colleagues from different countries around the world by using their mobile phones and computers [10]. It is considered as a platform that permits users to communicate with each other by creating connections between them [20].

Social Media has unfolded new potentials for organizations to involve their stakeholders by permitting them to transfer data and obtain instantaneous feedback from diverse stakeholders. A lot of researchers have done research on Twitter's effect in particular on organizations [15]. A study was performed in 2009 on five hundred Fortune companies to see whether their Twitter account can be classified as active, pending their accounts were updated in the last thirty days. The results of the study showed that out of the five hundred companies, only 35% of them had accounts classified as active and only 24% of them replied to their accounts' followers with new updated Tweets being posted on their accounts [1].

Text mining is the procedure of finding exciting and non-trivial patterns or facts from unstructured text files. It is also referred to as data mining of text and text database knowledge detection [21]. It is the discovery of relevant information in text [23]. It is seen that text mining is a more complicated assignment than regular data mining techniques since it handles text data that can be fundamentally unorganised and fuzzy. It is regarded as involving many academic disciplines such as text analysis, data withdrawal, data recovery, computational learning, classification, cluster analysis, visualisation, organizing data in databases and data mining [21].

Sentiment analysis can be recognized as a classification problem where the assignment is to categorize messages into two classes subject on whether they carry positive or negative feelings [2]. It is also defined as a natural language processing and data extracting tool whose object is to obtain the author's feelings conveyed in positive or negative remarks, inquiries and requests through analysing a big number of documents that contain text. Its purpose is to find the attitude of a speaker or an author concerning a specific topic or them or the complete tonality of a text file [18]. The feelings can also be false positive if they are incorrectly classified negative, and they can be false negative if they are incorrectly classified positive [6].

3 Methodology

For this work, the validation was run on data related to University of Derby as a business entity. This organization's Twitter account was selected to see if a positive perception from the public can be utilized to provide recommendations on how it can enhance its resources. The University of Derby's Twitter account was inspected to see how its community members communicate using it. This was done in order to explore the developing themes of discussions amongst University of Derby's online communities using a University of Derby's Twitter dataset.

The experimental environment used for this work is a Windows 10 Pro laptop with a processor of AMD E-300 APU with Radeon™ HD Graphics 1.3 GHz. The installed memory (RAM) is 6.00 GB. The System type is 64-bit Operating System, x64-based processor.

The data size for a data set can reach up to 3240 of the most recent tweets since this is the maximum number allowed using this method and Twitter API. This data set was selected because it represents an official Twitter account for a University of Derby online community. Thus, this data helps find developing themes of discussions by University of Derby students specifically.

A web-based system called Internet Community Text Analyzer (ICTA) on Netlytic [8] was used to analyse the data. ICTA uses two algorithms which are Name Network Algorithm and Name Alias Resolution Algorithm [5]. Name Network Algorithm is made up of node discovery and tie discovery. In node discovery, the algorithm eliminates "stop-words" and then normalises all remaining words by removing all "special symbols" from the start and the end of these words. For the rest of the words, the algorithm depends on a dictionary of names and a group of linguistic guidelines derived manually [5]. The algorithm depends on "context words" that frequently show personal names to discover names that might be found in the dictionary. It also eliminates word sequences of more than three capital words in order to eliminate words that are part of institutes [5]. Finally, after all relations have been generated with their total confidence levels, the algorithm goes through all the posts once again to change those names found in the body of the posts that have been related to an email at least. If in any case the name has more than one email, the algorithm will use the email with the maximum level of confidence.

3.1 Text and Social Network Analysis Using Netlytic

Following a suitable pre-processing to ensure the data followed a consistent format, the analysis via Netlytic was implemented. This is cloud-based text and social networks analyser [8] which was used to perform text analysis and network analysis for the data set that resulted from pre-processing. Key Extractor is a tool for text analysis, which helps to detect prevalent themes or topics in a dataset by calculating the word frequency in it [8]. In addition, this tool calculates the number of unique words found by eliminating all common words from a "list of stop-words" in eighteen languages. It will then calculate the total number of messages where every word exits. Upon picking a word or topic that was one of the top words or topics, Keyword Extractor allows the researcher to see that word highlighted in context of all the tweets that had it with the date and time it was posted. Another tool used is Manual Catagories for text analysis, which helps generate categories of words and phrases to symbolise wider notions like positive versus negative words [8]. The categories can be generated manually by a user, or automatically by Netlytic if this tool is being used for the first time for a designated dataset. It will automatically detect and calculate which tweets in a dataset belong to which category. This tool can be used for sentiment analysis since it extracts the positive and negative words from the tweets that can reveal if the attitude in the tweet is positive or negative. The results can be shown as a "Treemap visualization". Network Analysis to be used is a communication network constructed from detecting and extracting individual names in the tweets. It also allows the researcher to detect ties in name networks through linking a tweet sender to all the names existing in the sender's tweets. It calculates the number of nodes and the number of ties. It also calculates the number of names detected, which is equal to the number of single personal names that Netlytic detected in this dataset. The results can be shown in clusters.

4   Statement of Results and Analysis

Neylytic [8] was used where the dataset was imported to its cloud. The below portrays the results of text analysis and network analysis.

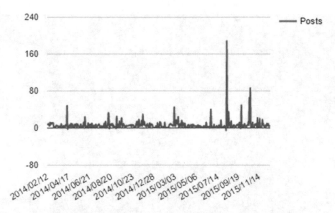

Fig. 1. Number of posts for DerbyUni between February 2014 and December 2015

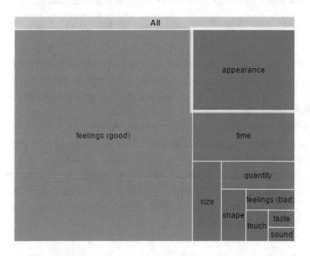

Fig. 2. Treemap of the DerbyUni tweets categories

The DerbyUni Twitter account discusses The University of Derby's open days as well as general information related to studying at University of Derby. Figure 1 shows that between February 2014 and December 2015, the highest number of posts recorded was 188 posts on August 13th, 2015. This can be justified since this month is right before the start of the academic year, which starts in September, and most university posts about open days will be close to that month. Some of the frequent words do not necessarily represent a topic, while others do. Subsequently, sentiment analysis was carried out based on the most common concepts as shown in Fig. 2. The positive tweets were 720 while the negative tweets were 20. 11 posts could not be categorised into to any of the categories. Some challenges were observed, for example some sentiments in "feeling (bad)" were not classified correctly due to negation in them. An example of this is tweets in "nervous", "bad", and "never" subcategories. Another example is the worried subcategory where the university was asking students to give

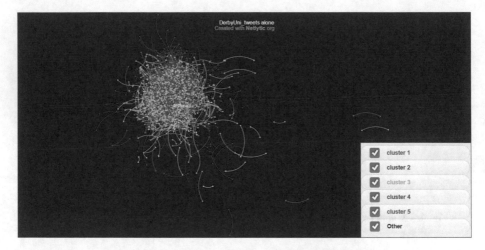

Fig. 3. Name network of DerbyUni

them a call if they are worried and it was incorrectly classified in the "feeling (bad)" category due to domain dependency. Another challenge is subjectivity detection when some subcategories contain tweets that include words that can be non-opinion but was put in one of the subcategories. In Fig. 3, 5 main clusters represent the name network for DerbyUni where each cluster contains tweets connected to each other and most of the tweets include a common name. Cluster 1 has 920 tweets that include "derbyuni", cluster 2 has 65 tweets that include "derbyunistudent", cluster 3 has 95 tweets that include "derbyunipress", cluster 4 has 59 tweets that include "teamderby", and cluster 5 has 26 tweets that include "derbyuniarts". Thus, it is observed that "derbyunistudent", "derbyunipress", "teamderby", and "derbyuniarts" are one of the frequent users who are included in posts by DerbyUni.

5   Conclusion

Throughout this work, web mining and opinion mining were applied to a University of Derby Twitter account for validation. The aim was to explore the developing themes of discussions amongst University of Derby's online communities through a Twitter account. It is observed that the highest amount of posts in the dataset was during or just before September, which means that most of the posts by students or the university occurred just before the start of a new academic year. It is recommended that there be a more constant level of activity on these account during other months of the year in order to discuss events, make suggestions, promote courses and organize meetings, which may encourage students to be more involved with their university. Finally, it observed that the dataset has networks of tweets that mostly contain other Derby university twitter accounts in their posts, which asserts that accounts tag each other in most of their posts. This helps the university's marketing campaigns for all events, course offerings, and announcements related to the university. It is recommended for university accounts to tag each other more in their posts for their posts to reach as many viewers as possible. Future work can be work on improving sentiment analysis to be more accurate so that it will produce results with less text that are incorrectly classified.

References

1. Barnes, N.G.: The Fortune 500 and social media: a longitudinal study of blogging, Twitter and Facebook usage by America's largest companies (2010). Retrieved from Society for New Communications Research on 6 March 2011
2. Bifet, A., Frank, E.: Sentiment knowledge discovery in Twitter streaming data. In: Pfahringer, B., Holmes, G., Hoffmann, A. (eds.) DS 2010. LNCS (LNAI), vol. 6332, pp. 1–15. Springer, Heidelberg (2010). https://doi.org/10.1007/978-3-642-16184-1_1
3. Bottles, K., Sherlock, T.: Who should manage your social media strategy. Physician Executive 37(2), 68–72 (2011)
4. Campbell, D.: The new ecology of information: how the social media revolution challenges the university. Environ. Plann. D: Soc. Space 28(2), 193–201 (2010)
5. Daniel, B.K. (ed.): Handbook of Research on Methods and Techniques for Studying Virtual Communities: Paradigms and Phenomena: Paradigms and Phenomena. IGI Global, Hershey (2010)
6. Elgamal, M.: Sentiment analysis methodology of Twitter data with an application on Hajj season. Int. J. Eng. Res. Sci. (IJOER) 2, 82–87 (2016)
7. Grosseck, G., Holotescu, C.: Can we use Twitter for educational activities. In: 4th International Scientific Conference, eLearning and Software for Education, Bucharest, Romania, April 2008
8. Gruzd, A.: Netlytic: Software for Automated Text and Social Network Analysis (2016). http://Netlytic.org
9. Holotescu, C., Grosseck, G.: An empirical analysis of the educational effects of social media in universities and colleges. Internet Learn. 2(1), 5 (2013)
10. Huberman, B.A., Romero, D.M., Wu, F.: Social networks that matter: Twitter under the microscope. arXiv preprint arXiv:0812.1045 (2008)
11. Hughes, A.: Higher education in a Web 2.0 world. JISC report (2009)
12. Johnston, R.: Social media strategy: follow the 6 P's for successful outreach. Alaska Bus. Mon. 27(12), 83–85 (2011)
13. Junco, R., Heiberger, G., Loken, E.: The effect of Twitter on college student engagement and grades. J. Comput. Assist. Learn. 27(2), 119–132 (2011)
14. Kietzmann, J.H., Hermkens, K., McCarthy, I.P., Silvestre, B.S.: Social media? Get serious! Understanding the functional building blocks of social media. Bus. Horiz. 54(3), 241–251 (2011)
15. Lovejoy, K., Waters, R.D., Saxton, G.D.: Engaging stakeholders through Twitter: how nonprofit organizations are getting more out of 140 characters or less. Public Relat. Rev. 38(2), 313–318 (2012)
16. Malita, L.: Social media time management tools and tips. Procedia Comput. Sci. 3, 747–753 (2011)
17. Markos-Kujbus, É., Gáti, M.: Social media's new role in marketing communication and its opportunities in online strategy building. BCE Marketing, Marketingkommunikáció és Telekommunikáció Tanszék, Budapest (2012)
18. Mukherjee, S., Bhattacharyya, P.: Sentiment analysis: a literature survey. arXiv preprint arXiv:1304.4520 (2013)

19. Piskorski, M.J.: Social strategies that work. Harvard Bus. Rev. **89**(11), 116–122 (2011)
20. Sultana, M., Paul, P.P., Gavrilova, M.: Identifying users from online interactions in Twitter. In: Gavrilova, M.L., Tan, C.J.K., Iglesias, A., Shinya, M., Galvez, A., Sourin, A. (eds.) Transactions on Computational Science XXVI. LNCS, vol. 9550, pp. 111–124. Springer, Heidelberg (2016). https://doi.org/10.1007/978-3-662-49247-5_7
21. Tan, A.H.: Text mining: the state of the art and the challenges. In: Proceedings of the PAKDD 1999 Workshop on Knowledge Disocovery from Advanced Databases, vol. 8, pp. 65–70, April 1999
22. Uzelac, E.: Mastering social media. Research **34**(8), 44–47 (2011)
23. Zhong, N., Li, Y., Wu, S.T.: Effective pattern discovery for text mining. IEEE Trans. Knowl. Data Eng. **24**(1), 30–44 (2010)

The 5th International Workshop on Collaborative e-Business Systems (e-Business-2019)

Bonus Programs for CRM in Retail Business

Wolfgang Neussner[1,2], Natalia Kryvinska[1,2(⊠)], and Erich Markl[2]

[1] Comenius University, Bratislava, Slovakia
neussner@stona.at, Natalia.Kryvinska@fm.uniba.sk
[2] FH Technikum Wien, Vienna, Austria

Abstract. This article does a literature research regarding the conception of bonus programs in scientific theory including concepts of customer value and loyalty. The starting point of the loyal customer as the basis of a successful customer value management in the context of maximizing enterprise value, the necessity of customer loyalty and emotional bonding supplemented by the development of purchase decision processes, the implementation in companies is quantitatively surveyed and compared with the theoretical concepts. In a saturated and highly competitive environment, the management of customer loyalty must become particularly important, as new customers can only be won at considerable expense and in manageable numbers. The aim is to strengthen customer loyalty to the company in order to ward off competitive marketing campaigns in the best possible way. For example, customer loyalty programs in which existing customers are rewarded (monetary or non-monetary) for their repurchasing behaviour are used to try to retain existing customers for the company.

1 Introduction

Customer loyalty programs can be understood as marketing programs designed to generate customer education by offering benefits to profitable customers. [1] whereas Henderson et al. regards it as an incentive system that influences customer behaviour in the sense that consumer behaviour is still improved without price or assortment changes [2]. Bonus programs are part of Customer Relationship Management (CRM). The goal is the long-term improvement of the profit potential. Key Performance Indicator is the Customer Lifetime Value (CLTV) [3]. This is calculated from the balance of all revenues and costs of a customer over the duration of the business relationship. Quantitative and qualitative components must be taken into account. The main qualitative factors include recommendation, cross-selling and up-selling potential. Since the duration of the customer relationship is a key input parameter, it makes more sense to calculate CLTV dynamically than statically [4]. The main factors influencing customer value are customer loyalty, cross-selling and up-selling potential, the reduction of customer service costs and other value-enhancing potential, such as recommendation behaviour. Customer loyalty is intended to generate loyal customers who deliver greater profits for companies, since acquisition costs are not incurred with every purchase. The increasing purchasing power in a person's life cycle can also have positive effects [3]. For a detailed calculation Kumar/Reinartz [5] are offering the following model in Fig. 1.

© Springer Nature Switzerland AG 2020
L. Barolli et al. (Eds.): INCoS 2019, AISC 1035, pp. 483–491, 2020.
https://doi.org/10.1007/978-3-030-29035-1_47

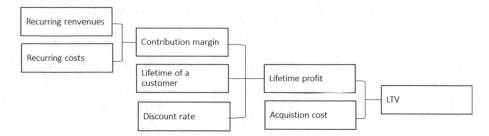

Fig. 1. Shows the calculation of the customer lifetime value Kumar and Reinartz [5]

Recurring revenues are reduced by recurring costs, which results in the contribution margin, which is discounted in connection with the customer duration and a discount factor to be selected, and the result is reduced by the acquisition costs. Customer Lifetime Value Management (hereinafter CLTV-M) has aligned all sales and marketing activities in the company with customer value. This serves to maximize the company's profit, based on a strategy of customer orientation, which through a conscious non-equal treatment of all customers "the right measure for the right customer". Higher expected profits also justify higher or non-standardized acquisition costs. The main task of the CLTV-M is therefore the adequate selection. Bolton et al. [6] see the use of marketing instruments (service quality, direct mailings, relationship instruments, reward programmes as well as advertising and distribution) as a way of perceiving the customer (price, satisfaction, commitment). If the instruments are chosen to suit the customer, this will be reflected on the one hand in the customer's behaviour (repetition, additional and crossbuying behaviour) and on the other hand in the resulting costs and revenues and thus ultimately in CLTV.

Figure 2 shows the demarcation between product, process and value orientation in the company. The process orientation complements the product orientation with the levels of customer marketing and customer care/retention, where the value-oriented approach has been expanded to include CLTV-M. The product orientation has been expanded to include CLTV-M as can be seen in Fig. 2.

This also shows that conventional (one-dimensional) strategies (cost leaders or quality leaders) must be developed in a value-oriented approach to multidimensionality in order to be able to live customer value in all dimensions. CLTV-M is therefore not an organisational unit, but a target for the creation and maintenance of profitable, sustainable customer relationships. In estimating the expected CLTV, valuable customers are treated individually and less valuable customers standardized. The assessment in valuable and less valuable is carried out by socio-demographic characteristics, the obtained turnover with the help of the analysis of the life cycle of the customer. The monitoring of a customer's intrinsic value is given very high priority, which is also measured continuously on the basis of the availability of data on individual customers. In this way only those customers are to be bound who are worth it to be bound [4].

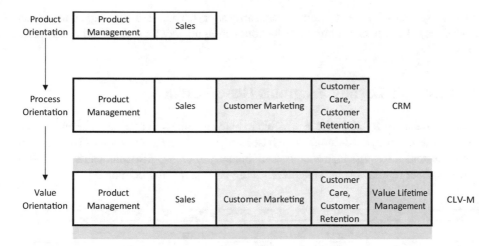

Fig. 2. Structure of customer lifetime value management (network consulting)

The challenge in the conception of such a bonus program is to bind the customer to the company and not to the bonus program. Even if companies regard the bonus program exclusively as part of the marketing mix, customers only see the totality. The consequence of this may be that a bonus programme that is not perceived as serious or the products and/or services offered as bonuses are not perceived as qualitatively meeting the requirements, and the products and services may also be perceived as inferior. For example, a bonus program that is perceived negatively can also have a negative impact on a company's sales [7].

The following distinctions regarding the degree of brand loyalty can be found in the literature.

"Hard-core Loyals" refers to customers who buy only one brand from a company at all times. In terms of customer loyalty, these are the target customers that all companies want. Spilt loyals' are those which consistently purchase two or three different brands from different dealers. This happens because no reason or obligation is seen to show exclusive loyalty to a brand [8]. Shifting loyals include those customers who now and then change brands from different companies [9], whereas "switchers" have no ties to any brand and very often switch between different brands, different businesses [9, 10]. If customers regularly buy from the same company and have little opportunity to switch to competing companies, this is referred to as "Behavioural loyalty" [11]. Attitudinal loyalty is the term used when a customer gives a strong preference to a brand [12]. "Attitudinal loyalty" leads to positive recommendations through "word-of-mouth propaganda" [11]. Customers with a "cognitive loyalty" are willing to pay higher prices for products or services compared to other providers) [11]. Customers can be either in connection with a company (the net benefit lies in the perception of the customer by the customer) or in a state of connection. In the second case, the condition is covered by a change barrier, which restricts customer freedom) [13]. However, customer loyalty can only be spoken of if net benefit aspects or change barriers are the trigger for repeat purchases [14]. A habitual buying behaviour can in no way be understood as customer loyalty [15].

2   Customer Loyalty Programs Classification

In order to maximize CLTV, it is possible to bind customers to one's own company and to increase it through additional purchases and repurchases. Ranzinger [16] differentiates customer loyalty instruments into bonus programs, discount cards, customer cards without regular incentives, point gluing campaigns and couponing.

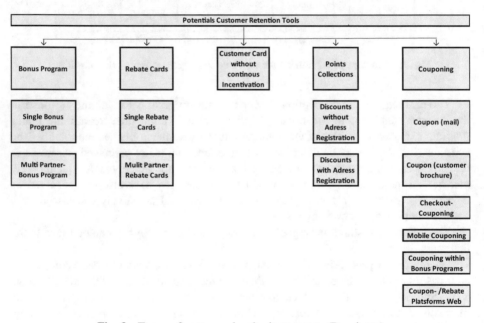

Fig. 3. Types of customer loyalty instruments (Ranzinger)

Bonus programs are divided into single and multi-partner programs. Thus advantages can be collected either only with one or several enterprises. Notwithstanding the single bonus program, products or services from other companies can also be offered as benefits. The multi-partner programs usually have only one player on the market, with no other competitors in the industry. Discount cards can be issued as single or multi-partner cards. It must also be distinguished whether the discount is an immediate discount at the cash desk or a refund at the end of the year. The refund at the end of the year can be in the form of money or vouchers of the company. The disadvantage of customer cards without regular incentives is that they are not shown at the checkout every time a customer makes a purchase due to the lack of an advantage, which means that data on customer behavior is not recorded even though the customer is aware of it. Irregular coupons are sent and can be earned without showing the customer card.

Point gluing actions can be carried out with or without address acquisition. From a predefined purchase value, customers receive adhesive dots. After the collection of a predefined number of adhesive dots, these can be redeemed partly with partly without additional payment. When redeeming, there is the possibility of collecting personal data in order to be able to use them for marketing purposes.

As shown in Fig. 3 in the case of couponing, a distinction is made according to the distribution channel (mailshot, customer magazine, checkout process, mobile, as part of bonus programs and platforms). Couponing can be implemented by industry funded discounts on special products or by pre-defined discounts on all products [16].

3 Bonus Schemes

Bonus programs serve as a means of customer retention. A clear definition is not given in the literature. For example, Lauer [17] defines bonus programs as "when a systematic offer is made by companies to customers to collect specifically created value units (bonus points) for certain behaviors, which can be converted into benefits (bonuses) above a certain size" (redemption threshold).

Künzel [18] defines the goal of a bonus program as a strategic marketing instrument for customer loyalty, whereby the rewards for the customers [18] are in proportion to their purchasing behaviour.

Diller [19], on the other hand, sees bonus programmes as rebate systems for customers who receive rebates in kind or cash bonuses after achieving predefined values (purchase quantities or points).

As different as the approaches to the definition of a bonus program may be, they all have four characteristics

- The bonus program runs over a relatively long period of time.
- Participants are rewarded for predefined behaviors.
- The participants of the bonus program collect units of value.
- The participants of the bonus program will exchange these value units for bonuses or rewards at a later date [3].

A distinction must be made between systems where participants can only collect and redeem points at one or more companies due to their purchasing behaviour. There is also the possibility of using the bonus systems Business to Business and Business to Customer [7].

The first bonus program to be mentioned in the literature was that of American Airlines in 1981 [20]. Bonus programs have their origin in the airline business and trade, although today other industries, such as banks, insurance companies or energy companies offer bonus programs. Bonus programs are strongly influenced in the retail sector by greater digitisation (e-commerce) and supra-regionalization of formerly individual businesses. The rapid progress of e-commerce providers is also leading to the digitalization of bonus programs [3].

According to Lauer [17] the basic mechanics of bonus programs work are presented in the Fig. 5.

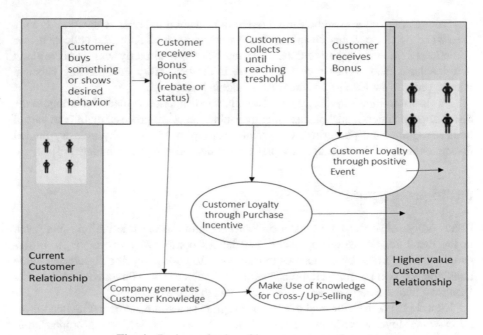

Fig. 4. Basic mechanics of bonus programs (Lauer)

As shown in Fig. 4 for a predefined behavior (e.g. turnover), the customer registered in the bonus program receives bonus points (discounts or status points), which he can exchange for a bonus performance when reaching a redemption threshold (minimum turnover or minimum number of points).

From the point of view of the company, the registration of the customer generates more knowledge about the individual customer, which leads to customer-specific offers and ideally ends in cross/upselling.

Functions of a bonus program are

- Identification function: the participants are identified and their buying habits are recorded. Sociodemographic data are recorded and a map is usually used for identification purposes.
- Bonus function: predefined behavior is rewarded.
- Interaction function: the identification also enables targeted interaction with the participant.
- Service function: Services for participants, such as free credit card or ticket service [7].

See the design possibilities of bonus programs [7] a combination of the following instruments. The target group is defined as business-to-business and/or business-to-customer. The extent and quality of the coverage must be determined and possibly also achieved in the form of cooperation. Should repeated shopping behaviour be positively rewarded and this be defined by entry barriers, rewarding the behaviour, defining point heights and the collection medium (card or app). Which channels should be used for communication with the participants (direct mailing/customer magazine/hotline/posters, TV). Should the redemption of points be made more difficult (bonus threshold/ gradations/expiration of points) or easier (co-payment, access). The benefit is offered in the form of money, vouchers, material prizes, experience prizes or a status program.

For bonus programs, a distinction must be made between multi-partner programs and bonus programs without partners. Cross-industry and therefore non-competing companies join together to form multi-partner programmes. This is usually handled by a specially founded company, which acts as a service provider and is financed by the participating partners. In return, the company takes over the entire processing, such as advertising, customer data management and the provision of premiums. The best-known multi-partner programs in Germany include Payback, HappyDigits, which was founded by Deutsche Telekom, DeutschlandCard and the bonus program of Lufthansa, Miles and More. Bonus programs without partners are offered by individual companies that do not have any cooperations or collection partners, which serves the purpose of clarity [7].

Lauer [3] sees only four ways to bonus credit. A distinction must be made between status points and discount points. Collection points may be provided by own products/services or by third parties. If the customers are not motivated by the bonus to the desired behavior, the bonus program loses its meaning.

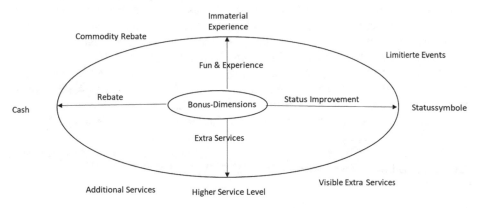

Fig. 5. Dimensions of bonification Lauer [3]

The first subdivision of bonus benefits distinguishes between economic and rational benefits. Economic benefit is understood to mean those products and services which could be acquired with the help of financial means even without membership, whereby the customer receives a measurable advantage (in monetary units). The (exclusively) emotional benefit, on the other hand, cannot be measured in monetary terms and cannot be acquired with money either. Since the monetary benefit can also lead to feelings, a clear separation is not possible and reasonable and leads Lauer to the following differentiation into four dimensions: Discount, extra services, status and fun & experience. A clear demarcation of these four models is often not given, whereby the best combination should be developed. By definition, the rebate is not granted immediately but only after thresholds have been reached. Since the discount is only paid out afterwards, it is also referred to as cashback. The advantage of the discount as a bonus option lies in the simplicity of the availability and logistics of the bonus (money) and practice over decades. The big disadvantage, however, is that little or no emotions can be aroused [3]. The dimensions of bonification are shown in Fig. 5.

The financial test carried out by Stiftung Warentest in August 2010 quoted the Gesellschaft für Konsumforschung (Gfk) as saying that cardholders of the cash back programme provider Payback in Germany made 25% more sales with their Payback card in the third year of their participation than in the first. ("Finanztest" [21]) Another example is ABCO, where the launch of a baby club generated 25% revenue growth in 6 months due to the consolidation of the share of total spending [22].

The meaningfulness of a bonus program for every company must be denied. If a company already has sufficient information about customers and their purchasing behavior or if there are insufficient resources available to analyze the generated data, a bonus program will not contribute to the positive development of the company (company characteristics). Bonus programs are designed to prevent customers from migrating to competitors. If there is a contractual or psychological dependence on the existing supplier or service provider, a bonus program will not increase customer loyalty (market characteristics). A high heterogeneity of the target customers leads to a high cross-selling potential, which can be exploited by a bonus program (customer characteristics). If the relevant competitors all have bonus programs, it must be considered whether the program will not (must) be aligned in the long term. If a bonus program is not an industry-specific minimum requirement, it should be dispensed with (competitive features) if it is not based on promising, affordable differentiation [7].

The fact that the conception of a bonus program must be thought through is also due to the fact that well-known companies such as Subway, American Airlines (note: reintroduced), eBay America, America Online have discontinued their bonus programs. The art lies in not being so generous that margins erode, but still generating additional sales and profit margins [22].

In 2010, Stiftung Warentest examined 29 bonus programs for their data protection conditions and rated four as acceptable [21] 2018, the DSGVO [23].

4   Conclusions

After an introduction to the essential concepts of bonus programs and customer loyalty, its influence on customer lifetime value was analyzed. The different types of loyalty have had an impact on the success of the bonus program. Furthermore, the possibilities of controlling customer behaviour and the ideal process were discussed within the framework of a bonus program. Opportunities and risks were presented. In a further step, the examples of successful and unsuccessful bonus programs available in the literature should be analysed and discussed, and a selection of bonus programs on the market should be surveyed in order to achieve a comparison between the concepts and the current situation.

References

1. Hoseong, J., Youjae, Y.: Effects of loyalty programs on value perception, program loyalty, and brand loyalty (2003)
2. Henderson, C., Beck, J., Palmatier, R., Henderson, C.M., Beck, J.T., Palmatier, R.W.: Review of the theoretical underpinnings of loyalty programs. J. Consum. Psychol. **21**(3), 256–276 (2011)
3. Lauer, T.: Bonusprogramme: Rabattsysteme für Kunden erfolgreich gestalten. Springer, Berlin (2011). https://doi.org/10.1007/978-3-642-19118-3
4. Hofmann, M., Mertiens, M. (eds.): Customer-Lifetime-Value-Management (2000)
5. Kumar, V., Reinartz, W.J.: Customer Relationship Management: A Databased Approach (2006)
6. Bolton, R., Lemon, K., Verhoef, P.: The theoretical underpinnings of customer asset management: a framework and propositions for future research. J. Acad. Mark. Sci. **32**(3), 271–292 (2004). https://doi.org/10.1177/0092070304263341AccesstoDocument
7. Musiol, G., Kühling, C.: Kundenbindung durch Bonusprogramme: erfolgreiche Konzeption und Umsetzung. Springer, Berlin (2009). https://doi.org/10.1007/978-3-540-87571-0
8. Maharaj, A.: Awareness, perceptions and effects of customer loyalty programmes within the retail sector of the Durban Metropolitan area (2008)
9. Kotler, P., Keller, K.L.: Marketing Management (2012)
10. Kasai, C., Chauke, M.X.D.: Investigating customer perceptions of loyalty cards and their influence on purchasing behaviour in major retail stores **5**, 23 (2017)
11. Srivastava, M., Rai, A.K.: Mechanics of engendering customer loyalty: a conceptual framework. IIMB Manag. Rev. **30**, 207–218 (2018). https://doi.org/10.1016/j.iimb.2018.05.002
12. Buttle, F.: Customer Relationship Management (2004)
13. Bliemel, F.W., Eggert, A.: Kundenbindung–die neue Sollstrategie? Mark. ZFP J. Res. Manag. **20**, 37–46 (1998)
14. Eggert, A.: Kundenbindung aus Kundensicht: Konzeptualisierung – Operationalisierung – Verhaltenswirksamkeit, Wiesbaden (1999)
15. Eggert, A.: Die zwei Perspektiven des Kundenwerts: Darstellung und Versuch einer Integration. In: Günter, B., Helm, S. (eds.) Kundenwert: Grundlagen – Innovative Konzepte – Praktische Umsetzungen, pp. 41–59. Gabler Verlag, Wiesbaden (2006). https://doi.org/10.1007/978-3-8349-9288-8_2
16. Ranzinger, A.: Praxiswissen Kundenbindungsprogramme: Konzeption und operative Umsetzung. Gabler, Wiesbaden (2011). https://doi.org/10.1007/978-3-8349-6734-3
17. Lauer, T.: Bonusprogramme richtig gestalten. Harvard Bus. Manag. **2002**(3), 98–106 (2002)
18. Künzel, S.: Das Bonusprogramm als Instrument zur Kundenbindung (2003)
19. Diller, H.: Vahlens großes Marketinglexikon (2001)
20. Glusac, N.: Bonusprogramme - ein wirkungsvolles Kundenbindungsinstrument? In: Hinterhuber, H.H., Matzler, K. (eds.) Kundenorientierte Unternehmensführung: Kundenorientierung — Kundenzufriedenheit — Kundenbindung, pp. 513–524. Gabler, Wiesbaden (2006). https://doi.org/10.1007/978-3-8349-9132-4_23
21. Finanztest. Stift. Warentest (2010)
22. Nunes, J.C., Drèze, X.: Your Loyalty Program Is Betraying You. Harv. Bus. Rev. (2006). https://hbr.org/2006/04/your-loyalty-program-is-betraying-you
23. DSGVO: Datenschutzgrundverordnung

Bus Ticket Reservation System Agile Methods of Projects Management

Mateusz Grzelak[1], Łukasz Napierała[1], Łukasz Napierała[1],
Vincent Karovič[2]([✉]), and Iryna Ivanochko[3]

[1] Lodz University of Technology, Lodz, Poland
[2] Faculty of Management, Comenius University, Bratislava, Slovakia
vincent.karovic@fm.uniba.sk
[3] School of Business, Economics and Statistics,
University of Vienna, Vienna, Austria
iryna.ivanochko@univie.ac.at

Abstract. This work concerns to a web application Bus Ticket Reservation System that can be used in a bus transportation system in TEWU to reserve seats, cancellation of reservation and different types of route enquiries used on securing quick reservations. Also contains questions about the software development methods and models and questions about the interface system for Internet applications. To develop Bus Ticket Reservation System it is going to choose XP (Extreme Programming) methodology. In the future it is possible make the system more functional and user-friendly, and as a result may bring benefits in many bus transport companies.

1 Introduction

Bus is an important public transport. Urbanization process becomes faster than usual. The level of urban public service also increases. That is why it is so important that the processes in bus transport companies run as optimally as possible and meet the customers' needs. In this paper, we investigate the problem of giving bus seat reservations online in the selected company.

Travel Everywhere With Us (TEWU) is a bus service company located in Lodz. For keep their customer's details and booking reservation records they usually do paper works. It takes a lot of time and causes many errors. To solve the above problem we want to propose a new online bus ticketing system which will be much easier for customers to book tickets as well as for the company to manage their overall business smoothly [1, 2, 6, 7].

Bus Ticket Reservation System is a web application that can be used in a bus transportation system in TEWU to reserve seats, cancellation of reservation and different types of route enquiries used on securing quick reservations. It will maintain all customer details, bus details, reservation details. Passengers may purchase the bus ticket or reserve their seats online without waiting on queue [3, 23, 25].

L. Barolli et al. (Eds.): INCoS 2019, AISC 1035, pp. 492–501, 2020.
https://doi.org/10.1007/978-3-030-29035-1_48

2 Problems of Current System

Currently the type of system being used in TEWU is a manual booking system. As a result they are facing some problems issuing booking requests of customers. All the necessary booking stuffs are being done in hard copy. It takes a lot of time and causes many errors. The problems facing the company are that customers have to go to the counter to buy bus ticket or ask for bus schedule. They will also need to pay cash when they buy the bus ticket. If customers need to reserve seat, they have to call them or walk in to their counter which is consider as wasting their valuable times. Sometimes the phone line also keep busy and they unable to reserve seats. Besides, TEWU need to keep records of the payments in papers and quite impossible for them to keep track on payment issues [4].

Considering the above problems, we suggest implementing a new system – online Bus Ticket Reservation System, which is developed to help passengers book their bus tickets online in a convenient, easy and fast way.

3 Objectives and Scope of Work

The main purpose of this work is to automate the manual activities of reservation a bus ticket for any journey made through Travel Everywhere With Us (TEWU). Using this system customers can select seats by themselves. Specifically, objectives of this project will consist of:

1. Providing a web-based bus ticket reservation function where a customer can buy bus ticket through the online system without a need to queue up at the counter to purchase a bus ticket.
2. Providing anytime anyplace service for the customer. Customer can buy bus ticket 24 h a day, 7 days a week over the Internet.
3. Enabling customers to check the availability and types of busses online. Customer can check the time departure for every TEWU bus.
4. Ability of customers to cancel their reservation.
5. Ability of customers to print bought bus ticket.
6. Admin user privileges in updating and canceling payment, route and vehicle records [5].

4 Software Development Model

A software development model in software engineering is a framework that is used to structure, plan as well as control the process of developing an information system. There are a few software development models which we usually follows. They are:

- Agile Software Development
- Extreme Programming
- JAD (Joint Application Development) Waterfall

- WSDM (Web Semantic Design Method)
- Design Improvement
- Small Releases

To develop Bus Ticket Reservation System we are going to choose XP (Extreme Programming) methodology.

Extreme Programming is one of the agile methodology for system development that aims to produce higher quality software, and higher quality of life for the development team. XP is the most specific of the agile frameworks regarding appropriate engineering practices for software development [8, 21].

There are some rules in XP methodology. Some of the basic activities that are followed during software development by using XP model are given below:

- Coding: The concept of coding which is used in XP model is slightly different from traditional coding. Here, coding activity includes drawing diagrams (modeling) that will be transformed into code, scripting a web-based system and choosing among several alternative solutions [9].
- Planning: The customer writes user stories, which define the functionality the customer would like to see, along with the business value and priority of each of those features. User stories don't need to be exhaustive or overly technical - they only need to provide enough detail to help the team determine how long it'll take to implement those features [10, 24].
- Testing: XP model gives high importance on testing and considers it be the primary factor to develop a fault-free software [11].
- Listening: The developers needs to carefully listen to the customers if they have to develop a good quality software. Sometimes programmers may not have the depth knowledge of the system to be developed. So, it is desirable for the programmers to understand properly the functionality of the system and they have to listen to the customers [12, 13].

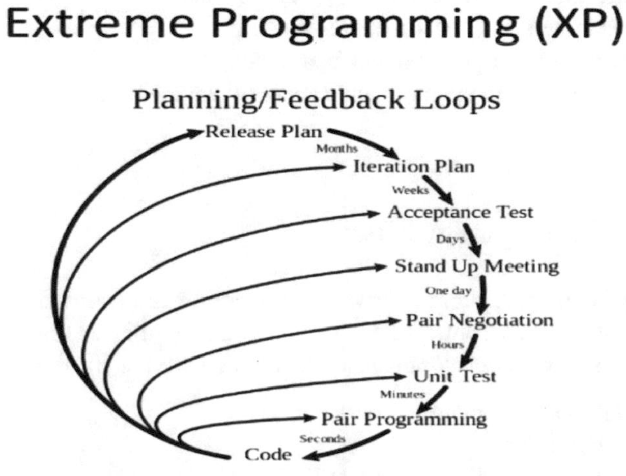

Extreme Programming (XP)

Planning/Feedback Loops

Fig. 1. Extreme Programming (XP) model

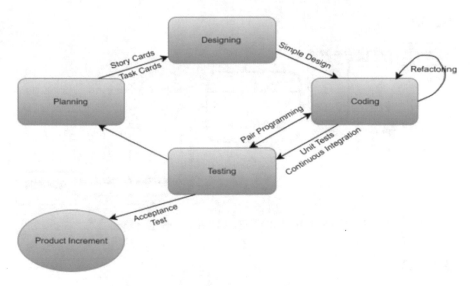

Fig. 2. Extreme Programming – basic activities

- Designing: Without a proper design, a system implementation becomes too complex and very difficult to understand the solution, thus it makes maintenance expensive. A good design results elimination of complex dependencies within a system. So, effective use of suitable design is emphasized [12–14, 22, 26] (Figs. 1 and 2).

4.1   Data Flow Diagram

Data flow diagram (DFD) is a graphical representation of the flow of data through an information system, modeling its process aspects. This DFD has been done in several levels. Each process in lower level diagrams can be broken down into a more detailed DFD in the next level. The top-level diagram is often called context diagram. It consist a single process bit, which plays vital role in studying the current system. The process in the context level diagram is exploded into other process at the first level DFD [15, 16].

Figures 3, 4 and 5 shows a data flow diagram in the system. We can see the context view, admin view and also user view of the described Bus Ticket Reservation System.

Level 0

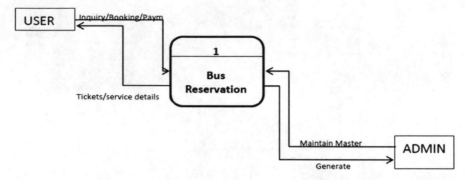

Fig. 3. Context view of Bus Ticket Reservation System

LEVEL 1

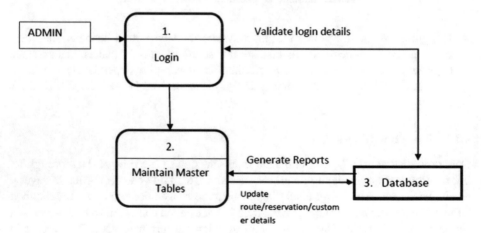

Fig. 4. Admin view of Bus Ticket Reservation System

Level 2

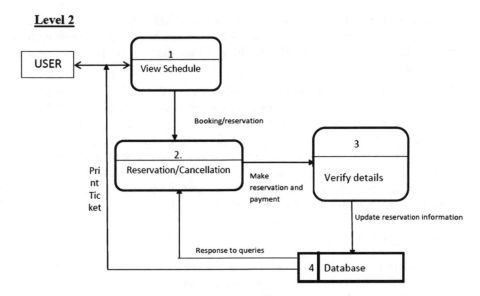

Fig. 5. User view of Bus Ticket Reservation System

4.2 Use Case Diagram

It is also called behavioral UML diagram. It gives a graphic overview of the actors involved in a system directly. It shows how different functions needed by the actors how they are interacted. In other words a use case describes "who" can do "what" with the system in question. The use case technique is used to capture a system's behavioural requirements by detailing scenario-driven threads through the functional requirements [17, 18].

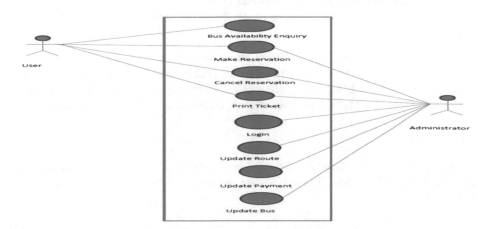

Fig. 6. Use case diagram for users and admin

There are two groups of users of our system: user (customer) and administrator. Depending on the user's membership in a given group, he has the right to perform certain application functions. In Fig. 6, the use case diagram of the new proposed system is shown [19, 20].

1. User: Customer - One of the main customer's capabilities is making ticket reservation. Every user can check the availability and types of busses online. They are able to select seats by themselves and then they may buy a ticket and print it. If something unexpected happens, they can cancel their reservation.
2. User: Admin - Admin is the key user of this entire web system. He is able to manage the ticket booking of the customers. Admin has access to every functionalities of the system. He can update and cancel the payment, route and bus if necessary. He can also delete users accounts.

4.3   User Interface

The user interface is very important part of the system. A good and user-friendly interface attracts the user toward it. Whereas a bad one makes the user experience bad and they never return to the system. We have shown several use cases and we have developed below interfaces to interact with the system [20].

Login Page:
Admin and customer need to log in using username and password. The system authenticates every user. The only valid user can access the data.

Admin Page:
This page is dedicated to an administrator of the system. This page contains the link to add buses and also the link to client's details. This page shows highlighted bookings, ongoing buses, newly added buses.

Customer Page:
This page shows the customer details. Clients can access recommended buses as per their last searched in this system. Clients can view the case update using this interface. This page would help the customer to check the ticket availability.

Reservation Page:
This page is used for making the online bus ticket reservation. The customer would have redirected to this page after checking the availability. Passenger need to fill the personal details. After filling the details, payment gateway page would come up. Once payment is done, user would get confirmation on email as well as on screen.

4.4   Test Cases

For the correct operation of the Bus Ticket Reservation System, it is necessary to perform the testing process. The application we propose should support the following test cases.

Test Case 1:

- Test Title: Travel Everywhere With Us Icon.
- Test Procedures: Click on the icon.
- Test Data: Users need to click on the icon bar.
- Expected Result: It will redirect to the system home page.

Test Case 2:

- Test Title: Customer and Admin Login.
- Test Procedures: Type username and password.
- Test Data: Username and password must be in alphanumeric. Otherwise system will show error (ex. Please enter valid alphanumeric data).
- Expected Result: It will redirect to login page.

Test Case 3:

- Test Title: Sign Up.
- Test Procedures: Click sign up page.
- Test Data: Input customer's information and click on the button "SIGN UP".
- Expected Result: It will register new customer.

Test Case 4:

- Test Title: Search.
- Test Procedures: Type destinations and bus name.
- Test Data: Valid destinations name with date.
- Expected Result: System will search according to customer's choice.

Test Case 5:

- Test Title: Make Reservation.
- Test Procedures: Click on the link "Make Reservation".
- Test Data: Add, Edit, Delete and Save button.
- Expected Result: Admin can edit, add and delete individual customers booking records.

Test Case 6:

- Test Title: Username and Password.
- Test Procedures: Enter customers or admin valid username (Upper Case and Lower Case) and password (Alphanumeric).
- Test Data: Invalid password will show the warning message (Please Enter Valid Password).
- Expected Result: The system will follow the validation pattern.

Test Case 7:

- Test Title: View Reservation Cancel Button.
- Test Procedures: Customers need to sign in and can cancel the reservation when they press cancel button.
- Test Data: Cancel the reservation record.
- Expected Result: The system will cancel client's reservation records from database.

Test Case 8:

- Test Title: Change Search.
- Test Procedures: Customers can change their search according to changing their destinations and bus.
- Test Data: Click on the button "Change Search" from bus availability page.
- Expected Result: Customers can select a new bus and date with destinations.

5   Conclusion

Bus Ticket Reservation System is developed to help passengers book their bus tickets online in a convenient, easy and fast way. Proposed application can provide benefits for the company and passengers. TEWU can run the business smoothly as there will be no hustle regarding managing customers details, booking details, payments and so on. This system will increase the productivity of their business. Using new system will also reduce the workload of the staff, reduce the time used for making reservation at the bus terminal and also increase efficiency. Transportation services may be available to users anywhere, anytime and any device for booking of bus tickets. Comparing to current system in this company which is just a manual book keeping, this online web system will automate their day to day business process.

References

1. AgileModelling: AgileModelling (2014). http://agilemodeling.com/artifacts/componentDia gram.htm. Accessed 13 Feb 2019
2. Buschmann, F., Meunier, R., Rohnert, H., Sommerlad, P., Stal, M.: Pattern-Oriented Software Architecture. SAGE Publication California, Mayfield Publishing Company, London (1996)
3. Extreme Programming. https://airbrake.io/blog/sdlc/extreme-programming. Accessed 13 Feb 2019
4. Extreme Programming. https://www.lucidchart.com/blog/what-is-extreme-programming. Accessed 13 Feb 2019
5. Kryvinska, N.: Building consistent formal specification for the service enterprise agility foundation. J. Serv. Sci. Res. **4**(2), 235–269 (2012)
6. Pedone, F.: Optimistic validation of electronic tickets. In: 20th IEEE Symposium on Reliable Distributed Systems (SRDS 2001) (2001)
7. Jakubauskas, G.: Improvement of urban passenger transport ticketing systems by deploying intelligent transport systems. Transport **4142**, 37–41 (2014). http://doi.org/10.1080/16484142.2006.9638075. Accessed 13 Feb 2019
8. Gregus, M., Kryvinska, N.: Service orientation of enterprises - aspects, dimensions, technologies. Comenius University in Bratislava (2015). (ISBN 9788022339780)
9. Kevin, O.C.: Web-based bus reservation and ticketing system: college of computer studies. Ateneo de Naga University, Naga City (2012)
10. Madden, A.D.: A definition of information. In: Aslib Proceedings, vol. 52(2000)
11. MakeMyTrip: MakeMyTrip (2016). https://www.makemytrip.com/bus-tickets/. Accessed 14 Feb 2019

12. March, S.T., Smith, G.F.: Design and natural science research on information technology. Decis. Support Syst. **15**(4), 251–266 (1995)
13. Oates, B.J.: Researching Information Systems and Computing. Sage, London (2005)
14. Kryvinska, N., Gregus, M.: SOA and its business value in requirements, features, practices and methodologies. Comenius University in Bratislava (2014). (ISBN 9788022337649)
15. Nichols, A.: Agile planning, estimation and tracking (2009). https://www.slideshare.net/andrewnichols/agile-planning-estimation-and-tracking. Accessed 13 Feb 2019
16. Online Bus Booking System. https://www.slideshare.net/stejinpaulson/online-bus-booking-system?related=2. Accessed 11 Feb 2019
17. Online Bus Ticketing Reservation System. https://www.slideshare.net/stejinpaulson/online-bus-booking-system?related=2. Accessed 11 Feb 2019
18. Wee, K.L.: Bus Reservation System: Faculty of Information and Communications Technology
19. Kaczor, S., Kryvinska, N.: It is all about services - fundamentals, drivers, and business models. J. Serv. Sci. Res. **5**(2), 125–154 (2013)
20. Molnár, E., Molnár, R., Kryvinska, N., Greguš, M.: Web intelligence in practice. J. Serv. Sci. Res. **6**(1), 149–172 (2014)
21. Poniszewska-Marańda, A.: Modeling and design of role engineering in development of access control for dynamic information systems. Bull. Polish Acad. Sci. Tech. Sci. **61**(3), 569–580 (2013)
22. Majchrzycka, A., Poniszewska-Marańda, A.: Secure development model for mobile applications. Bull. Polish Acad. Sci. Tech. Sci. **64**(3), 495–503 (2016)
23. Poniszewska-Marańda, A., Rutkowska, R.: Access control approach in public software as a service cloud. In: Zamojski, W., et al. (eds.) Theory and Engineering of Complex Systems and Dependability. Advances in Intelligent and Soft Computing, vol. 365, pp. 381–390. Springer, Heidelberg (2015). (ISSN 2194-5357, ISBN 978-3-319-19215-4)
24. Tkachenko, R., Izonin, I.: Model and principles for the implementation of neural-like structures based on geometric data transformations. In: Hu, Z., Petoukhov, S., Dychka, I., He, M. (eds.) Advances in Computer Science for Engineering and Education. ICCSEEA 2018. Advances in Intelligent Systems and Computing, vol. 754, pp. 578–587. Springer, Cham (2019)
25. Poniszewska-Marańda, A.: Selected aspects of security mechanisms for cloud computing – current solutions and development perspectives. J. Theor. Appl. Comput. Sci. **8**(1), 35–49 (2014)
26. Smoczyńska, A., Pawlak, M., Poniszewska-Maranda, A.: Hybrid agile method for management of software creation. In: Kosiuczenko, P., Zieliński, Z. (eds.): Engineering Software Systems: Research and Praxis. Advances in Intelligent Systems and Computing, vol. 830, pp. 101–118. Springer, Heidelberg (2019)

Virtual Teaching in an Engineering Context as Enabler for Internationalization Opportunities

Corinna Engelhardt-Nowitzki, Dominik Pospisil[✉],
Richard Otrebski, and Sabine Zangl

University of Applied Sciences Technikum Wien, Höchstädtplatz 6,
1200 Vienna, Austria
dominik.schremser@technikum-wien.at

Abstract. The mobility of students that would be necessary to gain intercultural internationalization experience during their studies is not available to the same extent for all students. In engineering it is partly possible to replace physical presence by virtualized technical arrangements – e.g. by web access to the sensors and actuators of a robot or machine or by using 3D printing technology. The project ENGINE of the University of Applied Sciences (UAS) Technikum Wien – Engineering goes International – develops such concepts from a technical and didactic point of view. This paper gives a brief overview of the experience gained in this case study with academic programs in e.g., mechatronics/robotics, mechanical engineering and international industrial engineering. The approach and achieved results can easily be applied as well in other engineering study programs – e.g., electronics, informatics and many more.

1 Virtual Teaching Concepts in Engineering as Enabler for Intensified Internationalization

In the engineering sciences, the curriculum of study programs such as mechatronics, robotics, mechanical engineering and similar courses includes both theoretical content and practical exercises – either in the form of computational or simulation tasks or in the form of practical laboratory exercises. In addition to the technical content, non-technical learning objectives are also becoming increasingly important in these study fields in the course of the globalized competition, especially internationalization. This is not limited to the ability to explain or negotiate complex technical content in English, but encompasses far more comprehensive cross-cultural experiences.

Typical full-time students typically have a high level of mobility due to their life situation: there is usually no dependent family situation (e.g. childcare obligations) that makes a stay abroad difficult or impossible. Typically, these students do not have a continuous professional obligation throughout the year, but can usually – insofar as employment relationships exist at all alongside their studies – reconcile studies and supplementary employment in such a way that travel opportunities are available. Funding programs such as ERASMUS provide substantial financial support.

© Springer Nature Switzerland AG 2020
L. Barolli et al. (Eds.): INCoS 2019, AISC 1035, pp. 502–512, 2020.
https://doi.org/10.1007/978-3-030-29035-1_49

This does not apply to part-time students. Since they combine their studies with a permanent employment contract, the possibility of going abroad exists only in exceptional cases or is at least easier to reconcile for those students who work in companies with foreign assignments (even in these cases, ongoing obligations or lack of approval from superiors can block this theoretically given mobility option). However, the proportion of students studying part-time and/or at an age when family, professional and financial dependencies typically make a stay abroad more difficult or impossible is increasing and is forecast to continue to rise [1]. According to the Austrian Ministry of Science's university mobility strategy, 30%–35% of annual university graduates should have completed a study-related stay abroad by 2025. However, the actual figures make it seem unlikely that this goal is achievable [2]. Therefore, further measures beyond travelling abroad are required.

In the field of engineering, the virtualization of learning tasks, which typically have to be accomplished in classroom teaching today, offers massive opportunities: If, for example, a robotics exercise can also be carried out virtually via the Internet instead of in the laboratory [3], or if students who have a project task to solve together can also do it from different locations via web communication and the exchange of files, then this opens up opportunities for cooperation with students from other countries and cultures who would have needed a trip without the virtualization.

The project ENGINE – Engineering goes international – aims to provide exactly such virtualized resources. For example, students from different universities will be able to work in a joint project team, communicate via the web, and exchange 3D print files or control programs, with the physical robot to be programmed or optimized being physically present at only one location. Sensors readable via the Internet, a webcam or similar simple means enable location-independent condition monitoring, control and manipulation. Or – due to the nowadays low material costs – each participant could afford to 3D-print a local prototype using jointly created designs, so that local tests and optimizations, purely based on file exchange are possible [4]. If case studies, projects or exercises do not require physical laboratory set-ups, but instead use VR (virtual reality) technologies, the need for physical devices is no longer required. These technological possibilities, in combination with suitable didactic settings, allow innovative solutions for the above-mentioned problem of limited student mobility as part of the essential internationalization. The project ENGINE therefore develops and tests an international case study (mixed student teams across universities) and virtual laboratory demonstrators for worldwide access via the Internet. The super-ordinate objective is the extended possibility of internationalization for students whose travel possibilities – for whatever reason (also e.g., a health impairment and many further occurrences can be such an obstacle) – are limited.

The remainder of the paper is as follows: section two provides a short overview on internationaliszation@home concepts. Section 3 discusses gender- and diversity aspects. Section 4 (empiric case study and project description) provides a brief overview of the engineering topics that the project ENGINE is currently working on at the University of Applied Sciences (UAS) Technikum Wien in concrete terms. This begins with data harmonization challenges and extends to the conception of concrete virtual laboratory demonstrators that can be used in class.

2   Internationalisation@Home Through Virtual Internationalization Concepts at the University of Applied Sciences Technikum Wien

Experience abroad, the development of intercultural skills, excellent foreign language skills and an understanding of international interrelations on the labor market are just some of the focal points that are not only gaining in importance in the course of globalization, but are increasingly already being assumed: human resources specialists increasingly expect these skills from career candidates [5].

In order to fulfill this international aspect, the conditions of the university have to be considered, in particular the situation of the students. Internationalization must be adapted to their living conditions, so that this offer for internationalization can actually be used. For example, it is not possible or only possible under certain conditions for working students to complete a semester abroad due to the compatibility of study, career and family. Internationalization at UAS Technikum Wien goes beyond the expansion of English-language courses. Of course, a corresponding range of courses is an essential aspect, because achieving a higher proportion of English-language courses within the curricula is a declared goal within the framework of the internationalization measures of the UAS Technikum Wien.

Currently, UAS Technikum Wien offers only a few study programs completely in English. However, all programs include singular courses in English as part of their curricula. The "International Office" acts as an integrated service point, coordinating mobility measures for students, teachers and staff. In the Campus International, which is managed by the International Office, students are supported in planning and organizing a semester abroad. In the course of this support, also incoming students are advised and accompanied. This range of courses is also available to domestic students, so that there is the possibility of experiencing international exchange impulses even without a stay abroad.

The model for the case study developed in the ENGINE project was previous cooperation activities in the course of the internationalization of the degree program "International Industrial Engineering". An awarded [6] international case study "Globally displaced workgroups: Creating a real world experience in the classroom", developed by the University of North Texas, was presented as part of the course "Supply Chain Management". A total of 15 universities (USA, Colombia, England, etc.) participated in this case study. In the case study currently in progress, 852 students were coordinated in 213 teams. The participants were assigned to the teams at random (but under the rule that a team had to be internationally diversified). The task was to develop a logistical solution for a company. Due to the different origins of the participants, these were virtual teams (communication and joint work across time zones via the Internet). The challenges of this project were not so much the solution of the logistical problem as the cooperation of the participants from different countries with different study programs, previous knowledge and cultural backgrounds. In the last run, for example, it was necessary to coordinate and communicate with team members from a total of 15 time zones.

ENGINE examines the question of whether such case studies can only be carried out meaningfully for logistic tasks, since these do not require any work on physical components, or whether subjects that require the development or modification of technical mechanisms (hardware) can also carry out a suitably adapted international case study in a meaningful way. This is intended to complement current mobility measures to the benefit of students with limited travel opportunities.

3   Gender and Diversity at UAS Technikum Wien – How to Substantiate These Concerns in Internationalized Engineering Teaching Formats

3.1   Gender and Diversity Principles in Relation to Internationalization

The UAS Technikum Wien sees itself not only as an internationally oriented university that promotes the intercultural competence of its students and staff, but also attaches great importance to diversity. This means that in all projects, but above all in the course of internationalization, the diversity of living conditions, cultural differences and the respective individual realities of life and needs must be taken into account. In particular, it is the responsibility of project managers and lecturers to prepare graduates in the best possible way and to align their training accordingly within highly diversified contexts [7].

Internationalization measures – no matter whether based on actual traveling or by means of internationalization@home concepts – are to initiate or enhance an exchange and a broad discussion between a variety of people: men and women, students, lecturers, industrial professionals or simply product customers and service users, i.e., humans of different ages, different nationalities and languages with a variety of social backgrounds, family commitments and working environments; as a consequence, individual needs and opinions, and also respective capabilities, skills and potentials may be extremely different.

Although there are in principle numerous interesting fields of activity open to women in industry and technology, they are still a minority in technically oriented courses of study. At the same time, society is becoming increasingly diverse and new demands are being made on the tertiary education sector. It is therefore important to get girls and women excited about technology and science at an early age and to increase the proportion of women in all degree programs [8]. The same applies to students from other backgrounds who, for example, have difficulties integrating themselves into such a course of study because of other language skills.

Thus, not only internationalization measures themselves, but as well gender and diversity issues must be thought by multiple means in order to strengthen the awareness of students and researchers for the importance of these topics. Innovative and forward-looking educational programs and lectures must therefore be internationalized on the one hand, and have to deal responsibly with the diversity of students and teachers at the universities and companies involved in a project on the other hand. In particular, this means harnessing the potential and opportunities resulting from university courses or projects for the benefit of all those involved.

3.2   Concretization of Virtual Training Concepts with Regard to Internationalization Within ENGINE

The internationalization project ENGINE pays respect to the aforementioned matters in manifold ways:

- in order to sensitize the university itself, and its researchers and teachers, to equal treatment and equality issues, training and evaluation activities are part of the project.
- for raising awareness among students, female characters or characters of different nationalities are also equally addressed – insofar as the technical exercise instructions and didactic materials show references to human persons. The second concerns in particular the team formation in the case study, which is carried out within the framework of ENGINE together with partner universities: student teams are formed in such a way that nobody works together with team members of their own nationality. Insofar as the gender ratio permits (in technical courses of study the proportion of male students typically dominates), attention is also paid to a balanced distribution of male and female students in the team composition.
- another challenge is the often unconscious attribution of gender or roles to technical objects by humans – here teachers and students: a robot has no gender per se, yet one might be tempted to attribute a male gender to a robot with a male voice and corresponding appearance, but to attribute a female gender to a service robot with female design attributes. In order to avoid this as far as possible, ENGINE deliberately avoids humanoid features of robots and has designed e.g. the virtual lab demonstrators rather as automation mechanisms (e.g. a ball labyrinth to be controlled) than as robots.
- The virtual laboratory arrangements developed within ENGINE allow for trying an engineering task as often as desired: in contrast to physical laboratory tests, there is no material consumption and unacceptable handling does not destroy the sensitive technical components: In the virtual laboratory, the student only receives an error message. In addition, the overriding of the time factor makes it easier for students with slower learning speeds to work out the problem: if a virtual task is interrupted, it can be resumed in the identical state without any problems. With physical laboratory equipment, material could have changed in the meantime (cooled solder, incorrect calibration of a robot, dusty mechanics, contacts covered with rust, plastic parts damaged by overheating or electronic components, …). The same applies to mechanical wear, e.g. due to abrasion: this does not occur in virtual laboratory exercises.

Manifold details of this kind have been incorporated into the didactic and technical project documentation and teaching materials of ENGINE. These small items may in themselves appear only slightly significant. Although this assumption is acceptable in principle, according to previous project experience with test students and in the research team itself, the impact of the concrete nature of these measures should not be underestimated: In addition to the important but nevertheless abstract declarations of principle concerning the consideration of gender mainstreaming and diversity, these small measures demonstrate very concrete ways of implementation. If attention is paid to some of these concrete situations (this is done consequently in ENGINE-inspired teaching and lab-processing), it is obvious and easy for teachers and learners to think of associations with other relevant situations – for example, with the safety relevance of religiously prescribed clothing, e.g. for women of Islamic religious belief: if a widely sewn piece of clothing or scarf gets caught in a robot, this can seriously injure the wearer. The assumption is: Whoever has once thought this risk and its fatal consequences might act with higher sensitivity in similar situations. The probability of this hypothesis would, however, have to be empirically tested in ENGINE follow-up projects which include social science components before this can be regarded as a confirmed finding. Likewise, future research projects should empirically investigate to what extent this awareness can be promoted even stronger if not only the didactic materials and technical apparatuses, but also the teaching of the lecturers is accordingly enriched – e.g. by means of the anecdotal integration of corresponding examples in the sense of "story telling".

4   ENGINE – Project Description and Virtualization Approach

4.1   Project Design

ENGINE serves the internationalization of the engineering courses of the UAS Technikum Wien. The challenge lies in preparing graduates of engineering courses for the international cooperation required in the future in this sector in view of the increasingly international division of labor in production and plant engineering and the rapidly advancing networking of production sites, resources and employees. Here, it is particularly important to train students in the courses of the study programs mechatronics/robotics, mechanical engineering, international industrial engineering, technology and innovation management and renewable energies to use key technologies of digital transformation to provide global services and to operate globally distributed production systems. On the other hand, aspiring engineers in preparation for their future professional environment need to acquire the competence for intercultural cooperation in the trade-off between technically complex solutions and efficient entrepreneurial or economic competition, political interests, regional influencing factors and, if necessary, trouble spots.

An important instrument for setting up teaching at a university of applied sciences internationally is media-supported teaching (electronic techniques for knowledge transfer and communication, e.g. online materials and instruments based on suitably prepared teaching materials, e.g. dynamic visualization of complex processes using video/audio clips and interactive game scenarios).

Especially part-time students (half of the study seats at UAS Technikum Wien) should be supported by Internationalization@home concepts. In cooperation with international partners, a cross-university case study will be developed (virtual mobility) and the curricula of the courses included will be internationalized. In addition, ENGINE creates the conditions for these courses to be an accepted partner in the International Network for Higher Education in Engineering (INHEE): such university partnerships are a compelling prerequisite for the substantial internationalization of degree programs. The focus of the first (completed) project phase was, on the one hand, on creating the technical prerequisites (data model, technical demonstrators) and, on the other hand, on the accompanying strengthening of both the university network and the international partnerships, which are particularly to be strengthened bilaterally through pure network membership. In particular, the university partners will in future be able to participate in an educational cross-university case study. Secondly, laboratory exercise units based on virtualized technical demonstrators are available via the Internet.

One of the main results of the initial literature research was the critical evaluation of the already mentioned case study of the University of North Texas with regard to the question of portability into an application that also includes physical prototypes. At its core, the finding is that ENGINE's future progress should include a precise segmentation: on the one hand, a didactic concept is being developed for the use of technical hardware for the case study on distributed cross-border application. The main topic "network technology" has proven to be particularly suitable here, as it is a part of numerous study programs and thus allows a broad integration of various universities and student classes. The use of physical prototypes within the framework of internationalization@home concepts as a "virtual lab" (for example, the automated marble maze, which has also already been mentioned) must be seen completely separately from this. The background to this distinction is both technical (different requirements for network access and Internet design) and didactic (different types of learning objectives, instructional documents and processes for use in teaching).

4.2   ENGINE Building Blocks (1) – The Modified Teaching Case Study

Under the title "Creation of a network in a digitalized factory", students from several universities are jointly developing a network concept for the "Digital Miniature Factory" of the UAS Technikum Wien using the Internet as a communication channel.

The Digital Miniature Factory consists of robots as well as automated manufacturing and assembly stations in small format. These are made of simple and inexpensive components, many of which can be easily produced by the students themselves using inexpensive 3D printing technology. The electronic components (sensors, drives, microprocessors) also come from the low-cost sector - often with a value of only a few cents. Therefore, it is unproblematic if components fail due to operating errors and student teams can afford to build several physical instances of the robot identically locally. The required design drawings of the components and the control programs for the robots can be easily exchanged via file transfer.

Figure 1 shows the Digital Miniature Factory. As part of the case study, each team is expected to solve the task of defining a network concept for this digital factory using predefined parameters within 7 (optionally 14) days. The task is technically easy, but (similar to the logistical task in the case study of the University of North Texas) solution-finding is nevertheless complex in cross-country, cross-cultural and cross-time zone communication. One of the learning effects of the case study is the need to include the resulting additional time and effort from the very beginning when planning international projects.

Fig. 1. The Digital Miniature Factory at UAS Technikum Vienna; e.g., the robot arm and other specifically shaped parts are 3D-printed (in the figure see filament colors red, blue, orange,…). The wood elements are simple laser-cut parts. All components are affordable also in less prosperous countries and don't require specific machinery, thus enabling barrier-free access.

The form of teaching is arbitrarily selectable – a university might participate in the course of a lecture, a seminar, an exercise or any other teaching format. This guarantees barrier-free access in only loose dependence on the respective curricula of the participating study programs. Each team consists of 3 students. The number of participating teams is arbitrary. Target students are bachelor students in engineering and bachelor students in business administrations. If the case includes interdisciplinary study programs, also the task assignments will contain interdisciplinary quests. In all participants derive from engineering programs, the managerial assignments could be skipped. In order to enable barrier-free access, also the required lecturer expertise is fundamental (basics of network technology).

On the first day all participants receive an e-mail describing the task, the timeline and the team assignment. Subsequently the teams can start solving the problem. Questions are answered by means of a FAQ-section on the case study website. The student teams have to enter their solution proposals on the website. On the last day they have to present their solution via a joint video conference, which is also graded by a lecturer who also attends the web meeting.

The learning objectives can easily be adjusted to the adaptive composition of the teams each year: based on the fundamental learning objectives further specific objectives can be added at any time. Technical targets are: upon completion, participants will be able to create simple communication structures in a digitized factory and to identify and explain relevant network elements and structures. In addition to the technical goals, the course leaders should also formulate learning goals related to internationalization, such as e.g., the ability to verbalize a technical problem solution developed jointly and transnationally (examination: questions form lecturers during the joint online presentation at last case study day).

4.3   ENGINE Building Blocks (2) – The Virtual Lab Concept

In the course of digitization, virtual tools have been developed for many engineering fields and have been widely tested in business practice [9]. Despite different technological approaches and applications, the basic idea is usually comparable: working in the virtual, i.e. in the reality simulated or recreated by means of powerful computers, should enable the user to execute his task or question without carrying out physical events. In this way, for example, the collision-free assembly of complex components can be investigated using VR [10], without having to manufacture these parts in reality. Another example is the simulation of processes, e.g. the analysis and optimization of manufacturing processes using discrete process simulation [11]. A further advantage of virtualization is safety [12]: there is no danger to humans, e.g. from explosions, collisions or contact with harmful substances, nor is physical damage caused by destruction or costly cleaning work a relevant issue. Valuable prototypes are only physically manufactured or tested after the virtual representation has reached operational readiness, which makes such problems unlikely or at least greatly reduces their effect.

This also offers immense opportunities for internationalization: if the virtual representation of an object or a process, e.g. a physical exercise, a real experiment or a measurement, makes it temporally and spatially independent from a fixed laboratory, this makes it possible to jump directly to another country: linguistic compatibility (possibly in the form of foreign or multilingual documents) and the IT-technical networking of the necessary technical components presupposed, it no longer plays a role from which location a student accesses a virtually available resource.

In the international network of the UAS Technikum Wien there are partner universities with which a mutual exchange within the framework of courses is conceivable. The Open Mint Lab [13] is a collection of laboratories which can be used online. This was developed by the universities of Kaiserslautern, Koblenz and Trier. Here laboratories are made available online on a website which can be called up at any time, by means of tools the tasks can be worked on and solved. The purpose of the Open Mint Lab is not only to work on the laboratories, but also to teach the basics and the correct procedure.

ENGINE takes up these ideas and thus opens up extended possibilities for its own students to combine technical education with points of contact in internationalization. At the same time, the possibility of mutual resource sharing arises within the network of several cooperating universities: Each university offers virtual laboratory offers in its own core competence fields and can make use of the offers of the partner universities.

Figures 2 shows two ENGINE demonstrators that can be integrated into such a virtual laboratory scenario: marble maze and candy grabber.

The marble maze facilitates the entry into data modeling, image recognition and the control of mechatronic systems. By means of reverse engineering the students grasp the structure of the data model and the mechanisms behind it (clusters of the data of the individual components, object-oriented data structure). They begin by controlling the maze task manually and subsequently carry out exercises on automated control. This provides insight into common image processing algorithms and search algorithms. The combination of these technologies enables the students to solve the labyrinth using a

Fig. 2. ENGINE Demonstrators: marble maze and candy grabber

self-developed algorithm. The candy grabber is the logical continuation of the maze assignment, as kinematics and exercises are more complex. VR-applications, website and exercise documents are accessible via QR-code.

5   Conclusion

Summarizing, the project ENGINE intends to give students from all over the world the opportunity to access laboratory resources online to perform active engineering exercises. They will also learn how to deal with intercultural communication in the case study described in the next sections.

Participation in the teaching case study also enables a large number of students to gain international experience in short format with comparatively little effort. The higher the number of universities and teams involved, the more diverse the learning experience, and the more complex the supervision.

From a technical and didactic point of view, the virtual lab exercises can be integrated into the classroom with little effort - both from the students' and teachers' point of view. However, such an arrangement, including ensuring functioning access (web communication, depending on the arrangement also a webcam, hardware maintenance, etc.), requires ongoing effort on the side of the offering university. In this respect, the mutual exchange with partner universities is recommended, each of which offers a few virtual lab resources in its own competence fields and brings these into a mutual exchange. This distributes the workload evenly among the universities involved.

Acknowledgments. We would like to thank the City of Vienna and especially the MA23 department for their friendly support of the project ENGINE. Thanks are also due to the company SMC for the friendly support with hardware components for the marble maze.

References

1. Sotz-Hollinger, G.: Karriereerwartungen berufsbegleitend Studierender. Zeitschrift für Hochschulentwicklung ZFHE **4**(2), 10–22 (2009)
2. BMBWF (previously BMWFW): Hochschulmobilitätsstrategie des BMWFW, Vienna (2016)
3. Candelas Herias, F.A., Puente, S., Torres, F., Ortiz, F.G., Gil, P., Pomares, J.: A virtual laboratory for teaching Robotics. Int. J. Eng. Educ. **19**(3), 363–370 (2003)
4. Despeisse, M., Minshall, T.: Skills and education for additive manufacturing: a review of emerging issues. In: Lödding, H., Riedel, R., Thoben, K.-D., von Cieminski, G., Kiritsis, D. (eds.) APMS 2017. IAICT, vol. 513, pp. 289–297. Springer, Cham (2017). https://doi.org/10.1007/978-3-319-66923-6_34
5. Bradford, H., Guzmán, A., Trujillo, M.-A.: Determinants of successful internationalisation processes in business schools. J. High. Educ. Policy Manag. **30**(4), 435–452 (2017)
6. https://northtexan.unt.edu/issues/unt-professor-ted-farris-earns-logistics-innovation-award. Accessed 10 June 2019
7. https://www.technikum-wien.at/en/about_us/gender___diversity/. Accessed 10 June 2019
8. https://www.technikum-wien.at/en/international/campus-international/. Accessed 15 June 2019
9. Schenk, M., Schumann, M. (eds.): Angewandte Virtuelle Techniken im Produkt-entstehungsprozess. Springer, Magdeburg (2016). https://doi.org/10.1007/978-3-662-49317-5
10. Mujber, T.S., Szecsi, T., Hashmi, M.S.: Virtual reality applications in manufacturing process simulation. J. Mater. Process. Technol. **155**, 1834–1838 (2004)
11. Nee, A.Y., Ong, S.K., Chryssolouris, G., Mourtzis, D.: Augmented reality applications in design and manufacturing. CIRP Ann. - Manuf. Technol. **61**, 657–679 (2012)
12. Grubert, J.: Die Zukunft sehen: Die Chancen und Herausforderungen der Erweiterten und Virtuellen Realität für industrielle Anwendungen, Coburg (2016)
13. https://www.openmintlabs.de/. Accessed 12 June 2019

Diagnosing the Administration Systems as a Prerequisite for Enterprises Business Processes Reengineering

Oleg Kuzmin[1], Vadym Ovcharuk[1], Volodymyr Zhezhukha[1(✉)],
Dhruv Mehta[2], and Jan Gregus[3]

[1] Lviv Polytechnic National University, Bandera Street, 12, Lviv, Ukraine
`{oleh.y.kuzmin, vadym.v.ovcharuk,`
`volodymyr.y.zhezhukha}@lpnu.ua`
[2] Imperial College London, London, UK
`d.mehta@aument.eu`
[3] Faculty of Management, Comenius University in Bratislava,
831 04 Bratislava, Slovakia
`Jan.Gregusml@fm.uniba.sk`

Abstract. This paper proposes the method for diagnosing the administration systems as a prerequisite for an enterprise's business processes reengineering. The method is based on identifying these systems' capacity to facilitate the achievement of the established purposes of the company ("target-means" principle) in the context of transposed projections of the Balance Scorecard (internal processes within administration systems; personnel learning and growth in them; "customers" satisfaction with administration systems; financial aspects of these systems).

1 Introduction

Effective administration systems in enterprises management enable managers of all levels to generate new business ideas, and to respond rapidly to the functional environmental changes by the relevant managerial decision-making. They also facilitate the establishment of efficient linkages, both forward and backward, within the management process, and provide balance to business processes at the enterprise [1–7]. It can diagnosed the level of efficiency in the administration systems in enterprise management, while taking into account the possibility of achieving various goals in an organization using these systems [8–14]. The review and synthesis of literary sources, as well as the research findings, allow us to justify use of the hierarchy analysis method for solving the problem of diagnosing administrative systems in enterprise management [15–20].

© Springer Nature Switzerland AG 2020
L. Barolli et al. (Eds.): INCoS 2019, AISC 1035, pp. 513–524, 2020.
https://doi.org/10.1007/978-3-030-29035-1_50

2 Diagnosing Administration Systems as a Prerequisite for Enterprises Business Processes Reengineering

The first stage envisages the assembly of experts to form an interim committee, distribution of responsibilities among its members, establishment of interaction, development and usage of documentary forms, establishment of a time schedule of work, and choice of the forms and methods of experts' work, etc. These experts must also be specialists in the subject area with a specific link of management technology, so various administration systems can be analyzed for effectiveness (for example, in supply chain, personnel management, sales, foreign trade, marketing, logistics, etc.).

The next stage of the method of diagnosing administration systems in enterprise management, considering the European integration processes, involves determining the levels of a generalized hierarchical model of such diagnostics. In this context, it should be noted that during such a diagnosis, two options are actually compared, namely:

$$\theta = \{n_1, n_2\}, \tag{1}$$

where θ – set of comparison options for diagnosing administration systems in enterprises;

n_1 – administration systems;

n_2 – goals of the enterprise.

Also, the advantage of considering four parameters of comparison m was substantiated above, namely: internal processes in administration systems; staff learning and growth in administration systems; "customers" satisfaction with administration systems; financial aspects of the administration systems.

At the next stage, it is necessary to diagnose administration systems according to certain parameters and identify the goals of the enterprise related to such systems. In other words, one should determine the list of such systems that would be diagnosed for efficiency, as well as relevant goals.

The next stage of the proposed method of diagnosing administration systems in enterprise management, considering the European integration processes, provides for the construction of matrix of pairwise comparisons based on a comparison of certain parameters of such systems with the goals of the enterprise. So, the pairwise comparison matrix $A = (a_{ij})$ will look like:

$$A = \begin{bmatrix} 1 & a_{12} & \cdots & a_{1j} & \cdots & a_{in} \\ a_{21} & 1 & \cdots & a_{2j} & \cdots & a_{2n} \\ \cdots & \cdots & 1 & \cdots & \cdots & \cdots \\ a_{i1} & a_{i2} & \cdots & 1 & \cdots & a_{in} \\ \cdots & \cdots & \cdots & \cdots & 1 & \cdots \\ a_{n1} & a_{n2} & \cdots & a_{nj} & \cdots & 1 \end{bmatrix}, \ a_{ij} = 1, \ a_{ji} = \frac{1}{a_{ij}}, \ i,j = 1, 2, \ldots, n, \tag{2}$$

where a_{ij} – elements of the inverse symmetric matrix A, which reflect the preference of option i over option j with respect to a certain comparison parameter (where the indices i and j refer to the row and column of the matrix, respectively); n is the number of comparison variants.

Table 1. The relative scale of comparing the parameters of administration systems with the favorableness of achieving the goals of the enterprise

Scores	Characteristics
1	The administration system has fully supported achieving the enterprise goals
3	The administration system has only marginally not supported achieving the enterprise goals
5	The administration system has largely not supported achieving the enterprise goals
7	The administration system has significantly not supported achieving the enterprise goals
9	The administration system has absolutely not supported achieving the enterprise goals
2, 4, 6, 8	Intermediate values between adjacent scale values

Note: developed by the authors based on [21, p. 111; 22, p. 53].

When constructing matrices of pairwise comparisons, it is advisable to pay attention to the fact that each their element a_{ij} has an exclusively positive value, that is, $a_{ij} > 0$ for all $i, j = 1, \ldots, n$. To provide a quantitative and qualitative assessment of compliance of option i with option j, it is proposed to use the well-known Saaty's nine-point scale (Table 1).

As it was mentioned above, in the context of diagnosing the administration systems effectiveness, the principle of pairwise comparison matrices inverse symmetry should be taken into account, according to which $a_{ji} = \frac{1}{a_{ij}}$. Therefore:

$$A = (a_{ij}), \; a_{ji} = \frac{1}{a_{ij}}, \; a_{ii} = 1, \; i,j = 1, 2, \ldots, n, \tag{3}$$

Next, within the proposed method of diagnosing administration systems in enterprise management, for each matrix of pairwise comparisons, eigenvectors and their normalized values should be established. As it is known from the theory of hierarchy analysis, the elements of these eigenvectors of matrices x should be calculated using the geometric average of the rows of matrix A:

$$x_i = \sqrt[n]{\prod_{j=1}^{n} a_{ij}}, \quad i,j = 1, 2, \ldots, n, \tag{4}$$

where x_i – the i-th value of the eigenvector element of pairwise comparisons matrix.

The normalized estimate of the i-th value of the eigenvector element of pairwise comparisons matrix y within the proposed method of diagnosing administration systems in enterprise management is calculated in this way:

$$y_i = \frac{x_i}{\sum_{i=1}^{n} x_i} \tag{5}$$

where y_i – is the normalized estimate of the i-th value of the priority vector element.

In this context, it should be noted that, under certain conditions, when diagnosing administration systems in enterprise management, it is necessary to calculate the index of the consistency of experts' opinions IU. From the hierarchy analysis theory, this is caused by a problem, when, in the opinion of one expert, alternative i is better than alternative j, and alternative j, in turn, is better than alternative k, while, in opinion of other experts, alternative k is dominant over alternative i. This inconsistency can exist, when there are at least three comparison options and at least three experts. The index of experts' opinions consistency IU is calculated as follows:

$$I_u = \frac{\gamma_{max} - n}{n - 1},$$

(6)

where γ_{max} – a maximum eigenvalue of any matrix of pairwise comparisons;
I_u – the index of consistency.

When identifying the consistency of experts' opinions within diagnosing administration systems in enterprise management, one should take into account the value of the random consistency index V_{iu} of pairwise comparison matrices (Table 2), justified by Saati [22, p. 25] on the basis of a large number of samples analysis at the Oak Ridge National Laboratory.

Table 2. A random consistency index V_{iu} of pairwise comparison matrices

Matrix dimension	1	2	3	4	5	6	7	8	9	10	11	12	13	14	15
Value of V_{iu}	0,00	0,00	0,58	0,90	1,12	1,24	1,32	1,41	1,45	1,49	1,51	1,48	1,56	1,57	1,59

Note: based on [22].

After that, within the framework of the proposed method of diagnosing administration systems in enterprise management, a generalized vector of priorities should be determined, which is calculated by multiplying the transposed row vector of the weighting factors by the combined normalized matrix of the estimates of the priority vectors elements. Such diagnostics complete the formation of findings, generalizations and recommendations.

3   Testing the Method of Administration Systems Diagnosis

The application of the method of diagnosing administration systems in enterprise management was carried out in the activities of a number of business entities. In particular, the corresponding calculations were carried out on the example of the Novator State Enterprise's (SE) systems of supply and foreign economic activities. The expert group in the first version included 5 specialists: 2 people - employees of the supply division of this enterprise, 2 people - representatives of the consulting company and the authors of the work.

The results of the preliminary discussion of the problem showed approximately the same level of expert competence, and as a result, it was considered not expedient to rank this level and take it into account in further calculations. The work of the experts involved periodic sessions, most often carried out in the form of a brainstorming session using expert focus. The method of averages was used to summarize the estimates obtained.

During the study, three levels of a generalized hierarchical model of diagnosing administration systems in enterprise management, presented in Fig. 6, were taken into account. Actualization of these levels at lower stages of the model was not carried out. Administration systems diagnostics were performed in the context of their internal processes, learning and growth of personnel in such systems, "clients" satisfaction with them, as well as in the financial aspects of their operation. To ensure a quantitative and qualitative assessment during such a diagnosis, the Novator SE used the nine-point Saaty scale represented in Table 1. Thus, at the first stage, four matrices of pairwise comparisons were obtained for each of the four parameters, the first of which is internal processes in the administration systems (Table 3).

Table 3. Matrix of pairwise comparisons for Novator SE regarding the parameter "internal processes within the administration system in supply"

Parameter "internal processes within administration system in supply"	Administration system in supply	Enterprise goals	Normalized estimate of the priority vector elements, y_i
Administration system in supply	1	1/5	0.167
Enterprise goals	5	1	0.833
$\gamma_{max} = 2.00$			

Considering formulas (4) and (5), we will get:

$$x_1 = \sqrt[2]{\frac{1}{5} \times 1} = 0.447;$$

$$x_2 = \sqrt[2]{1 \times 5} = 2.236;$$

$$\sum_{i=1}^{2} 0.447 + 2.236 = 2.683;$$

$$y_1 = \frac{0.447}{2.683} = 0.167;$$

$$y_2 = \frac{2.236}{2.683} = 0.833.$$

Maximal eigenvalue of the matrix of pairwise comparisons γ_{max} for Novator SE regarding the parameter "internal processes within the administration system in supply" will be equal to:

$$\gamma_{max} = 0.833 \times \left(1 + \frac{1}{5}\right) + 0.167 \times (5 + 1) = 2.00.$$

The matrix of pairwise comparisons for Novator SE regarding the parameter "personnel learning and growth within the administration system in supply" is presented in Table 4.

Table 4. Matrix of pairwise comparisons for Novator SE regarding the parameter "personnel learning and growth within the administration system in supply"

Parameter "personnel learning and growth within the administration system in supply"	Administration system in supply	Enterprise goals	Normalized estimate of the priority vector elements, y_i
Administration system in supply	1	1/3	0.250
Enterprise goals	3	1	0.750
$\gamma_{max} = 2.00$			

Considering formulas (4) and (5), we will get:

$$x_1 = \sqrt[2]{\frac{1}{3} \times 1} = 0.577;$$

$$x_2 = \sqrt[2]{1 \times 3} = 1.732;$$

$$\sum_{i=1}^{2} 1.732 + 0.577 = 2.309;$$

$$y_1 = \frac{0.577}{2.309} = 0.250;$$

$$y_2 = \frac{1.732}{2.309} = 0.750.$$

Maximal eigenvalue of the matrix of pairwise comparisons γ_{max} for Novator SE regarding the parameter "personnel learning and growth within the administration system in supply" will be as follows:

$$\gamma_{max} = 0.750 \times \left(1 + \frac{1}{3}\right) + 0.250 \times (3 + 1) = 2.00.$$

The matrix of pairwise comparisons for Novator SE regarding the parameter "clients" satisfaction with the administration system in supply" is represented in Table 5.

Table 5. Matrix of pairwise comparisons for Novator SE regarding the parameter "clients" satisfaction with the administration system in supply"

Parameter "clients" satisfaction with the administration system in supply"	Administration system in supply	Enterprise goals	Normalized estimate of the priority vector elements, y_i
Administration system in supply	1	1/7	0.125
Enterprise goals	7	1	0.875
$\gamma_{max} = 2.00$			

Considering formulas (4) and (5), we will get:

$$x_1 = \sqrt[2]{\frac{1}{7} \times 1} = 0.378;$$

$$x_2 = \sqrt[2]{1 \times 7} = 2.646;$$

$$\sum_{i=1}^{2} 2.646 + 0.378 = 3.024;$$

$$y_1 = \frac{0.378}{3.024} = 0.125;$$

$$y_2 = \frac{2.646}{3.024} = 0.875.$$

Maximal eigenvalue of the matrix of pairwise comparisons γ_{max} for Novator SE regarding the parameter "clients" satisfaction with the administration system in supply" will be equal to:

$$\gamma_{max} = 0,875 \times \left(1 + \frac{1}{7}\right) + 0,125 \times (7 + 1) = 2,00.$$

The matrix of pairwise comparisons for Novator SE regarding the parameter "financial aspects of the administration system functioning in supply" is represented in Table 6.

520   O. Kuzmin et al.

Table 6. Matrix of pairwise comparisons for Novator SE regarding the parameter "financial aspects of the administration system functioning in supply"

Parameter "financial aspects of the administration system functioning in supply"	Administration system in supply	Enterprise goals	Normalized estimate of the priority vector elements, y_i
Administration system in supply	1	1/7	0.125
Enterprise goals	7	1	0.875
$\gamma_{max} = 2.00$			

Considering formulas (4) and (5), we will have:

$$x_1 = \sqrt[2]{\frac{1}{7} \times 1} = 0.378;$$

$$x_2 = \sqrt[2]{1 \times 7} = 2.646;$$

$$\sum_{i=1}^{2} 2.646 + 0.378 = 3.024;$$

$$y_1 = \frac{0.378}{3.024} = 0.125;$$

$$y_2 = \frac{2.646}{3.024} = 0.875.$$

Maximal eigenvalue of the matrix of pairwise comparisons γ_{max} for Novator SE regarding the parameter "financial aspects of the administration system functioning in supply" will be equal to:

$$\gamma_{max} = 0.875 \times \left(1 + \frac{1}{7}\right) + 0.125 \times (7 + 1) = 2.00.$$

During the diagnosis of Novator SE administration system in supply, a second-level matrix of pairwise comparisons for four comparison parameters has been also constructed (Table 7).
Considering formulas (4) and (5), we will get:

$$x_1 = \sqrt[4]{1 \times \frac{1}{5} \times \frac{1}{3} \times \frac{1}{3}} = 0.386;$$

$$x_2 = \sqrt[4]{5 \times 1 \times 3 \times 3} = 2.590;$$

$$x_3 = \sqrt[4]{3 \times \frac{1}{3} \times 1 \times 3} = 1.316;$$

Table 7. Matrix of pairwise comparisons for Novator SE by the second level of generalized hierarchical model of diagnosing the administration system in supply

Parameters	Internal processes within the administration system in supply	Learning and growth within the administration system in supply	"Clients" satisfaction with the administration system in supply	Financial aspects of the administration system functioning in supply	Normalized estimate of the priority vector elements, y_i
Internal processes within the administration system in supply	1	1/5	1/3	1/3	0.076
Learning and growth within the administration system in supply	5	1	3	3	0.512
"Clients" satisfaction with the administration system in supply	3	1/3	1	3	0.260
Financial aspects of the administration system functioning in supply	3	1/3	1/3	1	0.150

$\gamma_{max} = 4.181$
$I_u = 0.06$

$$x_4 = \sqrt[4]{3 \times \frac{1}{3} \times \frac{1}{3} \times 1} = 0.760;$$

$$\sum_{i=1}^{4} 0.386 + 2.590 + 1.316 + 0.760 = 5.052;$$

$$y_1 = \frac{0.386}{5.052} = 0.076;$$

$$y_2 = \frac{2.590}{5.052} = 0.512;$$

$$y_3 = \frac{1.316}{5.052} = 0.260;$$

$$y_4 = \frac{0.760}{5.052} = 0.150.$$

Maximal eigenvalue of the matrix of pairwise comparisons γ_{max} for Novator SE regarding the second-level parameters will be equal to:

$$\gamma_{max} = 0.076 \times (1 + 5 + 3 + 3) + 0.512 \times \left(\frac{1}{5} + 1 + \frac{1}{3} + \frac{1}{3}\right) + 0.260$$

$$\times \left(\frac{1}{3} + 3 + 1 + \frac{1}{3}\right) + 0.150 \times \left(\frac{1}{3} + 3 + 3 + 1\right) = 4.181.$$

The index of the consistency of experts' opinions IU was calculated as follows:

$$I_u = \frac{\gamma_{max} - n}{n - 1} = \frac{4.181 - 4}{4 - 1} = 0.06.$$

Taking into account the fact that the random consistency index of pairwise comparison matrices V_{iu} for the matrix of the fourth dimension, according to Saaty scale, is equal to 0.90 (Table 2), we will get:

$$V_u = \frac{I_u}{V_{iu}} = \frac{0.06}{0.90} = 0.067.$$

The presented calculations indicate the consistency of experts' opinions in diagnosing the administration system in the supply of Novator SE, since the obtained value of $V_u \leq 0, 10$. Further, the generalized vector of priorities was calculated by multiplying the transposed vector the transposed row vector of the weighting factors by the combined normalized matrix of the estimates of the priority vectors elements:

$$\begin{bmatrix} 0.167 & 0.250 & 0.125 & 0.125 \\ 0.833 & 0.750 & 0.875 & 0.875 \end{bmatrix} \begin{bmatrix} 0.076 \\ 0.512 \\ 0.260 \\ 0.150 \end{bmatrix}.$$

Components of the priorities matrix-vector C will be equal to:

$$c_{11} = a_{11} \times b_{11} + a_{12} \times b_{21} + a_{13} \times b_{31} + a_{14} \times b_{41}$$
$$= 0.167 \times 0.076 + 0.250 \times 0.512 + 0.125 \times 0.260 + 0.125$$
$$\times 0.150 = 0.192;$$

$$c_{21} = a_{21} \times b_{11} + a_{22} \times b_{21} + a_{23} \times b_{31} + a_{24} \times b_{41}$$
$$= 0.833 \times 0.076 + 0.750 \times 0.512 + 0.875 \times 0.260 + 0.875$$
$$\times 0.150 = 0.808.$$

Generalized results of calculations are represented in Table 8.

Table 8. Generalized results of diagnosing the administration system of Novator SE in supply

Comparison variants	Values of comparison parameters and their weighting factors				Total score
	Internal processes within the administration system in supply	Personnel learning and growth within the administration system in supply	"Clients" satisfaction with the administration system in supply	Financial aspects of the administration system functioning in supply	
	0.076	0.512	0.260	0.150	
Administration system in supply	0.167	0.250	0.125	0.125	0.192
Enterprise goals	0.833	0.750	0.875	0.875	0.808

4 Conclusion

The results of the above mentioned analyses allow us to make a conclusion about the low level of the administration system favorability of Novator SE in the supply to achieve the goals of this enterprise (obviously, it is worth talking about the maximum possible benefit when the components of the priority matrix-vector (total scores) are 0.5; any decrease in the first component of the final matrix (and, accordingly, an increase in the second) indicates a deterioration in the favorability level). Thus, according to the experts, the administration system in the supply of Novator SE in general does not contribute to the achievement of the established goals of the enterprise by 30.8%. Identification of detected gaps for each of the comparison parameters and their study provides the ability to generate ideas, thoughts and suggestions in the direction of minimizing these gaps and improving the quality of administration systems in enterprise management.

References

1. Agrawal, S.: Competency based balanced scorecard model: an integrative perspective. Indian J. Ind. Relat. **44**(1), 24–34 (2008)
2. Kaplan, R., Norton, D.P.: The Balanced Scorecard: Translating Strategy into Action. Harvard Business Review Press, Boston (1996)
3. Kunz, G.: Zielvereinbarungen und Balanced Scorecard. Personal **51**(10), 488–493 (1999)
4. Shadbolt, N.M.: The balanced scorecard: a strategic management tool for ranchers. Rangelands **29**(2), 4–9 (2007)
5. Humphreys, K., Shayne Gary, M., Trotman, K.T.: Dynamic decision making using the balanced scorecard framework. Account. Rev.: J. Am. Account. Assoc. **91**(5), 1441–1465 (2016)
6. Kryvinska, N.: Building consistent formal specification for the service enterprise agility foundation. J. Serv. Sci. Res. **4**(2), 235–269 (2012)
7. Kaczor, S., Kryvinska, N.: It is all about services - fundamentals, drivers, and business models. J. Serv. Sci. Res. **5**(2), 125–154 (2013)

8. Kryvinska, N., Gregus, M.: SOA and Its Business Value in Requirements, Features, Practices and Methodologies. Comenius University in Bratislava (2014). (ISBN 9788022337649)
9. Molnár, E., Molnár, R., Kryvinska, N., Greguš, M.: Web Intelligence in practice. J. Serv. Sci. Res. **6**(1), 149–172 (2014)
10. Auger, N., Roy, D.: The balanced scorecard: a tool for health policy decision-making. Can. J. Public Health/Revue Canadienne De Sante'e Publique **95**(3), 233–234 (2004)
11. Burney, L., Swanson, N.: The relationship between balanced scorecard characteristics and managers' job satisfaction. J. Manag. Issues **22**(2), 166–181 (2010)
12. Bütikofer, P.: Balanced Scorecard als Instrument zur Steuerung eines IT-Unternehmens im Wandel – Ein Praxisbericht über die Einführung der Balanced Scorecard bei der Systor AG. Die Unternehmung **53**(5), 321–332 (1999)
13. Gardiner, C.: Balanced scorecard ethics. Bus. Prof. Ethics J. **21**(3/4), 129–150 (2002)
14. Kurylova, A.: Postroyeniye sbalansirovannoy sistemy pokazateley kak effektivnogo sredstva finansovogo mekhanizma upravleniya na predpriyatiyakh avtomobil'noy promyshlennosti. Korporativnyye finansy **1**, 55–67 (2011)
15. Kolesnykova, T., Martyshova, Y.: Issledovaniye vozmozhnostey vnedreniya sistemy elektronnogo dokumentooborota v organizatsii. Uspekhi v khimii i khimicheskoy tekhnologii **XII/10**, 22–25 (2007)
16. Dyer, J.: A clarification of "remarks on the analytic hierarchy process". Manag. Sci. **36**(3), 274–275 (1990)
17. Holder, R.: Some comments on the analytic hierarchy process. J. Oper. Res. Soc. **41**(11), 1073–1076 (1990)
18. Islam, R., Biswal, M., Alam, S.: Clusterization of alternatives in the analytic hierarchy process. Mil. Oper. Res. **3**(1), 69–78 (1997)
19. Gregus, M., Kryvinska, N.: Service Orientation of Enterprises - Aspects, Dimensions, Technologies. Comenius University in Bratislava (2015). (ISBN 9788022339780)
20. Lv, M., Chen, K., Xue, L., Su, Y., Li, R.: Hierarchy analysis for flow units division of low-permeability reservoir: a case study in Xifeng oilfield, Ordos basin. Energy Explor. Exploit. **28**(2), 71–86 (2010)
21. Korobov, V., Tutygin, A.: Preimushchestva i nedostatki metoda analiza iyerarkhiy. Izvestiya Rossiyskogo gosudarstvennogo pedagogicheskogo universiteta im. A.I. Gertsena **122**, 108–115(2010)
22. Saati, T: Prinyatiye resheniy. Radio i svyaz', Moskva (1993)

Product Lifecycle Management Service System

Dariusz Woźniak[1], Babak Gohardani[1], Emil Majchrzak[1],
Emiljana Hoti[2(✉)], and Oksana Urikova[3]

[1] Lodz University of Technology, Lodz, Poland
[2] Faculty of Management, Comenius University in Bratislava,
Bratislava, Slovakia
emiljanahoti@fm.uniba.sk
[3] Lviv Polytechnic National University, L'viv, Ukraine
mklimash@polynet.lviv.ua

Abstract. Product lifecycle management represents an all-encompassing vision for managing all data; its concepts were first introduced when safety and control were extremely important. Over the last ten years, benefits of PLM solutions were discovered and adopted efficient PLM software in increasing numbers. PLC is an assumption that every product goes through, which involves the same pattern of introduction into the market, growth, maturity, and decline. More time the product spends on market and through the cycle, the sales increase. Each product's PLC is different and at risk of not making to introduction phase. However, company's strategy should remain consistent throughout each of the phases. As in the market exist many PLM solutions this paper will review some prominent systems and come up with a system and solutions that combines the best features among them all. The focus of the system is for small and medium companies that are managed by either its owner or founder in medium companies or by a team leader. A multilevel list of all modules and components useful when managing a product through its whole creation process of exist is provided throughout the paper with all the indicators.

1 Introduction

Product lifecycle management represents an all-encompassing vision for managing all data related to the design, production, support and ultimate disposal of manufactured goods.

PLM concepts were first introduced where safety and control have been extremely important, notably the aerospace, medical device, military, and nuclear industries.

These industries originated the discipline of configuration management (CM), which evolved into electronic data management systems (EDMS), which then further evolved into product data management (PDM).

Over the last ten years, manufacturers of instrumentation, industrial machinery, consumer electronics, packaged goods, and other complex engineered products have discovered the benefits of PLM solutions and are adopting efficient PLM software in increasing numbers [2].

PLC is an assumption that every product goes through that involves the same pattern of introduction into the market, growth, maturity, and decline (Fig. 1). When

© Springer Nature Switzerland AG 2020
L. Barolli et al. (Eds.): INCoS 2019, AISC 1035, pp. 525–533, 2020.
https://doi.org/10.1007/978-3-030-29035-1_51

the product spends more time in the market and it makes its way through the cycle, its sales increase. Each product's PLC is different in the length of scope and duration, and each product is at risk of not making it out of the introduction phase. However, the company strategy should remain consistent throughout each of the phases [6].

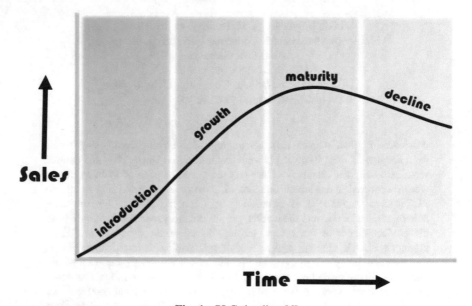

Fig. 1. PLC timeline [6].

Product lifecycle management consist of four stages:

1. Market development: In this stage new product is brought to the market, before there is even proven demand for it, and often before it has been proven out technically in all aspects. Sales are low and creep along slowly.
2. Market Growth: Demands for the product begins to accelerate and the size of the total market expands rapidly. Also called the takeoff stage.
3. Market Maturity: Demands levels off and grows, for the most part, only replacement and new family-formation rate.
4. Market Decline: Product begins to lose consumers appeal, and sales start to go down [1].
5. Additionally, the product life cycle affects the average selling price (ASP). The ASP is how much you generally sell your products or services for. When a product has many competitors or it is in the decline stage of its PLC, the ASP will be lower [6, 7].

2   Comparison of Existing Solutions

There are many different PLM solutions in the Market. Here we will review some of the prominent systems (Fig. 2).

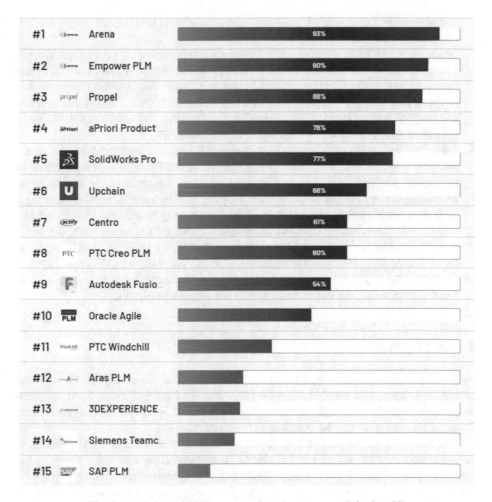

Fig. 2. Ordering of PLM systems based on users satisfaction [5]

1. Arena PLM: Arena Solution's Product Lifecycle Management (PLM) software is a cloud-based system designed for original equipment manufacturers (OEMs). It enables users to manage the design, production, and delivery of the products.
   Most of the users mention that this product is easy to deploy, maintain, and on cloud the software provider does maintenance [3].
2. Propel combines product lifecycle management (PLM), Product Information Management (PIM), and Quality Management Solutions (QMS) into one system natively built on the Salesforce platform. Its uses Salesforce cloud infrastructure and allows for Collaboration on customer design. Majority of the users liked the level of personalization allowed by the software.
3. Oracle Agile: Oracle Agile PLM is a comprehensive solution for enterprises to manage their product value chains and lifecycles. It is customizable and provides visibility into the product data. The integrated framework allows cross-functional

teams to work on co-related tasks collectively and synchronizes data with internal and external pools [9].

Agile PLM products include Agile Product Collaboration, Agile Product Quality Management, Agile Product Portfolio Management, Agile Product Cost Management, Agile Product Governance & Compliance, Agile Engineering Collaboration, Oracle PLM Cloud Strategies, and Oracle Product Lifecycle Analytics [8]. While this a quite comprehensive product many of the users complain about the complexity of use and steep learning curve of this PLM [11].

4. Autodesk fusion lifecycle: Autodesk Fusion Lifecycle is developed by renowned software manufacturer Autodesk. It has been designed to make the management of processes, projects, and people easy and effective by automating key tasks an delivering the right information to the right people at the right time, while making it such that people get to work closer to the community-based way people interact today. Autodesk Fusion Lifecycle represents a radical departure from the traditionally complicated, hardware-based systems that have kept PLM from living up to its full potential. Being a cloud-based system, the data is accessible anytime, anywhere, by anybody, on any platform. Similar to Oracle Agile, many users think the software would be too difficult to use for beginners [4, 10].

3 Our System

By assessing existing offers in the market of product lifecycle management, we came up with a solution that combines the best features among all of them. Some of these suppliers have been around for ages, but their product it is not polished to perfection. Some of them might even be outdated for the same reason.

We focus on developing a system for small and medium companies, which, is managed either by its owner and founder or in medium companies by a team leader. Such companies should not require a dedicated management team, but the system is flexible enough so if it were necessary a single manager would be enough to keep an eye on all teams and members. Other team members should have editing access rights limited only to the parts they are responsible for unless it is particularly needed otherwise.

Below we can see a multilevel list of all modules and their components we think are useful when managing a product throughout its whole creation process and its existence. It consists of modules and their parts [12–14].

System major components and their smaller modules (Fig. 3):

Administration

- Teams and members allocation
  - A view for creating teams, groups and managing people enrolment to groups, as well as their roles inside them.
- Portfolios

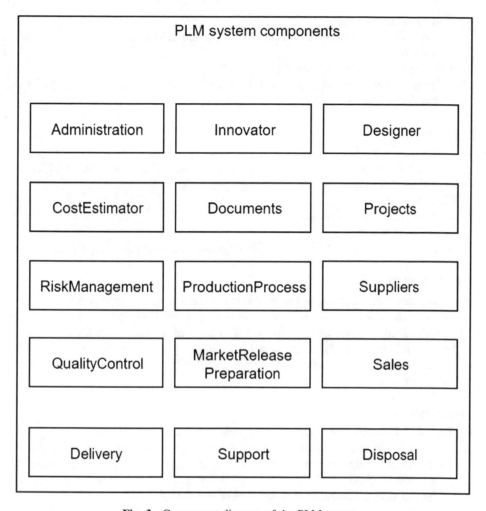

Fig. 3. Component diagram of the PLM system.

- A browser for people of a company. It stores currently hired, those previously working for it and some new applicants. It also contains logs of what each person was responsible for so far.
- Ranks, roles, permissions
  - Management of people status, their skills with corresponding levels, their permissions in the system, roles in a team, etc.

Innovator
- Brainstorming Platform
  - An online platform where all members can post their own ideas for others to comment and vote

- Idea Development Platform
  - An online platform where simple concepts are developed into more elaborate solutions
- Solution Verification and Opinions
  - This is where the concepts of solutions are verified and if accepted passed on to the design phase

Designer

- Design of the product parts
  - CAD file browser, viewer, manager, and editor. Displays thumbnails of all included files to help with searching.
- Final products
  - Combined designs of all parts.
- Products family
  - Quick access to all the products from the same product line or a product family.

Cost Estimator

- A complex calculator which helps with the estimates of how much each stage of a product life will cost. It can be used at any moment. Early on when a project is about to be developed the user has to fill in most of the input data for calculation, but it still helps to do not forget about anything important. During the product life cycle, the estimates get better and better as more of the data is precise and taken for the real world.

Documents

- Technical Documentation Creator (chapters > sections > documents)
  - It's a special kind of document editor which helps in creating a uniform and easy to browse the documentation. It allows for using any part of existing project data to create a schematic or a sketch. Any type of data from within the system can be imported into documentation, even only in selected parts.
- Templates
  - A template manager that allows preparing form type documents for other users to use and rely on or to lock some visuals on documents like backgrounds, borders, headers, company watermarks. Using it helps to keep all company documents uniform. This way documents look almost identical, even if they come from different people.

Projects

- A browser for all company projects. Those never put into production, those that are being developed and those were or are produced.

Risk Management

- Risk Assessment
  - Market analysis, the similarity of the project to other products existing on the market, demand verification, profitability estimation.
- Deadlines and dates
  - A timeline or a calendar with all events laid down chronologically.
- Progress
  - Shows what stages of a chosen project have been completed and what is next to come.

Production Process

- Components production
  - Detailed description and specification of the production process of parts
  - Supervision over the logistics and efficiency of production, handling potential breaks
- Assembly
  - Assembly instructions
  - Control over the product assembly
- Configurations
  - Matching various configurations with a specific customer order

Suppliers/Vendors

- Placing an order for parts or services required from third parties.

Quality Control

- Tests Specification
- Health and Safety Inspection
- Items Approval

Market Release Preparation

- Checklist to verify if everything was done and if there is enough of the product for a demand.

Sales

- Reports from all wholesalers and international distributors.

Delivery

- Collecting orders into packages, labeling, sending, and tracking. Integration with an API of a chosen delivery company.

Support

- Online Documentation
  - This is separate from the Technical Documentation Creator module as it is made for an end user, however, you can import parts of technical documentation that are shared.
- Maintenance
  - Includes both guides for user maintenance as well as professional support.
- Updates
  - A platform for distributing updates to already sold products.
- Recalls
  - Tracks all failures of products or specific parts.
  - Keeps tracks the status of warranty for each serial number. Also shows which type of warranty is assigned to a device.

Disposal/Recycling

- Collecting a product from a customer
  - Management of waste collection points
- Hazardous substances
  - It allows checking all company products on the matter of its components and used substances. Also, it allows for a search of a specific substance among all company projects and verifying how many of such devices are still out there.
- Disposal plans
  - For each device a list of steps that will be done to get rid of it.
- Waste collecting and recycling
  - Connection to other companies, which collect a different kind of waste and used them to extract raw materials.

4   Conclusions

As product lifecycle management represents all-encompassing vision for managing all data related to design, production, support and ultimate disposal of manufactured goods, throughout the paper different PLM solutions existing in the market are evaluated. This has led to the conception of a new solution, which performs better and combines the best features among them all by especially focusing in small and medium companies that are managed either by their owner or founder or by a team leader. Some of the suppliers have been around for ages, but their product was not polished to perfection and another part might be outdated for the same reason [15, 16]. A multi-level list of all modules and components useful when managing a product through its whole creation process of exist is provided that is believed to be the best solution while combining different system features. Some of them include; Administration, innovation, design and combined design of all parts, cost estimator, documents, projects, risk management, production process, suppliers, quality control, market release preparation, sales, delivery, support, disposal/recycling. All of the main components of the PLM system are represented at the end of the study by a diagram [17]. Such PLM system is believed to manage all data's related to the design, production, support and ultimate disposal of manufactured goods better and more efficient than the previous systems which lacked different aspects of practice or had disadvantages.

References

1. https://hbr.org/1965/11/exploit-the-product-life-cycle
2. https://www.product-lifecycle-management.com
3. https://www.softwareadvice.com/manufacturing/arena-plm-profile/
4. https://www.predictiveanalyticstoday.com/top-product-lifecycle-management-plm-software/
5. https://vendors.g2crowd.com/g2scoringmethodologies
6. https://www.smartsheet.com/product-life-cycle-management
7. Kryvinska, N.: Building consistent formal specification for the service enterprise agility foundation. J. Serv. Sci. Res. **4**(2), 235–269 (2012)
8. Kaczor, S., Kryvinska, N.: It is all about services - fundamentals, drivers, and business models. J. Serv. Sci. Res. **5**(2), 125–154 (2013)
9. Gregus, M., Kryvinska, N.: Service Orientation of Enterprises - Aspects, Dimensions, Technologies. Comenius University in Bratislava (2015). (ISBN 9788022339780)
10. Kryvinska, N., Gregus, M.: SOA and its business value in requirements, features, practices and methodologies. Comenius University in Bratislava (2014). (ISBN 9788022337649)
11. Molnár, E., Molnár, R., Kryvinska, N., Greguš, M.: Web intelligence in practice. J. Serv. Sci. Res. **6**(1), 149–172 (2014)
12. Poniszewska-Marańda, A.: Modeling and design of role engineering in development of access control for dynamic information systems. Bull. Polish Acad. Sci. Tech. Sci. **61**(3), 569–580 (2013)
13. Majchrzycka, A., Poniszewska-Marańda, A.: Secure development model for mobile applications. Bull. Polish Acad. Sci. Tech. Sci. **64**(3), 495–503 (2016)
14. Poniszewska-Marańda, A., Rutkowska, R.: Access control approach in public software as a service cloud. In: Zamojski, W., et al. (Eds.): Theory and Engineering of Complex Systems and Dependability. Advances in Intelligent and Soft Computing, vol. 365, pp. 381–390. Springer, Heidelberg (2015). (ISSN 2194–5357, ISBN 978-3-319-19215-4)
15. Tkachenko, R., Izonin, I.: Model and principles for the implementation of neural-like structures based on geometric data transformations. In: Hu, Z., Petoukhov, S., Dychka, I., He, M. (eds.) Advances in Computer Science for Engineering and Education. ICCSEEA 2018. Advances in Intelligent Systems and Computing, vol. 754, pp. 578–587. Springer, Cham (2018)
16. Poniszewska-Marańda, A.: Conception approach of access control in heterogeneous information systems using UML. J. Telecommun. Syst. **45**(2–3), 177–190 (2010)
17. Smoczyńska, A., Pawlak, M., Poniszewska-Maranda, A.: Hybrid agile method for management of software creation. In: Kosiuczenko, P., Zieliński, Z. (eds.) Engineering Software Systems: Research and Praxis. Advances in Intelligent Systems and Computing, vol. 830, pp. 101–118. Springer, Heidelberg (2019)

Building Microservices Architecture
for Smart Banking

Aneta Poniszewska-Marańda[1(✉)], Peter Vesely[2], Oksana Urikova[3],
and Iryna Ivanochko[3]

[1] Lodz University of Technology, Lodz, Poland
aneta.poniszewska-maranda@p.lodz.pl
[2] Faculty of Management, Comenius University, Bratislava, Slovakia
Peter.Vesely@fm.uniba.sk
[3] Lviv Polytechnic National University, L'viv, Ukraine
mklimash@polynet.lviv.ua

Abstract. We attempt to solve in this research a problem of efficiency and, at the same time, the reduction of the operating costs when creating complex banking IT systems. We propose an approach based on micro-service architecture. The article describes functionalities as well as methods and algorithms used during model development. The work emphasizes the importance of the methodically chosen system architecture to properly implement the user requirements and meet their expectations related to independence of scaling, ease of maintenance, and the reduced mutual blockages of websites.

Keywords: Microservices architecture · Banking systems ·
Economics of microservices · Web services ·
Large scale distributed architecture for smart banking

1 Introduction

The concept of micro-services is becoming more and more popular among software architecture designers, as an alternative approach to the monolithic building of the applications also in large information systems. This trend appears since internet and web browser becomes the integral and inseparable part of these systems. Nowadays, the applications are built from ready-made blocks, when each of them consists of one or more microservices.

Architecture is what allows systems to evolve and provide a certain level of service throughout their lifecycle. In software engineering, architecture is concerned with providing a bridge between system functionality and requirements for quality attributes that the system has to meet. Over the past several decades, software architecture has been thoroughly studied, and as a result software engineers have come up with different ways to compose systems that provide broad functionality and satisfy a wide range of requirements [1, 2, 19].

Microservice is a cohesive, independent process interacting via messages.

Microservice architecture is a distributed application where all its modules are microservices.

L. Barolli et al. (Eds.): INCoS 2019, AISC 1035, pp. 534–543, 2020.
https://doi.org/10.1007/978-3-030-29035-1_52

A service-oriented architecture (SOA) is a style of software design where services are provided to the other components by application components, through a communication protocol over a network.

The architecture of microservices is already used by such companies as we refer from articles [1–4]. In detail let mention at least couple of them: Netflix, Amazon, The Guardian, eBay, Twitter and many others. Let's have a closer look on some of them. Netflix, which is a very popular video streaming service that's responsible for up to 30% of Internet traffic, has a large scale, service-oriented architecture. It receives more than one billion calls every day, from more than 800 different types of devices, to its streaming-video API. Each API call then prompts around five additional calls to the backend service. Amazon has also migrated to microservices. They get countless calls from a variety of applications, including applications that manage the web service API as well as the website itself, which would have been simply impossible for their old, two-tiered architecture to handle.

Microservice architecture uses services to componentize and is usually organized around business capabilities; focuses on products instead of projects; has smart end points but not-so-smart info flow mechanisms; uses decentralized governance as well as decentralized data management; is designed to accommodate service interruptions; and last but not least, is an evolutionary model.

In further readings we were interested and focused on the idea of microservice architecture. In the article [5] presents the technique of identifying and defining microservices in monolithic enterprise systems. Thanks to the publication, we have the opportunity to learn about the assessment of the implementation of a banking system based on microservices.

In some other publications [6], presented an experience report of a real world case study in order to demonstrate how scalability is positively affected by re-implementing a monolithic architecture into microservices. The case study is based on the FX Core system, a mission critical system of Danske Bank, the largest bank in Denmark and one of the leading financial institutions in Northern Europe.

Inspiration could be taken also from some works around banking systems and electronic or mobile payment systems, based on services that fit in the definition of microservices [7] and [8].

Thus, we attempt in this paper to build/develop a generic model of microservices architecture for a smart banking system. Through research we will try to prove that the project designed and implemented by us in the architecture of microservices will give us many benefits. The implemented system will be more efficient, lightweight, and easier to maintain or repair.

2 Comparison Microservices with SOA Approach

Both Microservices Architecture (MSA) and Service-Oriented Architecture (SOA) rely on services as the main component. But they vary greatly in terms of service characteristics. We are presenting below in Table 1 comparing the SOA architecture with the microservice architecture [9, 13–15].

Table 1. Comparison of SOA architecture with microservice architecture

SOA	MSA
Service, service consumer provider contract pattern	Fine-grained service interfaces, independently deployable services, RESTful resources
More importance on business functionality reuse	More importance on the concept of "bounded context"
Common governance and standards	Relaxed governance, with more focus on people collaboration and freedom of choice
Uses enterprise service bus (ESB) for communication	Uses less elaborate and simple messaging system
Supports multiple message protocols	Uses lightweight protocols such as HTTP/REST & AMQP
Common platform for all services deployed to it	Application Servers not really used. Platforms such as Node.JS could be used
Multi-threaded with more overheads to handle I/O	Single-threaded usually with use of Event Loop features for non-locking I/O handling
Use of containers (Dockers, Linux Containers) less popular	Containers work very well in MSA
Maximizes application service reusability	More focused on decoupling
Uses traditional relational databases more often	Uses modern, non-relational databases
A systematic change requires modifying the monolith	A systematic change is to create a new service
DevOps/Continuous Delivery is becoming popular, but not yet mainstream	Strong focus on DevOps/Continuous Delivery

Now there will be presented advantages and disadvantages of microservices. This solution has a number of benefits:

- Each microservice is relatively small
- Easier for a developer to understand
- The IDE is faster which makes developers more productive
- The application starts faster, which makes developers more productive and speeds up deployments
- Each service can be deployed independently of other services - easier to deploy new versions of services frequently
- Easier to scale development. It enables you to organize the development effort around multiple teams. Each (two pizza) team is owns and is responsible for one or more single service. Each team can develop, deploy and scale their services independently of all of the other teams.
- Improved fault isolation. For example, if there is a memory leak in one service then only that service will be affected. The other services will continue to handle

requests. In comparison, one misbehaving component of a monolithic architecture can bring down the entire system.

- Each service can be developed and deployed independently.
- Eliminates any long-term commitment to a technology stack. When developing a new service, you can pick a new technology stack. Similarly, while making major changes to an existing service you can rewrite it using a new technology stack.

However, this solution has several drawbacks, too:

- Developers must deal with the additional complexity of creating a distributed system.
- Developer tools/IDEs are oriented on building monolithic applications.
- and do not provide explicit support for developing distributed applications.
- Testing is more difficult
- Developers must implement the inter-service communication mechanism.
- Implementing use cases that span multiple services without using distributed transactions is difficult
- Implementing use cases that span multiple services requires careful coordination between the teams
- Deployment complexity. In production, there is also the operational complexity of deploying and managing a system comprised of many different service types.
- Increased memory consumption. The microservice architecture replaces N monolithic application instances with NxM services instances. If each service runs in its own JVM (or equivalent), which is usually necessary to isolate the instances, then there is the overhead of M times as many JVM runtimes. Moreover, if each service runs on its own VM (e.g. EC2 instance), as is the case at Netflix, the overhead is even higher.

Both architectures have similar advantages and disadvantages but also some differences. In each of them we can identify services as we understand it in terms of technology and these services has a certain responsibilities, which is unlike a monolithic architecture. First and major difference is that in microservices, services can operate and be deployed independently of other services. It is strength in performance that doesn't affect other services but weakness that communication and cooperation between services are not so strong, thus productivity may be affected.

Last but not least main difference between SOA and microservices we can identify in the scope and size [10]. Microservice is/should be significantly smaller than what SOA usually are or to be and that leads to a smaller independently deployable service. However, an SOA can be a monolith architecture or even it can be an assembly of multiple microservices.

Finally, SOA typically evolve on existing system, trying absorbing and cover it at all, unlike microservices that usually appears as a "greenfield project" or a newcomer to service catalog.

3 Model Building

We describe here modules and functionalities for our smart system "e-bank", as well as algorithms and UML-diagrams as use-cases. Thus, our web application consists of the following modules:

- account support module - is responsible for the registration and signing in,
- authorization module - responsible for user authorization,
- payment system module - includes transaction realization, transaction history aggregation and creating incoming payments,
- loans and deposits module - makes possible to apply for a loan or to open a deposit by clients,
- correspondence module - allows the system administrator to contact with the bank clients.

Defined and potential system users are described in the Table 2.

Table 2. Potential banking system users

Name	Description
System administrator	Responsible for the information security management of the system. Particular emphasis is on the management of access control, user's rights and system limitations. Ensures that the system will be maintained all the time. The system administrator knows the functionality and structure of the web application. He defines the basic safety rules, that are results of the system specification
Client	A person using system for which access rules are defined. Uses all the functionalities that are available in the web application designed for the bank clients
Guest	Not loggedin user, possibly without bank account. Potential or future client

Further, we develop, explore, and analyze algorithm modules and diagrams that describe the system. Firstly, a conceptual model of use-cases that consists of a set of actors and use-cases assigned to them is analyzed.

The "system administrator - administrator" package includes use-cases responsible for managing access control, user rights and system limitations. Clients have no ability to change personal data without permission of the system administrator, e.g. phone number, identification number etc. In addition, they have no possibility to change own bank account number or making decision grant loan or not. This use-case diagram is shown in the Fig. 1.

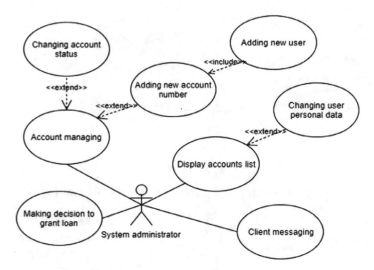

Fig. 1. Use-case diagram of a web-application for the "actor - system administrator".

This case should be in relation to other use-cases:

- by the relation ≪extend≫ with use case Account managing - Adding new account number, changing account status;
- by the relation ≪extend≫ with use case Display accounts list - Changing personal data;
- By the relation ≪include≫ base use case: Adding new account number is extended by Adding new user.

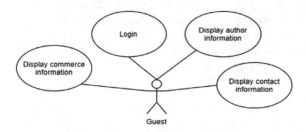

Fig. 2. Use-case diagram for the "actor - guest".

Figure 2 shows a use-case for an actor "Guest". The Fig. 2 does not contain any relations ≪include≫ and ≪extend≫. The client package includes use-case responsible for transfers' realization, displaying transaction history, opening deposits and applying for a loan. Figure 3 describes a use-case diagram for clients.

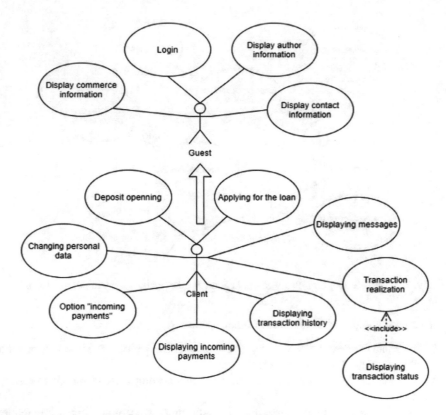

Fig. 3. Use cases diagram for "the actor - client"

The diagram in Fig. 3 has relation ≪include≫. The relation ≪include≫ base use-case: Transfer realization is extended by displaying transfer history. The diagram contains generalization. Generalization is a relationship between actors or use-cases in which one of them is general and other - detailed. In this current case, the Guest is a general one, and the Client is a detailed one.

Table 3. Technologies and methods used during "e-bank" system creation.

Module name	Programming language	Server	Database management
Authorization module	C#	IIS	MS SQL Server
Account service module	C#	IIS	MS SQL Server
Correspondence module	C#	IIS	MS SQL Server
Payment system module	C#	IIS	MS SQL Server
Deposit and loan module	Java	Tomcat	PostgreSQL

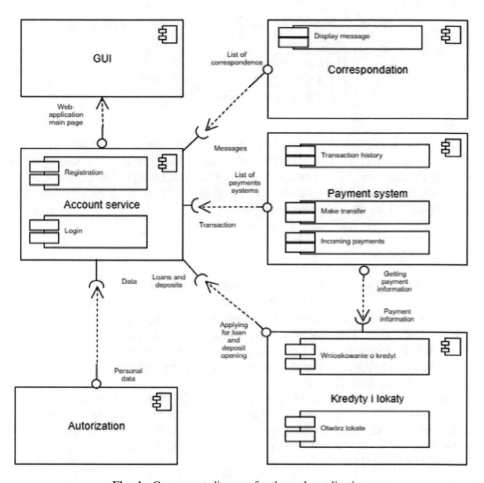

Fig. 4. Component diagram for the web-application.

Figure 4 shows components diagram for the smart banking system.

The methods used during the implementation have been selected to properly implement user's requirements. Technologies and methods used for the smart banking system creation are described in the Table 3.

Further, we assume that our system, based on the concept of building distributed applications composed of many components, should be equipped with the following mechanisms:

- horizontal scaling of services
- monitoring services
- debugging/logging services

In order to test the system, we use the tools presented in Table 4.

Table 4. Technologies used to test the IT system "e-Bank".

Mechanism	Technology
Horizontal scaling of services	Docker
Monitoring services	APImetrics
Debugging/logging services	Loggly

Since our research is still in progress - the verification and validation tests are on the primary stage. However, first results are promising - final requirements correspond to the original customer expectations. In practice, verification is a process that ends with a "zero-one" resolution - the product looks or behaves or is as in the specification. The operation of all use-cases has been checked.

4 Conclusion

The change of monolithic architecture to microservice one - despite many potential advantages - is not always a guarantee of benefits [11], [12]. Therefore, each case should be considered individually. If an enterprise decides to carry out the transformation, first it should remember to use potential of the change the best it can. Most likely, it will not only concern the product in the form of design and implementation of the software, but also it will concern changes in the processes and organization of the work of many IT cells [16–18].

In the next step we should provide testing. It will be necessary to select appropriate testing tools, set the measurement and evaluate the results of these tests. The results will show us performance ability, under different conditions, for example different traffic loads, or various requests. Finally tests and positive results will show us verification of presented model.

References

1. Shadija, D., Rezai, M., Hill, R.: Towards an understanding of microservices. In: 23rd International Conference on Automation and Computing (ICAC), pp. 1–6 (2017)
2. Baresi, L., Garriga, M., Derenzis, A.: Microservices identification through interface analysis. In: European Conference on Service-Oriented and Cloud Computing, pp. 19–33. Springer (2017)
3. Salah, T., Zemerly, M.J., Yeun, C.Y., Al-Qutayri, M., Al-Hammadi, Y.: The evolution of distributed systems towards microservices architecture. In: 11th International Conference for Internet Technology and Secured Transactions (ICITST), pp. 318–325 (2016)
4. Singleton, A.: The economics of microservices. IEEE Cloud Comput. **3**(5), 16–20 (2016)
5. Levcovitz, A., Terra, R., Valente, M.T.: Towards a technique for extracting microservices from monolithic enterprise systems. arXiv:160503175 Cs, Brazil (2016)
6. Bucchiarone, A., Dragoni, N., Dustdar, S., Larsen, S.T., Mazzara, M.: From Monolithic To Microservices: An Experience Report (2017)

7. Ivanochko, I., Greguš, M., Urikova, O., Masiuk, V.: mBusiness, mMarkets and mServices: exploration of opportunities. Int. J. Serv. Econ. Manag. **7**(1), 74–93 (2015). ISSN 1753-0822
8. Markoska, K., Ivanochko, I., Greguš, M.: Mobile banking services-business information management with mobile payments. In: Agile Information Business: Exploring Managerial Implications, pp. 125–175. Springer, Singapore (2018). (ISBN 978-981-10-3357-5)
9. Zimmermann, O.: Microservices tenets: agile approach to service development and deployment, overview and vision papier. In: Computer Science - Research and Development, Springer (2016)
10. Kryvinska, N., Greguš, M.: SOA and its business value in requirements, features, practices and methodologie, pp. 109–110. Univerzita Komenského, Bratislava (2014)
11. Dragoni, N., Dustdar, S., Larsen, S.T., Mazzara, M.: Microservices: migration of a mission critical system. arXiv:170404173 Cs, April 2017
12. Bossert, O.: A two-speed architecture for the digital enterprise. In: Emerging Trends in the Evolution of Service-Oriented and Enterprise Architectures, pp. 139–150. Springer, Cham (2016)
13. Kryvinska, N.: Building consistent formal specification for the service enterprise agility foundation. Soc. Serv. Sci. J. Serv. Sci. Res. **4**(2), 235–269 (2012)
14. Kaczor, S., Kryvinska, N.: It is all about services - fundamentals, drivers, and business models. Soc. Serv. Sci. J. Serv. Sci. Res. **5**(2), 125–154 (2013)
15. Gregus, M., Kryvinska, N.: Service Orientation of Enterprises - Aspects, Dimensions, Technologies. Comenius University in Bratislava (2015). (ISBN 9788022339780)
16. Molnár, E., Molnár, R., Kryvinska, N., Greguš, M.: Web intelligence in practice. J. Serv. Sci. Res. **6**(1), 149–172 (2014)
17. Poniszewska-Marańda, A., Rutkowska, R.: Access control approach in public software as a service cloud. In: Zamojski, W., et al. (eds.) Theory and Engineering of Complex Systems and Dependability. Advances in Intelligent and Soft Computing, vol. 365, pp. 381–390. Springer, Heidelberg (2015)
18. Tkachenko, R., Izonin, I.: Model and principles for the implementation of neural-like structures based on geometric data transformations. In: Hu, Z., Petoukhov, S., Dychka, I., He, M. (eds.) ICCSEEA 2018. Advances in Intelligent Systems and Computing, vol. 754, pp. 578–587. Springer, Cham (2019)
19. Semeniuk, M., Korpalski, E., Moradewicz, B., Moradewicz, M., Bartczak, M., Gomulak, P.: Microservice architecture by the example of a banking system. Term Paper, Pracownia problemowa, WiSe 17/18, Lodz University of Technology, Lodz, Poland

Crowdfunding – An Innovative Corporate Finance Method and Its Decision-Making Steps

Valerie Busse[1,2(✉)] and Michal Gregus[2]

[1] University of Vienna, Oskar-Morgenstern-Platz 1, 1090 Vienna, Austria
`valerie.busse@infinanz.de`
[2] Faculty of Management, Comenius University, Bratislava, Slovakia
`michal.gregusml@fm.uniba.sk`

Abstract. Crowdfunding represents a rather novel alternative of funding new ventures, and has gained increased attention during recent years. However, a distinct positioning of crowdfunding within the range of alternative financial methods and a systematization and preparation of complex decision-structures of the crowdfunding concept in terms of decision making aspects has remained largely unexplored. This research *(i)* gives a well-defined classification of crowdfunding compared to other financing methods and *(ii)* provides a clear picture of decision-making steps within the triadic relationship of involved actors.

1 Introduction

Since the increasing evolution of information and communication technologies (ICT), the integration of peoples' private, social and professional lives is no longer bound to a certain time or location, and is therefore nowadays often replaced through online interactions [21, 22]. Crowdfunding is a good example of this online interaction [28]. The phenomenon of crowdfunding emerged a decade ago and has grown in volume 1000% within the last four years [20]. It is a quickly expanding alternative way of funding for individuals or ventures through online platforms by collecting funds from a reasonable group of investors [29]. Thus, the basic idea of crowdfunding is to raise external funds from a large audience, i.e. "the crowd" [5]. Projects funded through the crowd may range from projects requiring only small amounts to large projects for entrepreneurs seeking hundreds of thousands of dollars [37]. Other scholars explain this enormous growth by referring to the low entry barriers compared to alternative funding methods [4]. In contrast to some highly regulated traditional financing methods, such as banks, and capital funding enterprises or less regulated traditional financing methods, for instance venture capital or business angels, the digital platforms for crowdfunding via the Internet are accessible for everyone and therefore easy to reach [7, 15, 20]. The term *crowdfunding* is based on the term crowdsourcing, which was first propagated in 1567 by Philipp of Spain, who proposed a reward to anyone, who finds a practical method for a precise determination of ship longitudinal [34]. Followed by Charles Babbage in the 19th century, who hired "the crowd" for an assistance in computing astronomical tables [2]. However, the use of the term "crowdfunding" has significantly increased since the financial crisis of 2008 when online platforms such as Kickstarter.com and IndieGoGo

© Springer Nature Switzerland AG 2020
L. Barolli et al. (Eds.): INCoS 2019, AISC 1035, pp. 544–555, 2020.
https://doi.org/10.1007/978-3-030-29035-1_53

were launched [9]. Whereas, certain studies examine parts of decision-making aspects in crowdfunding, there is a relative shortage of research providing a clear picture of how crowdfunding can be embedded in the overall financing system as well as confronting the overall complexity of decision structures made within the crowdfunding process considering the big picture. The need for a more systematic understanding of the complex concept of crowdfunding and particularly its decision-making components, including the differences between entrepreneur, intermediary and professional investors, has been highlighted by several authors [20, 30]. This research contributes to the complex phenomenon of crowdfunding as well as to the "the call for more research on network management" [33].

2   Theoretical Background

"How business start-ups are financed is one of the most fundamental questions of enterprise research" [10]. Financial capital, especially "seed capital", which is required in the beginning of a new venture is a necessary resource to form and subsequently operate an enterprise [10]. Amador and Landier highlight the spread of different forms of start-up financing caused by diverse sectors and regions. The financial decision subsequently impacts economic factors in terms of competition, innovation and employment rate [10]. The following sections define several financial methods in more detail. General financial methods for Start-ups follow the structure illustrated in Fig. 1. There is a differentiation of the criteria of capital inflow. Therefore, capital can be raised from inside or from outside the company. In the case when capital is raised from inside the company, the capital source is the cash flow, which is not included in the research object.

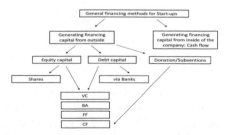

Legend: VC= Venture Capital , BA= Business Angel, FF= Family and Friends, CF= Crowdfunding

Fig. 1. Structure of general financing methods for Start-ups (based on [8])

By acquiring financing capital from outside, however, several methods are possible. Capital can be raised through equity-, debt-, or donation/subventions. Whereas, equity capital can be generated through shares (e.g. for small capitals via the new German stock-exchange segment "scale") or investments of stakeholders [42].

Debt capital is usually provided by banks in the form of bank loans. It has to be pointed out, that debt capital always requires predefined equity ratios.

A mixture of equity capital, debt capital and donation/subvention is given in form of venture capital (VC), business angel (BA), family and friends (FF), or/and crowd-funding (CF) [8, 10, 18].

2.1 Estimation of Evaluation Criteria of the Four Main Financial Decisions

By taking the four main financial decisions, the intensity of invested amount accessibility, involvement, investment motivation and risk vary in several ways. The following chart shows a first intensity estimation of these criteria rated on a scale from 0 to 5. The estimation is based on own long-time market observations in the crowdfunding market and corporate finance market and will be further explained in detail.

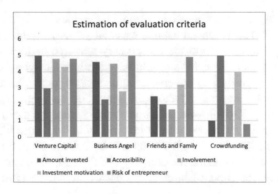

Fig. 2. Authors' estimation of varying degree of evaluation criteria depending on financing methods of Start-ups [e.g. 24, 45, 19].

2.2 Venture Capital

Venture capital (VC) has been an important financial source for innovative enterprises over the past 30 years [19]. VC is a form of financing which provides early stage companies with high potential growth predictions with required funds from investors known as venture capitalists [10]. Many successful companies including Tesla, Starbucks, Facebook, Apple, and Google used VC successfully as funding form [46]. Venture capitalists use their high "hands on" expertise to actively manage the organizations they finance and usually have broad control rights such as voting or board rights [10]. VC investors can act as groups, individuals, on behalf of a venture capital company, or as venture capital funds [19]. Due to the high risk of investing in start-ups, venture capitalists typically require various protections and possible returns [47]. Busse estimates that even if several venture capitalists experience major losses, they are usually wealthy enough to absorb the high risk of loss compared to other capital provider [8].

2.3   Business Angels

Business Angels or angel investors are "high net worth individuals who invest their personal capital in a small set of companies" [14]. Business angels typically invest "seed capital" which is required by firms in a very early stage of development [18]. Compared to other financial methods such as venture capital, business angel investments tend to be relatively small ranging from US$ 500,000 up to US$ 2 million [14]. According to Wong's research on angel investors, average companies are aged 10.5 months by the time of their first angel funding, while about 70% of companies receiving angel investments have not yet produced any revenue [47]. Angel investments are private transactions and therefore do not require any public disclosure [14]. The investment amount depends on how much the business angel is willing to provide. In return, the business angel has the opportunity to contribute and influence important company decisions [8]. However, according to Wong's study business angels provide less support in company decisions than venture capitals, as illustrated in Fig. 2 [47]. For example, only 24% of business angels assists in recruitment decisions concerning the top management team [14]. According to Wong [47] it is more likely of business angels to provide support if the business angel is geographically close to the company [44]. McKaskill claims the high rate of 50% of business angels' investments in projects, which turn out to be unsuccessful [27]. Additional evidence was submitted by Brettel who underlines this argument by quoting the high dependency on one person compared to other investment methods [7]. Despite the fact that research on angel investors indicates that business angels play an essential role in early stage start-up funding, which require smaller amounts of capital [14]. Business angels also play the role of the networker companies to receive subsequent funding. Thus, this form of investment provides rather a "bridge of financing", until the company is in a more advanced stage of development to receive venture capital [14].

2.4   Family and Friends

Generating capital through family and friends is the "most informal finance method" [25]. Several scholars highlight the cost advantages of using family and friends as a source to finance a business idea. Lee and Persson [25], however, argue that "borrowers often prefer formal finance" due to shadow costs such as intrafamily insurances and an underestimation of limited liabilities. McKaskill highlights the high risk-aversion of friends and family due to their awareness of the threat of not getting any return [27]. Other factors stated by Busse are the low accountability in terms of liquidity of family and friends [8].

2.5   Crowdfunding

One of the newer funding forms is called "crowdfunding" and this has developed into "the next big thing in entrepreneurial financing" within the last few years [26].

According to Poetz and Schreier the term "crowdfunding" derives from a wider concept of "crowdsourcing", and therefore can be described as "the outsourcing of problem-solving-tasks to a distributed network of individuals" [16, 38].

The phenomenon of crowdfunding provides start-ups with the possibility of generating feedback and funding through a certain crowd. Thus, the projects, which range from small ones to "seeking hundreds of thousands of dollars" ones, are provided by the entrepreneur [29]. The crowdfunding platform acts as an intermediary (and proves and provides the project on the platform for a certain return [9, 23]. Subsequently, the crowd is able to provide feedback and possible funding. This possible funding amount ranges from $5 up to $20 million depending on different platforms and projects.

Different platforms provide a wide range of funding-categories. Ranging from advertising and marketing, healthcare, finance, art, music, up to real estate projects [31, 32]. According to Cumming and Zhang, the phenomenon of crowdfunding has at least doubled within the last few years and is further rapidly increasing [13]. Catalini et al. [12] investigated a successful funding sum of $99 million in 2011 on Kickstarter.com. In 2014, $10.54 billion were counted in the market of Asia and Europe and up to $85.74 billion in North America [12, 13]. Crowdfunder.com estimated a raise of $4.1 billion from 2009 until 2019 on a single platform Kickstarter.com. The German market reports a financing volume of successfully funded projects of €91.6 million from January till March 2019. Due to the evolutionary expansion of crowdfunding, a concrete definition is still arguable [15, 29]. While Schwienbacher and Larralde define crowdfunding as an "An open call, essentially through the Internet, for the provision of financial resources either in form of donation or in exchange for some form of reward and/or voting rights in order to support initiatives for specific purposes" [40]. Bradford explains the term as "The use of the Internet to raise money through small contributions from a large number of investor" [6]. Other scholars such as Duoqi and Mingyu (2017) indicate "Crowdfunding consists of accumulating money from a group of people, typically comprising very small individual contribution, to support another's effort to achieve a specific goal" [15]. Summarizing the idea of crowdfunding allows individuals to request funding for cultural, social or for-profit projects by using a crowdfunding platform as an intermediary [29]. Thus, crowdfunding gives entrepreneurs with limited access to other financial sources such as venture capital, business angels and friends and family the opportunity to pursue their projects [16]. Beaulieu et al. proposes that even if other financial sources are available, crowdfunding is usually much quicker than taking advantage of other methods [3]. Other investigators such as Busse highlight the advantage of spreading the risk of failed projects among more individuals compared to other financing methods [9, 40]. Further advantages of crowdfunding are underlined by the generation of customer feedback from the crowd [15].

3   Analysis of General Types of Crowdfunding

By using crowdfunding as a financial method, the entrepreneur has multiple options to choose among the different forms of crowdfunding. As depicted in Fig. 2, options of crowdfunding types are reward-based platforms, donation-based platforms, equity-based platforms or lending-based platforms, which will be described in further detail in the following section.

3.1   Equity-Based

"Equity crowdfunding is a form of financing, in which entrepreneurs make an open call to sell a specified amount of equity or bond-like shares in a company on the internet, hoping to attract a large group of investors" [1]. Therefore, in equity-based crowd-funding, investors are able to receive equities or royalties in an organization [3].

3.2   Reward-Based

Reward-based crowdfunding offers investors non-monetary rewards such as discounts in the funded project, preordering of the funded project, or other tokens of appreciation, for example, the possibility of pre-ordering products and services [5, 20]. This type of crowdfunding is currently the most popular one compared to the other three types [13]. Reward-based crowdfunding appears in two different forms. First, in the "All or nothing (AON) model" which is used by platforms such as Kickstarter.com. The entrepreneur sets a funding goal and receives nothing until the goal is achieved. Thus, by not achieving the funding goal, the funds will be returned to the crowd [16]. Subsequently, according to Kickstarter.com, AON is less riskier for the crowd and motivates the entrepreneur to achieve the funding goal, as "All-or-nothing funding means that no one will be charged for a pledge towards a project unless it reaches its funding goal" [31]. Secondly, in "keep it all (KIA)" model, the entrepreneur collects the securities, even if the fundraising goal is not met. Therefore, if an unfunded project is proceeding the risk is maintained by the crowd [13]. Even if research found that the AON model is more likely to achieve the funding goal, due to the fact that KIA and "all and more" models provide the crowd with the risk that unfunded projects might lead to failure, it remains arguable which model is superior [13].

3.3   Lending-Based

Lending-based crowdfunding, which is also known as peer-to-peer lending (P2P) or debt-based crowdfunding, allows entrepreneurs to borrow money from the crowd by repaying in the form of interests [13, 35]. The lenders' goals vary between getting the interest and having intrinsic social goals of helping others [3]. Other scholars propose that lenders are more likely to respond positively to loan requests when they are framed as helping others rather than for some other occasion [35].

3.4   Donation-Based

Donation-based crowdfunding offers entrepreneurs the opportunity to raise money from the crowd in form of a donation. Thus, donation-based crowdfunding does not offer a certain reward, it uses crowdfunding as a collection of "social goods" and provides funding by the intrinsically or socially motivated crowd [9]. Donation campaigns range from educational topics up to medical subjects. GoFundMe is one of the most popular donation-based crowdfunding platforms and funded their biggest campaign of $2 million in November 2015 for a clinical treatment project called "Saving Eliza".

3.5   Advantages and Limitations of Crowdfunding

The general advantages of using crowdfunding as an investment method are identified by numerous scholars [4, 39, 40]. Scholars across a variety of disciplines have observed that crowdfunding can be used as a test of the market by getting a "real" market validation combined with a direct feedback from potential customers, and subsequently a risk reduction towards a non-acceptance of the product [45]. There is also a possible cashflow-optimization, as purchasing costs and production cost are more predictable and therefore easier to be anticipated. Additionally, the establishment of a crowd leads to deeper connections towards the customers as they are "part" of the projects. Hence, the crowd can be used as a multiplicator through social media or other channels [4]. Valenciene and Jegleviciute also mention the advantage of accessibility of capital compared to other funding methods [45]. Further positive effects of crowdfunding are proposed by Sigar [43]. The author mentions the capability of creating new jobs and fostering economic development as well as innovations [17, 41, 43]. Due to the fact that crowdfunding is mostly provided by online platforms, there could be a lack of guide and relevant personal advice for the customer [24]. Subsequently, scholars claim the administrative and accounting challenges arising through the crowdfunding process [42]. Moreover, due to the internet- based approach, both investors as well as entrepreneurs, face the risk of selecting a non- trustworthy crowdfunding platform as an intermediary [9, 43]. Consequences could be that the intermediary takes the invested money for his or her own use. Several authors quote the possibility of fraud due to weaker investor protection [9, 44]. Busse expresses risk associated with crowdfunding campaigns. Patents, ideas or products could be stolen or copied [9].

4   Analysis of Decision Making Processes in Crowdfunding

All participants in the process perform with different decisions. These decisions vary in their decision aims and decision drivers. Decision aims are for example from the viewpoint of the entrepreneur, crowdfunding or venture capital. Decision drivers are determinants of these decisions, for example, the costs of decision realization and the time of decision realization.

4.1   Triadic Interdependencies of Actors

Castelfranchi and Falcone explain the three-party-relationship between the client, contactor and authority [11]. The authors mentioned that it is often more likely that a party goes into a trust-relationship when the transaction is realized through a trustful intermediary. Thus, according to Peisl et al. the crowdfunding process can be described by a triadic relationship between the crowd, the web-portal and the entrepreneur [37]. The entrepreneur provides its ideas by a confident trust-relationship with the web-platform. According to Valenciene and Jegeleviciute entrepreneurs are often those who seek financing through the crowd due to limited access or failure of other financing methods [45]. Whereas, the web-portal can be defined as the intermediary [37]. The intermediary is described by Pavlou and Genfen as "a third-party institution that uses the internet infrastructure to facilitate transactions among buyers and sellers in its online marketplace" [36]. Valenciene and Jegeleviciute state that the main purpose of the intermediary is "to connect people" namely, the crowd and the entrepreneur. While the crowd - a large group of individuals - funds the entrepreneurs' idea by trusting the web- portal and provides feedback according to the marketability of the idea [13, 45]. According to Busse, all three actors within the triangle depicted in Fig. 3 have different intentions [9]. The entrepreneur aims to upload the idea, generate money, seeks feedback as well as a proof of competence, requires resources and is willing to be an early adaptor. Whereas the intermediary intents factors such as reputation, benevolence and integrity. While the crowds' intention is to invest, share personal and confidential data and seeks social-, reward- or financial compensations [9].

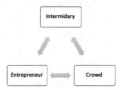

Fig. 3. Triadic relationship of crowdfunding [37]

As mentioned previously, the interdependences between the three actors, entrepreneur (E), intermediary (I) and crowd (C) have not yet been fully academically investigated. For that reason, the aim of this research is to analyze in an initial explorative step the entire decision-making process. In order to provide the basis for an analysis of the complexity and nature of the decision processes in between the three ideal-typical actors and their triadic relationship a detailed and systematized decomposition has been performed. Three decision diagrams shall provide a basic layout of fundamental decision alternatives for each actor, E, I and C, and provide an insight into the complexity of decision-making as well as the factors that influence the actors' decisions in crowdfunding.

4.2 Decisions from the Viewpoint of the Main Actors

In a first step basic decisions-making alternatives in the complex crowdfunding triangle are analyzed from the viewpoint of an entrepreneur. The following figures are not decision trees in the classical manner. They demonstrate basic decision alternatives not in a binary way, but in general terms. The decision structures of all three actors can be added with additional information in further research and provide the first basement of decision-making steps within the crowdfunding process.

4.2.1 Decision Analysis from the Viewpoint of the Entrepreneur
Figure 4 illustrates that the entrepreneur has, among others, general access to capital through four different channels. Namely, by crowdfunding, venture capital, family and friends or business angel.

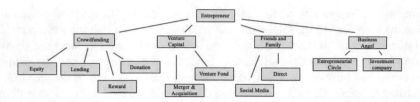

Fig. 4. Decision-making process from the viewpoint of the entrepreneur

In the case, the entrepreneur chooses crowdfunding he/she has the opportunity to decide which type. To be more specific, equity, reward, donation or lending. In any of these cases, there is the opportunity to use a platform (e.g. Crowdfunder.com), special fairs (e.g. Bits & Bretzels, Startup Camp Berlin), Social Media (e.g. Facebook) or TV shows (e.g. "Die Höhle der Löwen" in VOX). By using a crowdfunding platform several examples are given in Fig. 4, applied on each type. In the reward based type, the entrepreneur can decide between "All or nothing", "variable" or "both" as explained previously. Through considering venture capital, the entrepreneur can either use venture funds or mergers and acquisitions. By choosing friends and family as funding method, the entrepreneur has the opportunity to do this direct or indirect through social media. In the case the entrepreneur chooses a business angel he/she has further the opportunity to go through an investment company or an entrepreneurial circle.

4.2.2   Decision Analysis from the Viewpoint of the Crowd

As shown in Fig. 5, the crowd has, amongst others, also several opportunities when choosing to invest in an idea. First, the crowd chooses its drivers to invest. The most common driver is generally the return driver in the form of return on equity (ROE) or gaining a special interest rate if the crowd grants a loan. Other factors are the interest in a special company or product. The crowd could be interested in innovation, environmental advantage or social aspects. By being interested in the product, the crowd can be interested in one of the four crowdfunding forms (lending, reward, donation, equity). If the crowd wants to generate social help, it can do it either with cash or by providing real help.

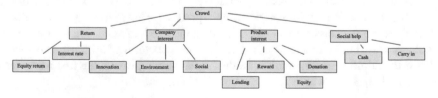

Fig. 5. The decision-making process from the viewpoint of the crowd

4.2.3 Decision Analysis from the Viewpoint of the Intermediary

Marking off the analysis from the viewpoint of entrepreneurs and crowd, the next step will be the decision analysis from the viewpoint of the intermediary, the platform provider, as shown in the following Figure.

The framework depicted in Fig. 6 illustrates the decision-making aspects of the intermediary, i.e. the crowdfunding platform provider. The crowdfunding platform provider has to decide whether to provide an equity-, reward-, lending-, donation or mixed-method approach. After defining the implemented approach, the platform provider has to decide further which structure or which visual appearance it might offer and how to select the crowd and the entrepreneurs.

Fig. 6. Decision-making process from the viewpoint of the intermediary

5 Conclusion

Due to a lack of analysis of decision making structures in crowdfunding research and particularly within the triadic relation of the players involved, several scholar claim the need for a more systematic exploration of decision-making components within crowdfunding [20]. However, to analyse complex decisions within the different participators, it is essential to clarify in a first step which decisions are necessary for every single component.

This paper provides the classification of crowdfunding into the overall financing methods and gives a detailed explanation of crowdfunding including its different forms. It analyzes the necessary decision-making steps of the involved actors as well as the interdependencies in the triadic relationships. Thus, it provides the basis for further research on decision-making in crowdfunding. In a next step further research is necessary to empirically underpin the results. Moreover, research could focus on specific analysis of decisions and empirically explore components and decision-drivers which support successful crowdfunding projects.

References

1. Ahlers, G., Cumming, D., Günther, C., Schweizer, D.: Signaling in equity crowdfunding. Entrepreneurship Theory Pract. **39**(4), 955–980 (2015)
2. Babbage, C.: On the Economy of Machinery and Manufactures. Lnd. Knight (1831)

3. Beaulieu, T., Saker, S.: A conceptual framework for understanding crowdfunding. Commun. Assoc. Inf. Syst. **37**, 1–31 (2015)
4. Bechter, C., Jentzsch, S.M.: From wisdom of the crowd to crowdfunding. J. Commun. Comput. **9**(1), 951–957 (2011)
5. Bellefamme, P.T.: Crowdfunding: tapping the right crowd. J. Bus. Ventur. **29**(5), 585–609 (2014)
6. Bradford, S.: Crowdfunding and the federal securities laws. Columbia Bus. Law Rev. **24**(4), 1–150 (2012)
7. Brettel, M.: Business angels in Germany: a research note. Int. J. Entrepreneurial Finan., 251–268 (2010)
8. Busse, F.-J.: Grundlagen der betrieblichen Finanzwirtschaft, 5th edn. Oldenbourg, München (2003)
9. Busse, V.: Crowdfunding - an empirical study on the entrepreneurial viewpoint. In: Fatos, X., Barolli, L., Gregus, M. (eds.) Advances in Intelligent Networking and Collaborative Systems: The 10th International Conference on Intelligent Networking and Collaborative Systems, pp. 306–318 (2018)
10. Cassar, G.: The financing of business start-ups. J. Bus. Ventur. **19**(2), 261–283 (2004)
11. Castelfranchi, C., Falcone, R.: Principles of Trust, pp. 55–99. Kluwer (2001)
12. Catalini, C., Fazio, C., Murray, F.: Can equity crowdfunding democratize. Innov. Sci. Rep., 1–16 (2016)
13. Cumming, D., Zhang, M.: Angel Investors Around the World. Elsevier, Amsterdam (2016)
14. Denis, D.: Entrepreneurial finance: an overview of the issues and evidence. J. Corp. Finan. **10**, 301–326 (2004)
15. Duoqi, X., Mingyu, G.E.: Equity-based crowdfuning in China: beginning with the first crowdfunding financing case. Asian J. Law Soc. **4**, 81–107 (2017)
16. Gerber, E., Hui, J.: Crowdfunding: motivations and derterrents for participation. ACM Trans. Comput.-Hum. Interact. **20**, 1–32 (2013)
17. Gobble, M.: Everyone is a venture capitalist. Res. Technol. Manag. **55**, 4 (2012)
18. Gompers, P.: Optimal investment, monitoring, and the staging of venture capital. J. Finan. **50**, 1461–1489 (1995)
19. Gompers, P., Gornall, W., Klaplan, S., Strbulaev, I.: How do venture capitalists make decisions. SSRN Electron. J. (2016)
20. Hoegen, A., Steininger, D.V.: How do investors decide? An interdisciplinary review of decision-making in crowdfunding. Electron. Markets **28**, 339–365 (2018)
21. Kaczor, S., Kryvinska, N.: It is all about services - fundamentals, drivers, and business models. Soc. Serv. Sci. J. Serv. Sci. Res. **5**(2), 125–154 (2013)
22. Kryvinska, N.: Building consistent formal specification for the service enterprise agility foundation. Soc. Serv. Sci. J. Serv. Sci. Res. **4**(2), 235–269 (2012)
23. Kryvinska, N., Auer, L., Strauss, C.: The place and value of SOA in building 2.0-generation enterprise unified vs. ubiquitous communication and collaboration platform. In: Mauri, J.L., Meloche, J.A., Balandin, S., Ibrohimova, M., Nakata, J. (eds.) The Third International Conference on Mobile Ubiquitous Computing, Systems, Services and Technologies. IEEE Press, Piscataway, pp. 305–310 (2009). https://doi.org/10.1109/ubicomm.2009.52
24. Kryvinska, N., Strauss, C., Zinterhof, P.: Migration strategies, planning methodologies, architectural design principles. In: Xhafa, N.B. (ed.) Next Generation Service Delivery Network as Enabler of Applicable Intelligence in Decision and Management Support Systems, pp. 473–502. Springer, Berlin (2011)
25. Lee, S., Persson, P.: Financing from family and friends. Rev. Finan. Stud. **29**, 2341–2386 (2016)

26. Li, Y., Rakesh, Reddy: Project success predicting crowdfunding. In: The ACM Guide to Computing Literature (2016)
27. McKaskill, T.: Raising Angel and Venture Capital Finance. Breakthrough Publications, Melbourne (2009)
28. Mladenow, A., Bauer, C., Strauss, C., Gregus, M.: Collaboration and locality in crowdsourcing. In: 7th International Conference on Intelligent Networking and Collaborative Systems (INCoS 2015), 2–4 September 2015, Taipei, Taiwan (2015). https://doi.org/10.1109/incos.2015.74
29. Mollick, E.: The dynamics of crowdfuning: an exploratory study. J. Bus. Ventur. **29**, 1–16 (2014)
30. Moritz, A., Block, J., Lutz, E.: Investor communication in equity-based crowdfuning: a qualitative empirical study. Qual. Res. Finan. Mark. **7**, 309–342 (2015)
31. n.d. (22 März 2019). *Kickstarter*. Von https://www.kickstarter.com abgerufen
32. n.d. (22 März 2019). *IndieGoGo*. Von https://www.indiegogo.com abgerufen
33. Naudé, P., Sutton-Brady, C.: Relationships and networks as examined in Industrial Marketing Management. Ind. Mark. Manag. **79**, 27–35 (2019)
34. O'Conner, J., Robertson, E.: Longitude and academie royale. In: Mac Tutor History of Mathematics (1997)
35. Paschen, J.: Choose wisely: Crowdfunding through the stages of the startup life cycle. Bus. Horiz. **60**, 179–188 (2017)
36. Pavlou, P., Gefen, D.: Building effective online marketplaces with institution-based trust. Inf. Syst. Res. **15**, 37–59 (2004)
37. Peisl, T., Raeside, R., Busse, V.: Predictive crowding: the role of trust in crowd selection. In: 3E Conference Ireland, pp. 1–19 (2017)
38. Poets, M., Schreier, M.: The value of crowdsourcing: can user really compete with professionals in generating new product ideas? J. Prod. Innov. Manag. **29**, 245–256 (2012)
39. Rossi, M., Thrassou, A., Vronits, D.: Open Innovation Systems and New Forms of Investment: Venture Capital's Role in Innovation. Cambridge Scholar Publishing, Newcastle upon Tyne (2013)
40. Schwienbacher, A., Laaralde, B.: Crowdfunding of Small Entrepreneurial Ventures. Oxford University Press, Oxford (2010)
41. Shirky, C.: Clay shirky talks JOBS act and the new business ecosystem (2012). (E. Blattberg, Interviewer)
42. Sickinger, M.D.: Überblick zu Börsensegmenten, Zugangsvoraussetzungen und Folgepflichten. *Münchner Kapitalmarkt Konfernez*. Heuking Kühn Wojtek, München (2017)
43. Sigar, K.: Fret no more: inapplicability of crowdfunding concerns the internet age and the JOBS Act's safeguards. Adm. Law Rev. **64**, 474–505 (2012)
44. Sullivan, B., Ma, S.: Crowdfunding: Potential Legal Disaster Waiting to Happen, pp. 474–505. Forbes (2012)
45. Valanciene, L., Jegeleviciute, S.: Valuation of crowdfuning: benefits and drawbacks. Econ. Manag. **18**, 39–48 (2013)
46. Vinturella, J., Erickson, S.: Raising Entrepreneurial Capital. Elsevier, Amsterdam (2013)
47. Wong, A.: Angel finance: the other venture capital. SSRN Electron. Bus., 1–66 (2002)

The 2nd International Workshop Machine Learning in Intelligent and Collaborative Systems (MaLICS-2019)

Survey on Blockchain-Based Electronic Voting

Shuai Xiao[1], Xu An Wang[1(✉)], Wei Wang[2], and Han Wang[1]

[1] Key Laboratory for Network and Information Security of the PAP,
Engineering University of the PAP, Xi'an 710086, Shaanxi, China
xs18829581835@163.com, Wangxazjd98@163.com
[2] Engineering University of the PAP, Xi'an 710086, Shaanxi, China

Abstract. With the progress of society and the improvement of people's democratic consciousness, voting, as a channel to fully develop democracy, is playing an increasingly important part in many application scenarios. In the era of Internet, electronic voting has replaced the traditional paper voting with the advantages of low cost, high efficiency and few mistakes. The fact that the data of electronic voting system is stored in the central database gives rise to the following problems: the voting data is not open and transparent enough, and it is easy to be tampered with and forged; the users' private information faces the risk of being leaked; voters can not verify the voting results. Fortunately blockchain technology can make up for the shortcomings of the current voting system, making the voting process open and transparent, preventing fraudulent votes, enhancing the security of voting data and verifying the voting results. The application of blockchain electronic voting system has very important significance and prospects. In this paper, we first introduce the development process of electronic voting system at home and abroad, then introduce the development status of blockchain-based electronic voting system, finally we compare and summarize several typical blockchain-based electronic voting schemes.

1 Introduction

With the progress of society and the great improvement of democratic consciousness, voting plays an increasingly important role in social life. At present, the main means of voting are: raising hands, paper-based voting, online voting and so on. The application scenarios of hands-up voting are extremely limited and have many limitations, such as the high degree of relevance between voting content and personal interests, voters are easily affected by the surrounding environment. Paper voting is simple and economic, but there are two main problems. On the one hand, paper voting is not suitable for larger voting, otherwise the accuracy of voting cannot be guaranteed; on the other hand, paper voting mainly depends on the procedural security of executing officials. In this process, collusion corruption easily occurs, which is contrary to the original intention of voting.

© Springer Nature Switzerland AG 2020
L. Barolli et al. (Eds.): INCoS 2019, AISC 1035, pp. 559–567, 2020.
https://doi.org/10.1007/978-3-030-29035-1_54

More or less, there exists some problems in the application scenarios of voting, such as the data is not open, opaque, votes fraudulent and arbitrary changes in the results.

Electronic voting is a new online voting system based on cryptography technology. Voters can vote conveniently on the Internet through computers or mobile devices. The final voting results can be automatically anonymously counted by the central server. These systems make the whole voting process interconnected, which greatly improves the efficiency of organizing, collecting ballots and counting the results of ballots compared with traditional voting, and ensures the fairness and openness of the voting process. But at present, the common electronic voting system also has the following drawbacks:

1. The security of data transmission cannot be guaranteed, and attackers can easily invade the system, tamper with or even destroy the voting results;
2. The risk of leakage of voters' personal privacy information;
3. Various risks such as data loss, document damage, official bribery and supplier's party-building and private operation in electronic voting system;
4. The voting results are totally centralized, and voters cannot verify whether their voting results are correct.

These problems have seriously hampered the construction of a more secure and efficient voting environment, which needs to be improved and solved urgently. The birth of Bitcoin in early 2009 has brought a new wave to the scientific and technological field in recent years. Its underlying technology-blockchain has gradually attracted people's attention, and became popular all over the world in early 2016. In particular, the technology of blockchain represented by Bitcoin and Ethereum (ETH) has attracted great attention from governments, financial institutions, technology enterprises, technology enthusiasts and the media. Blockchain is essentially a decentralized distributed ledger technology, which is composed of distributed data storage, peer-to-peer transmission, consensus mechanism, encryption algorithm and other technologies. Its technical feature is to build trust in a decentralized way, which has broad application prospects.

At present, there are many problems in electronic voting system, such as duplicate voting, fraudulent voting, privacy disclosure and data security. Combining with the characteristics of decentralization, traceability, untouchable modification and anonymity of blockchain technology, block chain technology is adopted as the accounting database of voting system in the voting system, and the pre-designed voting protocol can be automatically executed through intelligent contracts to ensure the electronic voting system. The traditional voting data is fair, transparent, verifiable and unalterable, which improves the accountability of the voting system and reduces the trust risk of the system. At the same time, it can effectively prevent illegal voters or malicious institutions from fraudulent voting, disrupt the voting process and interfere with the voting results. In addition, the anonymity algorithm protects the privacy of voters and allows anyone to query and verify the voting results without compromising the openness and fairness of the voting process. Therefore, the introduction of blockchain technology into electronic voting provides a new solution to the security problems encountered in the current voting system.

2 Development of Electronic Voting

2.1 General Electronic Voting

With the development of Internet technology and modern cryptography technology, electronic voting has become a new voting method, which solves many shortcomings of traditional voting methods, such as high cost, low efficiency and many mistakes. Since the first electronic voting protocol was proposed, after more than 30 years of development, electronic voting schemes based on different cryptosystems have been proposed by many cryptologists. Its goal is to provide a secure, convenient and efficient voting environment for the Internet. These mature solutions have also been applied to some government elections, corporate board voting and important decision-making voting.

Generally speaking, a secure electronic voting scheme should meet the following requirements:

1. Vote privacy: no one knows who the voters choose, and the content of the votes is hidden from the observer.
2. Personal verifiability: Voters can verify whether their votes are counted correctly after voting.
3. Qualification: Only legitimate voters can participate in voting activities.
4. Justice: Nothing can affect the result of voting. It is not allowed to divulge the result of voting or to add voters in the voting process. Otherwise, it will have an unfair impact on the result of voting.
5. Uniqueness: Each legitimate voter can vote only once.
6. Robustness: No one or any factor can affect or modify the final result of voting.
7. Integrity: Each ballot should be counted correctly.

At present, there are three main types of electronic voting schemes: electronic voting schemes based on blind signature and ring signature [1–5], electronic voting schemes based on homomorphic encryption [6–9], and electronic voting schemes based on hybrid network [10,11]. However, these three kinds of voting schemes have their own shortcomings (Table 1):

Chaum [12], an American cryptographer, proposed the first electronic voting scheme in 1981. The scheme transmits votes through anonymous channels and uses public key cryptosystem and digital pseudonym voting to hide voters' identity. Cohen and Fisher [13] proposed a secure voting protocol based on homomorphic encryption in 1985, but the protocol requires that all voting processes must be synchronized. Then Iverson [14], Benaloh [15] and Sako [16] proposed different voting protocols, in which Iverson's voting protocol could not effectively curb cheating, because only when all voters cooperated in the voting process can they vote effectively.

Table 1. Three types of electronic voting schemes

Voting schemes	Flaws
Electronic Voting Scheme Based on Blind Signature and Ring Signature	Generally, it is necessary to assume anonymous channels and trusted signature institutions
Electronic Voting Scheme Based on Homomorphic Encryption	Although ciphertext computation can be implemented to protect the privacy of ballot papers, homomorphic encryption is too complex to be practical
Electronic Voting Scheme Based on Hybrid Network	In theory, the public verifiability of decryption and counting can be achieved, but the algorithm is too complex and inefficient

In 1992, Fujioka, Okamoto and Ohta [17] proposed the first protocol (FOO) for large-scale voting scenarios. The voting mechanism consists of three parts: voters, voter sponsors and voter counters. It uses blind signature and bit commitment technology to encrypt ballot information and then sends it to the voting management agencies. It effectively protects the privacy of voters and guarantees the fairness of voting. But there are some problems in the scheme: voters cannot choose to abstain, ballot collision, cannot distinguish dishonest voting institutions and voters, and voter sponsors can get votes that voters have not voted out.

The electronic voting schemes mentioned above all need to be established in a secure and credible third party (TTP) voting agency to count votes, which poses a huge threat to voters' privacy. Usually, electronic voting protocols that protect voters' privacy rely on reliable authorities to decrypt and count votes in a verifiable manner. There are also many research institutions and scholars abroad who have conducted in-depth research on electronic voting schemes with self-counting votes. The electronic voting protocol (Helios) in document [18] usually uses threshold encryption to assign this trust to multiple statistical agencies. However, voters still need to believe that the statistical authorities will not fully collude, because once the statistical authorities fully collude, voters' privacy will be undermined.

Kiayias and Yung [19] first proposed a self-taffly voting protocol for board size scenarios, and then Groth [20] and Hao [21] proposed that the self-counting voting protocol transform the counting process into an open and verifiable process, allowing any voter or third-party observer to perform the counting process after all votes have been cast. This weakens the unique role of voting institutions in elections, because anyone can count the results of voting. These agreements provide voters with the greatest confidentiality and non-controversial, because only when the remaining voters are all in collusion will one person's ballot be disclosed and any third-party body be allowed to verify that voters are correctly following the voting agreement. However, the self-counting voting agreement also has a shortcoming of fairness, because the last voter can know the result of voting before others, which will cause the problem of destroying the result of voting. In other words, knowing the result of the vote in advance may affect the choice of the last voter.

2.2   Blockchain-Based Electronic Voting

In recent years, with the popularity of encrypted digital money, such as Bitcoin [22], many scholars have paid attention to blockchain and its underlying accounting technology. The essence of blockchain is an open and transparent database ledger technology, which records all transaction information. Its characteristic is that it can provide the characteristics of decentralization, non-tampering, open and transparent without third-party intermediaries [23]. At present, researchers at home and abroad use blockchain technology to carry out a lot of research and application. In the field of electronic voting, there are also many schemes combining blockchain technology. In 2015, Zhao and Chan [24] proposed an electronic voting protocol combining Bitcoin, which introduced a reward and penalty system for voters' voting behavior. Although the protocol has some limitations, it is the first attempt to combine electronic voting with blockchain.

In 2016, Lee, James, Ejeta and Kim [25] proposed another electronic voting protocol, which proposed using TTP (Trusted Third Party) to protect voters' votes in blockchain electronic voting protocol.

In 2017, Cruz et al. [26] proposed an electronic voting scheme based on blind signature technology and blockchain. The scheme is to write the blind voting content into the Bitcoin transaction with 80 bytes of additional information. In addition, the scheme also introduces a third-party voting agency to count the voting results.

The above schemes all have TTP's function of supervising the voting process, which requires voters to trust the TTP, but we cannot rule out the collusion attack between third-party voting agencies and voting sponsors, which results in the tampering of voters' votes and voting results in the voting process, as well as the leaking of voting results in advance by the voting agencies to control the whole voting results.

In 2017, Ayed [27] envisaged a new electronic voting system that could be used in local or national elections. This blockchain-based system would be secure, reliable and anonymous, and would increase the number of voters and the credibility of the government. Firstly, the author reviews the typical electronic voting systems currently used in several countries, analyses the shortcomings and security problems of these systems, and then introduces the new blockchain-based electronic voting system from definition to system requirements and implementation process. However, this system has limitations. The ideal state envisaged by the author is that voters use security devices to vote, but even if the system is very secure, hackers still have the ability to vote with malicious software installed on the voter's device beforehand or arbitrarily tamper with the voting results. In addition, if the user votes incorrectly, the system will not be able to modify. That is to say, users can only vote once.

In the same year, McCorry et al. [28] implemented a distributed and self-tally electronic voting scheme using the Ethernet blockchain, which maximized the protection of voter privacy. The voting protocol was written by ETH Smart Contract, which effectively replaced the third-party voting organization. In the voting process, the privacy of voters' voting information was protected by two rounds of zero knowledge proof, and the cost of contract execution was decomposed economically and numerically. A simulation experiment of 40 voters was carried out, but the voting scheme only allowed voters to choose two candidates (yes/no), which cannot satisfy the situation of multiple candidates in a single voting process.

In addition, domestic scholars also make use of the technical advantages of blockchain to make up for the shortcomings of electronic voting, and put forward many schemes. Among them, literature [29] proposes a board electronic voting system based on coalition blockchain to solve the anonymity of voting and the strict requirement of participant identity in small-scale voting scenarios. Smart contract is used to replace the traditional trusted third party institutions; the identity access mechanism of digital certificates is used to ensure the legitimate identity of the voters involved; and the voting protocol is designed by using elliptic curve blind signature technology to satisfy the anonymity of voting. In addition to these electronic voting protocols, many electronic voting applications on blockchain have been widely used. These applications basically use blockchain as a ballot box, so they still need to rely on third-party organizations to protect the privacy of voters.

3   Comparative Analysis of Blockchain-Based Electronic Voting Schemes

Literature [28] proposes a distributed blockchain electronic voting scheme with self-tally function by using blockchain. In the scheme, the voter's voting privacy information is protected by two rounds of zero knowledge proof protocol, but the voting scheme only allows voters to select two candidates (yes/no), namely 1-out-of-2 voting, which cannot satisfy the situation of multiple candidates in one voting process. Literature [30] proposes an electronic voting protocol based on blockchain by using ring signature algorithm. The security and privacy protection of the protocol is realized by ring signature. The voting process is realized by using the special 80- byte OP-RETURN script space of Bitcoin to write encrypted ballot information. However, the scheme has two apparent shortcomings: the ciphertext expansion of ring signature makes it impossible for the protocol to be applied to large-scale voting, and the amount of data written by OP-RETURN script is limited. If a certain amount of data is stored in each transaction, it will undoubtedly bring heavy burden to the Bitcoin network.

Hardwick et al. propose a voting scheme in the permission-based blockchain model, satisfying the basic notions fairness, eligibility, privacy and verifiability [31]. Their protocol uses the blockchain as a transparent ballot box. The system relies on a central certificate authority to authenticate voters and give permission to access the network. The bad thing is that, an authority, when byzantine, breaks the link between voter identity and casted vote, thus violating the ballot privacy (Table 2).

Table 2. Comparison of three types of voting schemes

Schemes	Encryption mode	Voting type	Vote counting method	Blockchain Technology
Hardwick et al. [31]	Blind signature	any	Self-tally	ETH
Wu [30]	Ring signature	1-out-of-m	Third party counting	Bitcoin
McCorry et al. [28]	2 Round-zero knowledge proof	1-out-of-2	Self-tally	ETH and Smart Contract

As can be seen from the table above, the scheme of literature [28] uses two rounds of zero knowledge proof protocol, but it can only achieve 1-out-of-2 type of voting. Although the scheme in literature [30] can achieve the 1-out-of-m type of voting, its encryption method is limited by the ring signature and is not flexible enough.

4 Conclusion

In this paper, we outline the development process of electronic voting. Through the comparative analysis of two typical schemes, we summarize their advantages and disadvantages. Generally speaking, the application scenarios of electronic voting based on blockchain are limited. To apply to large-scale scenarios, it is essential to improve the transaction throughput in blockchain. Because large-scale voting scenarios require high timeliness and throughput, once there is a large-scale voting, transactions increase sharply, the blockchain network is under great pressure. If we want to improve transaction throughput, we need to redesign and optimize consensus algorithm, block size, block generation time and transaction verification time to achieve better results.

Acknowledgements. This work was supported by the National Cryptography Development Fund of China (grant no. MMJJ20170112), Natural Science Basic Research Plan in Shaanxi Province of China (grant no. 2018JM6028), National Natural Science Foundation of China (grant no. 61772550, U1636114, and 61572521), and National Key Research and Development Program of China (grant no. 2017YFB0802000). This work is also supported by Engineering University of PAP's Funding for Scientific Research Innovation Team (grant no. KYTD201805).

References

1. Zhang, J., Li, Z., Liu, B., et al.: Multi-authorization electronic voting system based on group blind signature. Chin. Sci. Technol. (8) (2015). Pap. 980–983
2. Improvement of Yewei. FOO protocol and its application in electronic voting system. Wuhan University of Technology (2009)
3. Hua, R.: An electronic voting protocol for RSA signature system. J. Southwest Univ. Natly. (Nat. Sci. Ed.) **35**(5), 1091–1094 (2009)
4. Fan, A., Sun, Q., Zhang, Y.: Anonymous electronic voting scheme based on ring signature. Eng. Sci. Technol. **40**(1), 113–117 (2008)
5. Qiming, X.G.: A secret voting scheme suitable for large-scale electronic elections. J. Electron. Inf. Sci. **19**(5), 717–720 (1997)
6. Wang, Y., Xuchen, Chen, J., et al.: Safe electronic voting scheme based on HElib. Comput. Appl. Res. **34**(7), 2167–2171 (2017)
7. Chillotti, I., Gama, N., Georgieva, M., et al.: A homomorphic LWE based E-voting scheme. Springer (2016)
8. Zhu, Z.: An electronic voting scheme based on homomorphic encryption. Guangzhou University (2013)
9. Peng, K., Aditya, R., Boyd, C., et al.: Multiplicative homomorphic e-voting. In: Lecture Notes in Computer Science, vol. 3348, pp. 1403–1418 (2005)
10. Zhu, J., Fu, Y.: Advances in block chain application research. Sci. Technol. Bull. **35**(13), 70–76 (2017)
11. Huming, G., Jilin, W., Yumin, W.: An electronic voting scheme based on Mix Net. J. Electron. Sci. **32**(06), 1047–1049 (2004)
12. Chaum, D.L.: Untraceable electronic mail, return addresses, and digital pseudonyms. Commun. ACM **4**(2), 84–88 (1981)
13. Cohen, J.D., Fischer, M.J.: A Robust and Verifiable Cryptographic Secure Election Scheme, pp. 372–382. Mccarthy (1985)
14. Iversen, K.R.: A cryptographic scheme for computerized general elections. In: International Cryptology Conference, pp. 405–419 (1991)
15. Benaloh, J., Tuinstra, D.: Receipt-free secret-ballot elections. Proc. STOC **94**, 544–553 (1994)
16. Sako, K., Kilian, J.: Secure voting using partial compatible homomorphisms. In: Cryptology Conference on Advances in Cryptology, pp. 411–424 (1994)
17. Fujioka, A., Okamoto, T., Ohta, K.: A practical secret voting scheme for large scale elections. In: Auscrypt92 Gold Coast Queensland Australia, vol. 718, pp. 244-251, December 1992
18. Adida, B.: Helios: web-based open-audit voting. In: USENIX security symposium, pp. 335–348 (2008)
19. Kiayias, A., Yung, M.: Self-tallying elections and perfect ballot secrecy. In: International Workshop on Public Key Cryptography, pp. 141–158. Springer (2002)
20. Groth, J.: Efficient Maximal Privacy in Boardroom Voting and Anonymous Broadcast, vol. 3110, pp. 90–104 (2004)
21. Hao, F., Ryan, P.Y.A., Zielinski, P.: Anonymous voting by two-round public discussion. IET Inf. Secur. **4**(2), 62–67 (2010)
22. Nakamoto, S.: Bitcoin: a peer-to-peer electronic cash system (2008)
23. Yong, Y., Feiyue, W.: Development status and prospect of block chain technology. J. Autom. **42**(4), 481–494 (2016)
24. Zhao, Z., Chan, T.H.: How to Vote Privately Using Bitcoin. Springer (2015)

25. Lee, K., James, J.I., Ejeta, T.G., et al.: Electronic voting service using block-chain. J. Digit. Forensics Secur. Law: JDFSL **11**(2), 123 (2016)
26. Jason, P.C., Yuichi, K.: E-voting system based on the bitcoin protocol and blind signatures. Trans. Math. Model. Appl. **10**(1), 14–22 (2017)
27. Ayed, A.B.: A conceptual secure blockchain-based electronic voting system. Int. J. Netw. Secur. Appl. (IJNSA) **9**(3), 1–9 (2017)
28. McCorry, P., Shahandashti, S.F., Hao, F.: A smart contract for board room voting with maximum voter privacy. In: International Conference on Financial Cryptography and Data Security, pp. 357–375. Springer (2017)
29. Dong, Y., Zhang, D., Han, J., et al.: Board electronic voting system based on alliance blockchain. J. Netw. Inf. Secur. (12) (2017)
30. Wu, Y.: An e-voting system based on blockchain and ring signature. Master, University of Birmingham (2017)
31. Hardwick, F.S., Akram, R.N., Markantonakis, K.: E-voting with blockchain: an e-voting protocol with decentralisation and voter privacy. CoRR abs/1805.10258 (2018). arXiv: 1805.10258

Optimization of Maintenance Costs of Video Systems Based on Cloud Services

Dominika Karyś[1], Anna Pietrzyk[1], Rafał Kowalski[1], Peter Vesely[2], and Andrea Studenicova[2]([✉])

[1] Institute of Information Technology,
Lodz University of Technology, Lodz, Poland
[2] Faculty of Management, Comenius University, Bratislava, Slovakia
{Peter.Vesely,andrea.studenicova}@fm.uniba.sk

Abstract. The paper describes system architecture for storing and managing video resources. Presented architecture is based on the cloud. It will propose a way to optimize the costs associated with storing large files by using two types of storage. The first one will be dedicated to high-resolution files. The use of transcoder will be discussed including creation of a file of lower resolution. Such files will be stored in a second above-mentioned type of storage, which is available from the client application and is shared within end users. To illustrate the idea of cost optimization the services provided by Amazon were used. The paper describes a case study of highly scalable video archive with a dedicated web application. The architecture for this case is proposed and the basic cases of use are presented. However, the whole business analysis is omitted and the problem is treated at a high level of abstraction.

1 Introduction

Cloud computing is a computing paradigm, where a large pool of systems are connected in private or public networks, to provide dynamically scalable infrastructure for application, data and file storage. With the advent of this technology, the cost of computation, application hosting, content storage and delivery is reduced significantly. Cloud computing is a practical approach to experience direct cost benefits and it has the potential to transform a data center from a capital-intensive set up to a variable priced environment. The idea of cloud computing is based on a very fundamental principal of "reusability of IT capabilities". The difference that cloud computing brings compared to traditional concepts of "grid computing", "distributed computing", "utility computing", or "autonomic computing" is to broaden horizons across organizational boundaries [19,24,27].

Currently, mobile devices such as smart phones and portable computers are transforming our daily tasks, from the simple task of making calls to various applications such as sending emails, messaging, location services, multimedia,

© Springer Nature Switzerland AG 2020
L. Barolli et al. (Eds.): INCoS 2019, AISC 1035, pp. 568–578, 2020.
https://doi.org/10.1007/978-3-030-29035-1_55

banking, etc. Consequently, many challenges arise due to running such computationally high demanding applications on mobile devices. Though latest mobile devices use high-speed processors, with clock frequencies up to 1 GHz, power consumption is still a problem. Cloud computing offers a solution by making computations off line on the "cloud" thus reducing the power consumption on the device and allowing more elaborate and accurate algorithms to be performed on the server. It has been shown in that offloading is beneficial only when the amount of computation is fairly large with relatively small amounts of communication data which is the case in most image and video processing applications. It is assumed that the communication channel has enough upload bandwidth and the data to be transmitted is small while the operation to be performed on the image requires a lot of computational power [1,25,29].

The process of transcoding video to share video content imposes heavy strain on Internet infrastructure and computer resources. Recent video files have changed from low performance and definition to high capacity and definition. This was the reason that massive storage servers were required to store such files. In addition, all video transcoding processes consist of three vital sub-processes: decoding, resizing and encoding. This is a reason enormous computing power from processor resources is necessary and as a result creates need for an efficient approach to transcoding [23,26].

The paper describes a case study of highly scalable video archive with a dedicated web application. The architecture for this case is proposed and the basic cases of use are presented. However, the whole business analysis is omitted and the problem is treated at a high level of abstraction. The last section presents the comparison of application maintenance costs, focusing on the file storage. The comparison is based on the approach with the use of optimization through the use of two types of storages and the classic approach to file storage.

2   Cloud Computing, Its Services and Possibilities

Cloud Providers offer services that can be grouped into three categories:

- *Software as a Service (SaaS)*: in this model, a complete application is offered to the customer, as a service on demand. A single instance of the service runs on the cloud and multiple end users are serviced. On the customers' side, there is no need for upfront investment in servers or software licenses, while for the provider, the costs are lowered, since only a single application needs to be hosted and maintained. Today SaaS is offered by companies such as Google, Salesforce, Microsoft, Zoho, etc.
- *Platform as a Service (PaaS):* a layer of software, or development environment is encapsulated and offered as a service, upon which other higher levels of service can be built. The customer has the freedom to build his own applications, which run on the providers' infrastructure. To meet manageability and scalability requirements of the applications, PaaS providers offer a predefined combination of OS and application servers, such as LAMP platform

(Linux, Apache, MySql and PHP), restricted J2EE, Ruby etc. Google's App Engine, Force.com, etc.

- *Infrastructure as a Service (IaaS)*: IaaS provides basic storage and computing capabilities as standardized services over the network. Servers, storage systems, networking equipment, data centre space and others are pooled and made available to handle workloads. The customer would typically deploy his own software on the infrastructure. Some common examples are Amazon, GoGrid, 3 Tera, etc.

The features of Cloud Computing are that it offers enormous amounts of power in terms of computing and storage while offering improved scalability and elasticity. Moreover, with efficiency and economics of scale, Cloud Computing services are becoming not only a cheaper solution however much greener one to build and deploy IT services. The Cloud Computing distinguishes itself from other computing paradigms in the following aspects [20,21,28,32]: On-demand service, QoS guaranteed offer, Autonomous System, Scalability.

The process of high quality video encoding is usually very costly to the encoder and require a lot of production time. Considering situations where there are large content volumes, this is even more critical, since a single video may require the server's processing power for long time periods. Moreover, there are cases where the speed of publication is a critical point. Journalism and breaking news are typical applications in which the time-to-market video is very short, so that every second spent in video encoding may represent a loss of audience [1].

Wanting to ring the times of character coding into two solutions. With a view to increase their usability for reuse – only in peak pairs. The downside is that the infrastructure will remain inactive for the remaining time. The second solution is making and life-saving plant of resources. The ideal scenario would be to optimize using different solutions and evenly. In particular, new video coding, intuitive solutions for sharing into several different states in a cluster. The challenge of this problem is to split, as well as to defile video fragments without loss of synchronization [30,31].

Content Delivery Network (CDN) is an important aspect of data storage. They are provided by almost all providers. Their main task is to reduce access time to resources. This task is accomplished by creating a network of storage servers. When requesting access to resources, the main server redirects the request to another server that has the resource and that is closer to the client [17,33].

3   System Model for Storing and Processing Video Operating in the Cloud

This paper presents a model of a dedicated system for storing and processing video operating in the cloud. Amazon's services were used to design the architecture of the applications. These services are *storage* (S3 [1] and Glacier [2]) and *video processing* (Elastic Transcoder [3]). It is assumed that resource management is carried out through a web application written in any technology and hosted by another Amazon service: EC2 [4].

In addition, additional Amazon components supporting data management were used to build the system. They include services such as *database* (DynamoDB [5] or *relational database* such as PostgreSQL [6,15]), *mail server* (SES [7]) used to inform users about actions in the system, *serverless functions* supported by Lambda [8] or CloudFront [9] – Amazon Content Delivery Network. The last mentioned service can be triggered when adding or updating records in the database according to the business logic (Fig. 1).

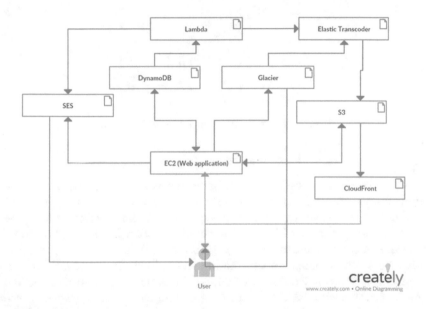

Fig. 1. Prototype of the system architecture to perform the described tasks

This architecture is the basis for all cloud-based systems whose task is to store, display, basic process and management video.

Basic Use Cases

To present sample usage of presented architecture, sample system is presented. This system is dedicated for video materials archiving (Fig. 2).

Storage for HR Files

The first problem in a system dedicated to file storage is their large size. In order to minimize this problem you can use storage, which is cheaper but the access time to it is much larger. Example of this is Glacier. In the presented model it will be used to store HR files that will not be available to users of the system, unless they order download actions in the original size. This will reduce network traffic and make the use of the application and downloading related resources faster.

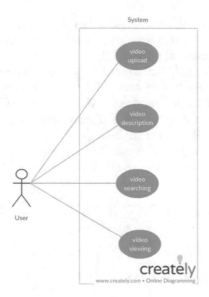

Fig. 2. Use case diagram for sample system

LR Files

In order to allow users to view existing resources on the system while the original files will be kept in storage with a long access time, it is necessary to transcode files to a lower quality. In order for this to be possible, it is necessary to use another service provided by the cloud provider. In the case of Amazon, this service is called Amazon Elastic Transcoder. The transcoding action can be initiated at the moment of completing the upload. Triggering, management and control of the transcoding task should be handled by a system hosted on a remote server such as EC2. To minimize number of user's requests, reduce network traffic and maximize access time to resources, only converted files will be available for the user (in the web application). Transcoded files will be stored in storage with low access time. The service that enables this is S3. HR files will be available only on demands.

Material Searching and Management

In order to enable users to manage and search for video files (which is the main use case), it is necessary to develop a customized system. It can be developed in any technology that enables the creation of web applications. This application should provide a user interface for uploading video files. It is also designed to allow advanced video searches for meta-data of various types. The application should be able to control access to resources for users with different permissions. Optionally, it is possible to consider automatic description of video material using video mining methods, which will be omitted in this paper.

Serverless Process Management

An interesting solution is the use of serverless technology to handle tasks related to video, such as order transcoding or changing location. Even approach creates a challenge such as function scaling and container discovery it offers interesting possibilities of use of resources [22]. Amazon provides a Lambda service that makes this possible. Calling functions in Lambda can be triggered e.g. by changing the state in the database. You can imagine situations when the upload of a video file ends, the management application changes the state in the database to uploaded which automatically triggers the function in lambda, which orders the transcoding of the HR file to LR. This approach will allow for a lot of flexibility and a clear division of responsibilities between individual components.

Content Delivery Network

The next component of presented model is CloudFront – Amazon Content Delivery Network. This component is necessary in case of application with many static resources. It allows to reduce latency of video downloading (including video preview in web application which actually is also downloading). CDN is a standard for applications containing static resources and its use is not a breakthrough.

Simple Email Service (SES)

The last component in the presented architecture is Amazon SES. It is cloud-based emailing sending service provided by Amazon. Its use is optional and can be replaced with any email server such as postfix. The one advantage of SES is that application DevOps does not have to take care of email server by itself. SES also allows to scale at the same time as web application.

Basic Processes of Stored Video Files

For optimization purpose required is design of customized processes of video processing. The first one is process of delivery video resource to system described by following steps:

- The user uploads original video file.
- Web application redirect file chunks directly into Glacier.
- When upload completed, web application mark state of resource as uploaded.
- Changing state of resource triggers lambda functions.
- Lambda function orders file transcoding providing original file location and destination – S3 storage.
- When transcoding is finished, transcoder triggers lambda function.
- Lambda function marks state of video as transcoded and, depends on business needed, sends email to user.

Another important process is video searching and preview with web application. This process could be designed as follow:

1. The user searches for video using application searcher.
2. The user selects video resource.
3. Application displays video overview with video player.
4. The user plays video with video player.
5. Application loads low resolution files.

From perspective of the end-user, application has to allow to access the original, high resolution files. Suggested way to ensure this need is very similar to previous process. On the video overview screen, the user should be allowed to download HR as well as LR files by clicking the button. Due to the Glacier, this is not possible download HR file immediately after clicking the button. Notification system informing about possibility to download should be implemented.

4   Comparison of Cloud Computing Providers from the Point of View of Service Price

This section presents the pricing of main components. Specify for presented model are two types of storage. Because of required file resizing, price of transcoding component is another important value. The last component which can affect total cost of maintaining the system is price of CloudFront. Another components are common in every application based on cloud so they do not change the total cost of maintaining regardless of architecture.

Table 1. Amazon elastic transcoder [10]

Output	Price per minute of video
Less than 720p	$0.017
720p and above	$0.034

Price of transcoding depends on source file duration and an output resolution (Tables 1, 2, 3). For example for 10 min video transcoded to 480p the price is:

$$10 \times \$0.017 = \$0.17 \tag{1}$$

Table 2. Glacier pricing [11]

Output	Price per minute of video
Storage	$0.004 per GB/Month
Retrieval pricing	$0.01 per GB
Retrieval request pricing	$0.055 per 1,000 requests
Upload request	$0.055 per 1,000 Requests
Data transfer out from glacier	$0.02 per GB

Price of upload depend on size and number of requests. Number of requests depend on chunking. The calculation assume that every chunk will be 10 MB big. The storage cost depends only on total size of file. Download cost depend on total size and number of GET request. Like previously, every chunk will be 10 MB big.

For 10 GB (which is equal to 1024 chunks) file the total price of one year storage with doubled download and upload is about $0.52. Below individual components of price are presented:

storage cost:

$$10 \times 12 \times \$0.004 = \$0.48 \tag{2}$$

request upload cost is:

$$1024 \times \$0.055/1000 = \$0.05632 \tag{3}$$

retrieval cost is:

$$10 \times \$0.01 = \$0.1 \tag{4}$$

data transfer is:

$$10 \times \$0.02 = \$0.2 \tag{5}$$

Table 3. S3 pricing [12]

Output	Price per minute of video	
Storage	First 50TB /Month	$0.023 per GB
	Next 450 TB/Month	$0.022 per GB
	Over 500 TB	$0.021 per GB
Upload request	$0.0004 per 1,000 requests	
GET request	$0.0004 per 1,000 requests	
Retrieval pricing	$0.01 per GB	

For 10 GB file to total price of one year storage with 2 downloads and upload is about $2.86. Below individual components of price are presented.

storage cost is:

$$10 \times 12 \times \$0.023 = \$2.76 \tag{6}$$

request upload cost is:

$$1024 \times \$0.0004/1000 = \$0.0004096 \tag{7}$$

retrieval cost is:

$$10 \times \$0.01 = \$0.1 \tag{8}$$

Table 4 presents the costs of Amazon Content Delivery Network. This component is optional.

Table 4. CloudFront pricing [13]

Output	Price (Europe)
Fist 10 TB	$0.085 per GB
Next 40 TB	$0.080 per GB
Next 100 TB	$0.060 per GB
Next 350 TB	$0.040 per GB
Next 524 TB	$0.030 per GB
Next 4 PB	$0.025 per GB
Over 5 PB	$0.020 per GB

5   Conclusions

The presented architecture allows to bring huge financial savings, especially in the case of long-term work. This has been achieved thanks to the introduction of a second, cheaper type of storage. Such an approach to building a system may work very well, e.g. for a film archive. When users have to work with high-resolution files, problems with architecture become apparent. The biggest problem is downloading HR files – they are not immediately available. For systems where response time is important, this will be a big problem. In addition, the financial benefits will not be so huge due to the high cost of preparing and downloading video files.

The proposed approach also requires a large additional workload. This generates additional software production costs. However, on the basis of the simulations presented, it can be seen that it will be an investment that will save money in the future. Another problem is that architecture is strongly based on the services of one provider and it is difficult to transfer it to another. That is why it is a certain risk. In the event of a change pricing or terms of use of services, the benefits of this optimization may be significantly reduced.

Despite many disadvantages, the system can bring huge savings for dedicated systems fulfilling selected needs. The most important case of use is the film archive. As simulations show, the archive, in which 100,000 min of video is added, savings are already counted in millions over a period of 5 years. For this reason, it is necessary to carefully examine the problem and consider the implementation of the proposed architecture.

References

1. Amazon S3. https://aws.amazon.com/s3. Accessed 15 Dec 2018
2. Amazon S3 Glacier. https://aws.amazon.com/glacier. Accessed 15 Dec 2018
3. Amazon Elastic Transcoder. https://aws.amazon.com/elastictranscoder. Accessed 15 Dec 2018
4. Amazon EC2. https://aws.amazon.com/ec2. Accessed 15 Dec 2018
5. Amazon DynamoDB. https://aws.amazon.com/dynamodb. Accessed 15 Dec 2018

6. Amazon RDS for PostgreSQL. https://aws.amazon.com/rds/postgresql. Accessed 15 Dec 2018
7. Amazon Simple Email Service. https://aws.amazon.com/ses. Accessed 15 Dec 2018
8. AWS Lambda. https://aws.amazon.com/lambda. Accessed 15 Dec 2018
9. Amazon CloudFront. https://aws.amazon.com/cloudfront. Accessed 15 Dec 2018
10. Amazon Elastic Transcoder Pricing. https://aws.amazon.com/elastictranscoder/pricing. Accessed 15 Dec 2018
11. Amazon S3 Glacier pricing. https://aws.amazon.com/glacier/pricing. Accessed 15 Dec 2018
12. Amazon S3 Pricing. https://aws.amazon.com/s3/pricing. Accessed 15 Dec 2018
13. Amazon CloudFront Pricing. https://aws.amazon.com/cloudfront/pricing. Accessed 15 Dec 2018
14. Video Space Calculator. https://www.digitalrebellion.com/webapps/videocalc. Accessed 15 Dec 2018
15. PostgreSQL: The World's Most Advanced Open Source Relational Database. https://www.postgresql.org. Accessed 15 Dec 2018
16. The difference between SaaS, PaaS and IaaS. https://www.computerweekly.com. Accessed 15 Dec 2018
17. Pallis, G., Vakali, A.: Content Delivery Networks: Status and Trends (2006)
18. Pallis, G., Vakali, A.: Insight and perspectives for content delivery networks (2006)
19. Voorsluys, W., Broberg, J., Buyya, R.: Introduction to Cloud Computing (2011)
20. Dillon, T., Wu, C., Chan, E.: Cloud Computing: Issues and Challenges (2010)
21. Foster, I., Zhao, Y., Raicu, I., Lu, S.: Cloud Computing and Grid Computing 360-Degree Compared (2008)
22. McGrath, G., Brenner, P.R.: Design, Implementation, and Performance (2017)
23. Velte, A.T., Velte, T.J., Elsenpeter, R.: Cloud Computing: A Practical Approach (2010)
24. Poniszewska-Maranda, A.: Modeling and design of role engineering in development of access control for dynamic information systems. Bull. Polish Acad. Sci. Tech. Sci. **61**(3), 569–580 (2013)
25. Majchrzycka, A., Poniszewska-Maranda, A.: Secure development model for mobile applications. Bull. Polish Acad. Sci. Tech. Sci. **64**(3), 495–503 (2016)
26. Poniszewska-Maranda, A., Majchrzycka, A.: Access control approach in development of mobile applications. In: Younas, M., et al. (eds.) Mobile Web and Intelligent Information Systems, MobiWIS 2016, LNCS, vol. 9847, pp. 149-162. Springer, Heidelberg (2016)
27. Kryvinska, N.: Building consistent formal specification for the service enterprise agility foundation. Soc. Serv. Sci. J. Serv. Sci. Res. **4**(2), 235–269 (2012)
28. Kaczor, S., Kryvinska, N.: It is all about services - fundamentals, drivers, and business models. Soc. Serv. Sci. J. Serv. Sci. Res. **5**(2), 125–154 (2013)
29. Gregus, M., Kryvinska, N.: Service Orientation of Enterprises – Aspects, Dimensions, Technologies. Comenius University in Bratislava (2015). (ISBN 9788022339780)
30. Poniszewska-Maranda, A.: Security constraints in access control of information system using UML language. In: Proceedings of 15th IEEE International Workshops on Enabling Technologies: Infrastructure for Collaborative Enterprises (WETICE-2006), UK (2006)
31. Poniszewska-Maranda, A.: Conception approach of access control in heterogeneous information systems using UML. J. Telecommun. Syst. **45**(2–3), 177–190 (2010)

32. Kryvinska, N., Gregus, M.: SOA and its business value in requirements, features, practices and methodologies. Comenius University in Bratislava (2014). (ISBN 9788022337649)
33. Molnar, E., Molnar, R., Kryvinska, N., Gregus, M.: Web Intelligence in practice. Soc. Serv. Sci. J. Serv. Sci. Res. **6**(1), 149–172 (2014)

Movies Recommendation System

Adrianna Frykowska[1], Izabela Zbieć[1], Patryk Kacperski[1], Peter Vesely[2],
and Andrea Studenicova[2(✉)]

[1] Institute of Information Technology,
Lodz University of Technology, Lodz, Poland
[2] Faculty of Management, Comenius University, Bratislava, Slovakia
{Peter.Vesely,andrea.studenicova}@fm.uniba.sk

Abstract. Due to ever increasing number of newly released movies, a recommendation system may be of use to majority of cinematography fans. This paper presents an approach to create such a system using existing database containing informations about movies and how they are rated by people. Features describing year of production, cast, director, genres and average rating are being extracted and then used with a kNN classifier to decide how much would someone rate any movie in the database. Based on that rating, a number of not yet seen movies is selected and recommended.

1 Introduction

Nowadays, more and more movies are created [16]. It is impossible to keep up with all the new releases without some sort of system that would inform about them. When choosing a new movie, people take a number of factors into consideration. They often enjoy movies of the same genres [5,23] or with actors they like. Person who watches a lot of films will likely have a favourite director and will often look forward to his new creations.

Our goal is to design and implement a system that would recommend movies based on user's preferences. After user rates a number of movies he has already seen, we will use that information to estimate how he would rate other movies and suggest him the ones with highest ratings.

Watching movies is currently one of the major source of entertainment among people all around the world. As more and more people become increasingly interested in films, more data is accumulated and freely accessible on the internet. For our system we will use The Movies Database (TMDB). It has a lot of data we can use about almost any movie that has been released and it has free to use, non comercial API that can be utilised to get access to their information.

From acquired data we will then extract features that will allow us to tell how similar or how different a pair of movies is. Work of Choi et al. [5,26] suggests that one of the major features that allow to distinguish films is a genre. TMDB has a set of genres associated with each movie, so for each pair we will see how many of them they have in common.

L. Barolli et al. (Eds.): INCoS 2019, AISC 1035, pp. 579–585, 2020.
https://doi.org/10.1007/978-3-030-29035-1_56

Other features that may help us in this task and are available on TMDB are director, year of production and cast. People, especially those who watch a lot of movies tend to have their favourite directors and actors. Therefore we will check, if two movies have the same director and how many actors they have in common. Year of production can sometimes also be an important feature. There are people that only like films they used to watch during their childhood and adolescence, whereas other like only new releases [22–25].

This paper presents an approach to create such a system using existing database containing informations about movies and how they are rated by people. Features describing year of production, cast, director, genres and average rating are being extracted and then used with a kNN classifier to decide how much would someone rate any movie in the database.

2   Recommendation Systems

The subject of movie recommendations systems is well studied and there is a lot of literature containing various methods. A brief review of modern recommendation algorithms can be found in the work of Asanov et al. [13]. Choi et al. [5] describe using correlation between movie genres with to achieve great results. We based most of our technique on this paper. Adeniyi et al. [12] show how to use kNN algorithm to create general purpose. Based on this work we decided to use this classifier to solve our problem.

There are many other approaches existing in literature that we reviewed, but in the end, decided not to use. Mukherjee et al. [11] use voting theory to create user model. Ahn and Shi [9] utilize cultural meta-data such as reviews, ratings and comments on blogs and social media sites to create their system. Chen and Aickelin [5] Artificial Immune Systems that model human immune system applied to standard Collaborative Filtering techniques.

Performed review allowed us to decide which method to use. One thing that most approaches had in common was using user preferences and data gathered from other people in recommending process, so we employed that by considering average rating of a movie during recommendation process [17, 18, 20, 21].

The basic features of a movie are as follows:

1. director(s),
2. cast – percent of cast members from unrate movie, which playing in both movies,
3. genres,
4. production year.

K-Nearest Neighbors algorithm is a non-parametric method of pattern classification. For new object it searches the training set for K objects closest to it by some metric. Then it classifies new object to a class which appears the highest number of times [12].

3   System Model for Movies Recommendation

To implement the system for movies recommendation we decided to use a Python language (version 3). We also used the following programming libraries:

- *json* – this library allows to serialize and deserialize data [14]
- *pandas* – this library allows to read CSV files [15].

Our recommendation system was created in Python language. We decided to split the whole system in smaller modules (Fig. 1):

1. *get_movies_features*: This script is responsible for getting all movies features from TMDB database. It creates a JSON with features which are used in recognition process (director(s), cast, genres and production year).
2. *get_movies_ratings*: This script gets all rates (for selected movies) from CSV file. It also counts mean rate value a and save it to JSON file.
3. *load_data*: There is a method, which is responsible for reading data form JSON files and writing data to array.
4. *gui*: This script gets attributes from user.

Figure 1 presents the activities of the recommendation system. It starts when system asking user for his id. Next, if user write a correct id, system asking for examples of watched movies, with rates. Finally, system is asking for number of suggestions that system should display. When user answer correctly for all questions, system loads movies data from JSON files. After that, system is getting movies features from JSON files and write them to the array. Next, using kNN algorithm to count potential rate for each movie, basing on movies rated by user. After that, system is getting this movies which has the highest potential rate and select from them m movies with the highest mean users rate. Finally, program displays selected movies in console.

The limitations of created recommendation systems are as follows:

- TMDB API does not allow to get a list of all movies in the database at once. We have to get each movie's data using its ID. It is possible to get the maximum movie's ID value. This value is already higher than half a million, so it would take a really long time to get data of all movies. For this reason, we decided to use only 200 movies. We think this will be enough to perform some recommendations.
- We assume a user logs in to our system only once. We don't store their movies' ratings nor user ID.
- Mean movies' ratings are calculated basing on different numbers of single ratings. Sometimes we get mean basing on for example 80 ratings and other time only 5 ratings.

4   The Results of Recommendation System Working

The features of movies provided by the user in the created movies recommendation system are presented in Fig. 2.

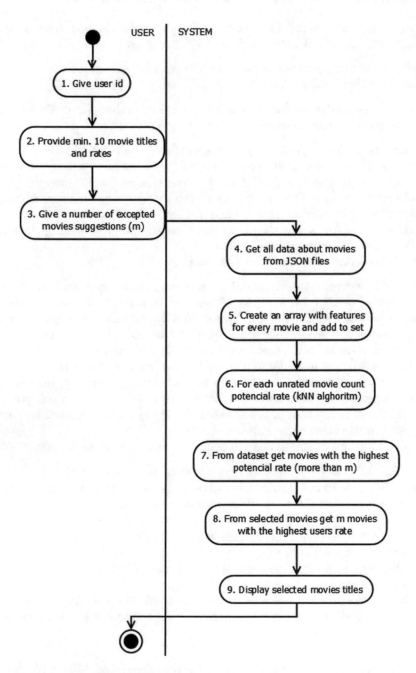

Fig. 1. Prototype of the system activities for movies recommendation

Title	Genres	Directors	Release year	Average rating	User rating
Alien	Horror, Science Fiction	Ridley Scott	1979	3.14	5
Blade Runner	Science Fiction, Drama, Thriller	Ridley Scott	1982	3.26	5
Finding Nemo	Animation, Family	Andrew Stanton	2003	3.2	4
Forrest Gump	Comedy, Drama, Romance	Robert Zemeckis	1994	3.2	3
Gone with the Wind	Drama, Romance, War	Victor Fleming	1939	3.01	3
Monty Python and the Holy Grail	Adventure, Comedy, Fantasy	Terry Gilliam, Terry Jones	1975	3.01	2
Return of the Jedi	Adventure, Action, Science Fiction	Richard Marquand	1983	3.07	2
The Exorcist	Drama, Horror, Thriller	William Friedkin	1973	3.24	0
The Godfather	Drama, Crime	Francis Ford Coppola	1972	3.3	1
The Green Mile	Fantasy, Drama, Crime	Frank Darabont	1999	3.55	4
The King's Speech	Drama, History	Tom Hooper	2010	3.4	1
The Intouchables	Drama, Comedy	Eric Toledano, Olivier Nakache	2011	3.5	0

Fig. 2. Features of movies provided by the user

Title	Genres	Directors	Release year	Average rating	Predicted rating
Terminator 2: Judgment Day	Action, Thriller, Science Fiction	James Cameron	1991	3.26	5
Die Hard	Action, Thriller	John McTiernan	1988	3.2	5
Gladiator	Action, Drama, Adventure	Ridley Scott	2000	3.16	5

Fig. 3. Features of movies provided by the system

The features of movies provided by the system in the created movies recommendation system are presented in Fig. 3.

The short description of all results from the created system is given below. They were compared with the movies that were rated 5 by the user.

Result 1: *Terminator 2: Judgment Day*:

- 2 of genres (thriller, science fiction) also appear in movies rated by the user,
- year of production is pretty close to both of them.

Result 2: *Die Hard*:

- 1 of genres (thriller) appear in one of movies rated by the user,
- year of production is very close to both of them.

Result 3: *Gladiator*:

- 1 of genres (drama) appear in one of movies rated by the user,
- director is the same as in both of them.

5   Conclusions

Recommending movies is not a trivial task as it's difficult to predict what causes people to like certain films and dislike others. Even when using as simple method as kNN it is possible to create system that allows to find similar movies.

We could store new user's ratings and take them into consideration during preparing recommendations for the next user. We could create a graphical user interface (GUI) and replace a command-line interface.

References

1. Lekakos, G., Caravelas, P.: A hybrid approach for movie recommendation. Multimedia Tools Appl. **36**, 55–70 (2008)
2. Arora, G., Kumar, A., Devre, G.S., Ghumare, A.: Movie recommendation system based on users' similarity. Int. J. Comput. Sci. Mobile Comput. **3**, 765–770 (2014)
3. Eyjolfsdottir, E.A., Tilak, G., Li, N.: MovieGEN: A Movie Recommendation System (2010)
4. Johansson, P.: MADFILM – A Multimodal Approach to Handle Search and Organization in a Movie Recommendation System (2003)
5. Choi, S.-M., Ko, S.-K., Han, Y.-S.: A movie recommendation algorithm based on genre correlations. Expert Syst. Appl. **39**(9), 8079–8085 (2012)
6. Lee, J.-S., Park, S.-D.: Performance improvement of a movie recommendation system using genre-wise collaborative filtering. J. Intell. Inf. Syst. **13**(4), 65–78 (2007)
7. Jeong, W.-H., Kim, S.-J., Park, D.-S., Kwak, J.: Performance improvement of a movie recommendation system based on personal propensity and secure collaborative filtering. J. Inf. Process. Syst. **1**(9), 157–172 (2013)
8. Szomszor, M., Cattuto, C., Alani, H., O'Hara, K., Baldassarri, A., Loreto, V., Servedio, V.: Folksonomies the Semantic Web and Movie Recommendation (2007)

9. Ahn, S., Shi, C.-K.: Exploring movie recommendation system using cultural meta-data. In: Transactions on Edutainment II, pp. 119–134. Springer, Berlin (2009)
10. Wang, Z., Yu, X., Feng, N., Wang, Z.: An improved collaborative movie recommendation system using computational intelligence. J. Vis. Lang. Comput. **25**(6), 667–675 (2014)
11. Mukherjee, R., Sajja, N., Sen, S.: A movie recommendation system - an application of voting theory in user modeling. User Model. User-Adap. Inter. **13**(1), 5–33 (2003)
12. Adeniyi, D.A., Wei, Z., Yongquan, Y.: Automated web usage data mining and recommendation system using K-Nearest Neighbor (KNN) classification method. Appl. Comput. Inform. **12**(1), 90–108 (2016)
13. Asanov, D.: Algorithms and Methods in Recommender Systems (2011)
14. json – Python 3.7.2rc1 documentation. https://docs.python.org/3/library/json.html. Accessed 17 Dec 2018
15. Pandas 0.23.4 documentation. https://pandas.pydata.org/pandas-docs/stable/. Accessed 17 Dec 2018
16. The Numbers – Top Movies of Each Year. https://www.the-numbers.com/movies/#tab=year. Accessed 18 Dec 2018
17. Poniszewska-Maranda, A.: Modeling and design of role engineering in development of access control for dynamic information systems. Bull. Polish Acad. Sci. Tech. Sci. **61**(3), 569–580 (2013)
18. Majchrzycka, A., Poniszewska-Maranda, A.: Secure development model for mobile applications. Bull. Polish Acad. Sci. Tech. Sci. **64**(3), 495–503 (2016)
19. Poniszewska-Maranda, A., Majchrzycka, A.: Access control approach in development of mobile applications. In: Younas, M., et al. (eds.) Mobile Web and Intelligent Information Systems, MobiWIS 2016, LNCS, vol. 9847, pp. 149-162. Springer, Heidelberg (2016)
20. Kryvinska, N.: Building consistent formal specification for the service enterprise agility foundation. Soc. Serv. Sci. J. Serv. Sci. Res. **4**(2), 235–269 (2012)
21. Kaczor, S., Kryvinska, N.: It is all about services - fundamentals, drivers, and business models. Soc. Serv. Sci. J. Serv. Sci. Res. **5**(2), 125–154 (2013)
22. Gregus, M., Kryvinska, N.: Service Orientation of Enterprises – Aspects, Dimensions, Technologies. Comenius University in Bratislava (2015). (ISBN 9788022339780)
23. Poniszewska-Maranda, A.: Security constraints in access control of information system using UML language. In: Proceedings of 15th IEEE International Workshops on Enabling Technologies: Infrastructure for Collaborative Enterprises (WETICE-2006), UK (2006)
24. Poniszewska-Maranda, A.: Conception approach of access control in heterogeneous information systems using UML. J. Telecommun. Syst. **45**(2–3), 177–190 (2010)
25. Kryvinska, N., Gregus, M.: SOA and Its Business Value in Requirements, Features, Practices and Methodologies. Comenius University in Bratislava (2014). (ISBN 9788022337649)
26. Molnar, E., Molnar, R., Kryvinska, N., Gregus, M.: Web intelligence in practice. J. Serv. Sci. Res. **6**(1), 149–172 (2014)

An Application-Driven Heterogeneous Internet of Things Integration Architecture

Changhao Wang[✉], Shining Li, Yan Pan, and Bingqi Li

School of Computer Science and Engineering,
Northwestern Polytechnical University,
Xi'an 710072, People's Republic of China
{wchanghao,panyan,libingqi}@mail.nwpu.edu.cn,
lishining@nwpu.edu.cn

Abstract. Smart city consists of heterogenous Internet of Things (IoTs) and urban application systems. To achieve smart city integration, One critical challenge is that these heterogenous systems are implemented independently, and hard to communicate with each other. To address this issue, the authors propose an application-driven heterogeneous IoTs architecture. Under this architecture, all kinds of resources are uniformly described by extracting the common characteristics of application requirements in smart cities to form independent and complete capability components, resources are reconfigured according to application requirements to provide adaptive resources and differentiated services for urban applications. The results show that our proposed architecture can effectively solve the problem of deep integration of heterogeneous Internet of things.

Keywords: Application-driven · Heterogenous architecture ·
Internet of Things

1 Introduction

The Internet of Things (IoT) connects the human world with the physical world through Internet technology. As an emerging industry, the Internet of things involves different fields and technologies. It is estimated that by 2020, there will be 50 billion devices connected to the Internet worldwide [1]. Real-time communication and data exchange are needed between multiple Internet of things, this poses a huge challenge to the carrying capacity of the existing IoT architecture. At present, various departments and industries in smart cities have established or are building their own IoT applications. These typical application systems are different in sensing terminal, network topology, access mode, communication protocol and application mode [2]. With a very significant heterogeneity, that is, a heterogeneous Internet of things. They operate independently, the lack of Internet sharing, it is difficult to interoperability, cannot be comprehensive to be effective, forming the islands of information.

The integrated service system of smart city with the concept of "sensingper-connection-recognition-application-integration" contains extensive requirements of extensive access of Internet of things and deep integration of urban system [3]. To

© Springer Nature Switzerland AG 2020
L. Barolli et al. (Eds.): INCoS 2019, AISC 1035, pp. 586–596, 2020.
https://doi.org/10.1007/978-3-030-29035-1_57

promote deep integration between urban IoT applications, Breaking the existing architecture of the Internet of things, we need to define a complete set of models, methods and independent innovation application-driven IoT architectures that support heterogeneous IoT interconnections. It can extend the interconnection and sharing of heterogeneous IoT resources in the vertical direction end-to-end with the application-driven configuration of IoT resources, and realize the integration of cloud, edge and end horizontally [4]. Through software definition, abstract description of heterogeneous IoT resources and encapsulation, cloud edge coordinated control and optimization technology, achieve unified abstract management and optimization of network, computing and storage, and sensing resources. Thus, it supports the cross-industry equipment connection management of wide-area access and provides support for massive smart city information and services.

The main idea of this paper is to design a new heterogeneous Internet of Things architecture that integrates network, computing, storage, sensing and control. It can allocate the resources needed by the application according to the application requirements: Including sensing devices, network resources, computing storage resources, etc., collaborative control, distribution of cloud, edge, and end devices, collaborative control and distribution of cloud, side and end devices, make the Internet of Things to respond flexibly to application requirements, optimize the allocation and utilization of resources.

2 Related Work

The earliest Internet of Things application architecture in foreign countries began with object identification and providing corresponding services. The proposed software-defined network (SDN) provides a solution for the flexible control and management of complex basic IoT hardware [5]. Google has also deployed an SDN-based B4 system in the global connected data center to achieve equal treatment for each private application. VMware's network virtualization platform provides business solutions based on SDN [6]. In China, after the Internet of Things industry was included in China's strategic emerging industries in the new period, the development of Internet of Things applications has steadily promoted.

However, with the continuous development of specific IoT application scenarios such as smart transportation, smart grid, and car networking, IoT device management and data processing have proliferated. The current IoT architecture is not enough to support the rapid development of the Internet of Things. Efficient device interconnection and resource interoperability between heterogeneous Internet of Things, differentiated resources required for dynamic allocation of devices across industries, and interconnection and interoperability between different vertical applications of IoT, and difficulty in sharing resources have become bottlenecks in the development of the Internet of Things. Researchers are looking for more flexible IoT architectures to meet the needs of heterogeneous IoT. Cheng etc. proposed a situation-aware dynamic IoT services coordination approach to address to heterogeneous physical devices into the context aware IoT infrastructure [7]. Zhang et al. gave a cache-enabled wireless heterogeneous network based on the control-plane and user-plane split to cooperated together in the backhaul scenario [8]. Qiao and others raises an event-driven, service-oriented IoT service providing method [9]. This method can conveniently support on-demand distribution and aggregation of perceived information and achieve event-driven dynamic collaboration of services across business domains and even across organizations.

These researches have solved the problem of resource allocation and dynamic coordination of heterogeneous Internet of things from different angles, but the overall architecture of heterogeneous Internet of things has not yet obtained a better way. Based on the typical application of IoT, this paper proposes an application-driven heterogeneous Internet of Things architecture for the difficult sharing and interconnection between typical IoT applications, it can dynamically arrange resources according to application needs, adjust allocation according to needs, provide differentiated resources for application needs, and realize efficient equipment interconnection and resource exchange among heterogeneous Internet of things.

3   An Application-Driven IoT Architecture

The traditional IoT system is chimney-like, each typical application scenario runs independently, and it is difficult to interconnect between systems due to interfaces, standards and other reasons, as a result, they do not extensions and scalability, and the application scenario is relatively closed, hardware configuration is not flexible, low rate of balancing load of the network, many infrastructure construction repeatedly, resulting in a large number of resources and equipment waste.

To tackle these issue, we propose an application-driven IoT architecture in this paper [10], we analyze the application requirements, business characteristics and deployment requirements of some typical application scenarios in the city evaluate its carrying capacity, extract common element requirements, and establish a set of meta-requirements such as network, processing, sensing, and control; Unified resource description of network, computing, storage, sensing and control resources distributed in smart cities, The network, computing, storage, sensing and control resources are virtualized into four resource pool, and abstract and encapsulate services to form independent and complete capability components. Trigger the rapid response of each layer of the IoT architecture through specific applications, resource allocation and capability matching are carried out based on application meta-requirements to provide adaptive resources and differentiated services for application requirements.

We name our architecture application-driven heterogeneous Internet of Things integration architecture. As shown in the Fig. 1, our architecture consists of four layers: application layer, control layer, network transport layer and sensing layer.

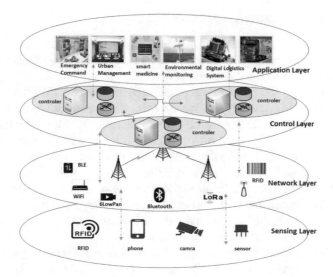

Fig. 1. Application-driven IoT architecture hierarchy model

3.1 Application Layer

The application layer is at the top of our architecture, at present, typical Internet of things systems in smart cities, such as smart transportation, smart medical care and smart power grid, are deployed and connected in this layer [11]. When intelligent transportation application system provides users with convenient navigation, path planning and other services, it needs to analyze and process road conditions, vehicle location, energy consumption and other information, so it needs data from wireless networks, sensors, GPS, cameras and so on. A smart medical system also needs to plan the first aid route and the corresponding road data when responding to the emergency of patients. For these application requirements, we can classify the data content and resource requirements required by application scenarios in accordance with the characteristics, set up network requirement set, process requirement set, perceive and control requirement set, then extract the minimum common requirements (we call them meta-demands), resource mappings are established for these meta-requirements in response to processing application requirements.

3.2 Control Layer

The control layer is the core layer of the architecture we designed, for the upper layer, it needs to respond to application requirements. For the lower layer, it can manage and control heterogeneous IoT system resources. In the control layer, we set up three control parallel controllers. SDN/NFV is responsible for the management, control and optimization of communication resources, the cloud side is responsible for computing the coordination and allocation of storage resources, sensing is responsible for the scheduling and control of perceived resources, they work together at the control layer to provide resource allocation and scheduling for business from the application layer.

3.3   Network and Transportion Layer

For heterogeneous IoT, different physical devices, different access networks, such as: WiFi, Bluetooth, zigbee, they are of different types and different protocols, which make it very difficult for the unified Internet transmission.

Cross-Technology Communication (CTC) technologies, e.g., WEBee [12], Lego-Fi [13] and BlueBee [14], provide a probability that messages can be transmitted across coexistent heterogenous IoT networks. For example, WEBee enables the direct transmission from the WiFi radio to the ZigBee radio with a high throughput of 126 bps; Lego-Fi builds the ZigBee-to-WiFi channel and achieves a throughput of 213.6 kbps. With the similar idea of WEBee, BlueBee builds the communication channel from the Bluetooth radio to the ZigBee radio. Inspired by these CTC technologiese, a framework for data transmission across heterogenous IoT networks is proposed. Please note although we take the WiFi and ZigBee as the examples, the framework can fit other heterogenous IoT networks.

With the help of the CTC technologies, the WiFi Access Point (AP) acts as not only the AP in the traditional WiFi network, but also the sink node in the ZigBee network. The workflow of the framework can be classified as the four parts:

- The WiFi AP communicates with the WiFi devices e.g., phones, cameras through the traditional WiFi-to-WiFi channel.
- The WiFi AP can disseminate the ZigBee data to the ZigBee network using the WEBee technology. Such downlink data dissemination can be used for routing formation, time synchronization, software updates [15, 16].
- Since the ZigBee nodes are energy constrained, the sensed data by ZigBee nodes usually needs to be relayed by multi-hop with the ZigBee-to-ZigBee communication channel.
- To aggregate the sensed data (by the ZigBee network) to the server or remote user, the WiFi AP utilizes the Lego-Fi to aggregate the sensed data.

With the proposed framework, the data transmission among heterogenous IoT networks can be effectively coordinated, which benefits the heterogenous IoT networks in reducing data transmission delay, improving data transmission throughput, reducing the energy consumption of IoT networks, etc.

3.4   Sensing Layer

The sensing layer integrates various physical sensing devices of heterogeneous Internet of things applications, the data, video, and various types of information required by the application system are obtained from the sensing layer, and the completion of the application task also requires the cooperation of the sensing device. At this level, sensing devices need to be arranged and abstractly described, the heterogeneous Internet of things system can dispatch sensing devices and respond to application tasks at any time forming a large cross-domain device connection management to support massive smart city information and services.

4   Methods and Procedures

In this section, we present a set of approaches to solve application-driven heterogeneous Internet of things systems, follow below Fig. 2:

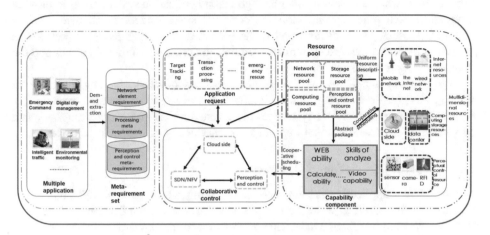

Fig. 2. Application-driven IoT architecture procedures

I.   The Ontology Web Language for Services (OWL-S) resource description method is used to describe the resources and devices in the sensing layer [17], including network resources, computing resources, storage resources, and sensing and control resources.

II.   Four resource pools are defined according to resource attributes, which are network resource pool, computing resource pool, storage resource pool, and sensing resource pool. The IoT device resources are virtualized and classified into corresponding resource pools according to the unified resource description results.

III.   From these virtualized resources, we abstract several capability components based on their service objects and service capabilities, such as: web capabilities, computing capabilities, analytical capabilities, heat capabilities, video capabilities, etc., and establish a mapping relationship between these capabilities and corresponding resources.

IV.   Sort out the application requirements of various application systems and extract their business characteristics, and decomposed into a single minimum element requirement, including network element requirements, processing meta-demands, perceptual control element requirements, and set up meta-demand set.

V.   When there is an application request, according to the application meta-requirement corresponding to the application request, under the collaborative management of the controller, the controller allocates the required capability components for the meta-requirement, according to the mapping relationship between these capability components and the resource pool, resource devices are scheduled to provide customized services for application requests and complete application tasks.

5   Application and Implement

In this part, we use the smart traffic application system and the smart medical system as an example to describe in detail the application-driven heterogeneous IoT architecture. For example, when a user in a smart transportation application system needs to provide navigation, path planning and other services, the system needs to allocate wireless network, sensors, GPS, camera and other resources. The emergency response of a patient in the smart medical system also requires the planning of first aid routes and the corresponding road resources.

We can see that these two independent systems have overlapping resources in sensing devices, network communication, data analysis, and application services. By connecting these independent systems to our heterogeneous Internet of Things, we can solve the waste of resources and optimize the services of the application system.

5.1   Uniform Description of Resources

When different application systems access to heterogeneous Internet of Things architecture, we first need to describe the resources in various application systems uniformly. We use OWL-S to describe these service resources for better semantic understanding and resource discovery. We describe each resource in the Internet of things as a quaternion relationship {name,sub,class,type}, among them:

Name: represents the name and identity of the resource; sub: represents the layer to which the resource belongs, which is one of the four layers of the heterogeneous IoT architecture; class: represents the category of the resource, such as: video, camera, RFID, etc. type: describe the specific parameters of the resource, such as the resolution of the camera, pixels, and parameters of the sensor, it can be a multi-group.

For a camera of a smart transportation system, we can describe it as {pho1,sensing,camera,{box,200W}}. For a medical system camera, we can describe it as {phon, application,camera,{globe,120W}}. In this way, we can describe each Internet of things device or resource. Then, according to the categories and attributes of the resources, we aggregate the resources into four resource pools, which are managed by the resource manager (RM).

In RM, we define the ability of a resource or device to perform one or more operations as a resource ability. In other words, one resource corresponds to multiple capabilities, thus, we can establish multiple mapping tables of resources and capabilities in RM, as shown in Table 1.

Through the resource mapping table, we can accurately pass out the operations that the resource can perform.

Table 1. Resource and capability mapping table

Resource name	Class of ability	Specific ability 1	Specific ability 2	... ...	Specific ability n
pho1	Video	Taking pictures	Recording video		Sensing
serv1	Computing	Analysis	Web		Computing
... ...	... ...	... ...	... ...	... ...	... ...
zigb1	Network transmission	Building network	Data transmission		Communication
sens1	Sensing	Speed sensing	Heartbeat		Heat rate

5.2 Meta-Demand

Our heterogeneous IoT architecture is a multi-user oriented application event that triggers the IoT platform to respond to it. This requires the decomposition of applications from the application layer.

We represent the application requirements of an application's system in a set, which is recorded as $Dn = \{R_1, R_2, \ldots \ldots R_i\}$; For one of the demand R_i, we can continue to decompose $R_i = \{E_1, E_2 \ldots E_j,\}$; Until E_j can no longer be decomposed, we call E_j a meta-demand, which is $E_j \in R_i \in D_n$.

Each application system requirement Dn can decompose the meta-demands and combine the same meta-demands.

Which is $Mt = R_1 \cup R_2 \cup R_3 \cup \ldots \ldots \cup Rn = \{Em, \ldots \ldots En\}$;

Mt is the set of meta-demands that we need. According to the attributes of meta-demands set, we subclassify it into network meta-demand set M_{net}, processing meta-demand set M_{pro} and sensing control meta-demand set M_{con}. After the meta-demand set is established, we can allocate IoT resources to the application according to the meta-demands, so that the application requirements can be quickly processed in the IoT platform.

5.3 Collaborative Allocation of Resources

After we connect different application systems to the heterogeneous IoT architecture, when users start an application request, we can allocate resources to meet the needs across the entire IoT platform. But it happens all the time, different application requests require the same resources. This requires our control layer to have a good scheduling policy.

In this paper, the heterogeneous Internet of things architecture we designed a center-free application-server cluster architecture. Under the controller of the control layer, resources are allocated on demand, process application requests in collaboration with cloud, edge, and end. To evaluate our proposed heterogeneous Internet of things (IOT) architecture, we use a center-free resource allocation algorithm [18].

There are N application requests, the set of N = {1, 2, ... N}.There are M servers whose collection is M = {1,2,...,M},Each application may be assigned to a server. Each application i can issue a request to each server j ∈ M to process x_{ij}, the algorithm is summarized as follows (Table 2):

Table 2. Center-free resource allocation algorithm

Algorithm 1. Center-free resource allocation algorithm

Input : x_i, z_j, p_j

Output : x_i^*,z_j^*,p_j^*

$\forall x_i$, z_j,p_j
for k=0 to max
    $\forall i \in N$, application i updates its own server request to get the corresponding resource x_i^{k+1}; application i sends x_{ij}^{k+1} to the corresponding server;
    $\forall j \in M$, the server j first calculates the required resource amount z_j and the number of resources p_j according to x_{ij}^{k+1};
    the server j allocates the number of available resources p_j to the application according to the resource amount z_j;
end

5.4   Performance of Architecture

According to the above method, we take the patient first aid event as an example to estimate the resources and system overhead required for this application-driven event. Compared with the resources and system overhead required by the independent intelligent transportation system and smart medical system. As is shown in Figs. 3 and 4.

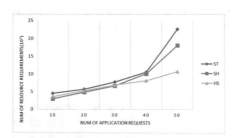
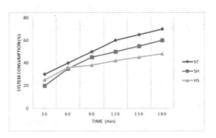

Fig. 3. Number of resources requirements   **Fig. 4.** System consumption proportion

In the above of Figs. 3 and 4, ST means smart transportation, SH means smart health, and HS means heterogeneous system. As you can see from the figures, at the beginning of the system operation, ST, SH and HS are close to the same sum of resource requirements and system consumptions. As some applications take over the system for a long time, ST and SH demand for resources and system consumption start to increase dramatically, however, HS's demand for resources and system consumption is not obvious increasing. The results show that the heterogeneous Internet of Things architecture can effectively save resource utilization and reduce system consumption.

6 Conclusion

In this paper, we review recent advances in architecture. We propose an application-driven architecture for heterogeneous Internet of Things. We detail its key components of application layer, control layer, net layer and sensing layer. We also presented specific methods and steps, and evaluated our architecture. Interconnection of heterogeneous Internet of things is an urgent problem to be solved in the future. How to reasonably allocate Internet of things resources according to application requirements, so that our architecture can be more scalable according to application slices, is our research work in the face of challenge in the future.

Acknowledgments. This work is supported by National Natural Science Foundation of China (NSFC) under Grant No. 61872434 and Key R&D Program of Shaanxi Province in 2017 (No. 2017ZDXM-GY-018).

References

1. Wang, J.: Research on routing in software defined internet of things. Huazhong University of Science & Technology, Wuhan, China (2016)
2. Wang, X.A., Liu, Y., Zhang, J., Yang, X., Zhang, M.: Improved group-oriented proofs of cloud storage in IoT setting. Concurr. Comput.: Pract. Exp. **30**(21), e4781 (2018)
3. Huo, R., et al.: Software defined networking, caching, and computing for green wireless networks. IEEE Commun. Mag. **54**(11), 185–193 (2016)
4. Wang, X.A., Yang, X., Li, C., Liu, Y., Ding, Y.: Improved functional proxy re-encryption schemes for secure cloud data sharing. Comput. Sci. Inf. Syst. **15**(3), 585–614 (2018)
5. Mckeown, N.: Software-defined networking. INFOCOM Keynote Talk **17**(2), 30–32 (2009)
6. Volpato, F., Silva, M.P.D., Dantas, M.A.R.: OFQuality: a quality of service management module for software-defined networking. Int. J. Grid Util. Comput. **10**(2), 187–198 (2019)
7. Cheng, B., et al.: Situation-aware dynamic service coordination in an IoT environment. IEEE/ACM Trans. Netw. (TON) **25**(4), 2082–2095 (2017)
8. Zhang, J., Zhang, X., Wang, W.: Cache-enabled software defined heterogeneous networks for green and flexible 5G networks. IEEE Access **4**, 3591–3604 (2016)
9. Qiao, X., Zhang, Y., Wu, B., et al.: Event-driven, service-oriented internet of things service delivery method. Sci. China Inf. Sci. **43**(10), 1219–1243 (2013)
10. Ye, Q., et al.: End-to-end quality of service in 5G networks: examining the effectiveness of a network slicing framework. IEEE Veh. Technol. Mag. **13**(2), 65–74 (2018)
11. Liu, J., et al.: Software-defined internet of things for smart urban sensing. IEEE Commun. Mag. **53**(9), 55–63 (2015)
12. Li, Z., He, T.: WEBee: Physical-layercross-technology communication via emulation. In: Proceedings of the 23rd Annual International Conference on Mobile Computing and Networking, pp. 2–14. ACM (2017)

13. Guo, X., He, Y., Zheng, X., et al.: LEGO-Fi: transmitter-transparent CTC with cross-demapping. In: Proceedings of IEEE INFOCOM (2019)
14. Jiang, W., Yin, Z., Liu, R., et al.: BlueBee: a 10,000 x faster cross-technology communication via phy emulation. In: Proceedings of the 15th ACM Conference on Embedded Network Sensor Systems, p. 3. ACM (2017)
15. Liu, J., Wang, S., Li, S., Cui, X., Pan, Y., Zhu, T.: MCTS: multi-channel transmission simultaneously using non-feedback fountain code. IEEE Access **6**, 58373–58382 (2018)
16. Wang, W., Liu, X., Yao, Y., Pan, Y., Chi, Z., Zhu, T.: CRF: coexistent routing and flooding using WiFi packets in heterogeneous IoT networks. In: IEEE IN-FOCOM 2019 - IEEE Conference on Computer Communications (INFOCOM 2019), Paris, France (2019)
17. Ling, J., Jiang, L.Y.: Semantic description of IoT services: a method of mapping WSDL to OWL-S. Comput. Sci. **4**, 89–94 (2019)
18. Huang, H.: Collaborative resource allocation algorithms over hybrid networks based on primal-dual method. University of Science and Technology of China (2017)

Author Index

Printed in the United States
By Bookmasters